042296
ROYAL AGRICULTURAL COLLEGE

D0613853

Enforcing the Common Fisheries Policy

Enforcing the Common Fisheries Policy

Ronán J. Long
Peter A. Curran

Fishing News Books
An imprint of Blackwell Science

Editorial Offices:
Osney Mead, Oxford OX2 0EL
25 John Street, London WC1N 2BL
23 Ainslie Place, Edinburgh EH3 6AJ
350 Main Street, Malden
MA 02148 5018, USA
54 University Street, Carlton
Victoria 3053, Australia
10, rue Casimir Delavigne
75006 Paris, France

Other Editorial Offices:

Blackwell Wissenschafts-Verlag GmbH
Kurfürstendamm 57
10707 Berlin, Germany

Blackwell Science KK
MG Kodenmacho Building
7-10 Kodenmacho Nihombashi
Chuo-ku, Tokyo, 104, Japan

First published 2000

Set in 9.5/12 pt Palatino
by DP Photosetting, Aylesbury, Bucks
Printed and bound in Great Britain by
MPG Books Ltd, Bodmin, Cornwall

DISTRIBUTORS

Marston Book Services Ltd
PO Box 269
Abingdon
Oxon OX14 4YN
(*Orders:* Tel: 01235 465500
Fax: 01235 465555)

USA
Blackwell Science, Inc.
Commerce Place
350 Main Street
Malden, MA 02148 5018
(*Orders:* Tel: 800 759 6102
781 388 8250
Fax: 781 388 8255)

Canada
Login Brothers Book Company
324 Saulteaux Crescent
Winnipeg, Manitoba R3J 3T2
(*Orders:* Tel: 204 837-2987
Fax: 204 837-3116)

Australia
Blackwell Science Pty Ltd
54 University Street
Carlton, Victoria 3053
(*Orders:* Tel: 03 9347 0300
Fax: 03 9347 5001)

A catalogue record for this title is available from the British Library

ISBN 0-85238-261-8

Library of Congress
Cataloging-in-Publication Data

Long, Ronan J.
Enforcing the common fisheries policy/
Ronan J. Long, Peter A. Curran.
p. cm.
Includes bibliographical references (p. 351) and index.
ISBN 0-85238-261-8 (hc)
1. Fishery policy – Europe. I. Curran, Peter A. II. Title.
SH253 .L66 1998
343.4'07692 – dc21 98-35425
CIP

For further information on
Blackwell Science, visit our website:
www.blackwell-science.com

Contents

List of Tables, Maps and Figures

Tables

Maps

Figures

Preface

This book assesses the present state of law enforcement in the common fisheries policy. At the time of writing, there is no authoritative work on the enforcement of Community law and the authors have thus drawn upon, in some instances, their first hand experience in fishery enforcement to fill the lacunae in the literature. In this endeavour there is an attempt to bridge the knowledge deficit between lawyers who may wish to know what the law is on this particular subject and law enforcement practitioners who are of course more interested in what States and the industry do in practice. The authors' understanding of fisheries law has been influenced by the views of Community fishermen who work and live in a most perilous environment. In this regard it is remarkable how the common fisheries policy transcends national boundaries to regulate the profession from the fog-bound Grand Banks of Newfoundland to the tropical climes of the Indian Ocean. It is hoped that what emerges from this book is a clearer understanding and more coherent picture of the remarkable developments in the enforcement of Community fishery law. A better understanding, the authors believe, is an essential element to ensure that the common fisheries policy works satisfactorily at a time when there is increasing concern regarding the sustainability of world fisheries.

The first author, Ronán Long, wishes to record a debt of gratitude to Professor William Binchy and to Alex Schuster BL, for providing him with the opportunity to undertake research for a PhD at the School of Law, Trinity College, Dublin. In Brussels, both authors wish to thank their colleagues in the European institutions who shared many of their insights into the complexities of the common fisheries policy. Special mention should be made of the staff who work in the information service of the Fisheries Directorate General, European Commission.

Any opinions or views expressed in this book are those of the authors and may not in any circumstances be regarded as statements of official policy of the European institutions.

Ronán J. Long
Peter A. Curran

Cases in the European Court of Justice

Case 25/59	*Kingdom of the Netherlands* v. *High Authority* [1960] ECR 355.
Case 26/62	*Van Gend en Loos* v. *Nederlandse Administratie der Belastingen* [1963] ECR 1.
Joined Cases 90/63 and 91/63	*Commission of the European Communities* v. *Luxembourg and Belgium* [1964] ECR 625.
Case 48/65	*Lütticke* v. *Commission of the European Communities* [1966] ECR 0019.
Case 29/69	*Stauder* v. *City of Ulm* [1969] ECR 419.
Case 9/70	*Grad* v. *Finanzamt Traunstein* [1970] ECR 825.
Case 22/70	*Commission of the European Communities* v. *Council of the European Communities* [1971] ECR 263.
Case 25/70	*Einfuhr- und Vorratsstelle für Getreide und Futtermittel* v. *Koester* [1970] ECR 1161.
Case 7/71	*Commission of the European Communities* v. *French Republic* [1971] ECR 1003.
JoinedCases 21–24/72	*International Fruit Co., N.V.* v. *Produktschap voor Groenten en Fruit* [1972] ECR 1219.
Case 39/72	*Commission of the European Communities* v. *Italy* [1973] ECR 101.
Case 5/73	*Balkan-Import-Export* v. *Hauptzollamt Berlin Packhof* [1973] ECR 1091.
Case 9/73	*Carl Schlüter* v. *Hauptzollamt Lörrach* [1973] ECR 1135.
Case 34/73	*Variola SPA* v. *Amministrazione Italiana Delle Finanze* [1973] ECR 981.
Case 167/73	*Commission of the European Communities* v. *French Republic* [1974] ECR 359.
Case 8/74	*Procureur du Roi* v. *Dassonville* [1974] ECR 837.
Case 34/74	*Van Duyn* v. *Home Office* [1974] ECR 1337.
Case 48/74	*Charmasson* v. *Minister for Economic Affairs and Finance* [1974] ECR 1383.
Case 43/75	*Defrenne* v. *Sabena* [1976] ECR 455.
Case 87/75	*Bresciani* v. *Amministrazione Italiana delle Finanze* [1976] ECR 129.
Joined Cases 3, 4 and 6/76	*Officier Van Justitie* v. *Cornelis Kramer and others* [1976] ECR 1279.
Case 50/76	*Amsterdam Bulb* v. *Produktschap voor Siergewassen* [1977] ECR 137.

Case 104/81	*Hauptzollamt Mainz* v. *C.A. Kupferberg & Cie KG a. A.* [1982] ECR 3641.
Joined Cases 138 and 139/81	*Directeur des Affaires Maritimes du Littoral du Sud* v. *Javier Marticorena-Otazo and Manuel Prego Parada* [1982] ECR 3819.
Joined Cases 137 and 140/81	*Directeur des Affaires Maritimes du Littoral du Sud-Oest and Procureur de la République* v. *Alfonso Campandeguy Sagarzazu* [1982] ECR 3847.
Case 230/81	*Luxembourg* v. *European Parliament* [1983] ECR 0255.
Case 281/81	*Anklagmyndigheden* v. *Jack Noble Kerr* [1982] ECR 4053.
Case 283/81	*Srl CILFIT and Lanificio di Gavardo SpA* v. *Ministry of Health* [1982] ECR 3415.
Joined Cases 13–28/82	*José Arantzamendi-Osa and Others* v. *Procureur de la République and Procureur Général* [1982] ECR 3927.
Joined Cases 50–58/82	*Administrateur des Affairs Maritimes, Bayonne and Procureur de la République* v. *José Dorca Marina and Others* [1982] ECR 3949.
Case 87/82	*Lt. Cdr. A.G. Rogers* v. *H.B.L. Darthenay* [1983] ECR 1579.
Case 14/83	*Sabine von Colson & Elizabeth Kamann* v. *Land Nordrhein-Westfalen* [1984] ECR 1891.
Case 24/83	*Gewiese and Mehlich* v. *Mackenzie* [1984] ECR 817.
Case 63/83	*Regina* v. *Kent Kirk* [1984] ECR 2689.
Case 106/83	*Sermide SpA* v. *Cassa Conguaglio Zucchero and others* [1984] ECR 4209.
Joined Cases 86–87/84	*Criminal Proceedings against I. Bout en Zonen BV* [1985] ECR 0941.
Case 100/84	*Commission of the European Communities* v. *United Kingdom,* [1985] ECR 1169.
Case 207/84	*Rederij L. De Boer en Zn. BV* v. *Produktschap voor Vis en Visprodukten* [1985] ECR 3203.
Case 234/84	*Belgium* v. *Commission of the European Communities* [1986] ECR 2263.
Case 40/85	*Belgium* v. *Commission of the European Communities* [1986] ECR 2321.
Joined Cases 89/85 114/85, 116–117/85 125–128/85	*Ahlström Osakeyhtiö and others* v. *Commission of the European Communities* [1988] ECR 5193.
Case 325/85	*Ireland* v. *Commission of the European Communities* [1987] ECR 5041.
Case 326/85	*Kingdom of the Netherlands* v. *Commission of the European Communities* [1987] ECR 5091.

Case 332/85	*Federal Republic of Germany* v. *Commission of the European Communities* [1987] ECR 5143.
Case 336/85	*French Republic* v. *Commission of the European Communities* [1987] ECR 5173.
Case 346/85	*United Kingdom* v. *Commission of the European Communities* [1987] ECR 5197.
Case 348/85	*Denmark* v. *Commission of the European Communities* [1987] ECR 5225.
Case 26/86	*Deutz und Geldermann, Sektkellerei Breisach (Baden) GmbH* v. *Council of the European Communities* [1987] ECR 941.
Case 34/86	*Council of the European Communities* v. *European Parliament* [1986] ECR 2155.
Case 46/86	*Albert Romkes* v. *Officier van Justitie for the District of Zwolle* [1987] ECR 2671.
Case 53/86	*Officier van Justitie for the District of Zwolle* v. *Albert Romkes* [1987] ECR 2691.
Case 55/86	*Arposol* v. *Council of the European Communities* [1988] ECR 0013.
Case 207/86	*APESCO* v. *Commission of the European Communities* [1988] ECR 2151.
Case 223/86	*Pesca Valentia Ltd.* v. *Minister for Fisheries and Forestries, Ireland and the AG* [1988] ECR 0083.
Case 237/86	*Kingdom of the Netherlands* v. *Commission of the European Communities* [1987] ECR 5251.
Case 239/86	*Ireland* v. *Commission of the European Communities* [1987] ECR 5271.
Case C-3/87	*The Queen* v. *Ministry of Agriculture, Fisheries and Food, ex parte Agegate Ltd.* [1989] I-4459.
Joined Cases 46/87 and 227/88	*Hoechst AG* v. *Commission of the European Communities* [1989] ECR 2859.
Case C-142/87	*Belgium* v. *Commission of the European Communities* [1990] ECR I-959.
Case 169/87	*Commission of the European Communities* v. *French Republic* [1988] ECR 4093.
Case C-216/87	*The Queen* v. *Ministry of Agriculture, Fisheries and Food, ex parte Jaderow Ltd.* [1989] ECR 4509.
Case 247/87	*Star Fruit* v. *Commission of the European Communities* [1988] ECR 291.
Case 262/87	*Kingdom of the Netherlands* v. *Commission of the European Communities* [1989] ECR 225.
Case C-290/87	*Commission of the European Communities* v. *Netherlands* [1989] ECR 3083.
Case 301/87	*French Republic* v. *Commission of the European Communities* [1990] ECR I-307.
Case 2/88	*J.J. Zwartveld and others* [1990] ECR I-4405.
Joined Cases 6/88 and 7/88	*Kingdom of Spain and French Republic* v. *Commission of the European Communities* [1989] ECR 3639.

Case 16/88	*Commission of the European Communities* v. *Council of the European Communities* [1989] ECR 3457.
Case C-39/88	*Commission of the European Communities* v. *Ireland* [1990] ECR I-4271.
Case C-64/88	*Commission of the European Communities* v. *French Republic* [1991] ECR I-2727.
Case C-68/88	*Commission of the European Communities* v. *Hellenic Republic* [1989] ECR 2965.
Case C-200/88	*Commission of the European Communities* v. *Hellenic Republic* [1990] ECR I-4299.
Case C-209/88	*Commission of the European Communities* v. *Italian Republic* [1990] ECR I-4313.
Joined Cases 46/87 and 227/88	*Hoechst AG* v. *Commission of the European Communities* [1989] ECR 2859.
Case C-301/88	*The Queen* v. *Intervention Board for Agriculture Produce ex parte The Fish Producers and Grimsby Fishery Producers' Organisation Ltd.* [1990] ECR I-3803.
Case C-331/88	*The Queen* v. *Minister of Agriculture, Fisheries and Food and Secretary of State for Health, ex parte Fedesa and others* [1990] ECR I-4023.
Case C-348/88	*Criminal Proceedings against Jelle Hakvoort* [1990] ECR I-1647.
Case C-358/88	*Oberhausener Kraftfutterwerk Wilhelm Hopermann GmbH* v. *Bundesanstalt fuer landwirtschaftliche Marktordnung* [1990] ECR I-1687.
Case C-370/88	*Procurator Fiscal* v. *Andrew Marshall* [1990] ECR I-4071.
Case C-9/89	*Kingdom of Spain* v. *Council of European Communities* [1990] ECR I-1383.
Case C-62/89	*Commission of the European Communities* v. *French Republic* [1990] ECR I-0925.
Case C-93/89	*Commission of the European Communities* v. *Ireland* [1991] ECR I-4569.
Case C-146/89	*Commission of the European Communities* v. *United Kingdom of Great Britain & N. Ireland* [1991] ECR I-3533.
Case C-213/89	*The Queen* v. *Secretary of State for Transport, ex parte Factortame Ltd. and others* [1990] ECR 2433.
Case C-221/89	*The Queen* v. *Secretary of State for Transport, ex parte Factortame Ltd. and others* [1991] ECR I-3905.
Case C-244/89	*Commission of the European Communities* v. *French Republic* [1991] ECR I-0163.
Case C-246/89 R.	*Commission of the European Communities* v. *United Kingdom of Great Britain & N. Ireland* [1989] ECR I-3125.
Case C-246/89	*Commission of the European Communities* v. *United Kingdom of Great Britain & N. Ireland* [1991] ECR I-4585.
Case C-258/89R	*Commission of the European Communities* v. *Spain* [1991] ECR I-3977.
Case C-280/89	*Commission of the European Communities* v. *Ireland* [1992] ECR I-6185.

Case C-35/5/89	*Department of Health and Social Security* v. *Christopher Stewart Barr and Montrose Holdings Ltd.* [1991] ECR I-3479.
Joined Cases C-6/90 and 9/90	*Andrea Francovitch and Danilia Bonifaci* v. *Italian Republic* [1991] ECR I-5357.
Case T-24/90	*Automec Srl* v. *Commission of the European Communities* [1992] ECR II-2223.
Joined Cases C-48/90 and C-66/90	*Kingdom of the Netherlands and Others* v. *Commission of the European Communities* [1992] ECR I-565.
Joined Cases C-73/90, C-71/90 and C-70/90	*Kingdom of Spain* v. *Council of the European Communities* [1992] ECR I-5191, I-5175, I-5159.
Joined Cases C-63/90 and C-67/90	*Portuguese Republic and Kingdom of Spain* v. *Council of the European Communities* [1992] ECR I-5073.
Case C-195/90R	*Commission of the European Communities* v. *Federal Republic of Germany* [1990] ECR I-3351.
Case C-240/90	*Federal Republic of Germany* v. *Commission of the European Communities* [1992] ECR I-5383.
Joined Cases 251/90 and C-252/90	*Procurator fiscal, Elgin* v. *Kenneth Gordon Wood and James Cowie* [1992] ECR I-2873.
Case C-286/90	*Anklagemydigheden* v. *Peter Michael Poulsen and Diva Navigation Corp.* [1992] ECR I-6019.
Case C-52/91	*Commission of the European Communities* v. *Kingdom of the Netherlands* [1993] ECR I-3069.
Case C-228/91	*Commission of the European Communities* v. *Italian Republic* [1993] ECR I-2701.
Case C-131/92	*Thierry Arnaud and Others* v. *Council of the European Communities* [1993] ECR I-2573.
Case C-135/92	*Fiskano AB* v. *Commission of the European Communities* [1993] ECR I-2885.
Case C-379/92	*Criminal Proceedings against Peralta* [1994] ECR I-3453.
Case C-405/92	*Etablissement Armand Mondiet SA* v. *Armement Islais SARL.* [1993] ECR I-6133.
Joined Cases C-46/93 and C-48/93	*Brasserie du Pêcheur Sa* v. *Bundesrepublik Deutschland and the Queen* v. *Secretary of State for Transport, ex parte Factortame Ltd and Others* [1996] ECR I-1029.
Case C-131/93	*Commission of the European Communities* v. *Federal Republic of Germany* [1994] ECR I-3303.
Case C-280/93	*Federal Republic of Germany* v. *Council of the European Communities* [1994] ECR I-4973.
Case T-493/93	*Hansa Fisch GmbH* v. *Commission of the European Communities* [1995] ECR II-0575.

Case T-572/93	*Odigitria AAE* v. *Council of the European Union and Commission of the European Communities* [1995] ECR II-2025.
Case C-25/94	*Commission of the European Communities* v. *Council of the European Communities* [1996] ECR I-1496.
Case C-44/94	*The Queen* v. *MAFF, ex parte National Federation of Fishermen's Organisations and Others and Federation of Highlands and Islands Fishermen and Others* [1995] ECR I-3115.
Joined Cases T-231/94 R T-232/94 R T-233/94 R T-234/94 R and T-551/93	*Transsacciones Maritimas SA, Recursos Marinos SA and Makuspesca SA* v. *Commission of the European Communities* [1994] ECR II-0885.
Case C-276/94	*Criminal proceedings against Finn Ohrt* [1996] ECR I-0119.
Case C-311/94	*Ijssel-Vliet Combinatie BV* v. *Minister van Economische Zaken,* [1996] ECR I-5023.
Case C-334/94	*Commission of the European Communities* v. *French Republic* [1996] ECR I-1307.
Case C-52/95	*Commission of the European Communities* v. *French Republic,* [1995] ECR I-4443.
Case T-194/95 Int.I	*Area Cova S.A. and Others* v. *Council of the European Community* [1996] ECR II-0591.
Case C-293/95 P.	*Odigitria AAE* v. *Council of the European Union and of the European Communities* [1996] ECR I-6129.
Case C-325/95	*Commission of the European Communities* v. *Ireland* [1996] ECR I-5615.
Case C-4/96	*Northern Ireland Fish Producers' Organisation Ltd (NIFPO) and Northern Ireland's Fishermen's Federation* v. *Department of Agriculture for Northern Ireland* [1998] ECR I-0681.
Case 151/96	*Commission of the European Communities* v. *Ireland* [1996] ECR I-5615.
Case C-316/96	*Commission of the European Communities* v. *Italian Republic* [1997] ECR I-7231.

International Agreements and Conventions

1. *International Agreements and Conventions*

Agreement for the Establishment of the Indian Ocean Tuna Commission, *OJ* L 236/25, 05.10.1995.
Agreement to Promote Compliance with International Conservation and Management Measures by Fishing Vessels on the High Seas, *OJ* L 177/24, 16.07.1996.
Agreement relating to the Implementation of Part IX of the UNCLOS of 10 December 1982, Council Decision of 25 July 1994, 94/562/EC, *OJ* L 215/9, 20.08.1994.
Convention on the Conservation of Antarctic Marine Living Resources *OJ* L 252/27.
Convention on Fishing and Conservation of the Living Resources of the High Seas (Geneva) 559 *UNTS* 285.
Convention on the Conservation of Salmon in the North Atlantic Ocean [1982] *OJ* L 378/25.
Convention on Fishing and Conservation of the Living Resources in the Baltic Sea and Belts (Gdansk Convention) [1983] *OJ* L 237/5, 26.08.1983, as last amended *OJ* L 021/70, 27.01.1996.
Convention on Future Multilateral Co-operation in the North-East Atlantic Fisheries [1981] *OJ* L 227/22, 12.08.1981.
Convention on the Future Multilateral Co-operation in the North-West Atlantic Fisheries [1978] *OJ* L 378/2, 30.12.1978.
Convention on the High Seas (Geneva) 450 *UNTS* 82.
Convention on the Territorial Sea and Contiguous Zone (Geneva) 516 *UNTS* 205.
Food and Agriculture Organisation Code of Conduct for Responsible Fisheries, Rome, 1995.
General Agreement on Tariffs and Trade 1994, 55 *UNTS* 194.
Helsinki Agreement for the Protection of the Baltic Sea [1994] *OJ* L 73/2, 16.03.1994.
International Convention for the Conservation of Atlantic Tunas [1986] *OJ* L 162/34, 16.06.1986.
International Convention on Safety of Fishing Vessels (Torremolinos 1977) Misc. 17 (1978) Cmnd. 7602.
International Convention on the Safety of Lives at Sea *UKTS* 46 (1980).
Lomé Convention *OJ* L 229/1.
OECD Convention 888 *UNTS* 179.
Tarawa Declaration on Driftnet Fishing, 1989, *LOS Bull* **14**, December 1989.
The Castries Declaration on Driftnet Fishing, 1989, *LOS Bull* **14**, December 1989.
United Nations Convention on the Law of the Sea III 1982, *OJ* L 179/3, 23.06.1998.
United Nations Agreement relating to the Conservation and Management of Straddling Stocks and Highly Migratory Fish, *OJ* L 189/17, 03.07.1998.
Vienna Convention on the Law of Fish Stocks Treaties 1969 *UKTS* 58 (1980).

Vienna Convention on the Law of Treaties between States and International Organisations and between International Organisations 1986 Misc. 11 (1987) Cmnd. 244.
Wellington Convention for the Prohibition of Fishing with Long Driftnets in the South Pacific *LOS Bull* **14**, December 1989.
World Trade Organisation Agreement.

2. *Fisheries Agreements between the European Community and Third Countries*

(i) Northern Hemisphere

Faeroes	*OJ* L 053/2, 22.02.1997.
Greenland	*OJ* L 351/16, 31.12.1994, Third Protocol *OJ* L 351, 31.12.1994.
Norway	*OJ* L 346/26, 31.12.1993.
	Additional Protocol consequent of the accession of Austria, Finland and Sweden to the European Union, *OJ* L 187/15, 08.08.1995.
Estonia	*OJ* L 332/17, 20.12.1996.
Latvia	*OJ* L 332/0002, 20.12.1996.
Lithuania	*OJ* L 332/7, 20.12.1996.
Iceland	*OJ* L 346/20, 31.12.1993.
	Council Regulation 1737/93 of 24 June 1993, *OJ* L 161/1, 2.07.1993.
Russia	Under negotiation.
Poland	Under negotiation.

(ii) Co-operation and Fisheries Agreements

Morocco	*OJ* L 306/07, 19.12.1995.
Mauritania	*OJ* L 334/19, 23.12.1996.

(iii) Latin America

Argentine Republic	Council Regulation No. 3447/93 of 28 September 1993, *OJ* L 318/1 of 20.12.1993
	Proposal for Council Regulation, COM/93/12final*/, *OJ* C 64/05, 06.03.93.

(iv) African and Indian Ocean Countries

Angola	*OJ* L 046/57, 17.02.1997, Protocol L 131/9, *OJ* 23.5.1997.
Cape Verde	*OJ* L 21/16, 28.01.1998, Protocol *OJ* L 21/01, 28.01.1998.
Comoros	*OJ* L 217/27, 05.08.1998, Protocol L 269/1, 06.10.1998.
Côte d'Ivoire	*OJ* L 25/81, 31.01.98, Protocol L 25/01, 31.01.1998.
Gambia	*OJ* L 079/2, 23.03.1994 (expired 30.06.1996).
Guinea-Bissau	*OJ* C 22/16, 27.01.1999 (proposal). Protocol *OJ* L 353/7, 24.12.1997.
Guinea	*OJ* L L 196, 14.07.1998, , Protocol *OJ* L 211, 29.07.1998.
Equatorial Guinea	*OJ* L 11/29, 17.01.1998, Protocol *OJ* L 11/1, 17.1.1998.
Madagascar	*OJ* L 295/31, 04.11.98, Protocol *OJ* L 324/14, 02.12.1998.
Mauritius	*OJ* L. 135/5, 27.05.1997, Protocol *OJ* L 278, 11.10.1997.
São Tomé	*OJ* L 279/30, 31.10.96, Protocol 164/1, 21.06.1997.

Senegal	*OJ* L 302/1, 5.11.1997, Protocol L 72/1, 11.03.1998.
Seychelles	*OJ* L 131/52, 27.05.1999, Protocol L 131/52, 27.05.1999.

3. International Schemes of Fishery Inspection and Enforcement

Joint International Scheme of Inspection in the North Atlantic Fisheries Organisation Regulatory Area, *OJ* L 175/1, 6.7.1988 (as amended by Council Regulation No. 436/92, *OJ* L 54/1, 28.2.1992).

System of Observation and Inspection under Article XXIV of the Convention for the Conservation of Antarctic Marine Living Resources, *OJ* L 252/26, 5.9.1981.

Draft Scheme of Inspection in the North East Atlantic Fisheries Regulatory Area (under negotiation).

4. Legal Instruments from other jurisdictions outside the European Union

(i) United States

Magnusson Fishery Conservation and Management Act	
Packard Amendment to the Magnusson Act,	16 USC § 1821 (1994).
Driftnet Impact Monitoring, Assessment, and Control Act of 1987,	codified at 16 USC § 1371–1407.
Pelly Amendment to the Fisheries Protection Amendment Act	22 USC § 1978 (1994).
Marine Mammals Protection Act	16 USC § 1371–1407.
Endangered Species Act	16 USC § 1537.
High Seas Driftnet Fisheries Enforcement Act	16 USC § 1826a–1826c (1994).

European Community Secondary Legislation

1. *Conservation*

(i) Council Regulation No.:

3760/92 of 20 December 1992, establishing a Community system for fisheries and aquaculture, *OJ* L 389/1, 31.12.1992.

1181/98 of 4 June 1998, amending Regulation No. 3760/92 establishing a Community system for fisheries and aquaculture, *OJ* L 164/1, 09.06.1998.

390/97 of 20 December 1996, fixing, for certain fish stocks and groups of fish stocks, the total allowable catches for 1997 and certain conditions under which they may be fished, *OJ* L 66/1, 06.03.1997.

48/1999 of 18 December 1998, fixing, for certain fish stocks and groups of fish stocks, the total allowable catches for 1999 and certain conditions under which they may be fished, *OJ* L 13, 18.01.1999.

(ii) Council Decision No.:

95/1/EC of 1 January 1995, adjusting the instruments concerning the accession of new Member States to the European Union, *OJ* L 1/1, 01.01.1995.

(iii) Commission Regulation No.:

3237/94 of 21 December 1994, laying down detailed rules for the application of the arrangements for access to waters as defined in the Act of Accession of Norway, Austria, Finland and Sweden, *OJ* L 338/20, 28.12.1994.

2. *Technical Conservation Measures*

(i) Council Regulation No.:

894/97 of 29 April 1997, laying down certain technical measures for the conservation of fishery resources, *OJ* L 132/1, 23.05.1997.

850/98 of 30 March 1998, for the conservation of fishery resources through technical measures for the protection of juveniles of marine organisms, *OJ* L 125, 27.04.1998. Corrigendum to Council Regulation No. 850/98, *OJ* L 318/63, 27.11.1998.

1239/98 of 8 June 1998, amending Regulation No. 894/97, *OJ* L 171/1, 17.06.1998.

Commission Regulation No.:

2108/84 of 23 July 1984, laying down detailed rules for determining the mesh size of fishing nets, *OJ* L 194, 24.07.1994. Corrigendum: *OJ* L 244/46, 14.09.1984.

2550/97 of 16 December 1997, amending Regulation No. 2108/84, *OJ* L 349, 19.12.1997. Corrigendum *OJ* L 054/10, 25.02.1998.

3440/84 of 6 December 1984, on the attachment of devices to trawls, Danish seines and similar nets, *OJ* L 318/23, 07.12.1984. Corrigendum *OJ* L 062/66, 01.03.1985.
955/87 of 1 April 1987, amending Regulation No. 3440/84, *OJ* L 090/29, 02.04.1987.
2122/89 of 14 July 1989, amending Regulation No. 3440/84, *OJ* L 203, 15.07.1989.

3. *Control*

(i) Council Regulation (EC) No.:
2847/93 of 12 October 1993, establishing a control system applicable to the Common Fishery Policy, *OJ* L 261/1, 20.10.1993.
2870/95 of 8 December 1995, amending Regulation. No. 2847/93, *OJ* L 301/1, 14.12.1995.
2489/96 of 20 December 1996, amending Regulation. No. 2847/93. *OJ* L 338/12, 28.12.1996.
686/97 of 14 April 1997, amending Regulation. No. 2847/93, *OJ* L 102/1, 19.04.1997.
2205/97 of 30 October 1997, amending Regulation. No. 2847/93, *OJ* L 304/1, 07.11.1997.
2635/97 of 18 December 1997, amending Regulation No. 2847/93, *OJ* L 356/14, 31.12.1997.
2846/98 of 17 December 1998, amending Regulation No. 2847/93, *OJ* L 358/5, 31.12.1998.
2241/87 of 23 July 1987, establishing certain control measures for fishing activities. *OJ* L 207/1, 29.07.1987.
3483/88 of 7 November 1988, amending Regulation. No. 2241/87, *OJ* L 306/2, 11.11.1988.
1381/87 of 20 May 1987, establishing detailed rules concerning the marking and documentation of fishing vessels, *OJ* L 131/10, 21.05.1987.
3690/93 of 20 December 1993, establishing a Community system laying down rules for the minimum information to be contained in fishing licences, *OJ* L 341/93, 31.12.1993.
1627/94 of 27 June 1994, laying down general provisions concerning special fishing permits, *OJ* L 171/7, 06.07.1994.
3317/94 of 22 December 1994, laying down general provision concerning the authorisation of fishing in the waters of a third country under a fisheries agreement, *OJ* L 350/13, 31.12.1994.

(ii) Council Decision No.:
95/524/EC of 5 December 1995, amending Regulation No. 2847/93 and Decision 89/631/EEC, *OJ* L 301/35, 14.12.1995.

(iii) Commission Regulation (EC) No.:
1381/87 of 20 May 1987, establishing detailed rules and documentation of fishing vessels, *OJ* L 132/11, 21.05.1987.
1382/87 of 20 May 1987, establishing detailed rules concerning the inspection of fishing vessels, *OJ* L 132/11, 21.05.1987.
3561/85 of 17 December 1985, concerning information about inspection of fishing activities carried out by national control authorities, *OJ* L 339/29, 18.12.1985.

2943/95 of 20 December 1995, setting out detailed rules for applying Regulation No. 1627/94 laying down general provisions concerning special fishing permits *OJ* L 308/15, 21.12.1995.

1292/97 of 3 July 1997, laying down pursuant to Article 10(2) of Regulation No. 2847/93 establishing a control system applicable to the common fisheries policy, notification deadlines for fishing vessels flying the flag of, or registered in, certain third countries, *OJ* L 176/21, 04.07.1997.

1489/97 of 29 July 1997, laying down detailed rules for the application of Regulation 2847/93 as regards satellite-based vessel monitoring systems, *OJ* L 202/18, 30.07.1997.

1449/98 of 7 July 1998, amending Regulation No. 2847/93, *OJ* L 192/4, 08.07.98.

435/98 of 24 February 1998, correcting Regulation No. 1489/97, *OJ* L 54/5, 25.02.98.

831/99 of 21 April 1999, amending Regulation No. 1489/97, *OJ* L 105/20, 22.04.1999.

4. *Fishing Effort*

(i) Council Regulation (EC) No.:

685/95 of 27 March 1995, establishing a system for the management of fishing effort relating to certain Community fishing areas and resources, *OJ* L 71/5, 31.03.1995.

2027/95 of 15 June 1995, establishing a system for the management of fishing effort relating to certain Community fishing areas and resources, *OJ* L 199/1, 24.08.1995.

149/1999 of 19 January 1999, amending Regulation No. 2027/95, *OJ* L 18/3, 23.01.1999.

2870/95 of 8 December 1995, amending Regulation. No. 2847/93, *OJ* L 301, 14.12.1995.

779/97 of 24 April 1997, introducing arrangements for the management of fishing effort in the Baltic Sea, *OJ* L 113/1, 30.04.1997.

(ii) Commission Regulation (EC) No.:

523/96 of 26 March 1996, adjusting the maximum annual fishing effort for certain fisheries, *OJ* L 77/12, 27.03.1996.

1059/97 of 11 June 1997, adjusting the maximum annual fishing effort for certain fisheries, *OJ* L 154/26, 12.06.1997.

2945/95 of 20 December 1995, amending Regulation. No. 2807/83 laying down detailed rules for recording information on Member States' catches of fish, *OJ* L 308/18, 21.12.1995.

2091/98 of 30 September 1998 concerning the segmentation of the Community fishing fleet and fishing effort in relation to multiannual guidance programmes, *OJ* L 266/36, 01/10/1998.

2092/98 of 30 September 1998, concerning the declaration of fishing effort relating to certain Community fishing effort areas and resources, *OJ* L 266/47, 01.10.1998.

5. *Statistics Logbook and Registration of catches*

(i) Council Regulation (EC) No.:

1382/91 of 21 May 1991, on the submission of data on the landings of fishery products in Member States, *OJ* L 133/1, 28.05.91.

3880/91 of 17 December 1991 on the submission of nominal catch statistics by Member States fishing in the north-east Atlantic, *OJ* L 365/1, 31.12.91

2018/93 of 30 June 1993, on the submission of catch data and activity statistics by Member States. fishing in the Northwest Atlantic, *OJ* L 186/1, 28.07.1993.
858/94 of 12 April 1994, introducing a system for the statistical monitoring of trade in bluefin tuna (Thunnus thynnus) within the Community, *OJ* L 9/1, 19.04.1994.
2104/93 of 22 July 1993 amending Regulation No. 1382/91, *OJ* L 191/1, 31.07.1993.
2597/95 of 23 October 1995, on the submission of nominal catch statistics by Member States fishing in certain areas other than those of the North Atlantic, *OJ* L 270/1, 13.11.1995.

(ii) Commission Regulation (EEC) No.:
2807/83 of 22 September 1983, laying down detailed rules for recording information on Member States' catches of fish, *OJ* L 276/1, 10.10.1983.
473/89 of 24 February 1989, amending Regulation No. 2807/83, *OJ* L 53/34, 25.02.1989.
2945/95 of 20 December 1995, amending Regulation No. 2807/83, *OJ* L 308/18, 21.12.1995.
137/79 of 19 December 1978, on the institution of a special method of administration co-operation for applying intra-Community treatment to the fishery catches of vessels of Member States, *OJ* L 20/1, 27.01.1979.
804/86 of 18 March 1986, amending Regulation No. 137/79, *OJ* L 75/14, 20.03.1986.
3634/89 of 4 December 1989, amending Regulation No. 137/79, *OJ* L 355/22, 05.12.1989.
3399/91 of 21 November 1991, amending Regulation No. 137/79, *OJ* L 320/19, 22.11.1991.
395/98 of 19 February 1998, amending Regulation No. 2807/83, *OJ* L 50/17, 20.02.1998.
1488/98 of 13 July 1998, amending Regulation No. 2807/83, *OJ* L 196/3, 14.07.1998.

(iii) Communication from the Commission No.:
85/C347/05, on the description of the ICES sub-areas and divisions used for the purpose of fishery statistics by Member States in the North-East Atlantic, *OJ* C 347/14, 31.12.1985.
85/C335/02, on the description of certain FAO areas, sub-areas and divisions used for the purpose of fishery statistics and regulations, *OJ* C 335/2, 24.12.1985.

6. *Baltic Sea*

Council Regulation (EEC) No.:
88/98 of 18 December 1997, laying down certain technical measures for the conservation of fishery resources in the waters of the Baltic Sea, the Belts and the Sound, *OJ* L 9/1, 15.01.1998.
414/96 of 4 March 1996, laying down certain monitoring measures applicable to fishing activities carried out in the waters of the Baltic Sea, the Belts and the Sound, *OJ* L 59/1, 08.03.1996.
1520/98 of 13 July 1998, amending Regulation No. 88/98, *OJ* L 201/1, 17.07.1998.

7. *Mediterranean Sea*

(i) Council Regulation (EC) No.:

1626/94 of 27 June 1994, laying down certain technical measures for the conservation of fishery resources in the Mediterranean, *OJ* L 171/1, 06.07.1994.

1075/96 of 10 June 1996, amending Regulation. No. 1626/94, *OJ* L 142/1, 15.06.1996.

782/98 of 7 April 1998, amending Regulation No. 1626/94, *OJ* L 113/6, 15.04.98.

(ii) Council Decision No.:

97/292/EC 28 April 1997, on a specific measure to encourage Italian fishermen to diversify out of certain fishing activities, *OJ* L 121/20, 13.05.1997.

(iii) Proposal for a Council Decision No.:

97/C 124/08, on the accession of the European Community to the General Fisheries Council of the Mediterranean, *OJ* C 124/61, 21.04.1997.

8. *North Atlantic Fisheries Organisation*

(i) Council Regulation No.:

1956/88 of 9 June 1988, adopting provisions for the application of the scheme of joint international inspection adopted by the Northwest Atlantic Fisheries Organisation, *OJ* L 175/1, 06.07.1988.

436/92, amending Regulation No. 1956/88, *OJ* L 54/1, 28.02.1992.

306/95, amending Regulation No. 1956/88, *OJ* L 329/1, 30.12.1995.

189/92, adopting certain provisions for the application of the Scheme of Joint International Inspection adopted by the Northwest Atlantic Fisheries Organisation, *OJ* L 21/4, 30.01.1992.

3068/95, adopting certain provisions for the application of the Scheme of Joint International Inspection adopted by the Northwest Atlantic Fisheries Organisation, *OJ* L 329/3, 30.12.1995.

3928/92 of 20 December 1992, establishing a NAFO pilot observer scheme applicable to Community fishing vessels operating in the Regulatory Area of the Northwest Atlantic Fisheries Organisation, *OJ* L 397/78, 31.12.1992.

1388/95 of 15 June 1995, amending Regulation No. 3928/95, *OJ* L 135/1, 21.06.1995.

3069/95 of 21 December 1995, establishing a Community observer scheme applicable to Community fishing vessels operating in the Regulatory Area of the Northwest Atlantic Fisheries Organisation, *OJ* L 329/5, 30.12.1995.

3070/95 of 21 December 1995, on the establishment of a pilot project on satellite tracking in the NAFO Regulatory Area, *OJ* L 329/11, 30.12.1995.

(ii) Council Decision

22 December 1995, concerning the conclusion of the Agreement constituted in the form of an agreed minute, an exchange of letters, an exchange of notes and the Annexes thereto between the European Community and Canada on fisheries in the context of the NAFO Convention, *OJ* L 327/35, 30.12.1997.

(iii) Commission Regulation No.:

2868/88 of 16 September 1988, laying down detailed rules for the application of the Scheme of Joint International Inspection adopted by the Northwest Atlantic Fisheries Organisation, *OJ* L 257/20, 17.09.1988.

9. *Conservation of Antarctic Marine Living Resources*

Council Decision No.:

81/691 of 4 September 1981, on the conclusion of the Convention on the Conservation of Antarctic Marine Living Resources, *OJ* L 252/26, 05.09.1981.

10. *Financial aid to the Member States*

(i) Council Decision No.:

89/631/EEC of 27 November 1989, on a Community financial contribution towards expenditure incurred by Member States for the purpose of ensuring compliance with the Community system for the conservation and management of fishery resources, *OJ* L 364/64, 14.12.1997.

92/393/EEC of 20 July 1992, amending Council Decision No. 89/631/EEC, *OJ* L 213/35, 29.07.1992.

94/207/EC of 12 April 1994, amending Council Decision No. 89/631/EEC, *OJ* L 101/9, 20.04.1994.

95/528/EC of 5 December 1995, amending Council Decision No. 89/631/EEC, *OJ* L 301/35, 14.12.1995.

95/527/EC of 8 December 1995, on a Community financial contribution towards certain expenditure incurred by the Member States implementing the monitoring and control systems applicable to the common fisheries policy, *OJ* L 301/30, 14.12.1995.

(ii) Commission Decision No.:

96/286/EC of 11 April 1996, laying down detailed rules for the application of Council Decision No. 95/527/EC, *OJ* L 106/37, 30.04.1996.

96/299/EC of 17 April 1996, on the eligibility of expenditure to be incurred by certain Member States in 1996 for the purpose of introducing monitoring and control systems applicable to the common fisheries policy, *OJ* L 114/35, 08.05.1996.

97/297/EC of 28 April 1997, on the eligibility of expenditure to be incurred by certain Member States in 1997 for the purpose of introducing monitoring and control systems applicable to the common fisheries policy, *OJ* L 122/24, 14.05.1997.

98/26/EC of 16 December 1997, amending Decision 97/297/EC, *OJ* L 8/30, 14.01.1998.

11. Common Organisation of the Market in Fishery and Aquaculture Products

(i) Council Regulation (EC) No.:

3759/92 of 17 December 1992, on the common organisation of the market in fishery and aquaculture products, *OJ* L 388/1, 31.12.1992.

697/93 of 17 March 93, amending Regulation. No. 3759/92, *OJ* L 76/12, 30.03.1993.

1891/93 of 12 July 93, amending Regulation. No. 3759/92, *OJ* L 172/1, 15.07.1993.

3318/94 of 22 December 1994, amending Regulation. No. 3759/92, *OJ* L 350/15, 31.12.1994.

2406/96 of 26 November 1996, laying down common marketing standards for certain fishery products, *OJ* L 334/1, 23.12.1996.

1093/94 of 6 May 1994, setting the terms under which fishing vessels of a third country may land directly and market their catches at Community ports, *OJ* L 121/3, 12.05.1994.

(ii) Council Directive No.:

91/493/EEC of 19 July 1991, laying down the health conditions for the production and the placing on the market of fishery products, *OJ* L 268/15, 24.09.1991.

92/48/EEC of 16 June 1992, laying down the minimum hygiene rules applicable to fishery products caught on board certain vessels in accordance with **Article 3 (1)(a)(i)** of Directive No. 91/493/EEC, *OJ* L 187/41, 07.07.92.

95/71/EC of 22 December 1995, amending the Annex to Directive No. 91/493/EEC, *OJ* L 332/40, 30.12.1995.

(iii) Commission Regulation No.:

3703/85 of 23 December 1985, laying down detailed rules for applying the common marketing standards for fresh or chilled fish, *OJ* L 351/63, 28.12.1985.

3506/89 of 23 November 1989, amending Regulation. No. 3703/85, *OJ* L 342/11, 24.11.1989.

323/97 of 21 February 1997, amending Regulation No. 2406/96, OJ L 52/8, 22.02.1997.

12. Structural Measures

(i) Council Regulation (EC) No.:

2080/93 of 20 July 1993, laying down provisions for implementing Regulation No. 2052 as regards the financial instrument of fisheries guidance, *OJ* L 193/1, 31.07.1993.

2930/86 of 22 September 1986, defining characteristics for fishing vessels, *OJ* L 274/1, 25.90.1986.

3259/94 of 22 December 1994, amending Regulation No. 2930/88, *OJ* L 339/11, 29.12.1994.

(ii) Commission Regulation No.:

109/94 of 19 January 1994, concerning the fishing vessel register of the Community, *OJ* L 19/5, 22.01.1994. (Repealed)

2090/98 of 30 September 1998, concerning the fishing vessel register of the Community, *OJ* L 266/27, 01.10.1998.

(ii) Commission Decision No.:

95/84/EC of 20 March 1995, implementing the Annex to Regulation No. 2930/86 *OJ* L 67/33, 25.03.1995.

97/259/EC of 1 April 1997, concerning the tonnage of the multiannual guidance programmes for the Community fishing fleets for the period 1992 to 1996 *OJ* L 104/28, 22.04.1997.

(iv) Communication from the Commission

COM(96)203 final of 30 May 1996, relating to management guidelines for the 4th generation of multiannual guidance programmes.

List of Abbreviations

ACP	African Caribbean Pacific (States)
AFDI	*Annuaire Français de Droit International*
AJIL	*American Journal of International Law*
Art	Article
Baltic States	Estonia, Latvia, Lithuania
BYIL	*British Yearbook of International Law*
CAP	Common Agriculture Policy
CCAMLR	Convention on the Conservation of Antarctic Marine Living Resources
CCP	Common Commercial Policy
CCT	Common Customs Tariff
CFP	Common Fisheries Policy
CFSP	Common Foreign and Security Policy
CIEM	Conseil International pour l'Exploration de la Mer
CMLR	*Common Market Law Reports*
CMLRev	*Common Market Law Review*
CLR	*Columbia Law Review*
COM	Communication from the Commission to the Council
COREPER	Comité des Représentants Permanents (Committee Permanent Representatives)
CYBIL	*Canadian Yearbook of International Law*
DG XIV	Directorate General Fisheries (European Commission)
EAGGF	European Agriculture Guidance and Guarantee Fund
EC	European Community
ECR	European Court Report
ECU	European Currency Unit (€)
EEC	European Economic Community
EEZ	Exclusive Economic Zone
EFTA	European Free Trade Trade Association
EFZ	Exclusive Fisheries Zone
ELJ	*European Law Journal*
ELRev	*European Law Review*
EP	European Parliament
EP. Doc	European Document
EU	European Union
EUROPOL	European Police Office
FAO	Food and Agriculture Organisation
FZ	Fisheries Zone
GATT	General Agreement on Tariffs and Trade

GIELR	*Georgetown International Environmental Law Review*
GNP	Gross National Product
GRT	Gross Registered Tonnage
GYIL	*German Yearbook of International Law*
HILJ	*Harvard International Law Journal*
IBSFC	International Baltic Sea Fisheries Commission
ICCAT	International Commission for the Conservation of Atlantic Tunas
ICES	International Council for Exploration of the Sea
ICJ	International Court of Justice
ICLQ	*International and Comparative Law Quarterly*
ICNAF	International Commission for the Northwest Atlantic Fisheries
IFREMER	Institut français de recherche pour l'exploitation de la mer
IGC	Intergovernmental Conference
IJECL	*International Journal of Estuarine and Coastal Law*
IJEL	*Irish Journal of European Law*
IJMCL	*International Journal of Marine and Coastal Law*
ILA	International Law Association
ILO	International Labour Organisation
IMO	United Nations International Maritime Organisation
INMARSAT	International Maritime Satellite Organisation
JCMS	*Journal of Common Market Studies*
JHA	Justice and Home Affairs
JMLC	*Journal of Maritime Law and Commerce*
LIEI	*Legal Issues of European Integration*
LJIL	*Leiden Journal of International Law*
LOS Bulletin	*Law of the Sea Bulletin*
LOS Convention	United Nations Law of the Sea Convention
MAGP	Multi Annual Guidance Programme
MARPOL	International Convention for the Prevention of Pollution from Ships
MLR	*Modern Law Review*
Mpolicy	Marine Policy
MSY	Maximum Sustainable Yield
NAFO	North Atlantic Fisheries Organisation
NASCO	North Atlantic Salmon Conservation Organisation
NEAFC	North East Atlantic Fisheries Commission
NYIL	*Netherlands Yearbook of International Law*
ODIL	*(Journal of) Ocean Development and International Law*
OECD	Organisation for the Economic Co-operation and Development
OJ	*Official Journal (of the European Communities)*
Reg	Regulation
RGDIP	*Revue Générale de Droit International Public*
RMC	*Revue du Marché Commun*
RTDE	*Revue Trimestrielle de Droit Européen*
SI	Statutory Instrument
SOLAS	International Convention for the Safety of Life at Sea
TAC	total allowable catch

TEU	Treaty of European Union
TJICL	*Tulane Journal of International and Comparative Law*
UKTS	United Kingdom Treaty Series
UNCED	United Nations Conference Environment and Development
UNCLOS	United Nations Conference on the Law of the Sea
UNCLOS III	Third United Nations Conference on the Law of the Sea (1973–1982)
UN Doc	United Nations Document
UNTS	*United Nations Treaty Series*
USC	United States Code
VMS	Vessel Monitoring System
PQ	Parliamentary Question (written) European Parliament

Introduction

Fishing activities throughout the world are subject to a good deal of regulation and are usually conducted within some type of management framework which is designed to conserve, manage and develop marine living resources.

Within the European Union fishing activities are governed by the Common Fisheries Policy (CFP). This policy is complex and its essential elements and underlying principles are often perceived as being incomprehensive and perhaps even contradictory. Moreover, the policy is frequently perceived as having failed to achieve its main objectives, the sustainable use and conservation of the the resource and the development of the fishing sector. The merits or otherwise of these perceptions are not explored in this book which has as its main focus the enforcement of the rules which underpin the CFP.

Enforcement in fisheries has now become a major priority throughout the Community. This has been brought about by a widespread realisation that enforcement is an essential element for the survival of the industry and that the absence of effective monitoring control and surveillance in the past has contributed to the recurring difficulties which have buffeted the fisheries sector over the last decade. Events and decisions outside the Community have also aided this process.

The aim of this book is to examine how the basic elements of the CFP are enforced, who conducts this task, the results attained and the probable shape of fisheries enforcement up to and beyond the year 2000. The context within which enforcement is conducted is explained in Chapter 1 which reviews the evolution of fisheries policy in the Community up to the adoption of the main elements of a definitive policy in 1983. This chapter also examines the detailed review of the role of enforcement which was conducted in the early 1990s.

Chapter 2 presents a profile of the industry which has undergone a major transformation over the last decade. Fleet size and employment levels in the industry have been reduced, access to traditional resources have become heavily curtailed and admission to new fishing opportunities is now more difficult to obtain. The second issue addressed in Chapter 2 concerns legal and enforcement competence in the Community – how and why it is largely retained in the hands of Member States. The division of competence and how it functions are crucial for an understanding of how enforcement is conducted in the fisheries sector.

Chapters 3–5 inclusive deal in depth with the Community control system as laid down in Council Regulation (EC) No. 2847/93 and subsequent amending legislation. This system covers key elements of the CFP, and its adoption in 1993 represents a major shift in enforcement policy in the Community.

The limited but crucial role played by the Commission and the Directorate General for Fisheries in the enforcement field is the subject of Chapter 6. The focus here is on how the Commission undertakes its task of monitoring enforcement by and in

Member States, the tools at its disposal and its capacity to bring about improvements. This Chapter concludes by examining the place of sanctions in the enforcement process. The vexed question of quota hopping and the attempts to curtail the activities of quota hoppers is addressed in Chapter 7 through a review of case law and national legislation. The enforcement problems with this phenomenon are assessed and possible solutions are explored. Chapter 8 looks at the procedure available under the Treaty for bringing Member States to court if there are manifest failures in enforcement. It also discusses some other options to the traditional processes and procedures in the enforcement of Community fishery law.

The costs incurred by and the financing of enforcement are examined in Chapter 9 where it is apparent that there has been a significant allocation of new resources since the mid 1990s. This highlights the seriousness which is now attached to the need for better enforcement compared to the early period of the CFP. The penultimate Chapter presents a case study on the interaction between the Commission and Member States when attempting to address a sensitive enforcement issue - the use of large scale driftnets by Community vessels. This example also serves to illustrate a number of crucial points about the enforcement task - the necessity and difficulty of encouraging national authorities to adopt appropriate control strategies, the capacity of the Commission to influence Member States and the role of environmental groups in focusing attention on deficiencies in control.

The potential presented by the use of modern technology in the enforcement field is examined in chapter 11. Here the Community has taken the initiative and adopted an ambitious plan to construct and operate a satellite-based tracking system covering the majority of the larger fishing vessels flying the flags of the Member States. In addition, the Community has supported the introduction of new technology in fisheries managed by international fisheries organisation. The second subject discussed in Chapter 11 relates to several recently concluded international legal instruments which will impact on the enforcement of fishery law in the global setting.

The book does not present any dramatic conclusions but seeks to stress that whilst much improvement is needed in the implementation of current rules, the framework for better control is in place, together with the financial resources and the best that modern technology can offer. Whether all these elements will be sufficient will be revealed following the next major review of the CFP in the run-up to the year 2002.

Chapter 1

The Context: Overview of the Common Fisheries Policy[1]

Introduction
The purpose of this chapter is to examine the evolution of the Common Fisheries Policy (hereinafter, referred to as the CFP in its abbreviated form) from the historical perspective.[2] After examining the Treaties of the European Union it is proposed to review the development of Community fisheries law in the context of three distinct time periods:

(i) The early period of the policy up to 1976.
(ii) Adoption of the CFP in 1983.
(iii) Revision of key elements of the CFP in the 1990s.

The Treaties

1. *The treaty establishing the European Community (EC Treaty)*[3]

The legal basis of Community fishery legislation can be traced to different articles of the EC Treaty. These can broadly be categorised for clarity, as the agriculture articles in Title II, and broader Treaty provisions such as **Articles 6** {12} and **235** {308}.

(i) Legal basis in the agriculture/fisheries ***Articles 38–47*** *{32–38}*[4]
In accordance with the original foundation treaty, the European Community shall:

> 'have as its task, by establishing a common market and progressively approximating the economic policies of Member States, to promote throughout the Community a harmonious development of economic activities, a continuous and balanced expansion, an increase in stability, an accelerating raising of living standards and closer relations between the States belonging to it.'[5]

In order to regulate and promote the free market which the EC Treaty created, it was necessary to adopt a number of common policies.[6] Interestingly, agriculture was virtually excluded from the initial draft of the Community written constitution.[7] However, by the time the Treaty was signed on 25 March 1957,[8] it was accepted by the negotiating states that there was a requirement for the adoption of a Common Agricultural Policy (hereinafter referred to as CAP).[9] **Articles 38–47** {32–38} of the EC Treaty set out the broad 'outline' for the development of the policy and provide a legal basis for the secondary implementation legislation.

By definition agricultural products means:

> 'the products of the soil, of stock farming and of *fisheries* and products of first stage processing directly related to these products.'[10] (emphasis added)

Furthermore, **Article 38(3)** {32(3)} states that the products subject to the provisions of **Articles 39** {33} to **46** {38} are listed in Annex 11 of the Treaty. These products include fish, crustaceans and molluscs. The common market therefore extended to fisheries and trade in fishery products, and **Article 38(4)** {32(4)} requires that the development of this market must be accompanied by the establishment of a common policy.

The foundation treaty explicitly deals with agriculture and only mentions fisheries specifically once - in the analogous context of agricultural products. However, it is within this context, that the EC Treaty, set down the legal requirement for the community to adopt a common fisheries policy. Since the EC Treaty entered into force in 1958 it is generally accepted that all the agriculture **Articles, 38** {32} *et seq.*, apply equally to fisheries.[11] Many of the original Treaty provisions however have since been amended by both the Treaty of European Union and the Treaty of Amsterdam.

The aims of the CAP are set out in **Article 39(1)** {33(1)} and **(2)** {33(2)}, and if the term *fisheries* is substituted for agriculture, and CFP for CAP, then the EC Treaty requires *inter alia*, the fisheries policy

(a) to increase *fishery* productivity by promoting technical progress and by ensuring the rational development of *fisheries* production and the optimum utilisation of the factors of production, in particular labour;
(b) thus to ensure a fair standard of living for the *fisheries* community, in particular to increase the individual earnings of persons engaged in *fisheries;*
(c) to stabilise markets;
(d) to ensure the availability of supplies;
(e) to ensure that supplies reach consumers at reasonable prices.

In formulating the CFP, and the special method for its application, **Article 39(2)** {33(2)} directs that account is to be taken of:

(a) the particular nature of *fisheries* activity, which results from the social structure of *fisheries* and from structural and natural disparities between the various *fisheries* regions;
(b) the need to effect the appropriate adjustments by degrees;
(c) the fact that in the Member States *fisheries* constitutes a sector closely linked with the economy as a whole.

In order to attain the objectives set out in **Article 39** {33}, **Article 41** {35} provides that provision may be made within the CFP for such measures as effective co-ordination of efforts in the spheres of vocational training, research and the dissemination of *fisheries* knowledge; this may include the joint financing of projects or institutions, and joint measures to promote the consumption of certain products.

The other important issues were markets, trade, and competition. The development of the common organisation of the market which was required by **Article 40** {34}, followed a European market organisation form.[12] Additional safeguards were provided for national markets under Article **43(3)** {37(3)}. Furthermore, **Article 42** {36} stipulated that EC competition and trade rules would apply to fishery products as determined by the Council.

Article 43(2) {37(2)} provides that the Council, acting on a proposal from the Commission after consulting the Economic and Social Committee and the European Parliament, shall implement the policy. There are no restrictions on the *ratione loci* of **Article 43** {37}.[13] The CFP should have been adopted by 31 December 1969, which was the end of the transitional period. However, as a result of numerous deliberations and *'twelfth hour'* negotiations, the Commission's proposals were delayed for nearly a year and were not finally adopted until October 1970.

(ii) Legal basis in other Treaty articles and the Act of Accession[15]

There are a number of provisions in the EC Treaty which have provided a legal basis for the development of specific aspects of the CFP. Specifically, **Article 6** {12} non discrimination on grounds of nationality; **Article 235** {308}, the 'catch all' provision, which is relied on in the absence of other appropriate legal basis for Community legislation; and, in cases of expediency such as arose on several occasions between 1977 and 1983, where Community legislation is required without the possibility of resorting to the procedure in **Articles 40** {34} and **43** {37}, then **Article 103** {99} ('conjunctural policy provisions') can be relied on as a legal basis. It has been suggested by some commentators that **Article 102** of the Act of Accession and **Article 100** {94} of the Treaty could alternatively have been used as a legal basis for fisheries legislation. The arguments are inconclusive.[16] It has also been argued that **Article 52** {43}, the right of establishment, and **Article 59** {49}, the freedom to supply services, are grounds for the equal access principle which, in theory, allows Community fishermen access to the fishing grounds in the waters under the sovereignty and jurisdiction of all Member States. The Court of Justice has ruled on the application of these principles in the colloquially referred to 'quota hopping cases' which are discussed in Chapter 10.

(iii) General application of Community Law

Unless stated otherwise, pursuant to **Article 38(2)** {32(2)}, then the general *corpus* of Community law applies to fisheries.[17] The legal instruments relied on by the Community institutions to implement the CFP are regulations, directives and decisions.[18] In practice, one of the distinctive features of Community fisheries law has been the reliance by the institutions on regulations as the predominant legislative tool in constructing the policy. In theory, this ought to facilitate the expeditious implementation of the policy in the Member States through the principles of direct applicability and direct effectiveness.

2. *The Single European Act (SEA)*[19]

The Single European Act (SEA) signed in 1986 made adjustments to the structure of the Community by expanding competencies and changing the institutional order and legislative procedure. This effected all policies including fisheries. Titles inserted into the Treaty included Economic and Social Cohesion, which emphasises the Community's role in promoting the social and economic welfare of its citizens and regions.

This had clear implications for the CFP as had the obligation on Member States to achieve the single market by 1992.

3. *The Treaty of European Union (TEU)*

The Treaty of European Union (TEU) was signed in Maastricht on 7 February 1992 and came into force on 1 November 1993. **Article 3(d)** of the EC Treaty has been renumbered and added to by a new article **3(e)** {3(e)}, which states that the activities of the Community shall include a common policy in agriculture and *fisheries* (emphasis added). Thus, the Community belatedly addresses, and acknowledges, the uniqueness of sea fisheries as distinct from agriculture, as well as providing a firm and clearly identifiable Treaty base for the CFP. The TEU sets out the framework and timetable for achieving the 'approximating of economic policies of Member States' required by the EC Treaty,[20] and the improvement of inter-Member State co-operation. Furthermore, the principle of subsidiarity as prescribed by the TEU (achieved through the inserting of a new **Article 3(b)** {5} in the EC Treaty) will have a profound impact on the development of the CFP and is now central to the division of competence between the Community institutions and the Member States. The principle is also very significant in the context of the Community fishery control and enforcement structure.[21] The second paragraph of **Article 3(b)** {5} reads as follows:

> 'In areas which do not fall within its exclusive competence, the Community shall take action, in accordance with the principle of subsidiarity, only if and in so far as the proposed action cannot be sufficiently achieved by the Member States and can therefore, by reason of the scale and effects of the proposed action, be better achieved by the Community.'

One of the significant provisions introduced by the TEU is the joint legislative procedure for legislation adopted jointly by the Council and the Parliament as laid down in **Article 189(b)** {247} of the EC Treaty.[22] However, because the majority of fisheries legislation is based on **Article 43** {37} the European Parliament has merely the right to be consulted.[23] In any case, it has been argued that the co-operation procedure,[24] and the joint legislative procedure are unsuitable for fisheries legislation on the grounds that conservation and enforcement measures by their nature require an expeditious legislative procedure to ensure the prompt adoption and implementation of regulatory measures.[25]

Co-operation and law enforcement in the fields of justice and home affairs are firmly anchored under Title VI of the TEU as an area of common interest for Member States to be dealt with on an intergovernmental basis. This extends to judicial co-operation in criminal matters which has several consequences for fishery law enforcement because in several Member States (United Kingdom, Ireland, the Netherlands) fisheries offences are dealt with through the criminal law process. While the TEU placed an obligation on Member States to consult, collaborate and co-ordinate their actions in this field, federal elements were lacking and the jurisdiction of the Court of Justice was severely limited. Experience in the first years of co-operation in the fields of justice and home affairs clearly demonstrated the shortcomings in Title VI {VI}. The deficiencies have been repeatedly criticised by the European Parliament and highlighted by each of the institutions of the European Union as well as by the refection groups in preparation of the Inter-Governmental Conferences (IGC) in 1996 and 1997.[26] Several of the concerns regarding the inadequacy of the TEU structures and procedures are addressed in the Treaty of Amsterdam.

4. *Treaty of Amsterdam*

Government leaders concluded a new Treaty for Europe in Amsterdam in June 1997. The Treaty of Amsterdam has four main objectives, which include the placing of employment and citizens' rights at the heart of the Union, the removal of remaining restrictions on freedom of movement but at the same time to augment internal security of the Union, to give the Union a stronger voice in world affairs, and finally to reinforce the Union's institutional structure to make it more efficient with a view to accommodating the accession of new Member States.

The Treaty of Amsterdam has several implications for the improvement and development of the law in the area of justice and home affairs. There has been a transfer of significant powers to the Community concerning internal and external borders, policies on visas, asylum and judicial co-operation in civil matters.[27] The transfer of powers to the Community in these areas will entail the use of Community legislative instruments (directives and regulations) as opposed to conventions, as well as the increased role for the Court of Justice (including the use of preliminary rulings, albeit restricted to last-instance courts and with the exception of issues concerning the abolition of border controls), and the exclusive right of the Commission in legislative initiative (following a five-year period of transition during which the requirement for Council unanimity continues). These are all major developments in the evolving dynamic of greater European integration. However, it needs to be stressed that co-operation between police and other agencies for criminal matters remains a matter for inter-governmental co-operation. Furthermore, and of major consequence for the enforcement of Community fisheries law, mutual co-operation in criminal cases remains exclusively within the remit of Member State competence, although it needs to be emphasised that, during the deliberations leading to agreement on the Amsterdam Treaty, Member States have recorded their agreement to improve the instruments and mechanisms of this co-operation in order to make it more efficient. Arguably, the proposed new structure which will be introduced by the Amsterdam Treaty will clarify the scope and ambit of Title VI in previous Treaties in so far as the new Treaty clearly circumscribes the normative limits of co-operation in the domain of police and criminal justice co-operation. This will be achieved by means of co-operation between police forces, customs authorities and other executive authorities in the Member States, either directly or via EUROPOL, improved co-operation between national authorities and the courts as well as the approximation of some criminal law provisions. Moreover, the third pillar has been considerably enhanced and now provides a role for the European Parliament. It replaces the existing norm-creating mechanism, the convention, with a new legal instrument, the 'framework decision', which is envisaged as operating in a similar *modus* as the directive in Community law. The proposed structure also extends the Commission's right of legislative initiative to cover all matters under Title VI of the Treaty and provides a limited role for the Court of Justice which entails reviewing the legality and interpretation of acts although its power to give preliminary rulings must be accepted by each Member State.

Section (i) The early period of the policy up to 1976

1. *The move towards a Common Fisheries Policy*[28]

In 1966, the Commission took the initial tentative step towards the creation of a common policy for fisheries by publishing a report on the situation in the fisheries sector of Member States and the basic principles for a common policy.[29] There were several obvious reasons for Community action on fisheries. Ostensibly there was increased exposure to competition for fisheries products as a result of the common market responding to the Common Customs Union and the GATT agreement. The consequent trade liberalisation placed strains on the traditionally protected French and Italian fishing industries.[30] Secondly, third countries were increasingly reluctant to allow Community vessels to fish in their coastal zones and in areas contiguous and adjacent to such zones. This increased the risk of greater fishing effort in the North Atlantic in waters adjacent to the coastal belts of Member States. Thirdly, the technical competitiveness of the European industry decreased in the face of rising competition at world level and catches were not sustainable as a result of the intensive exploitation of fishery resources. The challenge facing the Community was to restore the conditions for rational management and to ensure a socially-acceptable transition from the national management organisation to that of the Community. In common with many other Community policies this task has been pursued by way of legislative action.

Two years were to pass before the Commission presented two draft proposals for regulation to the Council.[31] After protracted Commission/Council dialogue the Council adopted two regulations which laid the cornerstones for the development of the CFP, namely:

(1) Council Regulation (EEC) No. 2141/70 of 20 October, 1970 on the establishment of a common structural policy for the fishing industry[32] (hereinafter referred to as the 1970 structural regulation).
(2) Council Regulation (EEC) No. 2142/70 of 20 October, 1970 on the common organisation of the market in fishery products[33] (hereinafter referred to as the 1970 market regulation).

The impetus for Council agreement on these regulations was provided by the doctrine that applicant states would have to accept the *acquis communautaire* as the basis of their terms of entry into the Community.[34] It was therefore essential for the Member States to agree on the principles of the CFP prior to the impending enlargement of the Community and thus avoid consultation with Ireland, the United Kingdom, Denmark, and Norway, whose interests were substantially different as these countries possessed significant fishery resources.[35] The *acquis* in fisheries was agreed on the 30 June 1970, which was one day prior to the commencement of accession negotiations with the new applicant States. As one commentator has noted, this ended any illusion that the new Member States might avert an unfavourable CFP settlement, but it also clarified their objectives in the accession negotiations.[36]

Council Regulations (EEC) Nos 2141/70 and 2142/70 laid the foundation for the development of the CFP and introduced principles and policies which have shaped the development of community fisheries law since. Firstly, the principle of equal conditions of access to fishery resources was established. Secondly, they provided the legal basis for the Council to adopt the necessary conservation measures to prevent over exploitation of certain fishery stocks. Thirdly, they introduced a financial mechanism to assist the fisheries sector through the implementation of regulations for the development of the structural and market areas of the industry.

2. *The principle of equal conditions of access*[37]

In discussing access, a distinction has to be made between access to fishing grounds of Member States by fishing vessels of a different Member State, and internal access to the fishing industry and resources within a Member State. Equal access refers to the former and it is the cornerstone of the common policy. National access, or internal access, falls within the general rubric of community law and featured in a number of fisheries cases before the Court of Justice in the 1980s and 1990s.[38]

Regulation No. 2141/70 was to become the *'bete noire'* of all subsequent accession negotiations because it laid down the principle of equal access for community fishing vessels to the fishing zones of other Member States. It is unlikely that the long term implications of the principle were fully foreseen in 1970 – bearing in mind that the majority of coastal states did not extend their jurisdiction or sovereignty to 200 miles until 1976. The equal access principle as subsequently carried over into **Article 2** paragraph 1 Council Regulation No. 101/76, laying down a common structural policy for the fishing industry, required that:

> 'Rules applied by each Member State in respect of fishing in the maritime waters coming under its sovereignty or within its jurisdiction shall not lead to differences of treatment of other Member States. Member States shall ensure in particular *equal conditions of access* to and the use of the fishing grounds situated in the waters referred to in the preceding subparagraph for all fishing vessels flying the flag of a Member State and registered in Community territory.'[39] (emphasis added)

There were three exceptions to the equal access principle. Firstly, a coastal band with a maximum limit of 12 nautical miles calculated from the baselines was reserved for local fishermen introduced by the 1972 Act of Accession and continued both by the 1983 and the 1993 Management Regulations (which in some instances was, and still is, subject to historic rights in the outer six-mile band);[40] secondly, the Orkney-Shetland box, a special sensitive fishery introduced by the 1983 Management Regulation;[41] and, thirdly, the special arrangements for Spain and Portugal under the Iberian Act of Accession.[42]

It has been suggested that the equal access principle became less important after 1983 with the introduction of Total Allowable Catches (TACs) and quotas, as well as the requirement for vessels to have a current Community fishing licence and fishing permit. However, it is still important in relation to the exceptions from the general principle (outlined above) and for the future development of the policy.[43]

3. *The enlargement of the Community*

In 1972 negotiations for the accession by Ireland, the United Kingdom, Denmark, and Norway to the European Economic Community hinged on achieving a solution to the issues relating to sea fisheries. The political compromise achieved in the accession negotiations was insufficient to placate a majority of Norwegians who subsequently rejected EC membership in a referendum even though their government had signed the accession Treaty. Fisheries was a major issue during the public debate in Norway (the industry lobbied against membership), in contrast to the other applicant states where there was a strong political desire that what was generally considered as a minor sector issue would not impede accession to the Community.[44]

The negotiations lead to some major amendments to the 1970 Structural Regulation. The principle of free access to all Community waters for fishermen of all Member States was unacceptable to the majority of applicant States.[45] The concession came in the form of four articles entitled 'Fishing Rights' in the Act of Accession 1972.

The Act of Accession 1972

Articles 98 and **99** of the Act of Accession 1972, which are the provisions relating to guide prices and the common organisation of the market, were accepted by the applicant States without much contention.

Articles 100 and **101** of the Act of Accession made provision for restricting access to coastal waters. Specifically, they allowed Member States to restrict fishing in waters under their sovereignty or jurisdiction, situated within a limit of six nautical miles, calculated from the baselines of coastal states, to vessels which fish traditionally in those waters and which operate from ports in that geographical area. Ireland, the United Kingdom and France received a further derogation under **Article 101** in that the 'limit of six nautical miles shall be extended to twelve nautical miles'. This derogation was to last until 31 December 1982, and then a decision was to be taken on the basis of a report from the Commission on the 'economic and social development of the coastal areas of the Member States and the state of the stocks'. The Council was to determine the provisions which were to follow.[46] In any event, the ten-year derogation period was extended by **Article 6** of the 1983 Management Regulation and again by **Article 6** of the 1993 Management Regulation.[47] The latter states that Member States are authorised to retain the arrangement defined in **Article 100** of the 1972 Act of Accession until 31 December 2002. Rights which other states had enjoyed under the 1964 London Convention and Bilateral Agreements thereunder were preserved by **Article 100(2)** of the Act of Accession.[48]

Article 102 called for agreement on a common conservation policies within six years.[49]

4. *International Developments in the Law of the Sea*[50]

In 1972 it was perhaps difficult to predict that, within four years, action by non-Member States was going to force the Community into revising the CFP. Following international momentum at the third United Nations Conference on the Law of the Sea (UNCLOS III) in May 1975 towards the establishment of a 200 nautical mile Exclusive Zones (hereinafter referred to as EZs), and the failure of two international fishery commissions (North East Atlantic Fisheries Commission, and The International Commission for the North West Atlantic Fisheries) to provide effective conservation measures,[51] Canada, the Faroes, USA, USSR, Iceland, and Norway extended their fishery limits to 200 miles.[52]

The exclusion of Community (and non-community vessels) from these North Atlantic grounds effectively forced the Commission and the Member States into considering a limit of 200 nautical miles for the sea areas adjacent to EC Member States. Otherwise, the Community faced the daunting prospect of large fishing fleets which were excluded from their traditional fishing grounds operating out of necessity in waters adjacent to Member States and threatening the sustainable yield of stocks, which were in many instances at the threshold of over-exploitation.

The Commission was under pressure from Ireland and the United Kingdom to introduce national exclusion zones to protect their inshore fishermen within any future 200-mile limit. Other Member States were fearful of losing their historic rights under the London Convention if these claims were successful. The Commission responded by producing a series of reports on these issues.[53] In order to achieve a balanced solution, the Commission proposed that it would thereafter be responsible for the conduct of external fisheries relations and that all 12-mile bands would be reserved for coastal fishermen. The proposals were not favourably received by Member States. In September 1976, the Commission proposed a package which included, *inter alia*:[54]

(1) that Member States extend fisheries zones to 200 miles as from 1 January 1977 in the North Sea and North Atlantic. The extension of areas referred to in **Articles 100** and **101** of the 1972 Act of Accession to 12 miles and the continuation of the derogation with respect to the access principle after the expired date in 1982, included the continuation of historic rights within the 12 miles;
(2) that the Commission would negotiate fishery agreements with third countries in order to obtain fishing opportunities for Community distant water fleets,[55] and that the Commission would represent Member States in international fishery organisations;
(3) a management system for the Community zone, including some specific conservation elements;
(4) measures to rationalise the community fleet, with special priority for the fishing industries in Ireland, the United Kingdom and Greenland.[56]

The Council had already considered an extension of the fisheries zones to 200 miles in a declaration on 27 July 1976, but, due to the disagreement and the delay which threatened the concerted action to extend the zones to 200 miles, the Foreign Ministers of the Member States, in default of a formal agreement, attended a special meeting in the Hague in October 1976.

5. *The Hague Resolution*

At the meeting in the Hague, the Council adopted a resolution, which is sometimes referred to as *the Hague Resolution* or *the Hague Agreement*.[57] The Hague Resolution laid down a series of guidelines as a matter of principle for the future development of the common fisheries policy. The Hague Resolution, which had eight annexes attached to it, was not published in the Official Journal of the European Communities.[58] The reason, according to the Council, was that at the time when the Resolution was adopted, it seemed inappropriate to publish some of its annexes which contained confidential instructions given by the Council to the Commission concerning future negotiations by the Community with non-member countries and international organisations. In the resolution the Foreign Ministers noted that:

> '... present circumstances and particularly the unilateral steps taken or about to be taken by certain third countries, warrants immediate action by the Community to protect its legitimate interests in the maritime regions most threatened by the consequences of these steps to extend fishing zones, and that the measures to be adopted to this end should be based on the guidelines which are emerging within the Third United Nations Conference on the Law of the Sea.'

The Hague Resolution was subsequently approved by the Council on 3 November 1976. In effect the Foreign Ministers agreed that by means of concerted action Member States would extend the limits of their fishing zones to 200 nautical miles off their North Sea and North Atlantic coasts, without prejudice to similar action being taken with respect to the jurisdiction of other fishing zones within their jurisdiction such as the Mediterranean, and as from 1 January 1977:

(1) that fishing in these zones would be governed by agreements between the Community and the countries concerned;
(2) that the Community was to obtain access to zones of third country and retain existing agreement by concluding appropriate agreements with the relevant third countries. To this end and irrespective of the action to be taken in the appropriate international bodies, the Council instructed the Commission to commence negotiations.

By the recommended date 1 January 1977, only four countries: Ireland, Denmark, the United Kingdom, and the Federal Republic of Germany, had enacted the necessary legislation to extend their fishing zones to 200 miles. In due course, and in accordance with international practice, all Member States in the North Atlantic and North Sea extended their coastal zones to 200 miles. This paved the path which led to the adoption of a common policy for fisheries.

One of the issues which is frequently subject to comment since 1976, is whether *the Hague Resolution* is legally binding and produces legal effects.[59] Following the scheme of Community law, Council resolutions are generally considered as expressions of the political will of the Council and thus do not have to be expressed in a particular form. The Court in the *ERTA* case ruled that a Council resolution constituted, what one eminent authority has classified as, a legal act *sui generis*.[60] The Court has since held that Annex VII of the Hague Resolution (which expresses the Council's political will to take account, in applying the future common fisheries policy, of the special needs of the regions in which the populations are particularly dependent on fishing and related activities), does not produce legal effects capable of limiting the Council's legislative powers.[61] The Council may adopt regulations or other measures which are legally binding to give effect to the guidelines contained in Annex VII of the Hague Resolution provided that these are validly adopted under **Article 43** {37} of the EC Treaty.

Section (ii) The adoption of the CFP in 1983

1. *The adoption of a common policy for fisheries*

Despite the initiatives taken by the Council at the meeting in the Hague regarding the extension of coastal Member State jurisdiction, progress in the development of a common policy proved to be dilatory. The procrastination of the Member States was overcome when, in a Council Declaration of 30 May 1980 the Council agreed that the completion of the CFP was a concomitant part of the solution to the problems with which the Community was confronted at that particular time.[62] The Council decided that a policy had to be put into effect by 1 January 1981. In order to comply with the treaties and to conform with the Hague Agreement the Council furthermore agreed that the policy should be based on the following guidelines:

(a) rational and non-discriminatory measures for the management of resources and conservation and reconstitution of stocks so as to ensure their exploitation on a lasting basis in the appropriate social and economic conditions;
(b) fair distribution of catches having regard, most particularly, to traditional fishing activities, to special needs of the regions where the local populations are particularly dependent upon fishing and industries allied thereto,[63] and the loss of catch potential in third-country waters;
(c) effective controls on the conditions applying to fisheries;
(d) adoption of structural measures which include a financial contribution by the Community;
(e) establishment of securely-based fisheries relations with third countries and implementation of agreements already negotiated. In addition endeavours should also be made to conclude further agreements on fishing possibilities, in which the Community, subject to the maintenance of stability on the community market, could also offer trade concessions.

The Council also agreed to examine the Commission proposal on technical conservation measures, control, and also to begin examination of a proposal on quotas for 1980. Although the Commission proposed a draft regulation establishing a community management system, agreement on the proposal led to protracted negotiation. It was not until December 1982 that there was agreement in the Council to adopt Council Regulation (EEC) No. 170/83 establishing a Community system for the conservation and management of fishery resources.[64] This regulation came into effect on 25 January 1983. As was the case of the accession of new Member States in 1972, it was political expediency which provided the necessary incentive for consensus in the Council. In this instance it was the imminent conclusion of the ten-year derogation period from the equal access principle in the 1972 Act of Accession. In 1983 the Council finally agreed the means by which the majority of stocks would be allocated to Member States. Interestingly, the method adopted in 1983 for the division and allocation of the total allowable catch between Member States has been retained ever since.[65] The essential management principles of the CFP were set out in Council Regulation (EEC) No. 170/83 (hereinafter referred to as the 1983 Management Regulation) and as many of the fundamental norms established by the regulation are still pertinent it is proposed to discuss some of the more significant provisions.[66]

2. *The 1983 Management Regulation*

Some of the important aspects of the 1983 Management Regulation related to:

(i) The identification of conservation as an objective of the CFP
As noted above, the EC Treaty defined the objectives of the CFP in the same context as the Treaty provisions relating to the common agriculture policy. Significantly, **Article 1** of the 1983 Management Regulation extended these objectives by providing that the aim of the community system for the management of fisheries resources was:

> 'to ensure the protection of the fisheries grounds, the conservation of the biological resources of the sea and their balanced exploitation on a lasting basis and in appropriate economic and social conditions.'

Moreover, and of particular importance to the peripheral regions within the community, the recital of the preamble stated that the Community system must safeguard the particular needs of regions where local populations are especially dependent on fisheries and related industries.

(ii) Community competence to adopt fisheries measures
The Community system for the conservation and management of fishery resources divides legal competence between the Community institutions and the Member States.[67] Essentially, the Community is legally competent to adopt fisheries measures. More specifically, the 1983 Management Regulation provided that the Council, acting on a proposal from the Commission and in the light of scientific advice, should adopt the conservation measures necessary to achieve the objectives of the Community fishery management plan.[68] Some of the measures detailed related to: the establishment of zones where fishing was prohibited or restricted; the setting of standards as regards fishing gear; the setting of minimum sizes of fish per species; and the restriction of fishing effort by limiting catches.[69] Moreover, conservation measures, total allowable catches (TACs), the adjustment of relative stability, controlling fishing effort by licences, and supervisory measures, were to be adopted by the Council acting by qualified majority.[70] The requirement for the Commission to consult the European Parliament, which has a specific fisheries committee, regarding fisheries legislative proposals will depend on which legislative base underpins the measure in question. Thus, for example, there is a requirement to consult the European Parliament if a proposal for legislation is based on **Article 43** {37} of the EC Treaty. The Parliament in the past has objected to such a limited role,[71] but the Court upheld the procedure in *Albert Romkes* v. *Officier van Justitie for the district of Zwolle.*[72] In this case the Court upheld the power of the Council to fix a total allowable catch each year for any species where it appears necessary to provide conservation measures.

(iii) Access to resources and the principle of relative stability
The 1983 Management Regulation formed the basis for the community system of Total Allowable Catches (TACs). TAC is the maximum amount of fish which is to be exploited from a particular stock during a given period. Scientists from scientific bodies such as the International Council for the Exploration of the Seas (ICES) propose certain TACs which in their view are the thresholds to which a fish stock may be

exploited without adversely affecting the means by which that stock rejuvenates.[73] The Commission on receipt of scientific advice turns to the Scientific, Technical and Economic Committee for Fisheries (STECF) for an opinion.[74] In the light of this opinion the Commission then draws up proposals for legislation which is forwarded to the Council and presented to the experts on the Advisory Committee for Fisheries (ACF). The quotas of fish allocated to each Member State is derived from the TAC by applying a fixed percentage (key) to each TAC.[75] Conveniently, the division of TACs into quotas and the allocation to individual states has followed the same pattern as that adopted *a posteriori* for 1982.[76] [The 1982 quota was used in 1983, as the Council did not achieve consensus until December of that year.] Since then there have been minor adjustments, but otherwise the Council has achieved agreement in time to prescribe quotas each year for certain species for the following year. Because the scheme of allocation has followed the same scheme to that adopted in 1983, this procedure is referred to as the application of the 1983 key. Annual temporary arrangements are contained in the Annual/TAC and Quota regulation. Since 1983 an allocation key has been established for other stocks on the basis of recent catches.

The principle of relative stability was the means by which fishing opportunities were allocated to Member Sates, that is to say, it was the method applied to the TAC to ascertain the division of quotas between Member States. In practice the principle guarantees each Member State a fixed proportion of each species. The principle may be traced back to **Article 4(1)** of the 1983 Management Regulation (repeated in **Article 8(4)** of the 1992 Management Regulation) which stated that:

> 'the volume of catches available to the Community ... shall be distributed between Member States in a manner which assures each Member State relative stability of fishing activities for each of the stocks considered.'

The notion of relative stability was defined as follows in the fifth, sixth and seventh recital in the preamble of the 1983 Management Regulation (later to be repeated in the 1992 Management Regulation):

> 'conservation and management of resources must contribute to a greater stability of fishing activities and must be appraised on the basis of a reference allocation reflecting the orientation given by the Council;
>
> ... in other respects, that stability, given the temporary biological situation of stocks, must safeguard the particular needs of the regions where local populations are especially dependent on fisheries and related industries as decided by the Council in its resolution of 3 November 1976, and in particular Annex VII thereto;
>
> ... therefore, it is in this sense that the notion of relative stability aimed at must be understood.'

The Court of Justice in *Romkes* held that the requirement of relative stability must be understood as meaning that, in the distribution of the total allowable catch to Member States, each Member State should retain a fixed percentage, unless there was an adjustment made by amending regulation adopted by the procedure provided for in the 1983 Management Regulation.[77] Despite the judicial clarification the principle, and its relationship with the Hague Preferences, has been the subject of frequent controversial debate. On occasion it has been asserted that the principle is contrary to some of the fundamental freedoms guaranteed by the EC Treaty.[78] The three criteria which were relied upon to deduce the principal of relative stability (the way the total allowable catch was divided up between Member States) namely: past fishing performance (evaluated on the basis of average catches landed by each Member State in the period 1973–1978); potential fishing losses suffered by Member States as a consequence of the 200-mile fishing limits and exclusive zones claimed by third countries (calculated for the reference period 1973–1976); and the special needs of the regions where the local population were particularly dependent on fisheries and related industries.[79] The system of Hague Preferences accommodated the third of these criteria and the regions in question were Greenland, certain northern parts of the United Kingdom and Northern Ireland, and Ireland. For each of these regions, the 'special needs' of the local population were considered as being represented by quantities of fish landed from certain stocks of fish of importance to the population. Occasionally, since 1983 when Member States' allocation of their share or quota of these particular stock falls below a certain threshold, then the issue and interpretation of the Hague Preferences is raised. This particular question has however been clarified by the Court of Justice in a preliminary ruling under **Article 177** {234(2)} of the EC Treaty on the validity of Annex VII of the Hague Resolution (this annex deals with the special needs of Ireland and other regions in applying the CFP). The Court held that the Council is not precluded from taking account of the Hague Preference system if a reduction in a TAC affects the vital interests of the communities which are particularly dependent upon fishing. Thus, by using the Hague Preference method to allocate quotas among the Member States, the Council cannot be regarded as committing a manifest error or manifestly exceeding the bounds of its discretion. The Hague Preference system thus provides a degree of protection to such communities if a reduction in a TAC affects their vital interests.

Spain and Portugal were not members of the Community in 1983 when the principle of relative stability was applied to allocate quota shares among Member States. In the intervening period between 1980 and 1986, Spain had a preferential fishing agreement with the Community which allowed a limited number of Spanish vessels access to Community resources. **Article 161** of the Iberian Act of Accession set out the percentages of the various TACs allocated to Spain until 2002. Subsequently, on their accession to the Community Spain and Portugal challenged the 1983 relative stability arrangements in the Court of Justice. The Court found that the requirement of relative stability in the allocation among Member States of the catches available to the Community, in the event of the limitation of fishing activities under **Article 4(1)** of the 1983 Management Regulation, must be understood as meaning: in that distribution each Member State is to retain a fixed percentage.[81] The distribution formula laid down under **Article 4(1)**, on the basis of **Article 11**, is to continue to apply until an amending regulation is adopted in accordance with the procedure laid down in **Article 43** {37} of the Treaty.[82] The principle of relative stability cannot be interpreted as placing the Council under the obligation to effect a fresh distribution whenever an increase of a particular stock is established, and where that stock was already covered by the initial allocation.[83] These cases established that the Council was not obliged to reassess the 1983 keys in order to take into account the fishing record of Spanish and Portuguese vessels in Community waters. This decision of the Court restricted Spain from gaining additional quota allocation and ultimately contributed to the phenomenon of quota hopping which is discussed in Chapter 7.

(iv) Historic Rights in coastal zones
Member States, pursuant to the 1983 Management Regulation, were allowed to extend from 6 to 12 miles the zones around their coasts in which access was limited to vessels which had fished traditionally in those waters and which operated from ports in those geographical coastal areas.[84] Essentially, this was just an extension of the ten years in the original derogation to the equal access principle.[85] Furthermore, the right of Community vessels to fish at will outside the 12-mile zones was restricted in order to protect certain fish stocks in certain biologically-sensitive areas (an example is the Shetland Box which restricts the rights of larger vessels to enter this zone to catch fish).[86] The licensing system necessary to control fishing effort in these areas was established.[87]

Finally, the 1983 Control Regulation contained provisions which provided scope for future review of the CFP after a ten-year period.[88]

(iv) Exchange and utilisation of quotas
The 1983 Management Regulation provided for the exchange of quotas between Member States, provided that the Commission has been given prior notice,[89] and this has since become a common feature of the CFP.

It is important to emphasise that Member States retained a degree of control in the management structure in so far as they were allowed to determine in accordance with the applicable Community provisions the detailed rules for the utilisation of quotas.[90] This is commonly referred to as national access. It allows Member States to determine who may fish against their quotas. The conditions under which this competence is exercised must however be in accordance with the general principles of community law.[91] On several occasions in the 1980s and 1990s, the Court invoked EC Treaty guarantees to curtail Member State measures which restricted access to national quotas. Thus, for example, in *The Queen* v. *Ministry of Agriculture, Fisheries and Food, ex parte Agegate Ltd.*, and *The Queen* v. *Ministry of Agriculture, Fisheries and Food, ex parte Jaderow Ltd. and others* the Court found that, under the national quota system, although a Member State can stipulate that vessels using its quota must have a genuine economic link with the flag State, it cannot hinder the normal activity of such vessels nor lay down particular restrictions regarding residency or nationality which are incompatible with some of the well-established general principles of Community law.[92] In *The Queen* v. *Secretary of State for Transport, ex parte Factortame Ltd and others* and in *Commission* v. *Ireland*, and in *Commission* v. *United Kingdom of Great Britain and Northern Ireland* the Court acknowledged that Member States, in this instance the United Kingdom and Ireland, retained competence regarding the conditions governing vessels sailing under their flag.[93] However, restrictions on access to the national register could not be based on conditions which are incompatible with certain principles in the EC Treaty, the Court cited **Articles 7** {12}, **52**{43}, and **221** {294} in these cases to support their view.[94]

3. *General comment on the 1983 Management Regulation*

The 1983 Management Regulation is the cornerstone of the CFP. While many of the provisions it introduced have since been amended, the regulation and the repealing regulation in 1992 continue to remain the guiding axiom for the sustainable development of Community resources. Indeed, it is generally accepted that if the TACs are properly managed and quotas complied with by the industry then these tools provide the basis for an effective fishery management system. However, as a management instrument the TAC and quota system has also been criticised on the grounds that it gives rise to the practice of fishermen discarding fish back into the sea – fish that may have been inadvertently captured when vessels target non TAC species, or when particular quotas for certain species have been exhausted.[95] In this respect the problem with the TAC/quota system is that it is difficult to assess how successful it has been as a conservation tool.

The management regulation was succeeded by many complementary regulations, most notably an array of technical conservation measures.

4. *Technical conservation measures*

Coinciding with the adoption of the 1983 Management Regulation, the Council agreed upon a package of technical conservation measures for the conservation of fishery resources. The 1983 Technical Regulation (Council Regulation No. 171/83) provided detailed rules for conservation in the Atlantic and the North Sea, in so far as it prescribes minimum mesh sizes, maximum bye-catches, minimum fish sizes and closed seasons. The aim of the technical measures is twofold: to protect small classes of fish size (juveniles) and to protect marine ecosystems by improving the inter-specie selectivity of fishing gear.

The Baltic Sea has its own technical regime which differs from that in the Atlantic and the North Sea. Council Regulation 88/98 lays down measures for the conservation of fishery resources in the waters of the Baltic Sea, the Belts and the Sound.[96] Similarly, a specific package of technical measures was adopted for the Mediterranean Sea in 1992.[97]

The original technical regulation for the Atlantic and the North Sea (Technical Regulation No. 171/83) was amended on six occasions and was ultimately repealed when the Council adopted Council Regulation 3094/86.[98] In the recital of the latter regulation it stated that because Regulation No. 171/83 had been amended six times, it was thus necessary to consolidate it for the proper understanding and for its effective enforcement. It was thus somewhat ironic that Regulation No. 3094/86 was itself subject to over 20 amendments,[99] before new proposals for technical measures were put forward by the Commission to clarify, simplify and remove some of the anomalies which exist in the 1986 Technical Regulation. After two years of debate in Council working groups, a new regulation was adopted by the Council on 30 March 1997.[100] The new measures, which will not be implemented until 2000 (in order to allow time for the industry to adjust to their new obligations), aim to reduce the exploitation of immature fish and to reduce the number of undersize fish discarded by fishermen at sea. The new regulation also simplifies previous rules. This, in theory, ought to lead to enhanced compliance by the industry and should also facilitate the verification of compliance by national inspection services. Significantly, the new regulation increases the mesh sizes that may be used in a number of fisheries in order to allow the smaller fish sizes to escape entrapment. It also regulates the selectivity of fishing gear with a view to providing better protection for the marine ecosystem. Importantly, the regulation prohibits fishing in particular areas which are associated with the reproductive cycles of stocks, areas identified by scientists as nursery grounds. The new measures also prohibit the sale and marketing of undersize fish.

Returning to the historical overview, the aforementioned management and technical regulations adopted in 1983 were preceded by the accession of Greece to the Community in 1981. These regulations were in the throes of implementation in the Member States when they had to be amended to allow for the accession of Spain and Portugal in 1986.

Section (iii) Revision of key elements of the CFP in the 1990s

1. Act of Accession of Spain and Portugal[101]

Under the Act of Accession, Spain and Portugal were not integrated fully into the Community system for the conservation and management of resources implemented by the 1983 Management Regulation.[102] Access to the Community fishery resources for Spanish and Portuguese vessels was restricted and the provisions concerning fisheries in the Act of Accession provided a transitional regime which was subject to review in 1996. In general it may be said that the overall framework was complex and only the principal elements in relation to Spanish and Portuguese fishing possibilities are summarised here.

The Act of Accession restricted fishing by Spanish vessels in the waters of the 'Ten' Community Member States.[103] In total a maximum of 300 Spanish vessels could have access to Community waters. The list containing the names of these vessels was referred to as the 'basic list'. Of the 300 vessels no more than 150 could fish at any one time. This was achieved through a 'periodic list'. All Iberian vessels were excluded from the North Sea and the Baltic Sea.

The Act also provided that Spanish vessels were not allowed to fish inside the 50 mile 'Irish Box' (this was defined as an area extending 50 miles from the Irish baseline on the North, West and South coasts, and extended across the Irish Sea to the Bristol Channel) for a 10-year period which commenced on 1 January 1986 and expired on 31 December 1995.[104]

Article 161 of the Act sets out the various proportion of the TACs allocated to Spain. The Council fixed the fishing possibilities for Spanish vessels each year on this basis. Furthermore, pursuant to **Articles 162** and **166** of the Act, these arrangements apply to the year 2002 with possible adjustments in 1992.

Fishing of the 'ten' Member States in Spanish waters was defined in **Article 164.**[105]

The detailed provisions with respect to Portuguese fishing activities were set out in **Articles 346–349** of the Act of Accession.[106] As Portugal was not allowed access to Community waters before accession only 11 vessels were on the list, and only six could fish at any one time. This number is not of great significance in terms of the fishery resources, nor were there fishing opportunities offered to the 'ten' Member States in Portuguese waters, pursuant to **Article 351**. Reciprocal arrangements between Spain and Portugal were fixed by the Act and also expired on 31 December 1995.[107]

The Act of Accession introduced special conditions with respect to licences. The system introduced was elaborate and is not examined here.[108] A special 'control' regime was implemented to monitor and manage the Spanish and Portuguese fleets.[109]

All Spanish and Portuguese vessels, since accession, have been obliged to comply with the *acquis communautaire* with respect to the conservation and management of fisheries resources as well as the Community control system. While the accession system posed a major challenge for the Community, the overall solution adopted in the form of the transitional arrangement created and highlighted some major shortcomings in the CFP. In this respect it is contended later in this book that one of the major weaknesses of the first phase of the CFP was the reliance on the exclusive competence of the national authorities to enforce the policy.[110]

After the third enlargement of the Community, the CFP entered a period of consolidation. The adjustment of the Spanish and Portuguese accession arrangements did not take place until 1996 and are examined below. Prior to this adjustment it was the mid-term review in 1992 which offered an opportunity to analyse the achievements of the CFP.

2. *Review of the effectiveness of the CFP during the period 1983–1992*[111]

The Community system for the management of marine resources which took effect on 1 January 1983, was established to last for a period of 20 years until 31 December 2002. It was the subject of a mid term review in 1992. The Commission prepared a report on the fisheries sector in the Community, which focused on, *inter alia*, the economic development of coastal areas, the state of the stocks, and the long-term sustainability of the resource.[112]

Other than the allocation of the available fish stocks, access rights to the coastal zones, and sensitive regions (i.e. the Shetland Box), the review (referred to as the '*1991 Report*') was broader than originally anticipated and highlighted the shortfall between exploitable resources and the fishing capacity of the Community fleet. The report examined the state of the industry and provided an outlook for the second half of the policy as well as the *post*-2002 period. The Commission, noting that profound changes had occurred in the sector since the inception of the original policy, recommended that the Council should adopt a new set of regulations to adjust the CFP. In particular the CFP needed to address the industry whose fishing capacity (the ability to catch fish) exceeded the available fishery resources in the Community zone. Furthermore, it was clearly apparent that although the 1983 regime had established a common policy, it had not resolved the latent problems in the sector. In 1992 the Commission also produced a special report reviewing the adjustment of the CFP as a consequence of Spanish/Portuguese accession.[113]

The common theme in these reports was that the sector faced a crisis unless the principal elements of the CFP were improved. It was evident that the TAC/quota management mechanism had not been a success, that fishing effort needed to be curtailed, and that the conservation policy had to be linked with the structure, and market policies. Importantly, one of the principal conclusions was that the control and monitoring of the CFP had been largely ineffective in the first half of the CFP.

In response to the Commission proposal,[114] the Council adopted in December 1992 a new basic regulation for a Community system for aquaculture and fisheries. (Hereinafter, referred to as the '1992 Management Regulation').

3. *The 1992 Management Regulation*[115]

The 1992 Management Regulation is all embracing. **Article 1** provides that:

> 'the common fisheries policy shall cover exploitation activities involving living aquatic resources, and aquaculture, as well as the processing and marketing of fishery and aquatic resources where practised on the territory of the Member States or in Community fishing waters or by Community fishing vessels.'[116]

Furthermore, **Article 2(1)** states:

> 'As concerns exploitation activities the general objectives of the common fisheries policy shall be to protect and conserve available and accessible living marine aquatic resources, and to provide for rational and responsible exploitation on a sustainable basis, in appropriate economic and social conditions for the sector, taking account of its implications for the marine eco-system, and in particular taking into account the needs of both producers and consumers.'

The aim of the 1992 Management Regulation is to provide basic guidelines for the establishment of a legislative framework for the CFP. This framework covers, *inter alia*: access to waters and the exploitation of resources; the harmonisation of the conservation policy and the restructuring of the Community fleet; and the essential requirement of an improved Community control system.[117]

It is proposed to examine the 1992 Management Regulation briefly with respect to each of these elements.[118]

(i) Access to waters and resources[119]

The Council retains exclusive competence in establishing the measures laying down the conditions for access to waters and resources, unless there is express provision otherwise in Community regulations.[120] There is nothing new in this requirement, nor the requirement for the Council to act in accordance with **Article 43** {37} of the EC Treaty. **Article 4** of the Management Regulation, continues the customary requirement for the Council to take account of reports from the Scientific Technical and Economic Committee for Fisheries and to act in the light of the available biological, socio-economic and technical analyses.[121] The type and ambit of the measures to be introduced are prescribed and include the broad range of provisions traditionally employed in limiting and controlling exploitation of fishery resources. These include, for example, technical measures for fishing gear, minimum sizes for particular species of fish and quantitative limits on catches.[122] It also introduces new controversial measures such as limiting the time vessels are able to spend at sea.[123] The latter is used as a means of limiting the fishing effort of fishing vessels. Fishing effort is defined in the regulation as: 'the product of its capacity and its activity, and fishing effort of a fleet or group of vessels is the sum of the fishing effort of each individual vessel.'[124] Fishing effort has been introduced to supplement the range of measures available to the Member States to attain a balance between the fishing capacity of their fleet and the resources available for exploitation.[125] In other words, these measures envisage vessels being restricted to port as a means of controlling fishing capacity and/or activity.

Member States are allowed to retain until December 2002 the derogation to the equal access principle defined in **Article 100** of the 1972 Act of Accession restricting access in the coastal 12-mile band.[126] Furthermore, they are also allowed to continue the neighbour access arrangements in the coastal band as defined in Annex 1 for the same period.[127] Effectively this is a continuation of the *status quo* with regard to the historic rights established since the 1977–1983 period. Furthermore, the restricted conservation region around the Shetland Islands is retained.

Also of major significance are the elaborate plans for controlling exploitation of resources through a Community system of licences and fishing permits which regulate the exploitation rate by restricting the volume of catches and, if necessary the fishing effort.[128] Each vessel was obliged to have a Community fishing licence by 31 December 1993.[129] The licence is administered by the flag Member State.[130] However, the flag Member State is obliged to provide the control authorities in a coastal Member States with the appropriate data on the identification, technical characteristics and equipment of vessels which are subject to inspection in the coastal Member State.[131] The data on the licence are required to correspond to both the data on the Community register and to that provided by the Member States in their multi-annual guidance programmes (MAGPs).[132] The licence requirement is supplemented by the requirement for fishing permits for specific stocks in certain sensitive fisheries.[133] A special fishing permit is a prior fishing authorisation issued to a Community fishing vessel to supplement its fishing licence, thereby enabling it to carry out fishing activities during a specified period in a given area for a particular fishery in accordance with the measures adopted by the Council.[134] Consequently, vessels which, for example, fish in areas such as western waters (the sea west of Ireland and the United Kingdom) are obliged to have a permit for this fishery. Access to resources is thus linked to the conservation policy and the limitation of fishing effort, belatedly addressing one of the oversights in the management regime introduced in 1983.

(ii) Harmonising of the conservation policy and the restructuring of the Community fleet

As previously stated, the conservation policy in the 1980s was based on the TAC/quota mechanism and supporting technical measures. While TACs and quotas were intended to control the rate of exploitation for certain stocks, they proved, however, to be an incentive to increase the catching capacity and efficiency of the Community fleet. This resulted in the Community's policy on structures and the policy on conservation being somewhat juxtaposed. That is to say, in order to sustain the increased fishing capacity, it was commercially necessary for the Community's fishermen to catch more fish. Unfortunately, there were not sufficient stocks in the Community fishing zone to withstand the increased exploitation rate. To redress this, the *1992 Management Regulation* aims to resolve the conservation problems by retaining the TAC/quota conservation mechanism while reducing and restructuring the catching capacity of the Community fleet.[135] In addition, the exploitation rate of certain stocks of fish may be regulated solely by a restriction of the fishing effort of the vessels which harvest these specific stocks.[136] In this regard the limitation of exploitation rates, where necessary, is to be undertaken within a planned community structured framework. Specifically, **Article 8** of the Management Regulation provides the Council with the appropriate legal base to establish management objectives on a multi-annual and, where appropriate, on a multi-species basis.[137] It is intended that these are to be updated at least one year before the end of the period fixed for each fishery or group of fisheries.[138] TACs and/or total allowable fishing effort (TAEs) are determined by the Council by qualified majority on a proposal by the Commission aimed at attaining global management objectives.[139] The principle of relative stability, so critical to the original management system, is retained in so far as the Council distributes fishing opportunities between Member States on that basis.[140] However, due cognisance is given to the development of mini-quotas and regular quota swaps since 1983.[141] Allowance is also made for the allocation of new fishing opportunities.[142]

Article 9 allows for the exchange of quotas, and requires Member States to inform the Commission of the criteria they have adopted for the distribution of the fishing possibilities allocated to them. In acknowledgement, perhaps, of the subsidiarity concept, Member States may take measures for the conservation and management of resources in waters under their sovereignty and jurisdiction, provided that such measures conform with Community law and do not transgress certain criteria prescribed in **Article 10**. Such measures must involve strictly local stocks which are of interest to fishermen from the Member State concerned, apply solely to fishermen from the relevant Member State, and are compatible with the objectives set out in the 1992 Management Regulation.

Pursuant to the Basic Regulation, the Commission proposed fixing a total allowable effort (TAE), 'days at sea', combined with or without a TAC/quota mechanism, allowing the Council to set these in a medium or multi-annual context.[143]

(iii) Improved control and enforcement of Community measures

As part of the review of the policy in 1992 the Commission submitted a detailed report to the Council and Parliament analysing the effectiveness of the Community fishery control and enforcement system during the first period of the policy.[144] This report is discussed in greater detail in Chapter 2. It is sufficient to note here that one of the principal recommendations in the report was the urgent need for the Community to adopt a new control system with extended scope to monitor all aspects of the fisheries industry. This recommendation was accepted by the Council and received legal form in Title III (under the inauspicious heading of 'General Provisions') of the 1992 Management Regulation which required the establishment of a Community control system.[145] The framework for the new structures and obligations is set out in Council Regulation No. 2847/93 (the 1993 Control Regulation).[146] Significantly, the Community system for control and enforcement established by this regulation is no longer limited to the technical aspect of conservation but also extends to the structural and marketing elements of the policy. The successful application of the Control Regulation in the Member States is now a prerequisite for the future of the CFP and the provisions it introduces are examined in Chapters 3–6.

4. *General comment on the redirection of the policy in 1992*

The 1992 Management Regulation lays out the strategy for the future of fisheries in the Community. As it is a basic regulation it requires to be supplemented by a significant number of implementation regulations. In this respect one of the essential requirements to ensure the transition to a well balanced policy is the political support of the Member States in completing the legislative framework.

Since 1993 several steps have been taken to achieve the improvements suggested in the 1991 report on the CFP. As noted above, various regulations have been adopted relating to licences and fishing permits in order to achieve the legal base to manage fishing effort. A new regulatory framework has also been adopted for the management of 'western fisheries' which integrates rules concerning TAC/quota mechanism with effort management.[147]

Coinciding with the mid-term review in 1992 was the review on the adjustment of the Spanish and Portuguese accession arrangements, and it could be argued that this was the first test to assess the transition to responsible fishery management in the *post*-1992 period of the policy.

5. *Integration of Spain and Portugal into the CFP*

Articles 162 and **350** of the Iberian Act of Accession required the Commission to publish a review report in December 1992 on the application of the Act of Accession in the fisheries sector (hereinafter referred to as the *Iberian Report*).[148] This was followed by a proposal to the Council on adjustments to the arrangements in the fisheries chapters of the Act of Accession of Spain and Portugal. The Council of Fisheries Ministers agreed that the revision of the CFP (following the *1992 Iberian Report*) required the integration of the Spanish and Portuguese into the CFP with effect from 1 January 1996. In accordance with the Commission's October 1992 proposal, this was achievable only if a number of conditions were satisfied. Firstly, the retention of the *acquis communautaire*, particularly the principle of relative stability so as to create a better balance between available resources and fishing effort; secondly, adherence to the concept of equal access, but allowing the derogations in the new Regulation for the Community system for fisheries and aquaculture;[149] and thirdly, maintaining the limits on fishing effort contained in the Iberian Act of Accession – i.e. no increase in the overall levels of fishing effort for the Spanish and Portuguese fleets.

The Council at its meeting in December 1994 agreed that Spain and Portugal would have to establish reference lists of all their named vessels which may have access to the Community area.[150] It was decided that capacity and catches were to be controlled in the zones covered by the 1985 Act of Accession. In this way access to resources was linked to the management of fishing effort. Each Member State was allowed to take up fully established fishing possibilities on the basis of the previous balance in exploitation by fishery and by zones – i.e. no increase in the overall levels of effort was allowed. In theory, this approach would not destabilise the principle of relative stability.

Member States were obliged to communicate to the Commission by 31 March 1995, the reference lists, the assessment and, where appropriate, the arrangements to regulate fishing effort. On the basis of this information, the Commission reported to the Council and proposed a regulation in 1 May 1995. On 15 June 1995, the Council adopted the first regulation establishing a system for the management of fishing effort relating to certain Community fishing areas and resources in ICES divisions Vb,VI,VII, VIII, IX, X and CECAF areas 34.1.1, 34.1.2 and 34.2.0(7) (see Map in Chapter 4, Section (ii), pp. 138–9).[151] The management scheme in these fisheries is distinguished in terms of gear deployed, the group of target species taken and the zone. For individual fleets, the maximum levels were set taking into account various factors having a bearing on their productivity. There was also a degree of flexibility built into these to allow for further adjustment. It was also agreed, after much contentious debate, that no access is allowed to vessels flying the flags of Spain and Portugal to the Irish Sea and the Bristol Channel. Furthermore, fishing by all Member State vessels, other than Spanish flagged vessels, was limited to previous activity in the area known as the 'Irish Box'. Forty vessels flying the flag of Spain were, however, granted access to this area.

Increased control measures in Member States were also introduced: these measures included *inter alia*, rules to guarantee that the requirements in relation to fishing effort are not exceeded; special fishing permits; and entry/exit communication in areas where effort is curtailed such as the 'Irish Box'.[152] The latter are to be supplemented by catch declarations from 1998. These requirements place additional responsibilities on the flag State and the coastal State and the significance of the new measures in the context of the evolving enforcement pattern in Community and international law are explored further below.[153]

At the June 1995 Council meeting several outstanding issues were resolved. It was clarified that the Commission could use the Management Committee procedure, laid down in **Article 18** of the Management Regulation, to reduce fishing effort as the need arose. This was an important development which allowed the Commission greater flexibility and autonomy in the management of Community fishery resources. Furthermore, there was agreement to review effort ceilings if they prejudice the taking of a quota share. France and Spain recorded their full commitment to the successful integration of Spain into the CFP. In order to improve bilateral relations they set up a 'French–Spanish Fisheries Committee'. Spain and Portugal reached an agreement with the aim of jointly regulating their respective fleets in each others waters. Spain, France, Portugal, and Belgium, negotiated a number of quota exchanges which give extended latitude to their national quota management mechanisms. Because the regulatory framework placed increased responsibility on the Irish inspection services, the Council noted that there would be additional financial Community aid for Ireland to improve national fishery enforcement structures. In a new departure, the Council gave a commitment that Community subvention would include covering the cost of operational expenditure incurred by fishery enforcement services.[154]

In 1995 the full integration of Spain and Portugal into the CFP was considered a priority for the future harmony of the CFP. In this regard one may deduce that the Community opposed a two-speed fisheries policy. It is contended, however, that the success of this approach is dependent on a similar adjustment to the control and enforcement regimes. One of the issues to be assessed in this book is whether the Community is capable of developing a truly uniform and integrated enforcement structure.[155]

6. *International Fishery Relations*[156]

The European Union plays a major role in world fisheries.[157] It is now proposed to mention briefly international relations under two separate headings. The first relates to bilateral agreements with third countries, the second concerns the Community's role in international fishery organisations.

The first role is a direct result of the Community's competence to negotiate bilateral fisheries agreements on behalf of the Member States.[158] This competence was granted as part of the Hague Resolution in 1976. The successful negotiation of these agreements is critical to ensuring that Community fishermen obtain fishing rights in the waters of third countries.[159] By 1998, the European Union had negotiated 26 fisheries agreements with third countries - 15 with African and Indian countries, ten with North Atlantic countries, and one with a Latin American country.[160] There is no single type of Agreement and the nature of individual Agreement depends on the objectives of the respective parties. The Commission has negotiated reciprocal agreements with Norway, the Faroes and the Baltic Republics. These Agreements entail access to Community waters by third-country vessels in return for access to third Country waters by Community vessels. It has negotiated access to stocks in return for payment and market access with Morocco, Mauritania and Greenland. In the past the Commission has obtained access to stocks in return for financial compensation with African, Caribbean and Pacific countries; and access to Canadian stocks was been obtained in return for *erga omnes* tariff concessions for Canadian imports to the Community. At the time of writing the Community does not have any fishery agreements with either Canada or the United States. More recently, however, the Community has concluded agreements with Argentina and Venezuela which are based on trade concessions with financial assistance for technical and scientific projects, and they offer special incentives for the establishment of joint enterprises. The success of these agreements has reduced the pressure on the European Union to redirect the over-capacity of the Community fleet towards the limited resources available in the Community zone.

The external strategy adopted by the Community in the negotiation of international agreements accords with the relevant provisions in United Nations Convention on the Law of the Sea (UNCLOS III) and in appropriate cases with the African Caribbean and Pacific European Community (ACP/EC) Lomé Conventions. In respect to the latter, fisheries agreements have to be consistent with the overall trust of the Community's development policy pursuant to **Article 130v** {178} of the EC Treaty. An interesting feature of recently concluded agreements, from the perspective of this book, is the emphasis which is placed on the promotion of an effective scheme of enforcement and the monitoring of fishing activities through observer and inspection programmes such as those that are included in the EU/Morocco Agreement and the EU/Mauritania Agreement.[161]

The second role of the Community in international fisheries is through its mandate to represent the interests of the Member States in international fishery organisations and to act as an observer in several organisations to which the Community has not yet acceded.[162] These organisations manage the exploitation of fishery resources on the high seas and allocate fishing possibilities among Contracting Parties. Traditionally, the success of international organisations in managing and conserving fishery resources was limited as their mandate did not allow them to monitor and control the activities of non-contracting parties. The latter are now subject to the extensive provisions of the United Nations Straddling and Migratory Stocks Agreement discussed in Chapter 11. The EU has supported the endeavours of the United Nations Food and Agriculture Organization (FAO) to foster better relations with non-contracting parties, and to promote the rational exploitation of stocks on the high seas. The role of the

Commission and the relationship between the Community institutions and the Member States in respect to membership of umbrella organisations such as the Food and Agriculture Organisation (FAO) has not always been harmonious.[163] General developments in the international law of fisheries influences the development of Community fisheries law as is evident from the discussion of the driftnet fishery in the Atlantic and Mediterranean which are the subject of a case study.[164]

European fisheries and global fisheries are closely interlinked. The Community plays an important role in the resolution of international fishery issues and provides support for the work undertaken by supra-national organisations. Typical of this role is the part played by the EU in the United Nations General Assembly to resolve outstanding issues arising out of UNCLOS provisions relating to straddling stocks and highly migratory species.[165] Similarly, the Community has played a significant part in supporting the negotiations which have led to the Code for Responsible Fisheries and the FAO Compliance Agreement.[166]

7. The Mediterranean[167]

Mediterranean fisheries differ from those of the Atlantic and the North Sea. The absence of a continental shelf has meant that fisheries resources are concentrated in narrow coastal bands. Coastal states have not sought extended exclusive fisheries areas with the same expediency as elsewhere.[168] Furthermore, although Italy, France, and Greece were Community Members, the Community management and conservation system, adopted in 1983, did not apply to Mediterranean fisheries.[169] However, as noted above, the Council adopted in June 1994 specific technical conservation measures for the Mediterranean Sea.[170]

8. Enlargement of the Union

With the fourth accession to the Community completed on 1 January 1995, the CFP faced a new challenge which diminished significantly with the non-accession of Norway to the Community. The Norwegian public rejected EU membership for a second time in a close-run referendum. Interestingly, negotiations with Norway had endeavoured to ease the full impact of the CFP on the fishing sectors by suggesting interim transitional arrangements similar to those that applied to Spain and Portugal in 1986. These were to restrict access to fishing zones and access to resources. Specifically access by Norwegian vessels to Community waters, west of 4° west, was to be managed by a basic and periodic list. Access to the North Sea and areas under Norwegian jurisdiction was to continue under the bilateral agreement which has been in place since 1981.[171] These arrangements are now of little significance and a special bilateral consultation takes place on an annual basis to decide mutual fishing rights and the management of common biological resources.[172]

There were special transitional arrangements for Swedish and Finnish vessels before they were absorbed into the general scheme of the CFP. In this regard the rules pertaining to the Community's fishing vessel register and licence scheme took effect as from 1 January 1995.[173] On acceding, Finland and Sweden were required to accept the *acquis communautaire* as it applies to the CFP.[174] Access for their respective fleets to Community resources is in accordance with the principle of relative stability and in line with previously established policy. Furthermore, the established norms governing the Community approach to external international relations in the domain of fisheries applies equally to the new Member States. In this regard of particular importance to Sweden and Finland is the special role that the Community plays in the International Baltic Fishery Commission which has a major input into the management of stocks in the Baltic.

9. *Structural measures to assist the fishing industry*

Introduction

The importance of the structural policy framework in the fisheries sector has varied since the adoption of the first structural regulation.[175] Essentially, the original objective of the structural policy was:

> 'to promote harmonious and balanced development of this industry within the general economy and to encourage rational use of the biological resources of the sea and of inland waters.'[176]

To achieve this objective the Community provided financial subventions to the fishing industry in the Member States. Early structural measures were used to develop, *inter alia*, Member State fleets; aquaculture; processing and marketing of products; and port facilities and shore-based infrastructure. This general approach could not be sustained in the context of the sweeping changes which would occur within a relatively short period of time. In particular it became clearly apparent after the extension of coastal state jurisdiction in 1976 that there were insufficient fishing opportunities for the Community fleet in the Community waters under the jurisdiction or sovereignty of the Member States. This excess imbalance between the fishery resources available and the fish catching capacity of the Community fleet is frequently referred to as *over-capacity* and is one of the principal problems which has beset the CFP from the start of the policy. The problem of over-capacity became more acute with the accession of Spain and Portugal in 1986. In order to deal with over-capacity the Community has introduced over a number of years several measures to reduce the number of vessels as well as to limit the activity of remaining vessels.

(i) Structural programmes

A succession of inconclusive structural programmes indicate that the critical significance of complementary synergy between the structural framework and the conservation policy was overlooked in the early period of the policy.[177] Thus, for example, from 1971 to 1978 there was no formal co-ordination of structural measures and the general framework of fishery management and conservation measures. The fact that aid to restructure fishing fleets was partly financed by the Guidance Section of the European Agricultural Guidance and Guarantee Fund (EAGGF) contributed to this segregation of two fundamental aspects of the CFP. In 1976, however, structural measures to rationalise the catching sector were included as part of the Commission's 'package' of proposals to deal with the problems created by the extension of coastal state jurisdiction. The structural scheme was modified in 1978 to include annual interim schemes for the inshore fleet and to cater for the growth of aquaculture.[178] The long-term proposals, however, were linked to the adoption of the Community management system, and so it was not until 1983 that Council adopted the first multi-annual guidance programmes (MAGP in its abbreviated form) to assist the restructuring of the industry.[179] Since 1983 there have been four MAGPs and it has become apparent that the poor implementation of these programmes has failed to check the fish catching capacity of Community vessels.

In order to reduce the complexity of the MAGPs and in response to the problem of increased efficiency of the Community fleet, a single regulatory framework had to be adopted in 1986 and it was also necessary to introduce a new programme for the period 1987 to 1991.[180] In the Structural Regulation, MAGP was defined as follows: 'a set of objectives, together with the statement of the means necessary for attaining them, for the development of the fisheries sector in the overall long-term context'.[181] The 1987–1991 programme aimed at reducing the fleet capacity by 2% in overall tonnage terms and by 3% in overall terms of engine power. This second generation programme is now referred to as MAGP II. It included financial aid to Member States for the scrapping or laying-up of vessels, grants for deploying excess capacity through exploratory voyages for species or in areas which had been previously under-utilised through joint ventures with third states; and financial aid for the modernisation and construction of vessels provided such aid formed part of a national program which was aimed at the long-term objective to balance capacity with catch potential.

A group of independent experts who prepared a report for the Commission concluded, however, that these modest reductions in MAGP II proved to be totally inadequate in reducing the over-capacity of the Community fleet.[182] This group also prepared guidelines for MAGPs for the fishing fleet for 1992–1996, and recommended a reduction of 40% in community fishing fleet.[183] The third MAGP (which had a scaled reduction in fishing effort) covered the period 1992–1996 and was only adopted after protracted and difficult negotiation at the Council.[184] This programme is referred to as MAGP III. Significantly, this programme necessitated an improvement of the Community register of fishing vessels in order to assist the Commission in monitoring the reductions in fishing effort, and in order to improve the veracity of the annual report to the Council.[185]

In 1995 a second group of independent experts (called after their chairperson, Mr Lassen) produced a report for the Commission on the fourth generation of MAGPs (referred to as MAGP IV) for the period 1997–2002. The report noted that many of the Community fish stocks are overexploited and significant reduction in fishing effort is required to remedy this situation. Indeed the group considered that a reduction of less than 20% in fishing effort on certain stocks to be inadequate. It was on the basis of the *Lassen Report* that the Commission proposed a major reduction in the capacity and effort of the Community fleet.[186] These proposals aimed to reduce the size of the Community fleet commensurate with the stocks available, but at the same time ensure the maintenance of a modern, competitive fleet capable of supplying the Community market.

(ii) Multi-Annual Guidance Programme (MAGP IV) for the period 1997–2002.

The Commission proposal to reduce the fishing capacity of the Community fleet, exceeded the recommendation of the *Lassen Expert Group,* entailed a 40% reduction in the number of vessels which fished the most exploited stocks between the years 1997 and 2002. This proposal was not well received by the industry. Significantly, both the United Kingdom (which at the time was facing a general election) and France voted against the Commission proposal on the fourth multi-annual guidance programme (MAGP IV) at the Council meeting on 15 April 1997.[187] The compromise proposal adopted at this Council meeting involves reducing the fleet capacity by 30% in the most vulnerable fisheries (that is to say in fisheries where there is serious risk of collapse of the stock), 20% reduction for the fleet which exploit stocks which are 'over-fished', and no increase in fishing effort for stocks which are defined as 'fully-exploited'. Member States may elect to select the means to achieve the reduction in fishing capacity and effort by their fleets. That is to say, Member States may choose, on the one hand, the permanent laying-up of vessels, or, on the other hand, to reduce the time spent at sea by particular vessels. It is most probable that Member States will opt for a combination of both of these methods.

The programme is scheduled to run for five years, from 1997 to 2002. An important element in the adopted programme for Member States is that the scheme only applies to vessels under 12 metres in length if they use static gear (that is vessels which use gear other than trawls or purse seines). Moreover, there is a provision in the legislative framework which allows vessel operators to increase the tonnage of particular vessels, without infringing the planned reductions in fishing capacity, in order to improve the safety and sea-worthiness of such vessels. Furthermore, agreement on MAGP IV means that the allocation of structural funds to aid fishing industry through the financial instrument for fisheries guidance to scrap vessels or to finance a change of activity, will continue to operate for the period 1997–2002. Significantly, the Community will not provide financial aid towards the construction or modernisation of vessels if the flag Member State has not achieved the global reduction in fishing capacity during the period of MAGP IV.

(iii) Structural measures in the processing and marketing sectors including financial aid

Originally the processing and marketing schemes for the fishery sector developed from the agricultural model and were covered by the same regulation.[188] In 1989, however, fisheries processing and marketing were integrated into the Community's structural funds. This is an important development because it results in the inter-linking of two policy elements which in the long-term will be critically affected by the restructuring of the catching sector.

In 1993 a new 'financial instrument for fisheries guidance' (FIFG in its abbreviated form) was integrated into the overall scheme for Community structural funds.[189] The FIFG, within the framework of the MAGP, contributes financial assistance to a number of operations including *inter alia*: the renewal and modernisation of the catching sector and facilities in ports; the improving of conditions for the processing and marketing of fishery products, and the modernisation of vessels and increasing gear selectivity; searching for new markets; and encouraging joint ventures and joint enterprises. Additional assistance is available to the fishery sector under the broader provisions pertaining to rural areas, and regions undergoing industrial conversion pursuant to the European Regional Development Fund and the European Social Fund. Finally, a recent Community initiative, referred to as PESCA, is specifically designated for assisting projects in coastal areas.[190]

10. The common organisation of the market in fishery products[191]

(i) The common market
As previously noted, **Articles 39–43** {33–37} of the EC Treaty, required the establishment of a common organisation of the market for fishery products. The Council adopted Council Regulation No. (EC) 2142/70[192] to establish this market. This had the dual aim of ensuring a reasonable income to the producers and stable supplies to the consumers. From the start it was modelled on the CAP and its structural side was geared principally to production. It sought to balance the different demands of each Member State and, like all good compromises, it contained something for each sector of the fishing industry. The market measures were extensive and went beyond the setting up of a Common Customs Tariff and abolition of internal customs. Like other elements of the CFP, the common organisation of the market has had to adapt to major changes since its creation in 1970.

It was inevitable that the extension of Member States fishing zones in 1976 necessitated a readjustment of the Community market structure for fishery products. The Commission, nonetheless, did not advance a proposal for the regulation of the common organisation of the market in fishery products until 1980.[193] This proposal was adopted by the Council in 1981 and took effect from 1 January 1983.[194] The operation of the market in fishery products during the period 1983–1993 is outside the remit of this book. It ought to be pointed out, however, that the review which led to the completion of the single market by 1 January 1993, was used as an opportunity to reform and introduce a new basic market regulation.[195] This regulation has four principal components to meet the objectives required by **Article 39** {33} of the EC Treaty. These included *inter alia*: a common price support system (internal market) which operates by either an intervention mechanism (i.e. withdrawal prices, carry-over aid and private storage aid), or a compensation mechanism; common marketing standards; the encouragement of the formation of Producers' Organisations; and rules on trade with third countries. In practice, the Commission monitors market trends and prepares price proposals for the following year on the basis of information supplied by the Member States.[196] These are the basis of guide prices which are published for certain fishery products such as the production price for tuna for canning. After consulting the Management Committee for fishery products, the Commission publishes the withdrawal and sale prices, the reference price, the amount of fixed premium, and the amount of carry-over aid.[197] To implement the market mechanism local Producers' Organisations (POs in their abbreviated form) representing fishermen or fish farmers are tasked with the implementation of the market mechanism.[198] While Producers' Organisations are partially financed by the Member States,[199] the ultimate objective is to make such organisations, which in 1998 numbered 160, self-financing. The importance of Producers' Organisations varies from country to country. In some Member States, they not only play a major part in adapting production structures to market demand, but are also involved in the day-to-day management of quotas. In the recent communication on the future of the market, the Commission suggested that

Producers' Organisations should be front-line players in the integrated management of resources and markets.[200]

In order to improve products quality and to make marketing easier for the benefit of both producers and consumers the Council adopted a Regulation in 1996 laying down common marketing standards for certain fishery products.[201]

The free circulation of fishery products in the Community has been achieved through the completion of the single market. In addition, there has been a continual trend towards liberalisation of international trade, through a succession of agreements under the General Agreement on Tariffs and Trade (GATT), which operates within the framework of the World Trade Organisation. The supply of fishery products by the Community fleet, however, has not served the market in terms of quantity, quality and regularity of supply. This has been partly due to the poor state of conservation of fish stocks. In many Member States, marketing structures have been adapted to serve the needs of Supermarket chains which are the main buyers of fishery products. The result has been a growing dependence on imported fisheries products, which now account for almost 60% of total consumption in the Community.[202] The Commission has emphasised the need for producers, merchants and processors to change their approach.[203] In particular it advocates:

- using the common organisation of the market to promote sustainable fishing and increasing producer involvement in market management;
- encouraging greater competitiveness inside the Community, especially in the domain of fresh fish products;
- promoting integration and transparency of the market through co-operation among players and better information regarding landings and quality;
- maintaining competition by keeping Community markets open to meet the needs of both consumers and processors which the Community cannot satisfy, provided rules regarding production safety and origin are complied with;
- finding new ways to increase the quality of fish products.

11. Concluding remarks on the evolution of a common policy for fisheries

Over the past 20 years Community fisheries legislation has been produced in considerable volume and has directly affected the lives of those who participate in the marine fishing industry. This chapter has been concerned with the development of the CFP and the complex route it has followed since its embryonic birth in the EC Treaty to its present status as an integrated and well-established Community policy. In this chapter, emphasis has been placed on the legislative architecture of the policy and in this regard it is apparent that the CFP has undergone important changes since 1983. One of the basic features of the policy is the wide legislative powers which **Article 43** {37} of the EC Treaty attributes to the Community legislature with respect to the choice of priorities and the means by which the objectives of the CFP can be obtained. This has resulted in a management plan for fisheries which is truly Community-based and is not defined, or constrained, by the individual interests of the Member States. Furthermore, it may be argued that the mandatory requirement for Member States to comply with Community regulations makes the CFP unique in the world of fishery management. This may be contrasted with the consentient approach to management which is a feature of many regional fishery organisations such as the North Atlantic Fisheries Organisation (NAFO). A second characteristic of Community fisheries legislation is the strong functional orientation of regulations. In this respect, the CFP is in many ways similar to the CAP.[204]

However, a common management plan is of no particular value unless there is effective supervision and enforcement of legislation. Indeed, outside the domain of fisheries, the exposure of the consumer sector in the United Kingdom to the Creutzfeldt–Jacob disease and to the *E. coli* (*Escherichia coli*) bacterium in Scotland clearly demonstrate the futility of regulation *per se* without adequate supervision and inspection procedures to ensure compliance with the rules to protect consumers. As law enforcement is at the core of all regulatory systems, it is now proposed to change the focus of the discussion and to give separate consideration to the subject of enforcement in the context of the CFP. In the following chapters it will be apparent that, while enforcement is of fundamental importance to the Community legal order, it is closely linked both to developments in the international legal order as well as the legal systems in the Member States.

References

1 This chapter gives an outline of the development of fisheries law in the European Community. For a comprehensive account, see, *inter alia*, Leigh, M., *European Integration and the Common Fisheries Policy*, (London, 1983); Wise, M., *The Common Fisheries Policy of the European Community*, (London, 1984); Farnell, J. and Elles, J., *In Search of a Common Fisheries Policy*, (Aldershot, 1984). The definitive authority on EC Fisheries Law, if not somewhat out of date, is still Churchill, R.R., *EC Fisheries Law*, (Amsterdam, 1986). This text provides valuable analysis of the legal issues pertaining to the fisheries policy. The most recent publication, which provides a critical insight into the political development of the conservation dimension of Community fisheries law, see Holden M., *The Common Fisheries Policy: Origin, Evaluation and Future*, (Oxford, 1994) (updated by Garrod, D., 1996). Hereinafter these authorities will be referred to as Leigh, Wise, Churchill, and Holden, respectively. For a discussion of the CFP from the United Kingdom perspective, see Munir, A.E., *Fisheries After Factortame*, (London, 1991). There has been a recent interdisciplinary study (economics and management) undertaken by Karagiannakos, A., *Fisheries Management in the European Union*, (Aldershot, 1995). The development of the policy has resulted in a sizeable library of periodical and journal articles; some of the most incisive include, *inter alia*: Song, Y.H., 'The EC's Common Fisheries Policy in the 1990s', *Ocean Development and International Law (ODIL)*, **26**, (1) 31–73; Churchill, R.R., 'EC Fisheries and an EZ - Easy', *ODIL*, **23**, 145–163; Freestone, D., 'Some Institutional Implications of the Establishment of Exclusive Economic Zones by EC Member States', *ODIL*, **23**, 97–114. For an appraisal of the performance of the Community fishery management system in the early period of the policy, see Churchill, R.R., 'The EEC's Fisheries Management System: A Review of the First Five Years of Operation', *Common Market Law Review (CML Rev.)*, **25**, 1988, 369–389.

2 The terms 'European Community' and 'European Union' are often used interchangeably in Community literature. The former will be used when discussing matters prior to The Treaty of Union (TEU), signed in Maastricht on 29 July 1992. Thereafter, in accordance with Article A {1}, Title 1, of that Treaty the term 'European Union' (EU in its abbreviated form) will be used where appropriate and principally in the context of the second and third

pillars of the Treaty. European institutions will be referred to by their short title, such as the Council, the Commission, the Parliament, and the Court for the European Court of Justice. The Council refers to the Council of Fisheries Ministers unless stated otherwise.

3 Hereinafter, where appropriate, referred to as the EC Treaty.

4 See, *inter alia:* Churchill, R.R., *EC Fisheries Law, op. cit.* fn 1, 23–31; Hiester, E. The Legal Position of the European Community with regard to the Conservation of the Living resources of the Sea, in *Legal Issues of European Integration* **55** (1976), 55–79.

5 EC Treaty, **Article 2** {2}. This Article has been substantially amended by the TEU, Article G(2) and by the Treaty of Amsterdam, **Article 2**. All Articles in the EC Treaty and the Treaty on European Union were renumbered by the Treaty of Amsterdam which was signed in 1997. In this book the new numbers are inserted in curly brackets (braces) after the old ones. Thus, for example, a reference to **Article 169** EC Treaty is recorded as **Article 169** {226}. The table of equivalence between the previous numbers and the new numbers is listed in the Annex to the Treaty of Amsterdam.

6 EC Treaty, **Article 3** {3}. This Article has been substantially amended by the TEU, Article G(3).

7 See Snyder, F., *Law of the Common Agricultural Policy*, (London, 1985), p. 7. On agriculture in general during the early phase of the policy, see, *inter alia*, Buckwell, Harvey, Thompson, and Parabou, *The Cost of the Common Agriculture Policy* (1982); Duchene, Szczepanik and Legge, *New Limits on European Agricultural Policy* (1985); Hill, *The Common Agricultural Policy: Past Present and Future* (1984). More recently, see *inter alia*, Barents, R., *The Agricultural Law of the EC: an inquiry into the administrative law of the European Community in the field of agriculture*, (Deventer, 1994); Rodgers, C.P., Margraves-Jones., *Agricultural Law*, (London, 1991); Blumann, C., *Politique agricole commune: Droit communautaire et agro-alimentaire*, (Paris, 1996).

8 Effective from 1 January 1958.

9 EC Treaty, **Article 3(d)**. This Article has been amended by the TEU, Article G (3), and is now Article {3(e)}.

10 EC Treaty, **Article 38(1)** {32(1)}.

11 This was acknowledged by the Court in Joined Cases 3, 4, and 6/76, *Officier van Justitie* v. *Kramer* [1976] ECR 1279 at 1309–1311. Hereinafter, where appropriate, referred to as the 'Kramer case'.

12 It has been defined by the European Court of Justice, in the agricultural context, in Cases 90 and 91/63, *Commission* v. *Luxembourg and Belgium* [1964] ECR 625 at 634; and Case 48/74, *Charmasson* v. *Minister for Economic Affairs and Finance* [1974] ECR 1383 at 1395.

13 See, '*Kramer*' fn 11, *supra*, and Case 61/77, *Commission* v. *Ireland* [1978] ECR 417 at 443. Case 141/78, *France* v. *United Kingdom* [1979] ECR 2923 at 2940. The scope of the fisheries policy is discussed in greater detail in Chapter 2, *post*.

14 EC Treaty, **Article 40(1)** {34(1)}.

15 The European Court has identified in several cases what the legal basis is for specific fisheries legislation. Most notably, *Kramer op.cit.* fn 11; Case 61/77, *Commission* v. *Ireland* [1978] ECR 417 at 443; Case 141/78, *France* v. *United Kingdom* [1979] ECR 2923 at 2940; Case 32/79 *Commission* v. *United Kingdom* [1980] ECR 2403; Case 804/79 *Commission* v. *United Kingdom* [1981] ECR 1045.

16 See, Churchill pp. 29–30 *op. cit.* fn 1, and the writers and cases quoted therein. **Article 102** was active for the transitional period of the Member States which acceded in 1972 only. This period terminated on 1 January 1979.

17 In the context of agricultural products this was decided by the court in Case 48/74, *Charmasson* v. *Minister for Economic Affairs and Finance* [1974] ECR 1383 paragraphs 7–20.

18 EC Treaty, **Article 189** {249}. As amended by the TEU, Article G(60).

19 The effect of the SEA on the fisheries market post 1992 is examined by Cuthbert, J.A.M.,

Scotland's fishing interests in the light of the common fisheries policy of the European Community (Glasgow 1991, unpublished thesis). On the judicial aspects of the SEA pertaining to agriculture and fisheries, see, opinion of the Economic and Social Committee, Flum, Pardon, Serra Caracciolo, *The Commission's agricultural proposals for implementing the Single Act. Own-initiative opinion of the section for Agriculture and Fisheries* (Brussels, EC, 1987), p. 19, Copy in the Commission Library, Brussels.

20 EC Treaty, **Article 3a** {4}. As inserted by TEU, Article G(4).

21 In relation to enforcement, see Chapter 2, *post.* On the general subject of subsidiarity, see Wilke, M. and Wallace, H., *Subsidiarity: Approaches to Power-Sharing in the European Community*, RIIA Discussion Papers **27**, London, Royal Institute of International Affairs, 1990. Since the Maastricht Treaty, the principle of subsidiarity and its application is referred to in all standard texts on the European Union and has resulted in a fount of academic literature, see, *inter alia,* Kapetyn, 'Community law and the principle of subsidiarity', *Revue des Affairs Européennes* (1991), **35**; Toth, A., 'Is subsidiarity justiciable?', *European Law Review*, Vol. **19** (3) (1994) 268–285; Cass, 'The word that saves Maastricht?: The principle of subsidiarity and the division of powers within the European Community', *CML Rev.* 29 (1992), 1107–1136; Peterson, J., 'Subsidiarity: a definition to suit any vision', *Political Affairs*, **47**, (1), (1994); Teasdale, A.L., 'Subsidiarity in Post Maastricht Europe', *Political Quarterly*, **64**, (2), (1993).

22 Inserted by TEU **Article G** (61).

23 EC Treaty, **Article 43(2)** {37(2)}. Furthermore, it ought to be pointed out that the Council may only conclude international fisheries agreements after consultation with the Parliament, pursuant to **Article 228** {300} of the EC Treaty. All standard texts on EC Law provide a description of the legislative procedures in the Community institutions, see, in particular, Hartley, T.C., *The Foundations of European Community Law*, (Oxford, 1998), pp. 37–48.

24 EC Treaty, **Article 189c** {252}, as inserted by Article G(61) TEU.

25 Holden, M., *The Common Fisheries Policy*, p.7, *op. cit.* fn 1.

26 See Resolution of the European Parliament on the progress made in respect to Title VI, *OJ* C 20/185, 20.01.1997. The Court of Justice, however, had limited jurisdiction to interpret provisions in Title VI, pursuant to the TEU, Article K 3(2)(*c*) However, see EC Treaty, Article {35}.

27 The new Community Title does not in principle apply to the United Kingdom and Ireland because of their special situation and a protocol also exists for Denmark. Ireland, however, has negotiated a clause which may revoke the protocol would she wish to participate fully at a later date.

28 A good account of the negotiation problems is discussed by Driscoll, D.J. and McKellar, N. 'The Changing regime of North Sea Fisheries' in Mason, C.M., (ed.) *The Effective Management of Resources. The international politics of the North Sea* (London, 1979), pp. 125–139. How the policy started and the marathon negotiations are discussed by Holden in Chapters 2 and 3, *op. cit.*, fn 1. See also Koers, A.W., 'The Fisheries Policy', in *Thirty Years of Community Law*, Commission of the European Communities, (Luxembourg, 1983), 467–475; and by the same author 'Participation of the European Economic Community in a New Law of the Sea Convention', *American Journal of International Law* **73** (1979), p. 426–443.

29 COM (66) 250, substantially reproduced in 1967, see, *Rapport sur la situation du secteur de la pêche dans les Etats Membres de la C.E.E. et les principes de base pour une politique Commune* (*OJ* EC 29.03.1967), pp. 861–900.

30 Wise, M. outlines the French fishing interests and demands during this period, *op. cit.*, fn 1, pp. 86–87.

31 Much regretted since is the omission from the CFP of the distinctive social policy references in the Report produced in 1966. Social policy has traditionally been a concern of the French fishing industry, but has never warranted much attention elsewhere in the Community. However by 1999, with increasing demand for large reductions in the European Union's

fishing capacity, this neglected dimension of the CFP is once again of significance. It was identified in a report completed by the Commission on the mid term review of the CFP in 1991 as an area that required to be urgently addressed, if the long-term sustainability of the fishing industry is to be guaranteed and to safeguard economic and social cohesion.

32 (*OJ* S Ed 1970 (111), pp. 703 and 707) and (*OJ* L 236, 27.10. 1970, p.5). Subsequently repealed by consolidated regulation in 1976, Council Regulation No. 101/76 of 19.01.1976, (*OJ* L 20, 19.01.1976, p. 19). Supplemented by Council Regulation No. 170/83, 25.01.1983, (*OJ* No L 24, 27.01.1983). Repealed by Council Regulation No. 3760/92, 20.12.1992 (*OJ* L 389/6, 31.12.92).

33 (*OJ* L 236, 27.10.1970) (repealed). Amended by Council Regulation No. 100/76 of 19.01.1976, (*OJ* L 20, 28.01.1976), and Council Regulation No. 3796/81 of 29.12.1981, (*OJ* L 379, 31.12.1981) (repealed). A new basic Regulation establishing the common organisation of the market in fishery and aquaculture products came into force on 1 January 1993, Council Regulation No. 3759/92 of 17.12.1992 (*OJ* L 388, 31.12.1992) amended for the third time on 22.12.1994.

34 The significance of the *acquis communataire* is examined by Kitzinger, *Diplomacy and Persuasion: How Britain Joined the Common Market*, (London, 1973), pp. 68–72.

35 Wise gives an informative although brief outline of French, West German, Italian, Belgian, Netherlands and Luxembourg objectives during the CFP negotiations prior to the enlargement of the Community, pp. 93–98, *op. cit.*, fn 1.

36 Leigh p. 39, *op. cit.*, fn 1.

37 Churchill, R.R., *EC Fisheries Law*, 124–133 *op. cit.*, fn 1, and the authorities cited therein, provide a comprehensive analysis of the legal basis and rationale of the Equal Access principle. There have been several articles published on the equal access principle, of particular interest see, *inter alia*, Churchill, R.R., 'The EC Fisheries Policy - towards a revision', *Marine Policy*, **1** (1977), 26–36, p.36. Olmi, G., 'Agriculture and Fisheries in the Treaty of Brussels of January 22, 1972', *Common Market Law Review* **9** (1972), pp. 293–321. Winkel, K., 'Equal Access Community Fishermen to the Member States Fishing Grounds', *Common Market Law Review*, **14** (1977), pp. 329–337. On the scope and origin of the regulation, see both Brown, E.D., 'British Fisheries and the Common Market', *Current Legal Problems* **25** (1972), 37–73, and Mensbrugghe, V. van der 'The Common Fisheries Policy and the Law of the Sea', *Netherlands Yearbook of International Law*, **6** (1975), 199–228.

38 See Chapter 6, *post*.

39 *OJ* L 20, 28.01.1976, p. 19.

40 Council Regulation No. 170/83, *OJ* L 24, 27.01.1983. Repealed by Council Regulation No. 3760/92, *OJ* L 389/6, 31.12.92. Article 6 of this regulation extends the derogation period until 31 December 2002. The access regime to follow this period will be decided by the Council, on the basis of a report submitted by the Commission. The precise access arrangements for the different Member States are set out in Annex II of Council Regulation No. 3760/92.

41 Council Regulation No. 170/83, **Article 7**, *op. cit.*, fn 40.

42 The Iberian Act of Accession is discussed in Section (iii), *infra*. Churchill, R.R., presents a detailed account of the derogations as they apply to the general principle, see *EC Fisheries Law*, *op. cit.*, fn 1, 133–139. Holden suggests that from the political perspective equal access has been a success, *op. cit.* fn 1, 117–118.

43 As is evident from the Council decision in 1994 regarding access to the sea area referred to as the 'Irish Box', see Section (iii), *infra*.

44 In relation to Norway and accession negotiations in 1995, see Section (iii) *infra*.

45 Denmark because of its own substantial industrial fishery interests in foreign waters supported equal access. The other applicant States felt that it was not for the Community but for their National Parliaments to decide on such matters. Their attitude thus appears to

call into question more than just the issue of fish, discussed by Kitzinger, *Diplomacy and Persuasion: How Britain Joined the Common Market, op. cit.*, fn 34.

46 Act of Accession, **Article 103**. This derogation period was similar to that in the European Fisheries Convention, however, it overturned the concept of progressive exclusion out to 12 miles.

47 Discussed in Section (iii), *infra.*

48 Council Regulation No. (EEC) 3760/92 of 20 December 1992, *OJ* L 389/1, 31.12.92. These rights were unsuccessfully invoked by Spain in Case 812/79 *AG* v. *Burgoa* (1980) ECR 2787.

49 Prior to the adoption by the Council of the necessary measures, Member States were empowered to adopt unilateral provisions to ensure protection of the fisheries resources in their own zones. This was acknowledged by the Court of Justice in the *Kramer* decision, *op.cit,* fn 11. The subsequent implications for Community competence are dealt with by Churchill, R.R., in an article entitled 'Revision of the EEC's Common Fisheries Policy-Part 11', *European Law Review*, **5** (1980), 3–37 and 95–111.

50 The international fisheries regime before and after UNCLOS III, has been the subject of a number of studies, including, *inter alia;* Carroz, J.E., 'Les Problèmes de la peche à la Confèrence sur le droit de la mer et dans la pratique des Etats', *Revue Générale de Droit International Public (RGDIP)* **84**, 705–1 (1980). Carroz, J.E. and Savini, M., 'The new international law of fisheries emerging from bilateral agreements', *Marine Policy*, **3**, 79–98 (1979). Moore, G., 'National legislation for the management of fisheries under extended coastal State jurisdiction', *JMLC* **11**, 153–182 (1980). Churchill, R.R., and Lowe, A.V., *The Law of the Sea*, (rev.ed. 1997), contains a succinct summary of international developments in fisheries law both prior to, and after the mid-1970s.

51 In relation to NEAFC see, *inter alia*, Driscoll, D.J. and McKellar, N., 'The Changing Regime of North Sea Fisheries', in Mason, C.H., (ed), *The Effective Management of Resources: The International Politics of the North Sea*, 125–167, *op.cit.* fn 28; Kwiatkowska, B., *The 200 Mile Exclusive Economic Zone in the New Law of the Sea*, (Dordrecht, 1989), 58–59, and 80.

52 At the 1960 Law of the Sea Conference the proposal to extend coastal states' exclusive jurisdiction to 12 nautical miles failed by one vote. In 1972 Iceland extended its exclusive zone to 50 miles, having previously in 1948 extended it to 12 miles. The exclusion of the United Kingdom from what was previously a lucrative cod fishing ground resulted in a succession of 'cod wars'. This unilateral action by Iceland resulted in the United Kingdom and the Federal Republic of Germany taking Iceland to the International Court of Justice. In an ambiguous judgment the court advocated that governments find an equitable solution having regard to both Iceland's coastal communities' preferential rights and other states' traditional rights in the same area, *The United Kingdom* v. *Iceland*, (1974) *ICJ* 3, 34; 175, 205–206. However, by the mid 1970s the Court's finding had been overtaken by political action. In particular developed states such as Iceland extended its jurisdiction to 200 nautical miles and this initiative was subsequently followed by the USA in 1976, the Fishery Conservation and Management Act 16 OSC 1801–1882.

In 1983 the United Nations concluded a new and elaborate convention to underpin the law of the sea (UNCLOS III). While many of the fisheries provisions in UNCLOS III, such as **Articles 61–73**, received widespread approval it was not until 1994 (after the General Assembly adopted a new resolution on deep-sea bed mining) that developed states ratified the convention.

Inevitably, because of the wealth of State practice many of the provisions in UNCLOS prior to ratification were considered as customary international law. In particular the provision enshrined in **Article 56 (1)**, which established that within 200 nautical miles the coastal state has sovereign rights for the purpose of exploring and exploiting conserving and managing the natural resources, whether living or non-living, became a well established and incontrovertible norm of international law.

Most of the literature on the law of the sea generally contains a review of the evolution of the guiding customary and convention principles. For a discussion of the issues prior to 1977, see Hollick, A., 'Origins of the 200 nautical mile offshore zone', **71**(3) *AJIL* 404 (1977). See also Buhl, J.F. 'The Third United Nations Conference on the Law of the Sea,' *CMLRev.*, **18**, 553-567, (1981). The colloquially referred to 'cod war' is subject to a study by Hart, J., *The Anglo Icelandic War of 1972-1973*, Research Series No. 29, (Berkley, 1976).

53 Sec (74) 4400; Sec (75) 3132; Sec (75) 4503: COM (76) 59, COM (76) 500. These reports examine the implications of 200 mile fishing limits from the Community perspective.

54 Proposal for a Council Regulation establishing a Community system for the conservation and management of resources. (*OJ* C 255, 28.10. 1976).

55 COM (80) 540 and 728.

56 The structural aspect of the CFP is dealt with in Section (iii) *infra*.

57 The Hague Agreement/Resolution refers to the entire agreement whereas the Hague preferences arose in the context of the establishment of relative stability whereby Community shares of the total allowable catch are allocated among Member States. The Hague preferences are discussed below in the context of the 1983 Management Regulation.

58 The first annex of the Hague resolution (Annex I) was published in the Official Journal in 1981. See, Council Resolution on 'Certain External Aspects of the Creation of the 200 mile fishing zone in the Community with effect from 1st January 1977', (*OJ* C 105/1, 07.05.1981). This annex had been previously published in International Legal Materials, **XV** (1976), p. 1425.

59 See discussion in Churchill, R.R. as to whether this resolution is legally binding, *op.cit.*, fn 1, pp. 70-71.

60 Case 22/70, *Commission* v. *Council*, [1971] ECR 263; see discussion of this case by Hartley, T.C., *The Foundations of European Community Law*, 4ed., (Oxford, 1998), 101-102. In regard to their content and other material factors resolutions may be legally binding, see the judgment in Case 9/73 *Schlüter* v. *Hauptzollamt Lörrach* [1973] ECR 1135, paragraph 40. Such resolutions must, however, be measures dealing with issues which the Council is required to resolve definitively, in the exercise of its own powers by way of a statement of position, rather than in the form or by means of one of the measures defined in **Article 189** {249} of the EC Treaty.

61 Case C-4/96, *Northern Ireland Fish Producers'Organisation Ltd. (NIFPO) and Northern Ireland's Fishermen's Federation* v. *Department of Agriculture for Northern Ireland*, [1998] ECR I-0681, paragraphs 26-39.

62 *OJ* NO. C 158/2, 27.06.1980.

63 Paragraphs 3 and 4 of Annex VII of Council Resolution 3 November 1973.

64 In the interim period, between the Commission proposal and Council adoption of Council Regulation No. 170/83, Member States were not to adopt unilateral measures pending the adoption of Community measures. In the absence of Community measures for 1977, however, Member States were authorised, having sought the approval of the Commission, to adopt non-discriminatory measures. Interestingly, one of the principal reason for the development of Community law in fisheries were several judgments of the Court in a series of cases upholding national measures in the absence of conservation measures adopted by the Council. Case 32/79, *Commission* v. *United Kingdom* [1980] ECR 2403, 2408-9; Joined Cases 185-204/78, *Officier van Justitie* v. *van Dam en Zonen* et al. [1979] ECR 2345 at 2360; Case 141/78, *France* v. *United Kingdom* [1979] ECR 2923 at 2940-2942. Case 287/81, *Anklagemyndigheden* v. *J. Noble Kerr* [1982] ECR 4053 at 4073.

65 The 1983 keys are explained in further detail below in the context of the 1983 Management Regulation.

66 Council Regulation No. 170/83 of 25.01.1983 establishing a Community system for the conservation and management of fishery resources *op.cit.* fn 32. Subsequently repealed by

Council Regulation 3760/92 of 20.12.92 establishing a Community system for fisheries and aquaculture, *op.cit.* fn 32.

67 The division of competence between the Member States and the Community is discussed in Chapter 2, *post.* At this point it is appropriate to note that the Court of Justice in the '*Irish Fisheries*' case acknowledged that the scope of Community authority extended to the 200-mile zone, Case 61/77, *Commission* v. *Ireland* (1978) ECR 417. In the earlier '*Kramer*' case the Court had found that the authority of the Community *ratione materiae* extended to fishing on the high seas, in so far as Member States had such competence under public international law. The Court interpreted **Article 227** {299} of the EC Treaty as applying to the Member States and not to their territory, paragraph 30 of the Judgement. The court also upheld the competence of the national authorities to adopt national measures, in the absence of regulation from the Community institutions. This was clear from the general scheme of Community law, Regulation No. 2141/70 and **Article 100** Act of Accession, *Kramer, op.cit.*, fn 11. The division of competence is discussed by Churchill, R.R., *EC Fisheries Law, op. cit.*, fn 1, 85–110, and has led to substantial case law and academic comment. See, Ribeiro, M., 'Compétence communautaire et compétence nationale dans le secteur de la pêche. Quelques considérations à propos de l'arrêt par la Cour de Justice dans l'affaire', *Cahiers de droit européen.* 1982, 144–185.

68 Council Regulation No. 170/83, **Article 2**, *op.cit.* fn 32.

69 See the discussion on technical conservation measures, *infra.*

70 Council Regulation No. 170/83, **Articles 10** and **11**, *op. cit.* fn 32.

71 Resolutions 16 March 1984, and 20 February 1986, *OJ* 1984, C 104/153 and 1986, C 68/108, 16.06.87. It should, nevertheless, be noted that a degree of parliamentary scrutiny was introduced by the 1983 Management Regulation in so far as the Commission was required to submit to the Parliament and the Council a report on the application of the measures implementing the management regulation, Council Regulation No. 170/83, **Article 9**, *op. cit.* fn 32.

72 Case 46/86, [1987] ECR 2671.

73 The International Council for the Exploration of the Sea (ICES), established in Copenhagen in 1902, is the oldest intergovernmental organisation in the world concerned with marine and fisheries science. ICES has 17 member countries on both sides of the Atlantic. The principal function of the organisation is to coordinate marine research, disseminate the results of research and to provide scientific advice to member governments and regulatory commissions. For a detailed account of how ICES operates and how it submits its recommendations to the community see Holden, M.J., 'Management of Fishery Resources: The EC Experience in OECD', *Experiences in the management of national fishing zones*, (Paris, 1984), 113–120. In relation to the procedures followed and the problems encountered by the EC in the implementation of the scientific recommendations of ICES on total allowable catches, see FAO, 'Expert Consultation on the Regulation of fishing effort', *FAO Fisheries Report No. 289*, **Suppl 3** (1985), 231–233.

74 Council Regulation No. 170/83, **Articles 12** and **13**, *op. cit.* fn 32, provide a legal basis for the Scientific and Technical Committee for Fisheries, and the establishment of the Management Committee for Fishery Resources.

75 Council Regulation No. 170/83, **Articles 3**, and **4**, *op.cit.* fn 32.

76 Council Regulation No. 170/83, **Article 4(2)**, *op. cit.*, fn 32. The key with minor adjustments is in percentage terms, Belgium 1.9%, Denmark 24.3%, France 11.7%, West Germany 12.1%, Ireland 4.7%, Netherlands 7.8%, United Kingdom 37.6%. The Mediterranean Member States do not have access to Community waters in the North Atlantic, and therefore Greece and Italy, (neither Luxembourg nor Austria) have not claimed an allocation. There was an adjustment for the accession of Spain and Portugal in 1986, discussed *infra* Section (iii). There has been considerable debate as to whether the quota allocation

conforms with Community law principles, Churchill, R., *EC Fisheries Law, op. cit.*, fn 1, 117–118.

77 Case 46/86, [1987] ECR 2671.

78 Holden, M.J, Chapter 7, and pp. 240–241, *op. cit.*, fn 1.

79 Council Regulation No. 172/83, *OJ* L 24/30, 1983. For a discussion of how these criteria were developed see, Farnell, J. and Elles, J., *In Search of the Common Fishery Policy*, (Aldershot, 1984), 107–118.

80 Case C-4/96, *Northern Ireland Fish Producers' Organisation Ltd. (NIFPO) and Northern Ireland's Fishermen's Federation* v. *Department of Agriculture for Northern Ireland*, [1998] ECR I-0681. See paragraphs 40–54.

81 Cases C-73/90, C-71/90, C-70/90, *Kingdom of Spain* v. *Council of the European Communities* [1992] ECR I- 5191, I-5175, I-5159; *Portuguese Republic and Kingdom of Spain* v. *Council of the European Communities* [1992] ECR I-5073.

82 *Ibid.*

83 *Ibid.*

84 Council Regulation No. 170/83, **Article 6**, *op. cit.*, fn 32.

85 On the extension of the derogation until the end of the present CFP in 2002. See section (iii), *infra.*

86 Council Regulation No. 170/83, **Article 7**, *op. cit.*, fn 32. These were listed in Annex II (A) of the Regulation.

87 Council Regulation No. 170/83, **Article 7**, paragraph 1(5), *op. cit.* fn 32.

88 *Ibid.*, **Article 8**.

89 *Ibid.*, **Article 5 (1)**.

90 *Ibid.*, **Article 5 (2)**.

91 The Court has upheld the TAC/quota mechanism in several cases, including *inter alia*: Case 46/86, *Albert Romkes* v. *Officier van Justitie for the district of Zwolle*, [1987] ECR 2671; Case 207/84, *DeBoer* v. *Produktschap voor Vis en Visprodukten*, [1985] ECR 3203.

92 Cases 3/87, *The Queen* v. *Ministry of Agriculture, Fisheries and Food, ex parte Agegate Ltd.*, [1989] I-4459 and Case 216/87, *The Queen* v. *Ministry of Agriculture, Fisheries and Food, ex parte Jaderow Ltd.* [1989] ECR 4509.

93 Case C-221/89, *R* v. *Secretary of State for Transport ex. parte Factortame* [1991] I-3905; Case 93/89, *Commission* v. *Ireland* [1991] ECR I-4569; Case 246/89, *Commission* v. *United Kingdom* [1991] ECR I-4585. See, discussion of the 'quota hopping cases', Chapter 7 *post.*

94 *Ibid.*

95 The effectiveness of TACs as a management system is much disputed, see Churchill, R.R., *op. cit.*, fn 1, pp. 112–115, and the writers cited therein. See Commission answers to questions in the European Parliament, PQ 171/88, *OJ* 1985 93/13 and PQ 1425/84, *OJ* 1985 C 135/13. See, Derham, 'The Problems of Quota Management in the European Community', in FAO, Expert consultation on the regulation of fishing Effort, *FAO Fisheries Report No. 289*, **Suppl. 3**, (1985) 241–250.

96 Council Regulation No. 88/98 of 18 December 1997 laying down certain technical measures for the conservation of fishery resources in the waters of the Baltic Sea, the Belts and the Sound, *OJ* L 9/1, 15.01.1998.

97 Council Regulation (EC) No. 1626/94 of 27.06.1994 laying down certain technical measures for the conservation of certain fishery resources in the Mediterranean, *OJ* L 171/1, 06.07.1994.

98 *OJ* L 288/1, 11.10.1986, p.1.

99 Council Regulation No. 3094/86, *OJ* L 288/1, 11.10.1986, p.1. As amended by Council Regulations Nos 4026/86, 2968/87, 3953/87, 1555/88, 2024/88, 3287/88, 4193/88, 2220/89, 4056/89, 3500/91, 345/92, 1456/92, 2120/92. See Holden *op.cit.*, fn 1, for an analysis of the amendments.

100 Council Regulation No. 850/98 of 30 March 1998 for the conservation of fishery resources through technical measures for the protection of juveniles of marine organisms, *OJ* L 125/1, 27.04.1998.
101 For a detailed account of Iberian accession, see Leigh, Chapter 7, *op. cit.*, fn 1. Holden, M. argues that the Act of Accession was poorly drafted and consequently was open to a number of interpretations, *op. cit.*, fn 1.
102 Council Regulation (EEC) No. 170/83, *op. cit* fn 32.
103 Act of Accession, **Articles 157–160**. Supplemented by Commission Regulation (EEC) No. 3531/85 of 12 December 1985; *OJ* L 336, 14.12.1985, p. 20.
104 Act of Accession, **Article 158(1)**. The box previously existed under the Spanish Community bilateral arrangements before accession. The 'Irish Box' derogation was subject to debate in 1994 as part of the review Spanish and Portuguese accession, discussed *infra*.
105 Supplemented by Commission Regulation No. 3716/85 and since 1986 by annual Council Regulations.
106 Supplemented by Commission Regulation (EEC) No. 3715/85, and since 1986 by annual Council Regulations.
107 Act of Accession, **Articles 165** and **352**.
108 The Commission report (1992) to the Council and the Parliament on the application of the Act of Accession of Spain and Portugal in the fisheries sector, contains a résumé of how the licensing system operated, SEC(92) 2340 final, p. 10.
109 Additional rules were laid down in Commission Regulation (EEC) No. 3531/85 of 12 December 1985, *OJ* L 336, 14.12.1985, p.20.
110 See Chapters 8, 10, 12, *post*.
111 On the enforcement issues pertaining to the Review see Chapter 2, *post*.
112 Report of the Commission to the Council and the Parliament on the common fisheries policy, SEC (91)2288 final, 04.12.1991. Other problems in the fishing sector were highlighted in the Commission Document 'Communication from the Commission to the Council and the European Parliament on the CFP', 06.12.1990. Discussed in Chapter 2, *post*. Member States also reviewed the effectiveness of the CFP, see, the House of Lords, Select Committee on the European Communities, *Review of the Common Fisheries Policy*, Session 1992–1993, 2nd. report (London: HMSO, 1992).
113 See further Chapter 2, *post*.
114 *OJ* C 311, 27.11.1992, p. 7.
115 Council Regulation No. 3760/92 of 20.12.1992, establishing a Community system for fisheries and aquaculture (*OJ* L 389,31.12.1992). This regulation was subject to criticism in the European parliament on the grounds that it did not adopt all the priorities for the CFP set out by the Parliament in their opinion of 15.12.1992 and their resolutions of 10.12.1991 and of the 15.5.1992, see PQ No. C 283/54, 20.10.93.
116 Council Regulation No. 3760/92, **Article 1**.
117 *Ibid.*, **Article 2**.
118 *Ibid.*, **Article 3** provides a list of definitions. This may avoid some of the contentious debate that heretofore was initiated by the use of fisheries terms in Community Regulations which were not consistent with the terms in use in the Member States.
119 *Ibid.*, Title 1, **Articles 4–10**.
120 *Ibid.*, **Article 4**.
121 *Ibid.*, **Article 16** specifically provides a legal basis and prescribes the terms of reference for the Committee which operates under the auspices of the Commission.
122 The list provided here is not exhaustive.
123 Council Regulation No. 3760/92, **Article 4**, subparagraph 2(d), *op. cit.* fn 115.
124 *Ibid.*, **Article 3**. For a discussion of the measures in relation to fishing effort, see Chapter 4, *post*.

125 Council Regulation No. 3946/92 of 19.12.1992, amending Council Regulation No. 4028/86 (*OJ* L 401, 1992).
126 Council Regulation No. 3760/92, **Article 6**, *op. cit.* fn 115.
127 Council Regulation No. 3760/92, **Article 6**, paragraph 2, *op. cit.* fn 115.
128 Council Regulation No. 3760/92, **Articles 5** and **7**, *op. cit.* fn 115.
129 Council Regulation No. 3760/92, **Article 5**, *op. cit.* fn 115.
130 Council Regulation (EC) No. 3690/93 of 20.12.1993 establishing a Community System laying down rules for the minimum information to be contained in fishing licenses (*OJ* L 341, 31.12.1993), **Articles 3–7**.
131 Council Regulation No (EC) No. 3690/83, **Article 8(2)**.
132 Council Regulation (EC) No. 3690/93, *op. cit.* fn 130. Commission Regulation (EEC) No. 109/94 of 19.01.1994 concerning the fishing vessel register of the Community (*OJ* L 19/5, 22.01.1994) as amended by Commission Regulation (EC) No. 493/96 of 20.03.1996 amending Regulation No. 109/94, (*OJ* L 72/12, 21.03.1996), both repealed by Commission Regulation (EC) No. 2090/98 of 30.09.1998 (*OJ* L 266), 01.10.1998.
133 Council Regulation (EC) No. 1627/94 of 27.06.1994 laying down general provisions concerning special fishing permits (*OJ* L 171, 06.07.1994).
134 Council Regulation No 1627/94, **Article 2**, *op. cit.*, fn 133.
135 In accordance with Council Regulation No. 3760/92 Title 11, **Article 11**, *op. cit.*, fn 32, the Council set the objectives and detailed rules, on a multiannual basis, for restructuring the fisheries sector. This was to take account of, on a case by case basis, the possible economic and social consequences and the special needs of the fisheries regions.
136 Council Regulation No. 3760/92, **Article 8**, paragraph 1, *op. cit.*, fn 32.
137 *Ibid.*, **Article 8**, paragraph 3(i) and (ii).
138 *Ibid.*, **Article 8**, paragraph 3(iii).
139 *Ibid.*, **Article 8**, paragraph 4(i)
140 *Ibid.*, **Article 8**, paragraph 4(ii)
141 *Ibid.*
142 *Ibid.*, **Article 8**, paragraph 8(iii)
143 For further information on how this is to be achieved, see, *inter alia*: Communication from the Commission to the Council, and the European Parliament, on the new components of the common fisheries policy and their practical implementation-Document COM (93) 664 final of 15.12.1993. Council Regulation No. 685/95 of 27 March 1995 on the management of the fishing effort relating to certain Community fishing areas and resources, *OJ* L 71/5, 31.03.95; Council Regulation No. 2027/95 of 15 June 1995 establishing a system for the management of fishing effort relating to certain Community fishing areas and resources, *OJ* L 199/1, 24.08.95. In relation to fishing effort see Chapter 4, *post.*
144 Report from the Commission to the Council and the European Parliament on monitoring implementation of the common fisheries policy-Document SEC(92)394 of 06.03.1992. See Chapter 2, *post.*
145 Council Regulation No. 3760/92 of 20.12.1992, establishing a Community system for fisheries and aquaculture. Article 12.(*OJ* L 389, 31.12.1992).
146 Council Regulation No. 2847/93 of 12.10.1993 establishing a control system applicable to the common fisheries policy. (*OJ* L 261, 20.10.1993). This Regulation was subject to major amendment in 1998, Council Regulation No. 2846/98, *OJ* L 358/5, 31.12.98. See Chapters 3–6, *post.*
147 Council Regulations No. 685/95 and 2027/95 *op. cit.* fn 143. See Chapter 4, *post.*
148 Report by the Commission to the Council and the European Parliament on the application of the Act of Accession of Spain and Portugal in the fisheries sector-Doc SEC (92) 2340 final of 23.12.1992.
149 Council Regulation (EEC) No. 3760/92 of 20.12.1992. **Articles 6** and **7**. (*OJ* L 389, 31.12.1992).

150 ICES divisions Vb, VI, VII, VIII, IX and X and the CECAF zones. These areas do not include the North Sea.

151 Council Regulation No. 2027/95 *op. cit.* fn 143.

152 Council Regulation No. 2870/95 of 08.12.1995 amending Regulation No. 2847/93 establishing a control system applicable to the common fisheries policy, *OJ* L 301/1, 14.12.95.

153 See Chapter 12, *post.*

154 See Chapter 9, *post.*

155 This point which is elaborated upon further in Chapters 8, 10, 12, *post.*

156 There have been several authoritative studies completed on international law of fisheries, including inter alia, Burke, T. *The new international law of fisheries: UNCLOS 1982 and beyond,* (Oxford, 1994). This book provides a remarkable analysis of international fishery law, it is used as a reference text in Chapters 2, 10 and 11, *post.* Symmons, C.R. *Ireland and the Law of the Sea,* (Dublin, 1993), examines the international regime as it pertains to Ireland, pp. 73–95 deals with fishery zones. Churchill, R.R. and Lowe, A., *The Law of the Sea,* (Manchester, rev.ed., 1997). Leonard, L., *International Regulation of Fisheries,* (Washington, 1994). O'Connell D.P., *International Law of the Sea,* 2 vols., (Oxford, 1982–1984). Fleischer, C.A., 'The new regime of international fisheries', *Collected Courses of the Hague Academy of International Law,* 1988, **2** (1988), T.209, *de la collection,* 95–222. Johnston, D.M., 'The international law of fisheries. A framework for policy, oriented inquiries', *New Haven studies in international law and public order,* 1987, 511–541; Churchill, R.R., Nordquist, N., Lay, R.R. and Simmonds, K.R. (eds), *New Directions in the Law of the Sea,* **I** to **XI**, (London, 1973–1981); Oda, S., 'Fisheries Under the United Nations Convention on Law of the Sea', *American Journal of International Law,* **77**(4) (1983), 739–755. Anand, R.P. 'The politics of a new legal order for fisheries', *Ocean Development and International Law,* **II** (3/4) (1982), 265–295. A recent contribution to the literature by the Spanish diplomat De Yturriaga, A., *The International Law of Fisheries from UNCLOS to the Presential Sea,* (The Hague, 1997); Orrego, F., 'The "Presential Sea": defining coastal states' special interests in high seas fisheries and other activities', *German Yearbook of International Laws* **35** (1992) 264–292.

157 See, *inter alia, Communication from the Commission to the Council and the European Parliament, Fisheries Agreements, Current Situation and Perspectives,* COM (96), Brussels, 30.10.1996; European Parliament Fisheries Seminar on Fisheries Agreements, Working Documents, (Brussels, 1996); Mathers, S., *European Community Fisheries Agreements with Third Countries and Participation in International Fisheries Agreements,* Working Papers produced for the European Parliament, Directorate General for Research, (Brussels, 1993); Duncan Crampton, P., *Report on International Fisheries Agreements,* (EP, Luxembourg, 1997).

158 The EU's competence to negotiate bilateral agreements does not prevent private parties within the Union negotiating access to third countries' resources. These arrangements cannot be concluded on a reciprocal basis, and the private parties must bear exclusive responsibility for their implementation and full liability in the event of their failure.

159 90% of global fish stocks now come under the exclusive jurisdiction of coastal States. Pêchestat, DG XIV, (Brussels 1994).

160 See Table of International Fishery Agreements and Conventions p. vii, *ante.*

161 *OJ* L 30, 31.12.1997, *OJ* L 334, 23.12.1996.

162 The Commission represents the EU in several international fishery organisations, including the North Atlantic Fisheries Organisation (NAFO), International Baltic Sea Fishery Commission (IBSFC), the International Commission for the Conservation of Atlantic Tunas (ICCAT), the North East Atlantic Fisheries Commission (NEAFC), the Food and Agriculture Organisation (FAO). See Table of International Fishery Agreements and Conventions pp. xxi–xxiii, *ante.*

163 On this point see, for example, Case 25/94 *Commission* v. *Council of the European Communities,* 1996 ECR I-1496. This case is discussed in the context of the division of competence in Chapter 2, *post.*

164 Chapter 10, *post*.

165 United Nations Resolution No. 47/192 of 22.12.1992 provided a mandate for the holding of an intergovernmental conference under the auspices of the UN to examine all aspects of the problem regarding these stocks and to propose appropriate recommendations. The conference which was convened for the first time in July of 1993 separated into two schools of interest. One school advocated the adoption of a convention which acknowledged the preferential rights of coastal States over straddling stocks or highly migrate species in areas adjacent to their jurisdictions. On the other hand, the second group of interests sought parity of esteem for the coastal and flag state in regard to the management and conservation of such stocks. The EU advocated a solution involving management by regional organisations based on international co-operation. See, United Nations Conference on straddling stocks and highly migratory species - New York, 12/30.7.93 - Negotiating document - A/Conf. 164/13. Conference, 14/31.3.94 - A/Conf. 164/Rev. 1.

166 See Chapter 11, *post*.

167 See, *inter alios*, Fiorentini L., Caddy J.F. and Leiva, J.L., *Long- and short-term trends of Mediterranean fishery resources*, (FAO, Rome, 1997); Caddy, J. and Griffiths, R., *Recent trends in the fisheries and environment in the General Fisheries Council for the Mediterranean area* (FAO, General Fisheries Council for the Mediterranean, 1990); and *Tendances récentes des pêches et de l'environnement dans la zone couverte par le Conseil général des pêches pour la Méditerranée* (FAO, Rome, 1991), which deals with statistics and the marine environment in the Mediterranean. For broad perspective on fisheries and aquaculture, Charbonnier, D., *Pêche et aquaculture en Méditerranée. Etat actuel et perspectives*, (Paris, 1990). International law, law of the sea and fisheries law as they apply to the Mediterranean, are among the topics dealt with by Leanza, U., *Le Régime juridique international de la mer Méditerranée*, Collected courses of the Hague Academy of International Law, **5** (1992), 127–460. See also the discussion of the issues pertaining to the driftnet fishery in the Mediterranean, Chapter 10, *post*.

168 The Member States acting in concert extended their fishing limits to 200 miles in 1977 did so without prejudice to action of the same kind in the Mediterranean Sea.

169 Council Regulation (EEC) No. 170/83, (repealed), *op. cit.*, fn 32.

170 Council Regulation (EC) No. 1626/94 of 27.06.1994 laying down certain technical measures for the conservation of certain fishery resources in the Mediterranean, *OJ* L 171/1, 06.07.1994.

171 Agreement on Fisheries between the European Economic Community and the Kingdom of Norway, *OJ* L 226/48, 29.08.1980.

172 See, for example, Council Regulation (EC) No. 46/98 of 19.12.1997 laying down, for 1998, certain measures for the conservation and management of fishery resources applicable to vessels flying the flag of Norway, (*OJ* L 012/50, 19.01.1998), and Council Regulation (EC) No 47/98 of 19.12.1997 allocating, for 1998, certain catch quotas between Member States for vessels fishing in the Norwegian exclusive economic zone and the fishing zone around Jan Mayen, *OJ* L 12/58, 19.01.1998. In 1993 the Council concluded a new agreement with Norway, 93/740/EC: *OJ* L 346/25, 31.12.1993; Agreement between the European Economic Community and Norway, *OJ* L 346/26, 31.12.1993.

173 Council Regulation (EEC) No. 163/89 of 24.01.1989 concerning the fishing vessel register of the Community (*OJ* L 20, 25.01.1989) (repealed, *op. cit.* 132) and Council Regulation (EC) No. 1627/94 of 27.06.1994 laying down general provisions concerning special fishing permits (*OJ* L 171, 06.07.1994).

174 Act of Accession of Austria, Finland and Sweden to the European Union (*OJ* C 241, 29.08.1994).

175 Council Regulation (EEC) No. 2141/70. For a comprehensive account of the structural policy, see, the Court of Auditors special report No. 3/93 concerning the implementation of the measures for the restructuring, modernisation and adaptation of the capacities of

fishing fleets in the Community together with the Commission reply, *OJ* 1994 C 2/01, 04.01.1994. On fishing capacity, see, Houghton, R.G., Nielsen, N.A. and Verdelhan, C. de, *Analysis of methods used to determine fishing capacity and establishment of a method suitable for Community needs,* (Brussels, 1981). A good account of the issues is presented by Song, Y., 'The Common Fisheries Policy of the European Union: Restructuring of the Fishing Fleet and the Financial Instrument for Fisheries Guidance', (1998) *IJMCL* **13** (4), 537–577. The difficulties associated with monitoring the implementation and application of structural measures are reviewed in Chapter 5, *post.*

176 Council Regulation (EEC) No. 101/76, *op. cit.*, fn 32.

177 Court of Auditors report, the place of structural policy in the CFP, paras 1.19–1.23, *op. cit.*, fn 175.

178 Council Regulation (EEC) No. 1852/78, (*OJ* L211/30, 1978).

179 Council Regulations (EC) No. 2908/83 and 2909/83, and Directive 83/515, (*OJ* L290/1, 1983, pp. 9 and 15). Repealed.

180 Council regulation No. 4028/86 of 1986 (*OJ* L 376, 31.12.1986). Regulation last amended by Council Regulation No. 3946/92 of 19.12.1992 on community measures to improve and adapt structures in the fisheries and aquaculture sector (*OJ* L 401, 31.12.1992).

181 *Ibid.*, **Article 2(1)**. This definition has been cited by the Court in Case C-44/94 *The Queen* v. *MAFF, ex parte National Federation of Fishermen's Organisations and others and the Federation of Highlands and Island Fishermen and Others,* [1995] ECR I-3115, paragraph 7.

182 Referred to as the 'Gulland report'.

183 Communication from the Commission to the Council and the Parliament (DOC SEC(90) 2244, 30.11.1990, P.18).

184 Commission Decision 92/598/EC of 21.12.1992 on a multi-annual guidance programme for the fishing fleet for three year period 1993 to 1996 pursuant to Council Regulation No. 4028/86 (*OJ* L 401, 31.12.1992).

185 The register which was introduced in 1989 does not replace the national registers in the Member States but may be used as a tool by the Commission in the management of fishing effort. Commission Regulation (EC) No. 2090/98, *op. cit.* fn 132. See Chapter 5, *post.*

186 Communication from the Commission relating to the management guidelines on the 4th generation of MAGPs, COM(96) 203 final, 30.05.1996. Proposal for a Council Decision concerning the objectives and detailed rules for restructuring the fisheries sector for the period 1 January 1997 to 31 December 2002 with a view to achieving sustainable basis between resources and their exploitation (COM(96) 237, 21.05.1996).

187 See, press statement Conclusions of Fisheries Council 14–15 April 1997, Reuter Business Alert, **http://www.cc:8081/inotes/rta/Eurofishing**. It is reported that the United Kingdom objected to the Commission proposal on the grounds that the issue of vessels engaged in quota hopping remained unresolved, and France objected on the basis that the period of the programme, five years, was too long.

188 Council Regulation No. 355/77 of 15.02.1977 on common measures to improve the conditions under which agricultural products are processed and marketed (*OJ* L 51, 23.02.1977). (Repealed).

189 Council Regulation (EEC) No. 2080/93 of 20.07.1993 laying down provisions for implementing Regulation (EEC) No. 2052/88 as regards the financial instrument for fisheries guidance (*OJ* L 31.07.1993), and Council Regulation (EC) No. 3699/93 of 21.12.1993 laying down the criteria and arrangements regarding Community structural assistance in the fisheries and aquaculture sector and the processing and the marketing of its products (*OJ* L 346, 31.12.1993).

190 Commission Communication on the future of Community initiatives under the structural funds - Document COM final/2 of 25.03.1994 - Communication on PESCA (*OJ* C 180, 1.07.1994).

191 See Churchill, R.R., Chapter 7 and 8, *op. cit.*, fn 1. On early reform see Wise pp. 211–214 and p. 254 and Leigh pp. 32–35, *op. cit.* fn 1. On the initial market policy, see *inter alia*, Garron, R., 'Le Marché Commun de la Pêche Maritime' (Paris, Librarie Techniques), **VII** (1971), *Collection de droit maritime et des transports*, Part III; Laing, A., 'The Common Fisheries Policy of the Six', *Fish Industry Review*, **1**(1971), 8–18.

192 Council Regulation No. 2141/70 and Council Regulation 2142/70 were replaced by consolidated Council Regulations No. 100/76 and 101/76 with the same title, (*OJ* L 20/1 and 19, 1976) and (*OJ* S ed 1970 (111) p. 707), as amended by Council Regulation No. 100/76, (*OJ* L 20/1, 1976). Replaced by Council Regulation No. 3796/81, (*OJ* L379/1, 1981) amended for the Iberian Act of Accession, (*OJ* L302/1, 1985).

193 COM (80)540 and 724.

194 Council Regulation No. 3796/81, (*OJ* L 379/1, 1981).

195 Council Regulation No. 3759/92 of 17.12.1992 on the common organisation of the market in fishery and aquaculture products. (*OJ* L 388, 31.12.1992) amended for the second time on 12.07.1993.

196 Commission Regulation No. 2210/93 of 26.07.1993 on the communication of information for the purposes of the common organisation of the market in fishery and aquaculture products, *OJ* L 197/8, 06.08.1993.

197 Council Regulation No. 103/76 laying down common market standards for certain chilled or fresh fish (*OJ* L 20, 28.01.1976) amended for the tenth time by Council Regulation No. 1935/93 of 12.07.1993. (*OJ* L 176, 20.07.1993).

198 Council Regulation (EEC) No. 105/76 of 19.01.1976 on the recognition of producers' organisations in the fishing industry, *OJ* L 20/39, 28.01.1976. For a list of the recognised producers' organisations in the fishery and aquaculture sector, see *OJ* C 92/3, 27.03.1998. On the role of Producers' Organisations in the decision making process in Denmark, see Nielsen, J.R., 'Participation in fishery management policy making. National and EC regulation of Danish fishermen', *Marine Policy*, **18**(1), 29–40, (1994). See also the authorities cited in Chapter 5, fn. 74, *post*. See also authorities cited in Chapter 12, *post*.

199 Council Regulation No. 3140/82 of 22.11.1982 on granting and financing aid by Member States to producers' organisations in the fishery products sector. (*OJ* L 331, 26.11.1982). See Chapter 9, *post*.

200 *Communication on the future of the market in fishery products in the European Union, reponsibility, partnership and competitiveness*, COM(97) 719, 12.1997.

201 Council Regulation No. 2406/96 of 26.11.1996 laying down common marketing standards for certain fishery products, *OJ* L 334/1, 23.12.1996.

202 See Chapter 2, *post*.

203 *Op. cit.*, fn 200.

204 On this point, see Barents, R., 'Recent Developments in Community Case Law in the Field of Agriculture', *CMLR* **34** (1997), 811–843.

Chapter 2
Fishery Enforcement and the CFP

Introduction
This chapter examines fishery law enforcement in the Community. It is divided into three sections. Section (i) presents some background information on the characteristics of the Community fishing industry and the resources available in the Member States to undertake the enforcement task. Section (ii) commences with a brief assessment of the Community's competence to regulate fisheries, this is followed by a discussion of some of the principles which underpin the enforcement of Community fishery law. In addition there are comments, from the general perspective, on the subject of Member States' enforcement obligations pursuant to the fisheries provisions in the United Nations Conference on the Law of the Sea 1982 (LOS Convention III).[1] On the other hand, Section (iii) focuses on the fishery enforcement weaknesses identified in a series of reports produced by the Commission in the latter part of the life of the first phase of the CFP regime (1986–1992). The reports are important as many of their proposals and suggestions were subsequently incorporated into the 1993 Control Regulation. They also mark the point of departure which ultimately led to the new scheme of enforcement which the Community adopted in 1993 and which is examined in detail below.

Section (i) The Community fishing industry

1. *The Sea Fisheries Sector in the EU*

The sector is composed of three elements: fishing fleets; fishermen; and the processing industry.[2]

(i) Productivity and zones of activity of the Community fleet
In 1994 the total catches of fish by the Community fleet (12 Member States) amounted to 6.7 million tonnes. If Sweden, Finland, and Austria are included, the quantity is 7.2 million tonnes. While the range of species harvested is extensive (30 varieties of species are used to compile the above figures) a small number dominate the catches taken annually. These are, principally, Atlantic mackerel, herring, sardine, horse mackerel, sprat, anchovy, tuna (skipjack and yellowfin) and cod.

One Member State, Denmark, contributed slightly over 25% of Community catches because its fleet targets and harvests large quantities of sandeels (839 000 tonnes), sprat (240 000 tonnes) and Norway Pout (166 000 tonnes). Spain is the next largest contributor to Community catches accounting for 17%, while the United Kingdom and France contributed 12 and 10%, respectively. Belgium contributed the smallest quantity, approximately, 35 000 tons or 0.48%. The Community fleet operate in eight main fishing zones. Table 2.1 indicates the eight zones which are classified by geographical ocean region and the catches of various Member States in each of these zones.

From the data presented in Table 2.1 it is evident that the catch of the Community fleet is predominantly taken from the Atlantic zone generally and on the Northeast Atlantic zone in particular. Catches in other coastal waters of Member States are modest, and contribute approximately 4% of all catches. The contribution of the Mediterranean Sea is not of great significance in the context of overall productivity. Spain takes catches in all fishing zones and up to 60% of its fleet is dependent on catches taken outside the Community zone. France, Italy and Portugal are present in most fishing zones, while Ireland, the Netherlands, Belgium, Finland, Sweden and Denmark confine their fleet activities to two zones.

If the internal fishing practices within the Community zone are examined, see data in Table 2.2, then it is evident that there are significant differences between the different fishing fleets of the Member States. The Irish and United Kingdom fleets fish predominantly in waters under their own jurisdiction and sovereignty. In comparison, vessels flying the flag of Spain depend largely on access to other Member States' fishery resources, third-country waters, and high seas fisheries. Of particular significance for these fleets is the international fishery agreement with Morocco and Mauritania, because it allows access for 550 Iberian vessels to third country waters.[3] Indeed, the agreements negotiated by the Commission with third countries, 19 of whom are African, have offset unemployment in the Community distant water sector.[4]

The value of catches into Community ports and non-Community ports by the European Union fleets in 1994 was €5.9 billion (12 Member States) or €6.1 billion (15 Member States). Spain accounted for approximately 26% of this figure while Italy and France account for approximately 20 and 13% respectively.

(ii) Fleet composition

The Community fleet (13 Member States) in 1997 consisted of 99 528 vessels. Of these, the bulk of the fleet is made up of 70 000 vessels which are under 12 metres in length and are mainly engaged in day trips or make trips of one or two days' duration. At the other end of the scale, there are approximately 2500 vessels in excess of 33 metres which are generally freezer trawlers or ocean purse seiners. The balance of the Community fleet are ± 20 000 vessels which undertake the major part of commercial fishing in Community waters. The fleet in the Mediterranean comprises 40 000 vessels of various lengths.

Table 2.3 gives figures for 1997 and is divided in the number of vessels per Member State by engine power (kW) and tonnage. It indicates that Spain represented almost one third of Community tonnage and one fifth of engine power (two of the criteria used to measure fishing capacity).

Table 2.4 gives an indication of the average tonnage and engine power per vessel per Member State in 1997.

The Member States with the highest average engine power and tonnage in 1997 were the Netherlands and Belgium. Both Member States have a relatively high percentage of beam trawlers in their respective fleets and this accounts for their position *vis-à-vis* other Member States. Both Mediterranean Member States have a preponderance of small vessels and this accounts for the low average tonnage noted in the table.

Table 2.1 Catches by country and the most important fishing zones in 1994: fish, crustaceans, molluscs, etc.[a] (Source: EUROSTAT.)

Fishing zones	Belgium	Denmark	Germany	Greece	Spain[b]	Finland	France	Ireland	Italy	Netherlands	Austria	Portugal	Sweden	United Kingdom	EUR 12	EUR 15
North-west Atlantic		245	305		54 053		0	0	0	0		30 157		49	84 809	84 809
North-east Atlantic	33 753	1 843 246	216 628	0	449 937	103 542	408 915	282 795	531	379 132		182 475	383 792	840 244	4 637 654	5 124 989
Central east Atlantic			0	9 317	340 000		78 865		43 578			24 524			496 284	496 284
Mediterranean				198 857	141 249		74 076		438 310			428			852 920	852 920
South-west Atlantic			0	0	72 071		0		5 810	0		3 888		1 256	83 025	83 025
South-east Atlantic			0		34 751		0		0			1 556			36 307	36 307
West Indian Ocean			0		111 746		96 592		8 717					0	217 055	217 055
Domestic waters	846	36 368	48 478	16 168	32 750	50 489	58 990	4 504	57 606	1 636	4 548	2 219	5 202	18 118	277 684	337 923
Total	34 599	1 879 859	265 411	224 342	1 136 557	154 031	717 437	287 300	554 552	380 768	4 548	245 246	388 994	859 667	6 685 738	7 233 311

[a] Nominal catches (live weight equivalent of quantities landed) in metric tonnes, including aquaculture production.
[b] Estimated by the FAO.

Table 2.2 Flag Member States with vessels operating in the waters of other coastal Member States.

		BEL	DNK	GER	GRC	ESP	FRA	IRL	ITA	NLD	PRT	FIN	SWE	UK	NAFO
F															
L	BEL	X	X	X			X	X		X				X	
A	DNK	X	X	X			X	X		X		X	X	X	X
G	GER	X	X	X			X	X		X			X	X	X
	GRC				X										
	ESP					X	X	X			X			X	X
	FRA	X	X	X		X	X	X		X	X			X	
	IRL						X	X						X	
S	ITA								X						
T	NLD	X	X	X			X	X		X				X	
A	PRT					X					X				X
T	FIN		X	X								X	X		
E	SWE		X	X								X	X	X	
	UK	X	X	X			X	X		X				X	
		BEL	DNK	GER	GRC	ESP	FRA	IRL	ITA	NLD	PRT	FIN	SWE	UK	

'X' signifies which flag Member State has vessels operating in the waters of each Coastal State.
Greek fishing vessels operate in Greek territorial waters, in international waters and in the coastal areas of west Africa.
Italian vessels operate in Italian territorial waters, in international waters of the Mediterranean and in the waters of certain African countries.

Table 2.3 Number of vessels per Member State by engine power (kW) and tonnage in 1997.

Member State	Number of vessels	% of Community fleet	kW% of Community fleet	Tonnage % of Community fleet
Belgium	148	0.12	0.8	1.1
Denmark	4 648	4.7	4.8	4.8
Germany	2 373	2.4	2.1	3.7
Greece	20 243	20.3	8.2	5.5
Spain	17 972	18.1	18.4	28.7
France	8 836	8.9	14.3	10.2
Ireland	1 246	1.3	2.4	3.0
Italy	16 325	16.4	18.9	12.7
Netherlands	1 040	1.0	6.0	8.5
Portugal	11 579	11.6	4.9	6.0
Finland	3 979	4.0	2.7	1.2
Sweden	2 481	2.5	3.2	2.4
United Kingdom	8 658	8.7	13.1	12.3

(iii) Employment in fishing activity

The number of persons employed as fishermen has been in decline in most Member States since the mid 1980s. Approximately 260 000 were employed either part-time or full-time in 1997. Table 2.5 summarises the situation in 1995 and in the nearest corresponding year.

At the level of the Member States, Spain has the highest number of fishermen ± 75 000 followed by Italy (45 000) and Greece (40 000). The largest concentration of

Table 2.4 Average tonnage and engine power per vessel per Member State in 1997.

Member State	Average kW	Average tonnage
Belgium	438	156
Denmark	81.9	21
Germany	72.2	32
Greece	32.3	5.5
Spain	88.29	32.79
France	129.1	23.7
Ireland	152.9	49
Italy	92.7	15.96
Netherlands	463.7	167.6
Portugal	33.9	10.7
Finland	55.2	6.0
Sweden	103.4	19.6
United Kingdom	121	29.2

Table 2.5 Number of fishermen per Member State.

Member State	1986	1990	1994	1995
Belgium	896	845	652	624
Denmark	9 000	6 945	5 275	5 055
Germany	3 089	4 812	4 979 (part-time excluded)	4979
Greece	39 237	39 124	40 164 (1993 figure)	40 164
Spain	94 246	87 351	77 962	75 009
France	30 036	32 622	27 598	26 879
Ireland			7 700 (approximately)	5 500
Italy	44 925	41 429	45 000 (1993 figure)	45 000
Netherlands		3 502	2 834 (1993 figure)	2 752
Portugal	41 775	40 610	31 721	30 937
Finland	6 950	3 046	2 372	2 792
Sweden	4 678	3 823	3 500	3 400
United Kingdom	22 224	24 230	20 766	19 928

fishermen is to be found in Pontevedra, and La Coruna in Spain, with approximately 31 500 fishermen. Fishermen as a proportion of the working population only exceeds 1% of one Member State – Greece. Fishermen, nonetheless, represented 3.7% of persons employed in the primary sector jobs in the Community.

As noted above, the majority of vessels in the Community are under 12 metres in length and because of their size do not require more than two or three crew members. Table 2.6 indicates the average number of jobs per vessel per Member State in 1994.

The table indicates that Irish vessels (average tonnage 49 tonnes) generates the highest number of jobs while Swedish vessels (average tonnage 5.93) maintain the lowest number. In terms of manpower requirements, per tonne, the Dutch fleet is the most efficient, while the Spanish, French, Irish and Portuguese are the least efficient from this perspective.

(iv) Trade in fisheries products

Over the lifetime of the CFP (post 1983) the Community has had a substantial trade imbalance in the fisheries sector. The imbalance is generated by the absence of sufficient resources in Community waters, consumer preferences for species caught in

Table 2.6 Average number of jobs per vessel per Member State in 1994.

Member State	Average number of jobs	Average tonnage of vessels
Belgium	4.2	156
Denmark	1	21
Germany	2	32
Greece	2	5.5
Spain	4.11	32.79
France	3	23.7
Ireland	4.4	49
Italy	2.75	15.9
Netherlands	2.6	167.6
Portugal	2.67	10.7
Finland	1.4	6.0
Sweden	1	5.93
United Kingdom	2.3	29.2

non-Community waters and the requirements of the processing sector for supplies of pelagic and demersal species. In 1995 the trade balance was as follows:

- Quantity of Imports: 4.3 million tonnes
- Quantity of Exports: 1.50 million tonnes

- Value of Imports: €8.6 billion
- Value of Exports: €1.69 billion

The main Member States which imported fishery products in 1996 were:

Spain 681 000 tonnes, value €1419 million
United Kingdom 728 000 tonnes, value €1327 million
Germany 756 000 tonnes, value €1218 million

The main Member States which export fishery products outside the European Union:

Denmark 407 000 tonnes, value €370 million
Spain 134 000 tonnes, value €295 million
Netherlands 385 000 tonnes, value €233 million

The number of Europeans who consume fish varies considerably from Member to Member State (see Table 2.7).

(v) Financial aid to the fishing industry and the processing sector

The fishing industry receives financial aid from the Community and the Member States. Table 2.8 covers the period 1997 and divides expenditure under a number of headings:

Table 2.7 Consumption levels of fish per head of population 1991–1993.

Member State	kg/head
Portugal	58.8
Spain	36.3
Finland	33
France	27.1
Sweden	28.5
Italy	22
Greece	25
Denmark	20.5
Belgium/U. Kingdom	19.2
Ireland	15
Netherlands	14
Germany	12.7
Austria	9.7

- Fisheries Guidance Fund (EFGF) for non-aquaculture investment in fleet development and port facilities which accounts for approximately 60% of the aid;
- EAGGF Guidance for investment in the processing and marketing of fisheries products (20% of aid);
- aid for aquaculture and the development of coastal areas (10% of the aid);

Table 2.8 Distribution of Community financial aid in 1997 and 1990–1997.

	Financial aid 1997		Financial aid 1990–1997
Member State	€ '000	Percentage	€ '000
Spain	220 869	52.1	1 177 019
Italy	34 542	8.15	353 155
Portugal	45 409	10.71	339 430
France	44 581	10.52	331 255
Greece	33 801	7.97	461 148
Denmark	7 493	1.77	201 359
Germany	1 944	0.46	186 087
United Kingdom	18 263	4.31	178 252
Ireland	13 275	3.13	93 816
Netherlands	246	0.06	63 486
Belgium	1 059	0.25	44 216
Finland	800	0.19	27 910
Sweden	1 525	0.36	47 934
Others	129	0.03	4 390
Total budget	423 936	100	3 509 457

- EAGGF Guarantee Sections intended to provide market support for fish withdrawal and storage (5–10% of aid).

At the beginning of 1994, the two EFGF and EAGGF Guidance Sections were merged into one financial instrument, Financial Instrument Fisheries Guidance (FIFG). The latter now accounts for approximately 92% of Community aid while the EAGGF Guarantee section represents 8%. Aid committed through the Community budgets over the period 1990–1997 increased from €250 million to €423 million. The distribution of financial aid for the period 1990–1997 and in 1997 are summarised iin Table 2.8.

2. *Agencies responsible for the enforcement of Community fisheries law*

The issue of a Member State's exclusive competence to enforce Community fishery law is a central tenet of the Community management and enforcement system and is discussed in Section (ii) below.[5] The actual resources available to undertake the enforcement task vary considerably from one Member State to another. All are affected by the control and enforcement system adopted in 1993. The Commission has published detailed data on the resources available for enforcement[6] and, pursuant to a provision in the Control Regulation, each Member State is obliged annually to furnish the Commission with an assessment of the technical and human resources used in applying the regulation.[7] The Commission, in turn, is obliged to publish a factual report every year and an assessment report every three years, based on their evaluation of the Member State Reports. The first report was made available in 1995.[8] In its preparation for the report, the Commission requested Member States to structure their reports around certain crucial aspects of enforcement: organisation, resources, and enforcement activities at sea and in port. In their submission, the Member States provided information to a greater or lesser extent on these three aspects, and this forms a basis for comparison between Member States. Table 2.9 of the report, reproduced here, contains a number of interesting points of comparison.

With respect to the means theoretically available for fisheries control, the table shows that for example, Greece has more inspectors than the United Kingdom while

Table 2.9 Inspection/control resources in the Member States of the European Union.

Member States	Inspectors	Vessels	Planes	Helicopters
Belgium	8	3[9]	1	–
Denmark	143	5	–	–
Germany	30	21[10]	–	–
Greece	180[11]	30	4	–
Spain	30[12]	31	–	2
France	144[13]	35	–	–
Ireland	18	10	1	–
Italy	–[14]	300	16	–
Netherlands	58	3	–	–
Portugal	12	71	3	–
United Kingdom	169	17	5	–

Italy has the largest number of personnel engaged in control. Clearly these figures need to be interpreted with considerable care. In the UK, the inspection service is port based and devotes a considerable amount (but not all) of its time to actual inspection of landings and catch registration generally. The data presented in the table do not take account of the considerable personnel resources of the Royal Navy which assists the Sea Fisheries Inspectorate in the enforcement task at sea. In Greece on the other hand, where a roughly comparable number of personnel is available, these are under the control of the Coastguard Service and undertake a range of tasks which go far beyond traditional inspection. The situation in Italy is more complex as responsibility for control is shared by at least six public organisations all of which undertake security and administrative tasks which cannot realistically be classified as traditional control functions.

The same table also indicates the number of patrol vessels and aircraft available for enforcement purposes. Once again the information needs to be treated with caution as it indicates that, for example, Germany, which has a relatively short/limited coastline, has a greater number of vessels available for control than the UK, which has an extensive coastline and exclusive economic zone (some 725 000 km^2 of sea area, with a coastline of some 16 000 km^2 and over 450 locations on which fish may be landed at shore). Moreover, Germany reported that its services in 1995 had conducted 4488 sea inspections while the UK reported a similar number (4509).

A word of explanation is also necessary for the resources reported by Spain, which has the largest fishing fleet of all Member States. A large number of personnel in the inspectorate are based in the Ministry of Agriculture, Food and Fisheries in Madrid. The inspectorate visits the coastal ports particularly those located in Galicia and the Basque region where the bulk of the Spanish vessels land their catches. In addition, there are local inspectors under the control of the autonomous region who perform control tasks. The latter have limited jurisdiction and deal with locally-adopted conservation/management rules governing fishing activities inside the 12-mile control zone. Their rôle was extended in 1995 to cover the control of the first sale of fresh fish in auction centres. In real terms, the Madrid based inspectorate has the task of inspecting those Spanish vessels authorised to fish in Community waters and those vessels engaged in fishing activities in the context of fisheries agreements between the European Union and third countries (principally Morocco and West African coastal states). In addition, the inspectorate is also charged with the task of controlling the Anglo-Spanish fleet which lands the majority of its catches into Spain. Since 1997 the Madrid inspectorate is supported by a permanent cadre of inspectors deployed in the important fishing ports.

Table 2 of the report (reproduced here as Table 2.10), indicating both the number of competent authorities responsible for fisheries and whether they are centrally- or regionally-based, demonstrates that in all Member States a variety of national authorities are involved. Thus, for example, in the United Kingdom responsibility for fishery enforcement is shared between four Government Departments.[15] Other examples include France, where the *Affairs Maritimes* (AFFMAR), a component of the *Ministère de l'Équipement*, is responsible for the inspection of vessels engaged in fishing activities. Inspections in port are undertaken by the *Gendarmerie Maritime* (GM) which is under the responsibility of the Minister of Defence and the French Navy. The French customs service is responsible for enforcing aspects of the structural policy such as fishing

Table 2.10 Number of authorities, regional authorities and specialised services in the Member States.

Member State	Total number of competent authorities	Regional authorities	Specialised service competent up to first sale
Belgium	5	no	yes
Denmark	3	no	yes
Germany	7	yes	yes
Greece	3	no	no
Spain	2 + autonomous	yes	yes
France	5	no	no
Ireland	3	no	yes
Italy	6	no	no
Netherlands	6	no	yes
Portugal	7	no	no
United Kingdom	6	yes	yes

vessel engine power and tonnage. As noted above, there is also an elaborate national fishery enforcement organisation in Italy, where inspections are conducted by several authorities which include the *Guardia di Finanza* (Ministry of Finance), *Polizia di Stato* (Ministry of the Interior), *Arma dei Carabinieri* (Ministry of Defence), *Unita Sanitarie Locali* (Ministry of Health), and many municipal authorities.

This finding regarding the diverse number of organisations is not really a surprise, as fishery enforcement covers a number of specialised sectors: markets, fleet registration or vessel licences etc., and often requires the support of specialised services, for example, the Navy and Air Force. In addition, one should recall that the range of enforcement tasks is continuing to expand and that many are highly specialised (determining engine power for instance) and consequently cannot be simply grafted onto existing more narrowly-focused inspection services.

As the Commission notes in the 1995 Control Report, the number of inspectors in each port differs substantially from one Member State to another. In some ports there may be up to four inspectors, while in others only one inspector is present. Although the number of inspectors in port will be determined by the level and timing of landings which are expected to occur, a single inspector in a large fishery port may not have the capacity to fully implement an integrated control and enforcement regime. On the other hand, it is equally clear that inspections cannot be conducted on all landings and that priority fleets (i.e. those harvesting particularly sensitive quota species, such as flat fish), have to be identified for ongoing enforcement purposes during the fishing season. In practice, national enforcement authorities have to strike a balance and seek to ensure that physical inspections are carried on a representative sample of the fleets landings into their ports. This will offer the opportunity of confirming that the quantities landed are accurately recorded in landing declarations and that subsequent cross-checking of eventual sales notes is based on reliable data.

Inspection activity generally follows the 'swings' common place in fishing activity which is highly seasonal in character. The enforcement resources in some Member States are thus marshalled to cope with particularly intense series of landings which are usually associated with the periods when quotas are close to exhaustion. In addition, seasonal activity such as 'klondyking', that is the practice of selling pelagic species (herring, mackerel in particular) 'over the side' to 'factory' vessels predominantly from eastern Europe and the Russian Federation make heavy demands on the national control services in Scotland and Ireland.

In addition to the resources available in the Member States the Commission has a team of 24 inspectors employed in Directorate General XIV (the Fisheries Directorate) which undertake a monitoring and verification role of Member States' efforts to implement and comply with the Control Regulation. The function of the Community Inspectors and their interrelationship with Member State enforcement agencies is examined in Chapter 6.

Section (ii) Community legislative competence to regulate fisheries and the issue of enforcement

Introduction

In this section there is a brief review of Community legislative competence to regulate fisheries. This topic is examined both from the viewpoint of legal competence within the Community as well as the external competence of the Community in relation to international fishery relations and agreements. Competence refers to the Community's law making powers. The term competence is derived from the French legal term *la compétence* and is frequently used in Community literature to express the concept of the Community's legal power to act in a particular domain.

1. *Community legislative competence to regulate fisheries*

(i) The internal perspective[16]

The Community has the capacity, as a manifestation of its unique legal order,[17] to prescribe fisheries law.[18] The Community right to exercise this competence on both the external plane in relation to international fishery agreements and on the internal plane to regulate the fisheries has been enunciated by the Court of Justice in several cases dealing with fisheries issues.[19] According to the Court in the *Kramer* case, it followed from the provisions of the Act of Accession, the Treaties, and the various measures adopted by the Council, taken as a whole, that the Community has at its disposal, on the internal level, the power to take any measures for the conservation of the biological resources of the sea.[20] In the seminal case on the subject of exclusive competence, *Commission* v. *United Kingdom*, the Court unequivocally concluded that 'since the expiration on 1 January 1979 of the transitional period laid down in the Act of Accession, power to adopt, as part of the CFP, measures relating to the conservation of the resources of sea belonged fully and definitively to the Community'.[21] The principal consequence of this judgement is that Member States may no longer act unilaterally in this field. In the words of the Court:

> 'the power to adopt measures ... has belonged fully and definitively to the Community. Member States are no longer entitled to exercise any power of their own in [these matters]. The adoption of ... measures is a matter of Community law. The transfer to the Community of powers in this matter being total and definitive, ... a failure [of the Council] to act could not in any case restore to the Member States the power and freedom to act unilaterally in this field.'[22]

Nevertheless, the Court recognised that the Member States might be required to legislate at national level to conserve fish stocks in order to protect the Community's interest until the Community had taken action.[23] It was made clear, however, that the Member States' powers to act only continued until the Community had itself acted; that the Member States' action should only be limited to what was strictly necessary for the conservation and exploitation of the sea's resources in the common interest of the Community as a whole, and that any national action had to be undertaken in close consultation with the Commission.[24] An example of this arose in *Anklagemyndigheden* v. *Noble Kerr* when the Court held that Member State legislation which had as its purpose the fixing of a total catch quota during the transitional period in which the Council had not exercised its power, did not impugn Community law if the measure was justified by objectives relating to the protection of the needs of the coastal population concerned and the maintenance of a situation temporarily created in the area in question.[25] The Court has also held that competence in certain areas which impact on fishing activities remain within the remit of the Member States include, *inter alia*: the power, in principle, to decide which vessels fly their flag[26] and the power to set national boundaries.[27] In such cases the Court has strictly or narrowly defined the exclusive competence of the Community and limited it to the regulation of fishing activity.

In some areas, nevertheless, the Community has delegated some of the prescriptive competence back to the Member States. For example, Member States may take emergency fishery conservation measures where urgency is required because of a serious and unexpected upheaval liable to jeopardise conservation, subject to a power of review by the Commission and/or the Council.[28] It is also the Member States' responsibility to manage the quotas allocated to them each year under the annual TAC and quota regulations.[29] Member States can take national control measures which go beyond the minimum requirement in Community legislation provided they comply with Community law and conform with the CFP and are communicated to the Commission.[30] The Court has also been active in this area and has identified in *Rogers* v. *Dartheney*,[31] and in *Officier Van Justitie* v. *Bout*[32] that Member States may adopt national measures when there are lacunae in Community law.

(ii) Community competence to prescribe fisheries law on the external plane

The Community's powers in external fisheries relations are derived expressly or by implication from the EC Treaty as amended by the TEU and the Treaty of Amsterdam.[33] Surprisingly, the only express provision in the text of the Treaties which refers to external fishery competence is **Article 113** {133} of the EC Treaty which gives the Community the competence to enter into agreements relating to trade in fishery products and other aspects of the common commercial policy. In contrast, the doctrine of implied powers, developed in the jurisprudence of the Court, is considered to flow by implication from the internal practice of the Community in relation to fisheries.[34]

The Court in *Kramer* pointed to 'the general system of Community law in the sphere of the external relations of the Community'. **Article 210** {281} of the EC Treaty means that the Community has capacity to enter into international commitments over the whole field of its objectives as defined in the EC Treaty. The Court held:

> 'To establish in a particular case whether the Community has authority to enter into international commitments, regard must be had to the whole scheme of Community law no less than its substantive provisions. Some authority derives not only from an express conferment by the Treaty, but may equally flow implicitly from other provisions of the Treaty, from the Act of Accession, and from the measures adopted, within the framework of those provisions, by the Community institutions.'[35]

As a result of the exclusive powers vested in the Community for the conservation and management of marine biological resources the Council gave the Commission responsibility in 1976 (the Hague Resolution) for negotiating and concluding fisheries agreements with relevant third countries and for representing the Community in international fishery management organisations.[36] The general practice is for the Commission to negotiate bilateral agreements in line with negotiating mandates received from the Council. Before taking effect the agreements must be adopted by the Council in the form of a regulation based on **Articles 43** {37} and **228** {300} of the EC Treaty (which require prior consultation with the Parliament). Since 1976 the Community has extensively relied on its implied treaty-making powers. Moreover, the Court has held that Member States lose their competence to enter into agreements with third states once the Community has taken action on the internal or external plane.[37] It has also been consistently held by the Court of Justice that as regards the high seas, the Community has the same regulatory powers in areas falling within its authority, as those recognised under international law, ascribed to States. This has been upheld by the Court in *Mondiet* v. *Islais* which is discussed below in the context of the implementation of Community rules on the use of driftnets.[38] Although it is now well established that the Community has exclusive competence in the negotiation of international fishery agreements and representing the Member States in international fishery organisations and commissions,[39] Member States nonetheless retain the right, under general Community law, to remain party to a treaty when the treaty covers matters which fall outside Community competence or where the treaty covers a range of issues where competence is divided between the Community and Member States. The latter are referred to as mixed agreements.

Although the competence of the Community to enter into negotiations to establish new or amend existing fisheries organisations and to become a party to any resulting treaty is well established this has never received the unequivocal acceptance of the Member States.

An illustrative example of the inherent tension that exists as a result of the Community's membership in fisheries organisations and some of the procedural difficulties which may arise is evident in the recent case of the *Commission of the European Communities* v. *Council of the European Union supported by United Kingdom of Great Britain and Northern Ireland.*[40]

In this case the Commission applied to the Court to annul the decision of the 'Fisheries' Council of 22 November 1993 giving the Member States the right to vote in the United Nations Food and Agriculture Organisation (FAO) for the adoption of the

'Agreement to Promote Compliance with International Conservation and Management Measures by fishing vessels on the High Seas' (FAO Draft Agreement).[41] The Court ruled in favour of the Commission and held that the Council had erroneously granted the Member States the right to vote for the adoption of the FAO Draft Agreement. The Court was of the opinion that this was a clear breach of an '*Arrangement*' concluded in 1991 between the Council and the Commission regarding preparations for FAO meetings, statements and voting, which all parties were required to observe. The Court drew attention to the General Rules of the FAO which provides for a system of alternative exercise of rights attached to membership, which allow the Commission to speak and vote where an agenda item is within exclusive competence of the Community. Where an agenda item contains matters containing elements both of national and of Community competence, however, the Commission is obliged to present the thrust of the issues which fall within the exclusive competence of the Community.[42]

In annulling the Council decision the Court was aware that to conclude otherwise would mislead third countries as to the competence of the Community in the context of its membership of international organisations such as the FAO. Moreover, the Council by granting the right to vote to the Member States would prevent the Community from having any effective say in the deliberations which would precede the final decision on the text of the agreement. Finally, by virtue of the *Arrangement* concluded in 1991 the fact that the Member States voted in accordance with the common position gave third countries and the FAO the impression that the thrust of the Agreement did not fall within the exclusive competence of the Community. As the Court pointed out the Council's vote therefore had legal effects as regards the relations between several parties, including: the institutions of the Community and the Member States, between the institutions of the Community and, finally, between the Community and its Member States on the one hand, and between the other subjects of international law, especially the FAO and its members, on the other.

(iii) Enforcement competence

Although it is clearly established that the power to prescribe fisheries law on both the internal and external plane rests with the Community, the competence to enforce such measures, however, remains almost exclusively with the Member States. In this regard there is no express or implied powers derived from the Treaties which vest autonomous inspection and enforcement powers in the Commission's fisheries inspectorate which is based in Brussels. There are, however, two instances where the Council has adopted regulations vesting inspectors with independent inspection powers in order to comply with international schemes of enforcement for high seas areas. The first one is the Joint International Scheme of Inspection adopted by the NAFO Fisheries Commission on 10 February 1988 and adopted by the Council on 9 June 1988 which authorises the Commission to assign Community inspectors to the scheme of inspection in the NAFO Regulatory Area.[43] Interestingly, this instrument also classifies such an inspector as any inspector appointed by the Commission or the Member States.[44] The second example is the system of observation and inspection established under Article XXIV of the Convention on the Conservation of Antarctic Marine Living Resources.[45] The system of observation and inspection pursuant to this Convention applies to the Community and allows Community inspectors to inspect Community vessels operating in the Convention area (Antarctica) for compliance with Community conservation and control measures as well as the appropriate Convention measures.[46] To date, these are the only two examples of Community legislative instruments which vests identical inspection competence in both Commission and Member State inspectors. Examples of the competence which remains almost exclusively with the Member States is the power to arrest or detain fishing vessels, seizures of gear and catches and imposition of fines and other penalties.[47] The latter is discussed in some detail in Chapter 6.

Community fishery law, similar to many other sector policies, is enforced by the Member States. The requirement for Member States to enforce Community law has its origin in **Article 5** {10} EC Treaty which provides, *inter alia* that: 'Member States shall take all appropriate measures, whether general or particular, to ensure fulfilment of the obligations arising out of this Treaty or resulting from actions taken by the institutions of the Community'.[48] The scope and extent of the procedures and sanctions relied upon by the Member States to discharge this obligation vary considerably.[49] In some Member States (Ireland, the United Kingdom and the Netherlands) for instance fisheries enforcement comes within the scope of the criminal legal system, whereas in other Member States (Germany, Spain and Portugal), fisheries offences are dealt with through administrative procedures and may result in the imposition of administrative sanctions. The use and rôle of criminal proceedings in the Member States in enforcing Community law was accepted by the Court in the *Amsterdam Bulb Case*,[50] and the Court further elaborated in the *Von Colson Case* that if criminal sanctions were relied upon that they must be effective, dissuasive and proportionate.[51] More recently, the Court in the *Greek Tobacco Case* has advanced the principle of assimilation - that there is an obligation to use criminal enforcement in cases where a similar infringement of national rules would be considered a criminal offence.[52]

Enforcement jurisdiction at national level may be exercised at two levels. On the one hand, there is the competence to inspect and collect evidence of violations of fishery law which in normal circumstances is comparable to a policing task and may be referred to as 'inspection jurisdiction' and, on the other hand, the final competence of courts to deal with alleged violations of the law which may be referred to as 'judicial jurisdiction'. The former, 'inspection jurisdiction', is a fundamental component in the scheme of fishery law and there is a clear symbiotic relationship between 'inspection jurisdiction' and 'judicial jurisdiction'. It is 'inspection jurisdiction' which delivers the elements of proof or evidence which are essential to allow the Courts to exercise their 'judicial jurisdiction' and evaluate whether there is compliance or non compliance with the regulatory framework. Indeed, one of the themes which is evident in Section (iii) of this chapter is the failure of the 'inspection jurisdiction' during the period 1983–1992 to ensure a satisfactory level of compliance with the CFP.

Despite the almost exclusive competence of the Member States in the domain of enforcement, fisheries law, nonetheless, shares the principle of 'dual vigilance' with Community law in general. In this regard, there are two levels of judicial enforcement: the higher at Community level, leading if necessary to **Article 169** {226} EC Treaty proceedings;[53] the lower at national level, based on the principle of direct effect which allows national courts to bring the full remit of their powers to bear to uphold Community law.[54] Unusually, the Community fishery enforcement system is almost unique in the world in so far as there is also 'dual vigilance' with respect to 'inspection jurisdiction'. The higher is at Community institution level where a Commission inspection unit is charged with monitoring the activities of national enforcement authorities, and the other, which is the norm, in the Member States where national inspectors bear the responsibility of monitoring and ensuring compliance with fisheries legislation.[55] The Commission of course has a legal duty in order to ensure the proper functioning and development of the common market to ensure that the provisions of the Treaties are applied.[56] Furthermore, the obligation for the Member States and the Commission to co-operate is a reciprocal obligation underlying the duty of sincere co-operation as expressed in **Article 5** {10} EC Treaty.[57] A point which the Court emphasised in the *EC Fisheries Inspector Report Case* when it noted that it was incumbent on every Community institution to give active assistance in national proceedings and in this case making available Commission documents to be used in criminal proceedings and making Commission Officials available to testify at such proceedings.[58]

Finally, it ought to be pointed out that the approach to the enforcement of Community fishery law conforms with established practice in the international law of fisheries.

(iv) The international law model – the Law of the Sea perspective

The approach to fishery law enforcement in the CFP conforms with State practice in both customary and convention law. The provisions in the United Nations Law of the Sea Convention (LOS Convention) relating to fishery enforcement reveal that there is an enforcement cleavage which separates the responsibilities of coastal State from those of the flag State. Indeed, the Law of the Sea Convention is remarkable in that it contains relatively few provisions on enforcement within national jurisdiction (**Articles 73, 49, 21,** and **25**), and even less with regard to enforcement outside the coastal State jurisdiction (**Articles 66, 110–111,** and perhaps **116**).

Coastal State competence to enforce fishery law within the Exclusive Economic Zone/ Exclusive Fishery Zone In order to ensure compliance with its fishery laws and regulations applicable in the EEZ, the coastal State is empowered pursuant to **Article 62**, paragraph 4(k) and **Article 73** LOS Convention to take such measures as may be necessary, including boarding, inspection, arrest and judicial proceedings.[59] Interestingly, coastal State penalties for violations of fisheries laws and regulations in the EEZ may not include imprisonment, in the absence of Agreements to the contrary by the States concerned, or any form of corporal punishment.[60] Moreover, arrested vessels and their crews have to be released promptly upon the posting of a reasonable bond or other security[61] and, in cases of arrest or detention of foreign vessels, the flag State has to be notified promptly of the action taken and of any penalties imposed.[62] It should also be mentioned that **Article 58** LOS Convention details the conditions limiting coastal state enforcement affecting freedom of navigation.

State practice in relation to coastal State competence to enforce fishery law within the EEZ/EFZ has developed in line with **Article 73**, and it has been suggested by several commentators that its provisions were accepted as part of customary international law, prior to the ratification of the Law of the Sea Convention in 1994.[63] Nationals of other States fishing in the EEZ are obliged to comply with the conservation measures and with the other terms and conditions established in the laws and regulations of the coastal State.[64] Such laws and regulations include, *inter alia:* licensing conditions, the determination of species to be caught, regulations in relation to gear etc.[65] Importantly, nationals are obliged to comply with the enforcement procedures of the coastal State which are consistent with the provisions in LOS Convention.[66]

One of the significant features of fisheries law in the European Community is that there are major differences between the obligations placed on third-country vessels operating in the Community zone and the obligations imposed on vessels flying the flag of a Member State which operate in the coastal zones of another Member State. These differences are examined in greater detail in Chapter 3.

The inadequacy of the flag–coastal Member State enforcement model to regulate foreign fishing in the European Community has been compounded by the fact that fishery enforcement is a difficult task to undertake within the 200-mile zone. One authority has cited two reasons which have accentuated the difficulties encountered by coastal States.[67] First, the absence of sufficient resources such as patrol vessels, aircraft, and appropriately trained personnel to undertake the surveillance and inspection task. The second relates to the EEZ/EFZ which covers an area which was previously governed by the freedom of fisheries and there may thus be a tradition of non-compliance in the zone in question. Outside the European Union, the most frequently adopted solution to facilitate enforcement is the granting of foreign access to the EEZ/EFZ not in terms of fishing quotas but in terms of number and size of vessels. The latter is also a more efficient means to control the level of fishing effort. An example of this approach is evident in the EU–Morocco and EU–Mauritania fishery agreements. The advantage of this approach is that it does not require intensive at-sea inspection to verify quota uptake.

While there has been some academic comment as to whether **Article 110** LOS Convention curtails coastal State authority to use naval vessels and military aircraft for fishery enforcement, there is little evidence to support this interpretation and the practice in the European Union is that Member States deploy naval vessels as well as ships that are clearly marked and identifiable as being on government service to enforce Community fishery law.[68]

Flag State competence to enforce fishery law in respect of vessels flying its flag The jurisdiction of a State over vessels flying its flag is well established in customary and convention law, unless such jurisdiction is conceded by a rule of law or by treaty. Specifically, the exclusive right of the flag State to exercise legislative authority and enforcement authority over its vessels on the high seas is codified in both **Article 6(1)** of the Geneva Convention on the High Seas and of **Article 92(1)** LOS Convention. However, when vessels operate in waters which are under the sovereignty or jurisdiction of the coastal State there is almost universal acceptance of coastal State authority to undertake the fishery enforcement function.[69] In this respect, the rule is that where there is legislative competence to prescribe regulation then the nature and degree of enforcement power is determined by legislative jurisdiction.[70] In the case of the European Union, while it is well-established that the Community has almost exclusive powers to prescribe legislation in the context of the CFP, there is unquestionable acceptance that the enforcement responsibility of the flag Member State relates to all fishing vessels operating in the waters under flag Member State sovereignty or jurisdiction, and to vessels flying the flag of the Member State in question when such vessels operate on the high seas. In other words, unless there is a joint scheme of enforcement between two or more Member States, such as exists between the United Kingdom and Belgium, then the flag Member State does not enforce Community fishery law in the EEZ/EFZ of other Member States.

The 1995 United Nations Agreement for the Conservation and Management of Straddling Fish Stocks and Highly Migratory Species has several extensive provisions which impinge on the enforcement competence of the flag State and some of these are examined in Chapter 11.

Concluding comments on Community and Member State legislative competence to regulate and enforce fishery law

The Community power to prescribe and regulate fisheries is well settled and is now rarely disputed. This state of affairs is mainly due to the jurisprudence of the Court which has clearly defined the prescriptive limits and the division of competence between Community institutions and the Member States in the domain of marine fisheries. Moreover, the power to enforce Community fishery law remains with the Member States and despite the evolution of a more collaborative approach, or a joint action approach, to fisheries enforcement, there is little argument regarding the *ratione* of Member State enforcement jurisdiction.

It may be helpful to summarise the position as shown in Table 2.11.

Table 2.11 Division of legislative and enforcement responsibility in the European Union.

Community	Member State
Prescribe fisheries law Exclusive competence	Prescribe fisheries law Limited competence: • delegated competence • national measures
Enforcement (i) Court of Justice Judicial Jurisdiction – **Articles 169–171** {226–228} EC Treaty	Enforcement Criminal/Administrative Courts/ Tribunals Judicial jurisdiction through criminal and administrative process – **Article 5** {10} EC Treaty
(ii) European Commission Supervisory role – **Article 155** {212} EC Treaty Delegated authority: such as CCAMLR and the NAFO Scheme of Joint International Inspection and Surveillance	Enforcement/Inspection Agencies Enforcement/inspection jurisdiction – **Article 5** {10} EC Treaty

Section (iii) Compliance with the CFP during the period 1983–1992

Introduction

In discussing the background of the Control Regulation it is helpful to recall briefly that the CFP has a life span of 20 years, 1982–2002.[71] The first control regulation which was adopted in 1983 establishing certain control measures for fishing activities by vessels of Member States[72] (the 1983 Control Regulation), was later repealed and replaced by a new consolidated regulation in 1987.[73] It is not proposed to discuss these regulations as they have since been repealed. A mid-term review of the CFP was undertaken in 1992, ten years after the inception of the policy.[74] The review was published in a report by the European Commission (the *1991 CFP Report*),[75] and this was followed shortly thereafter by a detailed report on enforcement in March 1992 (the 1992 Control Report).[76] This resulted in the Council adopting new regulatory provisions for the fishing sector,[77] which, in turn, anticipated the adoption of a new Community control system.[78] The new measures introduced by the Council sought to address the challenge which the CFP faces in the latter part of its regime and to provide the necessary basis for a successful and balanced fisheries sector in the period *post* 2002, when the life of the present policy expires.

Review of enforcement during the period 1983–1992

The Community enforcement framework and legislative regime was gradually reinforced in the 1980s. This was achieved, largely, on the Commission's own initiative and with Parliament's support. It was necessary because the Commission was devoid of enforcement clout in the initial regime, and the legal framework had to be adjusted to tackle some of the more flagrant weaknesses in the Member State enforcement policy. The 1983 control regulation was amended to strengthen the control of vessels transhipping fishery products in the waters of other Member States. It had also been amended to allow for obligatory quota transfers in the instances where Member States overfishing penalised other Member States. The 1987 consolidated regulation was amended to make it more difficult for flag vessels landing outside their flag state to circumvent the TAC and quota regulations. This was ostensibly to curtail the activities of the Spanish 'quota hopping' vessels which were registered in the UK and Ireland.[79] This more stringent approach to enforcement which started with the amendments to the 1983 and 1987 regulations continues in the 1993 Control Regulation. It was the natural, if not the only, response to the problems identified in the aforementioned reports and the perceived failures of the CFP in the period 1983–1991. The reports are essential reading and made a valuable contribution to the reorientation of the policy during the first phase of the policy as well as at the mid-term stage. In this section the following reports are examined:

(1) 1986 Report – Report by the Commission to the Council on the Enforcement of the CFP.[80]

(2) Communication from the Commission to the Council and the European Parliament on the CFP.[81]

(3) Report from the Commission to the Council and the European Parliament on the CFP (the 1991 CFP Report).[82]
(4) Report from the Commission to the Council and the European Parliament on the monitoring and implementation of the CFP (the 1992 Control Report).[83]

The discussion, which follows, is limited where possible to the control and enforcement issues identified in these reports.

1. *The 1986 Report – Report by the Commission to the Council on the enforcement of the CFP*[84]

The 1986 report examines the enforcement of the CFP in the first three years of the operation of the Community system. It dwells, mainly, on the conservation provisions of the CFP. Although the report received a muted response from the industry, there was considerable surprise in the national administrative and enforcement bodies with the direct approach taken by the Commission in assessing Member States' achievements in enforcing Community law. Not only did the report draw attention to the general dereliction of Community obligations but it also detailed the specific failures of Member States. It should be noted, however, that the Commission added the caveat that the conclusions presented in 1986 were provisional and subject to review.

The report notes that not all Member States had adopted the necessary national measures to give full effect to Community legislation.[85]

The Commission report also assessed the performance and the failures of the national enforcement bodies in all relevant Member States.[86] As a general observation, the Commission recorded that there was a lack of uniformity in regard to the catch figures supplied by Member States and that there was continued overfishing of national quotas.[87] This indicated inherent weaknesses in national catch registration systems. The Commission also expressed concern about the enforcement of the logbook/landing declarations.[88] Moreover, the inappropriateness and inadequacy of certain Member States' sea-based inspection resources such as patrol vessels contributed to the non-enforcement of technical measures. The Commission further observed that the lack of uniformity in applying the technical conservation measures was endangering the overall objectives of the CFP. Finally, the Commission pointed out that the lack of uniformity among Member States was not an acceptable reason for not improving national standards, or to neglect the harmonisation of the enforcement efforts undertaken by the national authorities.

The report examined the role of the Commission and drew attention to the fact that the powers and competence of the Commission inspectorate needed to be expanded.[89] The Commission declared that, during the period under review it had eschewed the pursuit of procedures pursuant to **Article 169** {226} of the EC Treaty (that is taking Member States to the European Court of Justice) in all but the most flagrant violations of Community fisheries law.[90] Furthermore, the Commission expressed the intention, thereafter, to have recourse to the full range of measures available for ensuring a greater level of enforcement with a view to establishing greater credibility for the CFP. The options open to the Commission, in order of severity, included conducting administrative enquiries into how Member States carries out their enforcement tasks pursuant to **Article 12** of the 1983 Control Regulation, the presentation of proposals to the Council in relation to new powers of enforcement for the Commission, and resorting to **Article 169** {226} proceedings against the Member State.[91]

The report gave a clear indication of the preferred approach of the Commission in dealing with Member States for non-compliance with their Community obligations. That is to say, the Commission started with a persuasive approach to improve enforcement in the Member States and then aimed to resort to the more conventional remedies such as **Article 169** {226} proceedings.[92] Indeed this approach has since become typical of the Commission's *modus operandi*.[93] Furthermore, the 1986 Report led to some legislative amendments which were later carried over into the 1993 Control Regulation. With hindsight, possibly the most important observation of the Commission in the 1986 Report, particularly in the context of what has occurred in the intervening period, is the warning that without proper enforcement, conservation would be threatened and the Community's international fisheries relations would suffer. Thus enforcement was not only a legal obligation for Member States, but also a political necessity as its absence would undermine the credibility of the CFP.

2. *Communication from the Commission to the Council and the European Parliament on the CFP*[94]

The Commission, in its communication to the Council and the Parliament presented in 1990, reviews the CFP. It deals with the exploitation and management of fisheries resources, and, also of major importance, with the structural and market aspects of the policy. It identifies the main problem in the fishing industry as being the imbalance between fishing resources and fishing capacity. Overfishing had undermined the conservation policy. The Commission expressed the opinion that there was an urgent necessity to ensure the management and control of fishing activities, particularly by restricting access to stocks. The need for improved synergy between the different components of the CFP was also identified as requiring attention.

Although the control and enforcement aspect of the policy is only briefly discussed, the communication reiterates several of the concerns identified previously by the Commission and repeated in greater detail in both the 1991 CFP Report and again in the 1992 Control Report.[95] For the first time, the Commission advocates a new approach based on the principle of subsidiarity to tackle the entire fishery sector. It was the Commission's view that apart from the excessive catching capacity of the Community fleet, enforcement difficulties are exacerbated because of the heterogeneity of the fishing industry, the vast extent of the areas to be supervised, the mobility of fishing vessels and the multiplicity of landing places.[96] Furthermore, the Commission unequivocally cited the lack of commitment by the political authorities to take difficult decisions as a major contributory reason why conservation measures were being flouted.[97]

For the first time, the Commission also advocates the introduction of new technology to improve surveillance and inspection procedures of fishing vessels. Specifically, the issue of monitoring vessels by using satellite technology was raised in the Report.[98]

The Communication was followed by a detailed report on the CFP from the Commission to the Council and the European Parliament in 1991.

3. *Report from the Commission to the Council and the European Parliament on the CFP (the 1991 CFP Report)*

This report was presented by the Commission to the Council and the European Parliament in December 1991. Member States were invited to submit their views on the operation of the CFP. Both Ireland and the United Kingdom completed reports.[99] Subsequent to the receipt of the national reports the Commission drafted a detailed Report (the 1991 CFP Report) which was a direct follow-up to the previous Commission communication to the same institutions, and provides an extensive analysis of the fishing industry and proposes a series of guidelines for the post 1992 period.[100]

In the 1991 CFP Report the Commission took the opportunity to undertake a broader and more complete review of the CFP, over the period 1983–1990, than was originally anticipated in the 1983 terms of reference. The review mapped out the approach the Commission deemed most appropriate to safeguard the future of the fishing industry and the ancillary sectors, particularly in the latter half of the CFP regime. Furthermore, special consideration was given to the problems that needed to be surmounted if the policy was to prepare the sector for the post-2002 period. Interestingly, the report contained no formal proposals. Its stated purpose was to provide debate in the various Community institutions, so as to prepare the ground for Commission legislative proposals which (to paraphrase the report) would set out the 'new rules of the game' for the entire fisheries sector for the 1993–2002 period and improve the overall functioning of the policy.

The facts presented by the report were uncompromising. The Community fishery resources were overfished and the sector faced a crisis.[101] The economic situation was described, at best, as bleak. The rapid decline in the number of fish landed and the corresponding reduction in the incomes of fishermen had been economically off-set by the increased demand for high value species.

From the enforcement perspective, it is interesting to note that there were several contributory reasons identified by the report as leading to the sector crises. First and foremost of these was the absence of any form of control over fishing capacity.[102] Secondly, the complex resource management model which was necessitated by the Community fisheries sector required major surveillance and enforcement mechanisms and inter-Member State co-ordination and procedures. Moreover, the existing system was unable to ensure compliance with the rules given the inadequacy of the coercive measures at Community level and the lack of political will in the Member States. Thirdly, the compartmentalisation of the CFP measures and lack of coherence between the different elements of the policy, especially between the market mechanisms and structural policy, was aggravated by failure to apply sanctions against illegal practices.

Among the solutions for redressing the problems in the fisheries sector, the Commission advocated the urgent redirection of the policy. Particularly, it recommended the redistribution of responsibility at all levels in accordance with the subsidiarity principle, conferring certain management responsibilities on fishermen's organisations.

4. *Report on the monitoring and implementation of the CFP (the 1992 Control Report)*[103]

The 1992 Control Report was presented by the Commission to the Council on March 6, 1992. The 1992 Control Report tackles major issues as opposed to identifying specific failures of the Member States.

The 1992 Control Report is divided up into six sections which endeavour, *inter alia*, to examine the basic principles of the organisation of fisheries surveillance in the Community; discuss the practical organisation of both the national and Community inspectorates; describe the operation of and compliance with certain conservation measures; highlight some of the major shortcomings in the implementation of inspection legislation; provide a detailed summary of how the Commission planned to table certain proposals for the amendment and improvement of the Community rules pertaining to enforcement and inspection along the lines identified in the report.[104]

In factual terms, the '1992 Control Report' revealed that the Community control system had failed to attain its modest objectives in the first half of the life of the CFP. The report identifies many inherent weaknesses in the legal framework and the lack of the necessary political commitment on the part of Member States to implement Community legislation. In particular, the Report identified three contributory factors to poor enforcement.

(i) Lack of understanding regarding the purpose of the policy

The Commission makes the revealing observation that the lack of understanding by the entire sector, from producers to consumers, of the purpose of the conservation and control legislation has exacerbated other inherent policy failures. This manifested itself in the frequently-cited problems of fishermen frequently opting for financial gain in disregard of Community regulations, and the continued demand by the consumer for undersize species of fish.[105] Furthermore, in certain jurisdictions, national inspectors were unable to keep abreast of Community legislation.[106] The report also asserts that judicial bodies failed to appreciate the gravity of fishery offences when prosecuting offenders of Community fisheries' law.

(ii) Inadequate legal framework

Failures were not limited to human elements but extended to the control framework. Of particular concern were the lacunae in Community legal instruments, such as the lack of power for the Commission Inspectors to undertake autonomous inspections. This issue, which is central to the division of legal competence between Member States and the Commission, is examined in Chapter 6.

(iii) Inadequate national structures and little inter-Member State enforcement co-ordination

National legal systems failed in several respects. The 1992 Control Report lists among the failures, *inter alia*: inadequate penalties and sanctions in proportion to the economic gain from the infringement; delays in prosecuting suspected offenders which reduced the deterrent effect of instituting legal proceedings for fisheries offences; the persistent infringement of access rules, particularly by multiple offenders, due to the lacunae in the procedure for implementing Council Regulation No. 3781/85;[107] and, ultimately, the fact that many of the parties who gained from the violation of Community regulations, such as Producers' Organisations, bore no financial responsibility where conservation measures were not respected.

Enforcement authorities with responsibility for inspection at sea were frequently operating from vessels unsuited to the task of fishery inspection.

National control authorities failed to implement and monitor the application of vital components of the CFP. The report states that this was probably at its most serious with respect to checking the landings of catches and the transhipment of fishery products. Furthermore, at Member State level, few countries complied with the closures of fisheries and there were inadequate measures to prevent the exhaustion of quotas. Even in the event of closures, producers continued to fish in closed areas and, in any case, the penalties for landing undersize species, if existent and applied, were modest in many jurisdictions. The '1992 Control Report' identifies the two crucial elements on which fishery enforcement depends: the possibility of inspection; followed by the dissuasive nature of the penalties invoked for breach of Community fishery law.[108] Fraud in fisheries is linked to the absence of competent and sufficient enforcement resources. The report notes that it has been established elsewhere that such fraud is lucrative and generates profits exceeding the penalties. It cites the example of the George Bank fishery in the United States, where it is said that illegal landings make up from 11% to 25% of catches; fishermen who are alleged to regularly commit infringement make up between 25% and 49% of the fishing population.[109]

In general, the report focuses on the theme of co-operation and co-ordination and draws attention to the fact that national enforcement bodies had little, if any, liaison with other enforcement authorities, particularly those in adjacent jurisdictions. The report stresses that there were no joint inspection arrangements, other than the international law right of hot pursuit, for pursuing suspected violators of Community fisheries law from one jurisdiction to another. The Report also notes that the monitoring of access to fishing zones was on a random basis and there was little enforcement over the activities of third country vessels operating in Community waters.

Some conclusions regarding the issues identified in the reports

Perhaps the most valuable contribution made by the reports, other than stimulating dialogue, was that they identified those issues which needed to be addressed in the context of the adjustment of the CFP in 1992, and in particular by the new control system for the entire fishery sector. The reports are unerringly consistent in expressing the view that the CFP has failed in many respects and is unlikely to attain anything other than some of its most modest objectives. Furthermore, they pointed out that unless all parties gave due cognisance to the need for reform, a sector crisis in the industry was inevitable. Clearly, the Commission believed that the plethora of regulations amounted to nothing more than platitudes unless there was effective enforcement. In addition it pointed to the inadequate power of the Commission to contribute to more effective enforcement and the need for the harmonisation of all aspects of the policy as well as the necessity for all Member States to co-operate in pursuing the aims of the control policy.

It was clearly evident by 1992 that the CFP had not been properly implemented in the first half of its life. It was equally clear that the future credibility and sustainability of the CFP hinged largely on the control and enforcement dimension of the policy.[110] Many of the conclusions on the role of enforcement reached in the series of reports dating back to 1986 may be summarised as follows:

> 'Without proper enforcement, conservation would be threatened and the Community's international fishery relations would suffer. Moreover, the many restrictions which the Communities fisheries legislation places on fishing activities in order to conserve resources in the general interest will only be accepted by fishermen in the longer term, on the understanding that such restrictions will be equally enforced by all Member States concerned. A proper level of enforcement is not only a legal duty laid upon Member States but also a political necessity without which the conservation component of the common fisheries policy would lose credibility and respect.'[111]

The next task which faced the Community was the adoption of legislation which would reflect the concerns and solutions highlighted during the lengthy review period.

References

1 This chapter does not examine the relationship between enforcement and extended jurisdiction. Law of the Sea obligations, however, are occasionally mentioned for comparative purposes, or, in some instances, they are discussed when they directly impinge on a particular area in the Community enforcement framework. In this regard, the Community enforcement problem in the high seas driftnet fishery in the North-East Atlantic and the Mediterranean discussed in Chapter 10 is relevant.

2 For detailed information of the sea fisheries sector in the EU see, *inter alia*, the 21 *Regional, Socio-Economic Studies in the Fisheries Sector*, which were completed in 1992 for use within the services of the Commission COM XIV/243/93; annual statistical bulletin produced by the Directorate General of Fisheries (DG XIV); *Review of Fisheries in OECD Countries* (Paris, 1997); *Towards sustainable fisheries: Country Reports*, (Paris, 1997).

3 Council Regulation No. 150/97 of 12.12.1996 on the conclusion of an Agreement on co-operation in the sea fisheries sector between the European Community and the Kingdom of Morocco and laying down provisions for its implementation, *OJ* L 30, Vol. **40**, 31.01.1997.

4 This aspect of the policy has been expensive, costing the Community budget on average €68 million a year in the period 1983–1990, or 29% of the fishery budget. It has been questioned whether the value of the fish caught exceeds the cost of the agreements. Holden, M., *The Common Fisheries Policy: origin, evaluation and future*, (Oxford, 1994), 37–38. For further discussion of the Community budget as it applies to the CFP see Chapter 9, *post*.

5 It is sufficient to note at this point that each Member State is responsible within its territory and within the maritime waters subject to its sovereignty or jurisdiction to monitor fishing activity and related activities. The Commission draws attention to this issue in reply to practically every question asked on the subject of monitoring and enforcement which is raised in the European Parliament, see, *inter alia*, answer to Parliamentary Question (hereinafter referred to as PQ) No. 1813/92, *OJ* C 47/1, 18.02.93.

6 COM, SEC(90) 2244.

7 Council Regulation No. 2847/93, **Article 35**, discussed in Chapters 3 and 8, *post*.

8 *Monitoring the Common Fisheries Policy, Commission Report*, COM(96) 100 final of 18.03.1996,

examined in Chapter 8, *post*. See also Commission reply PQ No. E-337/94, *OJ* C 376/44, 30.12.94.

9 Vessels belonging to the Navy.

10 Some of these vessels belong to the Ministry of Environment.

11 Greece. Primary rôle played by the Coast Guard Service which is based in all ports. In theory one person could conduct inspections.

12 Spain. This figure only includes the national Madrid-based inspections. In addition there are 232 inspectors employed by the various authorities in the Autonomous Regions.

13 France. Figure is based on:

(1) Unités Littorales as Affaires Maritimes (88)
(2) Gendarmes Maritime (44)
(3) Seagoing inspectors from Affaires Maritimes (12)

14 Italy: Potentially there are 1200 persons engaged in some form of fisheries inspection work. The real figure is probably much smaller.

15 Ministry of Agriculture, Fisheries and Food (MAFF), Welsh Office Agriculture Department (WOAD), Scottish Office Agriculture Department (SOAFD), Department of Agriculture for Northern Ireland (DANI). These offices have been renamed.

16 Churchill, R.R., 'EC Fisheries and an EZ – Easy', *Ocean Development and International Law*, **23**, (1992) 145–163.

17 Case 26/62 *Van Gend en Loos* v. *Nederlandse Administratie der Belastingen* [1963] ECR 1.

18 See discussion of the Treaties, Council Regulations Nos 2141/70 and 2142/70 and the Hague Agreement, in Chapter 1, *ante*.

19 See Churchill, R.R., *EC Fisheries Law*, *passim*. The subject of Community competence to regulate fisheries on the external plane is also discussed in Chapter 10, *post*.

20 Joined Cases 3, 4, and 6/76, [1976] ECR 1279, paragraphs 11–14.

21 Case 804/79, *Commission* v. *United Kingdom*, [1981] ECR 1045.

22 *Ibid*. at paragraphs 17, 18 and 20.

23 See Chapter 3, *post*.

24 Case 804/79, paragraphs 22–27, *op. cit.*, fn 21.

25 Case 287/81, [1982] ECR 4053.

26 Subject to the Community laws on the rights of establishment, see Case C-221/89 *Regina* v. *Secretary of State for Transport, ex parte Factortame* [1991] ECR I-3905; Case C-246/89 *Commission* v. *United Kingdom* [1991] ECR I-3533.

27 Subject to the duties incumbent on States in Community law: Case C-146/89 *Re Territorial Sea: Commission* v. *United Kingdom* [1991] ECR I-3533.

28 Council Regulation No. (EC) 3760/92, Article 15, *OJ* L 20, 28.01.1976, p. 19.

29 See for example Council Regulation (EC) No. 45/98, *OJ* L 12/1, 19.01.1998, see Chapter 1, Section (iii), *ante*.

30 Article 38 Council Regulation No. 2847/93 No. L 261/16, 20.10.93, discussed in Chapter 7 *post*. See Case C-3/87 *The Queen* v. *Ministry of Agriculture, Fisheries and Food, ex parte Agegate Ltd*. [1989] I-4459; Case 216/87 *The Queen* v. *Ministry of Agriculture, Fisheries and Food, ex parte Jaderow Ltd. and others* [1989] ECR 4509; Case C-221/89, *The Queen* v. *Secretary of State for Transport, ex parte Factortame Ltd and others*, [1991] ECR I-3905; Case C-93/89; *Commission* v. *Ireland*, [1991] ECR I-4569; Case C-246/89, *Commission* v. *United Kingdom of Great Britain and Northern Ireland* [1989] ECR 3125; regarding utilisation of national quotas, discussed in Chapter 7 *post*.

31 Case 87/82, *Lt. Cdr. A.G. Rogers* v. *H.B.L. Darthenay* [1983] ECR 1579, [1984] 1 CMLR 135.

32 Cases 86-87/84, *Criminal Proceedings against I. Bout en Zonen BV*., [1985] ECR 941, [1985] 3 CMLR 218.

33 In *Opinion 2/91 (Re ILO Convention 170)* [1993] ECR I-1061, the Court provides a useful

summary of the law on external competence and distinguishes between the source of Community competence and the nature of that competence.

34 See, in particular, Case 22/70 *Commission* v. *Council* ('the AERT case') [1971] ECR 263; and on fisheries specifically see Joined Cases 3, 4, and 6/76, *Kramer* [1976] ECR 1279, paragraphs 19–20. Although previously considered unclear and controversial, recent case law has clarified to some degree the exact scope of the implied powers, see, *Opinion 2/91 (Re ILO Convention 170)* [1993] ECR I-1061; *Opinion 1/94 (Re the WTO Agreement)*, [1995] 1 CMLR 205; *Opinion 2/92 (Re OECD National Treatment Instrument)* 25 March 1995.

35 Joined Cases 3, 4, and 6/76, [1976] ECR 1279, paragraphs 19–20.

36 See Chapter 1 *ante*.

37 *Op. cit.*, fn 34. There are a few bilateral agreements which have remained in place after Member States have acceded to the Community. See, for example, the agreements between Sweden and Russia and between Portugal and South Africa.

38 Case 405/92 *Mondiet* v. *Islais* [1993] ECR 1-6133, paragraph 12.

39 It is beyond the scope of the present study to analyse in detail the participation of the Community in external fishery organisations, it should nevertheless be pointed out that this an area in which the Court has played an active rôle in defining Community legislative competence. For a more detailed discussion see, *inter alia*, Churchill, R.R., *EEC Fisheries Law*, Chapter 5 on external relations in fisheries management matters; and Frid, R., *The Relations Between the EC and International Organizations* (The Hague, 1995), Chapter 7, and in particular the discussion of some of the consequence for the Member States of the Community's membership in fisheries organisations, 332–335; MacLeod, I., Hendry, I.D. and Hyett, S., *The External Relations of the European Communities*, (Oxford, 1996), Chapter 10.

40 Case C-25/94 *Commission of the European Communities* v. *Council of the European Union*, [1996] ECR I-1469.

41 This Agreement is discussed in Chapter 11, *post*.

42 In the case in question the Court noted that when the Council adopted the contested decision the essential object of the draft Agreement submitted for adoption by the FAO was compliance with international conservation and management measures by fishing vessels on the high seas. In particular the Draft Agreement no longer contained provisions on flagging on which the Council based its erroneous conclusion. Furthermore, the Court held that the Council was wrong in maintaining that the granting of a licence to fish on the high seas, issued by Member States subject to compliance with conservation and management measures, is comparable to the power to authorise a vessel to fly a particular flag. It also upheld the Commission's contention that fishing licences are a traditional means of managing fishing resources which, *inter alia*, give fishing vessels access to waters and resources and thus differed fundamentally from the general conditions which the Member States may lay down, under international law, in allowing ships of all categories to fly their flags.

43 Council Regulation No. 1956/88 of 09.06.1988 adopting provisions for the scheme of joint international inspection adopted by the Northwest Atlantic Fisheries Organisation, *OJ* L 175/1, 06.07.88. As amended by Council Regulation No. 436/92 of 10 February 1992, *OJ* L 54/1, 28.02.92.

44 **Article 2.1**, *ibid*. Furthermore, a Community inspector may be placed on board any Member State vessel engaged in or about to be engaged in inspection duties in the North Atlantic Fisheries Organisation Regulatory Area. Community inspectors are authorised to inspect all vessels to which the scheme applies (there are 16 Contracting parties). Inspectors may check Community vessels for compliance with NAFO regulations and other Community conservation and control measures applying to vessels flying the flags of the Member States while these vessels are operating in the NAFO Regulatory Area.

45 Approved by the Council in Decision No. 81/691/EEC, *OJ* L 252, 05.09.1981, p. 26.

46 Council Regulation EEC No. 3943/90 of 19.12.1990, *OJ* L 379/45, 31.12.90.

47 The Commission, however, may in some instances decide to refuse to issue a fishing licence to vessels in instances of recorded infringements. The Court of Justice has ruled on the legal nature of this sanction, see Case 55/86, *Arposol* ECR [1988] I-13; Case 135/92, *Fisakano AB* v. *Commission* ECR [1994] ECR I-2885, discussed in Chapter 7, *post.*

48 On **Article 5** {10} of the EC Treaty, see, Temple Lang, J., 'Community Constitutional Law: Article 5 EEC Treaty', *Common Market Law Review (CMLRev.)*, **27** (4) (1990), 645-681.

49 See discussion of sanctions and authorities cited in Chapter 6, Section (ii), *post.*

50 Case 50/76, *Amsterdam Bulb* v. *Produktschap voor Siergewassen* [1977] ECR 137.

51 Case 14/83, *Von Kolson and Kamann* v. *Land Nordrhein-Westfalen* [1984] ECR 323.

52 Case 68/88, *Commission* v. *Greece* [1989] ECR 2965.

53 See Chapter 8, *post.*

54 The issue of direct effect is discussed in relation to the Control Regulation in Chapter 3, *post.*

55 Other examples of this approach are the specialised agencies in the olive oil sector, the establishment of a Community fraud preventive unit (UCLAF) and the extended role of the Community in monitoring the viniculture sector. One authority considers this division of enforcement competence between the Member States and Community institutions to be inappropriate on the grounds that in some cases it mixes criminal procedures and administrative procedures and that it obfuscates two legal orders, the Member States and the Communities. See Harding (eds.), *Enforcement of Community Rules*, (Aldershot, 1996), Chapters 2 and 10.

56 EEC Treaty, **Article 155** {211}.

57 EEC Treaty, **Article 5** {10}.

58 The duty to co-operate is examined in Chapter 3, *post.* For a discussion of Case C-2/88 *Zwartveld and others* [1990] ECR I-3365, see Chapter 6, *post.*

59 LOS Convention, **Article 73(1)**.

60 LOS Convention, **Article 73(3)**.

61 LOS Convention, **Article 73(2)**.

62 LOS Convention, **Article 73(4)**.

63 See Browne, E.D., *The International Law of the Sea*, (Aldershot, 1994), p. 233; Attard, D. *The Exclusive Economic Zone in International Law*, (1987), p. 180; Dahmani, M., *The Fisheries regime of the Exclusive Economic Zone*, (Dordrecht, 1987), 83–92.

64 LOS Convention, **Article 62(4)**.

65 *Ibid.*

66 LOS Convention 62(4)(k).

67 Kwiatkowska, B., *The 200 Mile Exclusive Economic Zone in the New Law of the Sea*, (Dordrecht, 1989), p. 87.

68 See Burke, W. *The New International Law of Fisheries*, (Oxford, 1994), 312–314; Adams, M., *The Role of the Navy in Domestic Law Enforcement*, (Naval War College paper, 1984); Fidell, 'Fisheries Legislation: Naval Enforcement', *Journal Maritime Law and Commerce* **7**, p. 351–366, (1976).

69 Vessels not engaged in fishing operations have the rights and obligations stemming from the rights of innocent passage and the freedom of navigation.

70 See Dahmani, M., *The Fisheries Regime of the Exclusive Economic Zone, op. cit.*, fn 63, p. 84.

71 An overview of the historical development of the CFP is presented in Chapter 1, *ante.*

72 Council Regulation No. 2057/82, *OJ* L 220, 29.07.1982, p. 1.

73 Council Regulation No. 2241/87, *OJ* L 207/1, 29.07.87, (repealed).

74 Council Regulation No. 170/83, **Article 8**, See Chapter 1, *ante.*

75 Report from the Commission to the Council and the European Parliament on the CFP, Document SEC (91) 2288 final 04.12.1991.

76 Report from the Commission to the Council and the European Parliament on monitoring and implementation of the CFP, SEC (92) 394 of 06.03.1992.
77 Council Regulation (EEC) No. 3760/92 of 20.12.1992, establishing a Community system for fisheries and aquaculture, *OJ* L 389, 31.12.1992.
78 Council Regulation (EEC) No. 2847/93 of 12.10.1993, establishing a control system applicable to the CFP, *OJ* L 261, 20.10.1993.
79 Holden states that the United Kingdom authorities had been infuriated that the landings of these vessels had not been reported in time to the United Kingdom which resulted in the national quotas being overfished, *op. cit.* fn 4, p. 86. The enforcement difficulties associated with the phenomenon of 'quota hopping' are examined in Chapter 7, *post.*
80 COM(86) 301 final, 09.06.1986.
81 SEC (90) 2244.
82 *Op. cit.* fn 75.
83 *Op. cit.*, fn 76.
84 COM(86) 301 final, 09.06.1986.
85 *Ibid.*, p. 3.
86 Belgium, Denmark, Germany, Spain (only tentative comments as Spain had only acceded to the Community), France, Ireland, Netherlands, Portugal (as with Spain) and the United Kingdom. There are two Annexes in the 1986 report. One details the resources devoted to enforcement in the Member States. The second gives a brief account of the Commission inspectorate.
87 COM (86) 301 final, *op. cit.*, fn 84, 19–20.
88 Council Regulation No. 2807/83, *OJ* L 276/1, 10.10.83. This regulation was subject to a separate review in 1987, so the Commission did not wish to pre-empt this in the 1986 report.
89 See Chapter 6, *post.*
90 See Chapter 8, *post.*
91 These are examined in the context of the 1993 Control Regulation, Chapter 6, *post.*
92 See Chapter 8, *post.*
93 *Ibid.*
94 SEC (90) 2244.
95 *Ibid.*, pp. 19–22.
96 *Ibid.*, p. 20.
97 *Ibid.*, p. 21.
98 See Chapter 11, *post.*
99 In particular the House of Lords Select Committee of the European Communities undertook an extensive review, and published a report in which enforcement is discussed in a rather diffuse manner, see *Review of the Common Fisheries Policy*, House of Lords Paper **9** (1992–1993), published in June 1992.
100 This communication was adopted in November 1990, SEC (90) 2244.
101 The crises in the Community market sector were the subject of a detailed communication from the Commission to the Council and the European Parliament entitled, *The Crises in the Community Fishing Industry*, COM (94) 336 final, Brussels, 19.07.1994, discussed in Chapter 1, *ante.*
102 The definition of fishing capacity has led to much discussion and is examined in Chapter 5, *post.*
103 Report from the Commission to the Council and the European Parliament on monitoring and implementation of the CFP, SEC (92) 394 of 06.03.1992.
104 In addition there are six Annexes: **Annex I** looks at the Community financial contribution to control costs. **Annexes II** and **III** describe the resources available for fisheries inspection in eight Member States Belgium, Denmark, Germany, Spain, France, Ireland, Netherlands,

Portugal, United Kingdom. **Annex IV** compares the delay by national authorities in transmitting catch data to the Commission. **Annex V** examines the concept of satellite technology for fishing surveillance. Finally, **Annex VI** tabulates, for comparative purposes, the maximum fines available under national legislation for certain infringements concerning sea fisheries, and compares the fines in other Member States for the same infringements.

105 The example of the juvenile hake market is cited as an example, *op. cit.* fn 103, p. 36.

106 This was partly due to the large volumes of short term legislation and the frequent amendment to technical regulation as noted in Chapter 1, *ante.*

107 *OJ* L 363/26, 31.12.1985 (corrigendum in *OJ* L 167/63, 26.06.1987). This regulation laid down the measures to be taken in respect of operators who do not comply with certain provisions relating to fishing contained in the Act of Accession of Spain and Portugal. In accordance with this regulation, the Commission, on the basis of a judicial decision notified by a Member State, or in any other case of recorded infringement of the rules, the vessel could be removed from the basic list of licensed vessels for a given period. One authority has pointed out, that whereas removal from the list after a judicial hearing is acceptable, there is a doubt whether the removal at the request of a national authority is in accordance with the general principle of Community law, which entitles a party to a fair hearing if subject to the adjudication by an administrative body.

108 Report from the Commission to the Council and the European Parliament on the CFP, Annex V, pp. 18–21 of annexes, Document SEC (91) 2288 final 4.12.1991.

109 Sutinen, J.G., Rieser, A., Gauvin, J.R., 'Measuring and Explaining Non-compliance in Federally Managed Fisheries', *ODIL* **21** (1990), 335–372.

110 This is a common theme in many of the statements issued by political leaders to justify a more robust approach to enforcement. See for example, Commissioner Bonino press release, and speech to European Parliament, regarding her objectives as Fishery Commissioner, 31.09.1994; for a similar view on enforcement see Mr Michael Jack, Minister of Agriculture, Fisheries and Food, speech to the House of Commons on the introduction to Parliament of the Sea Fish (Enforcement of Community Control Measures) Order 1994 (SI 1994/451) which provided for the enforcement of Council Regulation No. 2847/93 establishing a Community control system applicable to the CFP, as reported by *Reuter* copy dated 01.03.1994.

111 COM (86) 301, p. 1.

Chapter 3
A New Approach to Enforcement – The Control Regulation

Introduction
On 12 October 1993 the Council adopted Council Regulation (EEC) No. 2847/93 establishing a control system applicable to the CFP[1] (hereinafter referred to as the 'Control Regulation' or as the '1993 Control Regulation'). This regulation as amended is the keystone of fishery enforcement in the European Union. It provides the means for implementing the new integrated approach to the control and enforcement of the Community's fishery policy. In its many provisions one can note a departure from the traditional *laissez-faire* view of enforcement to a more proactive all embracing control and enforcement structure which was recommended by the Commission in the reports presented in the early 1990s. Section (i) reviews the legislative history, the general principles and the scope of the Control Regulation. Section (ii) examines a number of the general provisions which are dealt with in Title IX which is the final title in the Control Regulation. These include, the Management Committee procedure and the adoption of implementation rules to complete the regulatory framework. Section (iii) examines the frequently litigated area of national enforcement and conservation measures.

Section (i) Legislative history and introductory comments on the Control Regulation

1. *Legislative history*

The origin of the Control Regulation can be traced back to the first control regulation which was adopted by the Council in 1982,[2] and later consolidated into Council Regulation No. 2241/87.[3] Experience gained in the application of these regulations clearly indicated that there was a need to strengthen enforcement. The legislative base for the Control Regulation lies in the revised 'Basic Regulation', establishing a Community system for fisheries and aquaculture, which required the introduction of a new control regime.[4] **Article 12** of the regulation states:

> 'To ensure compliance with this Regulation, the Council, acting in accordance with the procedure laid down in **Article 43** of the treaty, shall install a Community control system which shall apply to the entire sector.'[5]

The administrative preparatory work on the new Control Regulation began in 1992 although there had been general political consensus about the need to strengthen the monitoring and control of the application of the rules governing the CFP, particularly after the Commission had presented the *1992 Control Report*. Agreement on the Commission proposal was reached only after about one year of thorough and often controversial discussions in Council Working Groups and after numerous meetings of legal and technical experts. In common with standard Community legislative procedure, the Commission proposal for a new control system was amended in accordance with the institutional consultation procedure, to take account of the opinions of the European Parliament and the Economic and Social Committee.[6] The proposal was duly presented to the Council in June 1993 in the form of a proposal (Council Regulation No. 2847/93) establishing a control system applicable to the CFP, and, having acquired political consent by qualified majority, it was formally adopted by the Council, without further debate, at a Fisheries Council meeting in Luxembourg.[7]

In the period 1993–1998 the Control Regulation was subject to six major amendments.[8]

2. *Some general comments on the Control Regulation*

The Control Regulation is comprised of 40 articles providing for Community-wide structures and procedures for control and enforcement.

Although the Control Regulation does not diverge from its predecessor[9] in regard to the principle of Member States' competence to undertake control activities, its scope is, nonetheless, greatly expanded and includes new areas such as the enforcement of measures pertaining to the market and structural element of the policy. This, and the many technical differences between the old and the new regulation, places additional responsibilities on Member State authorities. It introduces new control tools whose potential for success will require major adjustment and acceptance by the industry and whose full potential will only be realised if there is rigorous implementation in the Member States.

The Control Regulation is binding in its entirety and directly applicable in all Member States.[10] It has the same effect as if it were national law. Rights, obligations, and duties are placed on Community Institutions, Member States and individuals (including artificial and legal persons). The Control Regulation is binding automatically on individual fishermen and confers rights and obligations on them which can be invoked in national courts.[11] Although the Control Regulation, in common with all Community regulations, does not require national implementing legislation,[12] it may be supplemented by national legislation on such matters as criminal enforcement.[13] The provisions in the Control Regulation which apply to fisheries agreements entered into by the Commission on behalf of the European Union are also binding on Community fishermen and may similarly confer rights and obligations on them which may be invoked in national courts.[14] However, these rights and obligations do not arise if the agreement provision stipulates otherwise or is imprecise and conditional, pursuant to the decision of the Court of Justice in the *International Fruit Co Case* and a long line of subsequent cases.[15] The significance of the '*direct effect*' doctrine, which has been developed by the Court, is that it has provided a novel and extremely effective method of policing the implementation of both Community law and also international treaties within the national courts of the Member States. In areas such as the law of fisheries and indeed the international law of the sea, where enforcement of international obligations can present problems, the doctrine of direct effectiveness has considerable power and potential.

All the courts in the Member States are obliged to apply Community law.[16] Each Member State is required to have a legislative, judicial and administrative structure enabling it to meet the obligations imposed by the Community legislation.[17] Each Member State is also obliged to devote an adequate amount of human and material resources to the execution and enforcement of its obligations. The administrative, legal, and judicial structures in the Member States, as well as the physical resources available to undertake fishery enforcement, are occasionally referred to in later chapters.

3. *Commencement date*

Article 39(1) of the Control Regulation states that Council Regulation No. 2241/87 (the '1987 Control Regulation') is repealed on 1 January 1994, with the exception of **Article 5** (which provides that Member State vessels shall keep a logbook for TAC and quota species), which shall continue to apply until 1 January 2000.[18] Furthermore, references to the repealed '1987 Control Regulation' in other legislation shall thereafter be construed as references to the 1993 Control Regulation.

The Control Regulation took effect from 1 January 1994, pursuant to **Article 40**. A number of exemptions were provided, however, for a period of two years, until 1 January 1996 from the obligation to apply the provisions of **Articles 9** (sales note), **15** (quota uptake), **18** (notification to the Commission) in so far as they concern computer transmission of sales notes and landing registration. Further exemptions apply to the Mediterranean Sea until 1999, **Article 8** (landing declarations), **Article 19** (Data bases) to fishing operations undertaken in this sea area. The 1998 amendment to the Control Regulation took effect from the date of publication in the Official Journal, 31 December 1998. The revised provisions on the transhipments and fishing operations involving two or more vessels (**Articles 11** and **28b2**) shall apply from the date of the entry into force of the relevant detailing implementing rules referred to in Council Regulation No. 2846/98. Furthermore, Member States are not obliged to apply **Article 6** (logbooks) to the Mediterranean Sea until the year 2000.

Section (ii) Legal basis, recital and legal competence

1. *Legal basis*

The legal basis of the Control Regulation is rooted in **Article 43** {37} of the EC Treaty. As noted above, the Commission was bound to consult both the European Parliament and the Social and Economic Committee[19] and, on receiving the opinion of each institution, there was no re-consultation.[20] In common with other Community regulations, the Control Regulation is presumed to be valid until struck down by the Court of Justice. Furthermore, any national legislation which conflicts with the Control Regulation is automatically invalid and of no effect.

2. Preamble

The preamble of the Control Regulation is worthy of consideration as it provides an introduction to the new measures and provides additional impetus to existing provisions in the control framework. It acknowledges that the success of the CFP requires the implementation of an effective system of control, covering all aspects of the policy and the entire fishery sector. This is to include provisions relating to the effectiveness of sanctions to be applied where the aforementioned measures are not observed (this is also the aim of the Control Regulation pursuant to **Article 1(1)** where the subject of sanctions is again explicitly raised). It is interesting to note that the recital stipulates this in a negative sense, i.e. the rules must include provisions to deal with failure to carry out the required measures. Perhaps this is an early acknowledgement or reminder that the Community control system must focus on the *direct* action to be taken when Community measures are transgressed, as well as just monitoring their observance.[21] This could be construed as an admission, with a certain air of inevitability, that fishery regulations will not be adhered to and, therefore, that a major issue to be emphasised in the new control framework is the sanction for non-compliance.[22]

The reality of fishery enforcement is acknowledged by virtue of the fact that the recital also records that the control system can only achieve the desired result if the industry recognises that it is justified.

3. Division of competence

The global issue of competence in the domain of prescribing and enforcing Community fisheries law has already been examined,[23] and consequently the following discussion is restricted to the issue of competence in the context of the Control Regulation.

Although competence to adopt the new control framework introduced by the Control Regulation rested with the Community, the competence to enforce such measures remains, almost exclusively, with the Member States. The legal competence of the Member States to undertake control and adopt appropriate measures is acknowledged in **Article 1(2)** which obliges Member States to adopt measures to ensure the effectiveness of the control system. Such measures have to conform with Community rules. The necessity of stating the obvious, (i.e. in accordance with Community rules) in **Article 1(2)** is due, possibly, to the propensity of Member States in the 1980s to introduce arbitrary and discriminatory measures, particularly in regard to non-nationals fishing other Member State quotas under flag of convenience arrangements. This frequently resulted in the intervention of the Court to censure such discriminatory provisions.[24] The competence of Member States to undertake the control function is clearly recognised and established in **Article 2** which states:

> 1. 'In order to ensure compliance with all the rules in force, *each Member State* within its territory and within maritime waters subject to its sovereignty or jurisdiction, shall monitor, inspect and maintain surveillance of all activities in the fisheries sector, particularly fishing itself, transhipment, landing, marketing, transport and storage of fisheries products and the recordings of landing and sales. *The Member State shall* take the necessary measures to ensure the best possible control within

> their territory and within maritime waters subject to their sovereignty or jurisdiction, taking into account their particular situation.' (Italics and emphasis added.)
> 2. '*Each Member State shall* ensure that the activities of its vessels outside the Community fishery zone are subject to proper monitoring and, where such Community obligations exist, to inspections and surveillance, in order to ensure compliance with Community rules applicable in those waters.' (Italics and emphasis added).

Arguably, **Article 4(1)** is more precise in defining legislative competence in so far as it states, that: *the inspection and monitoring specified in* ***Article 2*** *shall be carried out by each Member State on its own account by means of a system decided by the Member State.* Thus the competence for undertaking inspection is the prerogative of the Member State. However, the second subparagraph in **Article 4(1)** stipulates that the Member State must ensure compliance with requirements of **Article 2** (i.e. this is to ensure that **Article 4(1)** is not a loophole or an escape clause by which Member States can avoid their obligation pursuant to Community law to undertake the fishery inspection and control responsibilities). The national competence to undertake control is also qualified by the requirement, noted both in the preamble and in Title VII, that the Commission should also seek to ensure that the Member States monitor and prevent infringements in an equitable manner.[25] The financial and legislative means to undertake this are guaranteed, thus providing a proper legal basis to make available sufficient resources to carry out the new and additional enforcement tasks. This is to ensure that the agencies and organisations vested with the enforcement function are not circumvented by depriving them of the necessary financial resources to undertake their assigned tasks effectively.[26]

4. *The* ratione materae *and* ratione loci *of the Control Regulation*

Article 1(3) sets out the *ratione materae* and *ratione loci* of the control system, which applies:

> 'To all fishing and associated activities carried out within the territory and within the maritime waters subject to the sovereignty or jurisdiction of Member States including those exercised by vessels flying the flag of, or registered in, a third country, without prejudice of the right of innocent passage in the territorial sea and freedom of navigation in the 200-mile zone; it shall also apply to the activities of Community fishing vessels which operate in the waters of non member countries or on the high seas, without prejudice to the special provisions contained in fisheries agreements concluded between the Community and third countries or in international conventions to which the Community is a party.'

While the precise scope of the CFP is clearly not the focus of this book,[27] nevertheless the wide scope of **Article 1(3)** is interesting for two reasons. Firstly, the CFP has traditionally been associated with sea fisheries. However, the new provision clearly extends the *ratione materae* of the control system to all fishing and *associated* activities within the territory of Member States and, in theory, may allow the Community control system to apply to inland fisheries. The latter is unlikely to occur, as the management and control of inland fisheries is an obvious area for the application of the subsidiarity principle. Furthermore, because inland/onshore fisheries activities are not within the remit of the Community management policy there are consequently no Community regulations to enforce. Secondly, and most importantly, Member States are obliged to extend the scope of enforcement measures, from the traditional areas such as the inspection of vessels, to include the verification of all activities, including landing, selling, transporting and storing of fish, and recording landings and sales of fish.[28] In this regard, it is instructive to recall that, prior to the introduction of the Control Regulation, in most Member States, inspection of vessels at sea and in ports and of fishery products after first sale were the responsibility of different authorities who co-ordinated their inspection effort. Violations of the conservation regulations went unchecked simply because there was no combined effort between the different responsible bodies to assemble incriminating evidence with the requisite expediency to initiate legal or administrative proceedings. In order to close the many loopholes in the control framework, the Control Regulation places the onus squarely on Member States to detect and contain unauthorised fishing activities. It encourages co-ordination particularly between Member States, as well as between the respective responsible bodies in Member States.[29] The 1998 amendment to the Control Regulation introduced several substantive provisions into the legal framework to improve co-operation among the authorities responsible for monitoring in the Member States and with the Commission.

In general, maritime waters coming within the sovereignty or within the jurisdiction of the Member States, are those waters so described by the laws in force in each Member State. The issue of 'maritime waters', which is not a recognised term in international law and the precise territorial scope to which the Community fishery law applies has been comprehensively examined elsewhere.[30] Here, it is proposed to limit the discussion to the maritime waters subject to the sovereignty or jurisdiction of Member States followed by a brief examination of some of the enforcement issues that have arisen from the rights of vessels flying the flag of, or registered in, a third country, when exercising the right of innocent passage in the territorial sea and freedom of navigation in the 200-mile zone, and operating in the Community zone.

There is no declaration in the Treaties regarding the geographical scope of Community fisheries legislation. **Article 299(1)** of the EC Treaty states that the Treaty applies to 'the Kingdom of Belgium, the Kingdom of Denmark . . .', but does not offer any indication if the EC Treaty applies to maritime zones beyond the territorial sea. There is, therefore, the presumption, supported by the judgement of the Court in several fisheries cases that the scope of fisheries law is the same as that of the Treaties in general.[31] Indeed, in determining the precise scope of Community powers the Court of Justice has tended to adopt a purposive, rather than a literal, approach.[32] Nevertheless, several writers have expressed opposing views on the geographical scope of Community fishery law.[33] For example, one view is that the EC Treaty has the same geographical scope as the jurisdiction of Member States in relation to the differing subject matter of the Treaty, unless the scope is expressly restricted.[34] There is, however, a considerable body of case law, which has approved the extraterritorial scope of other policies, notably competition, and has also found that as the maritime zones increase in size, so the existing body of the relevant fisheries law is automatically extended in scope.[35]

The Court of Justice has confirmed that in respect of fisheries, the Community has competence to adopt measures which apply to activities within the exclusive economic zone, the territorial sea, internal waters, and ports of the Member States,[36] and in so far as the Member States have similar authority under international law, to activities on the high seas.[37] In maritime zones outside Member State sovereignty or jurisdiction, the Community has, in matters relating to the competences attributed to it, the same legislative powers at international level as that accorded to the flag State, or the State of registration of a vessel. In this regard the Communities' competence must be exercised in accordance with the applicable rules of international law.[38]

The unique legal position of certain parts of Member States' European territories (such as the Channel Islands) and Member States' territories outside Europe (such as French Guyana) have not created any particular enforcement problem.[39]

Since 1979, the regulation of fisheries has been the competence of the Community,[40] the establishment of maritime zones (and the baselines from which zones are measured) remains, however, *ex hypothesi*, to be determined by each Member State. The Treaty does not empower the Community to alter the maritime zones of its Member States. On the contrary, **Article 2(3)** of Council Regulation No. 101/1976[41] (the original structural regulation) provides that it is up to each Member State to define in its legislation the maritime waters coming under its sovereignty or within its jurisdiction. This requirement has been accepted implicitly by the European Court in several cases including the aforementioned *Irish Fisheries Case*[42] and more recently in *Commission* v. *United Kingdom*.[43] In the latter, there is clear evidence that the Court would not be persuaded by arguments that are solely based on the practical difficulties or expediency with respect to the activities of the law enforcement agencies. For example, the Court rejected the argument that two sets of 12-mile limits (one for the purposes of Council Regulation No. 170/83 and the second for other purposes such as the Technical Regulation - salmon fishing and beam trawling) would not lead to practical enforcement difficulties. In particular the Court noted that 'coastal surveillance activities require in any case, on account of the existence of different fishing arrangements depending on the zone, a very great degree of accuracy on the part of the competent authorities in determining the zones in which the vessels subject to control are situated'.[44] Furthermore, arguments that the presentation of maritime charts would be greatly complicated and that other practical difficulties such as defining the Orkney-Shetland Box were deemed to be without foundation. The decision in the *United Kingdom* base line case clarified that the drawing of baselines is within Member States' competence albeit constrained by Community law. The Court did not express any doubt as to the ability of fishermen and fisheries enforcement officials to cope with the existence of two 12-mile zones. The ruling created a further zone to be charted and enforced. The United Kingdom is in a unique position in this respect within the EC as no other Member State has low tide elevations beyond 3 nautical miles of the coast.[45]

The *ratione materae* and *ratione loci* of the Control Regulation also cover the activities of third-country vessels fishing in the Community zone.

5. *Third-country vessels fishing in the Community zone*

It is generally accepted that pursuant to custom and convention in international law, prior to the ratification of UNCLOS III, that the coastal State may, in the exercise of the sovereign rights to explore, exploit, conserve and manage the marine resources in the exclusive economic zone, take such measures, including boarding, inspection, arrest and judicial proceedings, as may be necessary to ensure compliance with the laws and regulations adopted by it in conformity with the LOS Convention.[46] Thus, it is not surprising that **Article 1(3)** also includes within the scope of the Control Regulation, the activities of vessels flying the flag of, or registered in, a third country, who fish in the Community zone, without prejudice to the rights of these vessels under international law.

For the purpose of the Control Regulation 'third-country fishing vessels' is given a broad definition and means:[47]

- a vessel, whatever its dimensions, used primarily or secondarily to take fisheries products;
- a vessel that, even if not used to make catches by its own means, takes the fisheries products by transhipment from other vessels;
- a vessel aboard which fisheries products are subject to one or more of the following operations prior to packaging: filleting or slicing, skinning, mincing, freezing and/or processing;

and flying the flag of, and registered in, a third country.

The scale and size of third-county fishing activity, and the several special requirements which these vessels have to fulfil while operating in the Community zone, are examined in Chapter 4.[48] At this point it is sufficient to note that prior to the amendment of the Control Regulation in 1998 there were major differences between the obligations placed on third-country vessels operating in the Community zone and the obligations imposed on vessels flying the flag of one of the Member States. It is a useful exercise for the purpose of illustrating the extent of the coastal State enforcement model to tabulate the control and reporting conditions imposed on third-country vessels operating in the coastal zones of the European Union prior to the entry of new Community measures adopted in 1998. It is evident from the data presented in Table 3.1 that there were minimum requirements such as the obligation to retain a licence on board as well the requirement to complete a fishing logbook.[49] There were significant differences, however, if the coastal State enforcement model relied upon in the Community is compared with the obligations imposed by New Zealand or Australia on foreign vessels operating in their respective zones during the same period.[50] For example, in addition to the requirements listed in Table 3.1, New Zealand imposes additional obligations, which include the requirement for foreign vessels to embark observers; to submit a fishing plan; to undergo mandatory port inspection on entry/departure from the fishing zone; and to appoint a national representative in New Zealand. In some instances, there is a requirement for vessels to install satellite transponders; to carry interpreters; and land/process the catch in New Zealand ports. In addition, there is also a much broader range of obligations regarding reporting requirements placed on vessels prior to and during fishing operations. Furthermore, because both New Zealand and Australia are parties to the South Pacific Forum Fisheries Agency (FFA), there is an arrangement which stipulates that licences will not be issued to foreign fishing vessels unless those vessels are listed in the FFA-maintained regional register of fishing vessels.

In view of the disparity which existed between Community rules and coastal State

Table 3.1 Compliance, control and reporting conditions imposed by coastal Member States on third-country vessels operating in the coastal zones of the European Union.

Coastal Member State	A	B	C	D	E	F	G	H	I	J	K	L	M	N	O	P
Belgium	x	x	x	x	x	x %	x %		x op	x						
Denmark	x	x	x	x	x	x %	x %	x	x op							
Germany	x	x	x	x	x	x %	x %		x op	x	x					
Greece	x	x	x	x	x	x	x		x	x						
Spain	x	x	x	x	x	x	x		x op	x		x				
France	x	x	x	x	x	x w	x w		x	x						
Ireland	x	x	x	x	x	x %	x %		x op							
Italy	x	x	x	x	x	x %	x %		x op							
Netherlands	x	x	x	x	x	x %	x %		x op							
Portugal	x	x	x	x	x	x	x %		x op				x	x	x	
Finland	x	x	x	x	x	x w	x w		x							
Sweden	x	x	x	x	x	x w	x w	x	x				x		x	x
United Kingdom	x	x	x	x	x	x %	x %		x op							

A vessel to bear specified identification marks;
B requirement regarding protection of local fisheries/gear;
C reporting requirement on, or prior to, entry into the zone;
D reporting requirement on, or prior to, exit of the zone;
E reporting prior to entry into coastal state port;
F timely report on position;
G timely report on catch and fishing effort;
H vessel to stow gear while in area while not authorised to fish;
I maintenance of fishing logbook;
J vessel permit to be carried on board;
K vessel to put into port for entry/departure from zone;
L controls over transiting of zone by unlicensed foreign fishing vessels;
M vessels to board observers as required;
N reporting on commencement or cessation of fishing operation;
O submission of logbooks/catch reports: on request/as required/at specified intervals;
P vessel to seek prior authorisation prior to transhipment;
% reports must be made every three days for vessels from Norway, Sweden, Faeroes when fishing for herring, weekly when fishing for other species;
op logbook to be filled out at the end of every fishing operation or at the end of every 24-hour period;
w weekly.

legislation elsewhere, the Council adopted in 1998 an amendment to the Control Regulation, Title VIa, which strengthens the control provisions which apply to third-country vessels operating in the Community zone. In particular third-country vessels are obliged to have a licence as well as a fishing permit and are restricted from transhipment operations without prior authorisation from the relevant Member State. Significantly, all third-country vessels over 24 metres overall will be obliged to be equipped with position monitoring system from 1 January 2000. Prior to this date such vessels will have to comply with a system for reporting movement and catches retained on board as well as the rules on the marking and identification of fishing vessels and their gear. Third-country vessels are also obliged to comply with the instructions of the authorities responsible for monitoring, particularly as regards for inspections prior to leaving the Community fishing zone. Stricter rules pertaining to third-country vessels which utilise the landing facilities in Community ports have been adopted. There are also specific rules for fishing vessels which have catches on board taken on the high seas. In this regard the Control Regulation follows the steps taken by certain regional fisheries organisations to increase the effectiveness of measures for the conservation of fishery resources on the high seas.[51]

6. *Third-country vessels on passage or in port in the Community zone*

For the purpose of this discussion the term *on passage* is intended to mean vessels *in transit* and not engaged in fishing activity. All third-country vessels have in principle freedom of navigation through the exclusive economic zone of the coastal state[52] and the right of innocent passage through the territorial sea although the latter falls under the sovereignty of the coastal state.[53] The application and enforcement of Community fisheries law to third-country vessels which are not engaged in fishing activity in the Community zone raises some interesting questions regarding the relationship between international law and Community law in general and more specifically the need in Community law to distinguish jurisdiction to prescribe law from jurisdiction to enforce regulatory measures. This distinction is best understood if one examines a practical example such as the one that arose in *Anklagemyndigheden (Public Prosecutor)* v. *P.M. Poulsen and Diva Navigation.*[54]

In this case the international law rights accorded to third-country vessels operating in the maritime waters of the Community and in the area referred to as the high seas were examined by the Court. The Court's judgement has been hailed as containing the most important pronouncement on the relationship between Community law and international law since *Wood Pulp.*[55] The Court's decision also highlighted many issues which stem from the interaction between the Community law prescriptive and enforcement provisions when applied in the context of third-country vessels. The case resulted from a preliminary ruling from the Danish Court on the interpretation of **Article 6(1)(b)** of Council Regulation (EEC) No. 3094/86 (the technical regulation)[56] which prohibited the retention on board, the transhipment and the storage of salmon fished in regions 1, 2, 3 and 4 of the North Atlantic (which include waters outside the sovereignty or jurisdiction of the Member States - see Map in Chapter 4 pp. 138–9). This provision implemented the Reykjavik Convention for the Conservation of Salmon in the North Atlantic Ocean, under which the EC undertook to prohibit salmon fishing on the high seas by Community fishing vessels.

The facts of the case are typical of the issues which occasionally arise from the routine inspection of fishing vessels. In this instance, it was a routine inspection carried out by Danish inspectors in a Danish port of the vessel *Onkel Sam* which was registered in Panama (flew the Panamanian flag). The vessel was owned by Diva Navigation, a company incorporated under Panamanian law, wholly owned by a Danish national. The Master of the vessel and the crew of the vessel were Danish and were paid in Danish kroner. Between voyages the vessel was normally berthed in a Danish port. The *Onkel Sam* had 22,332 kg of salmon on board which had been caught in the North Atlantic in waters outside Member State sovereignty or jurisdiction. The vessel had been on passage to Poland but owing to mechanical difficulties and inclement weather conditions sought refuge in a Danish port to carry out repairs. As a result of their inspection, the Danish fishery officers seized the catch. Subsequently, the skipper and the owner were indicted before a Danish Court for violation of **Article 6(1)b** of Regulation 3094/86.

In answering the Danish preliminary questions, the Court of Justice held that the Community must respect international law in the exercise of its powers and that, consequently, **Article 6(1)b** must be interpreted, and its scope limited, in the light of the relevant rules of the international law of the sea, as reflected by the 1958 Geneva Convention on the Territorial Sea and the 1982 UN Convention on the Law of the Sea.[57] The Court noted that the purpose of **Article 6(1)b** was the conservation of protected species and thus met the obligation to co-operate in the conservation and management of biological resources in the high seas as provided for in **Article 118** of the LOS Convention. **Article 6(1)b** thus had to be interpreted to 'give the greatest practical effect' (*effet utile*), within the limits of international law.

On the issue of nationality of the vessel, the Court held that:

> 'a vessel registered in a non-member country may not be treated, for the purpose of **Article 6(1)b** ..., as a vessel possessing the nationality of a Member State on the grounds that it has a genuine link with that Member State.'[58]

On the effect of the nationality of the crew of the vessel, the Court found that **Article 6(1)b** may not be applied to the Master and other crew members *qua* nationals of a Member State, irrespective of the State in which the vessel is registered and sea area in which the vessel is located.[59] In practice, this means that the nationality of the Master and the crew of a vessel have no bearing on the applicability or enforceability of Community regulations.

The Court then examined the geographical scope of applicability of **Article 6(1)b** in the different sea areas. It was the Court's view that the provision may not be applied to a vessel on the high seas registered in a non-member country, since in principle such a vessel is governed by the law of its flag. The Community has the power to adopt rules for the Exclusive Economic Zone, the Territorial Sea, inland waters and ports of the Member States. The jurisdiction of the coastal State in those areas in relation to vessels which fly the flag of a third country is not absolute, however, and must respect the right to innocent passage in the territorial sea (LOS Convention **Articles 17** to **32**, Geneva Convention on the Territorial Sea and the Contiguous Zone 1958, **Articles 14** to **23**) and the freedom of navigation in the exclusive economic zone (**Article 58(1)** LOS Convention).[60] By contrast, the Court concluded that **Article 6(1)b** may in principle be applied to a vessel registered in a non-member country only when that vessel is in the inland waters, or more especially, when the vessel is in the port of a Member State, where in the opinion of the Court it is subject to the unlimited jurisdiction of the State.[61] (This is a major departure from the opinion of Advocate General Tesauro who suggested that vessels not landing their cargo should not be subject to Community jurisdiction).[62]

On the penultimate issue regarding the seizure of the vessel's catch, the Court pointed out that:

> 'The confiscation of a cargo of fish forms part of a panoply of measures that Member States are bound to provide for in order to ensure that Community legislation is observed and to deprive those who contravene it of the financial gain from such contravention. Confiscation is thus an ancillary measure which may be ordered when there is an infringement of Community legislation.'[63]

Although in principle a Member State measure, this seizure sought to enforce Community law. Since neither the nationality of the vessel's owner and the temporary nature of the cargo's presence in waters under Community jurisdiction had any material effect on the illegality of the transport, **Article 6(1)b** applied and the cargo could be seized.[64]

Finally, the Court noted that as the relevant provisions of Community law did not contain any rules on distress, it was for the national court to assess, in accordance with international law, the legal consequences which flow from such a situation.[65] It has been suggested by one commentator that, as the Court was leaving room for an equitable solution on the grounds that Community law contains no legal concept of distress, assessment of its effect is thus within the competence of the Member States.[66] The Court concluded that the Community lacked jurisdiction and left the decision whether, under international law, distress can excuse an otherwise illegal transport, to the national courts. This approach allowed the Court to affirm the full scope of Community legislation but to leave the more intricate and difficult problem of enforcement to the Member States. On the other hand, this approach has been criticised on the grounds that there are international rules concerning vessels in distress and it would appear that such rules limit the scope of Community legislation as much as the rules of international law which the Court does apply.[67] That it is to say that the Court by avoiding this issue seemed not to adhere to its own finding that the Community must respect international law in the exercise of its powers. Extraordinarily, when the national court in Denmark subsequently adjudicated on the distress issue, the State was unable to produce the requisite evidence to rebut the claim that the vessel was legitimately exercising this right. The national Court thus held for the defendant on this point and ordered the Ministry of Fisheries to compensate the vessel owner for the seizure of the catch and related expenses incurred as a result of the detention of the vessel.

Other than the rejection of the 'genuine link' doctrine claimed by Denmark and the concept of 'personal jurisdiction' of the crew, the Court's decision in *Anklagemyndigheden (Public Prosecutor)* v. *P.M. Poulsen and Diva Navigation* accentuated that Community powers must be exercised in accordance with international law and that Community jurisdictional limits are the same as those of the Member States.[68] In this regard the Court established the different basis of jurisdiction (territoriality principle, personality principle), their modification by international law (exclusive flag state jurisdiction on the high seas) and the exemption from coastal state's territorial jurisdiction (innocent passage, freedom of navigation) in the traditional manner.

From the enforcement perspective, although the Court distinguished between jurisdiction to prescribe and jurisdiction to enforce, there is little in the way of reasoning to support the conclusion that the Community has jurisdiction over vessels registered in third countries while they are in the internal waters and ports of the Member States, particularly as it is a fundamental assumption in all schemes of law that enforcement jurisdiction only follows in the wake of legislative jurisdiction. The Court's judgement would be more satisfactory if it had established clear grounds on which to base the enforcement jurisdiction of the Community. The argument of Advocate General Tesauro is more persuasive on this point. It was his view that the Community only has jurisdiction over foreign vessels when the catch is actually landed, displayed or offered for sale. This is based on two premises: Firstly, the prohibition to retain salmon on board cannot be separated from the prohibition to fish. Since the prohibition to fish on the high seas did not extend to Panamanian vessels, then these vessels have not committed an offence under Community law and therefore cannot be sanctioned. Secondly, the Advocate General cited case law in the USA, France and Italy to support the contention that ancient and constant practice does not create complete and unconditional Port State jurisdiction. If the Court had distinguished the offences (into a fishing offence and a transport offence) and clearly linked prescriptive jurisdiction to enforcement jurisdiction, rather than justifying the legality of the catch's seizure on the grounds that this was a Member State measure, then it would be easier to accept the finding that **Article 6** applied to a third-country vessel when in a Member State's port or internal waters. One commentator has noted that there was a second way open to the Court to establish legislative jurisdiction and that was to invoke a universal duty to protect endangered species. Early on in the judgement, the Court makes an assumption regarding the obligation to conserve and manage the high seas resources, but this remained an *obiter dictum,* and stops short of applying the universality principle for asserting legislative jurisdiction.[69] This authority points to several international conventions including SOLAS and MARPOL to indicate that favourable treatment is not given in ports to foreign ships which fly the flag of a State which is not a party to these Conventions.[70] Non-convention flagged vessels are inspected with the object of ensuring that all vessels comply with the convention regardless of whether they have ratified the convention. In this regard developments in the Memorandum of Understanding on Port State Control allow extended powers of enforcement to the coastal state.[71]

As a final point, it would appear that enforcement jurisdiction arising from the judgement in *Anklagemyndigheden (Public Prosecutor)* v. *P.M. Poulsen and Diva Navigation* is not general in application, but rather appears to be limited to giving '*effet utile*' to Community fishery conservation measures. The issues raised in *Anklagemyndigheden (Public Prosecutor)* v. *P.M. Poulsen and Diva Navigation* are significant in the context of the many international fisheries regulated by regional organisations. They are particularly important in regard to the high seas driftnet fishery discussed in Chapter 10, where the possibility of flag of convenience registration may be viewed as a means to circumvent Community legislation. The decision of the Court in *Anklagemyndigheden (Public Prosecutor)* v. *P.M. Poulsen and Diva Navigation* must now be viewed in the context of the 1998 amendment to the Control Regulation and several international legal instruments (the Compliance Agreement, the Code of Conduct for Responsible Fisheries, the Straddling Fish Stocks and Highly Migratory Species Agreement) which have been ratified by the European Community in the early 1990s.[72] In general it may now be said that the Community is obliged to monitor carefully and to control the activities of vessels flying flags of convenience which utilise Community ports.[73]

6a. Additional requirements adopted in 1998 to monitor the activities of third-country vessels utilising Community ports

The 1998 amendment to the Control Regulation (Title VIa) requires the masters of third-country vessels to give 72 hours advance notification to the authorities of the Member States whose ports or landing facilities they wish to use and vessels may only put in at the ports designated by the Member State.[74] The latter requirement does not apply in cases of *force majeure* or where a vessel is in distress. It also needs to be emphasised that as a result of the 1998 amendment to the Control Regulation, **Article 28g** requires where the master or his representative of a third-country vessel declares that catches have been taken on the high seas, the competent authorities shall authorise landing only if it has been proved to their satisfaction by the master or his representative that:

- the species retained on board have been caught outside the regulatory areas of any competent international organisation of which the Community is a member; or
- the species retained on board have been caught in compliance with the conservation and management measures adopted by the competent regional organisation of which the Community is a member.

7. Community vessels fishing outside the Community fishery zone

There are many reasons why Community fishing vessels fish outside the Community zone. First and foremost it is the Community fleet that has explored, discovered and developed many of the world's commercial fisheries. Secondly, in recent times there have not been sufficient natural fish resources in the Community zone to sustain the fishing capacity of the Community fleet. As a consequence of the progressive introduction of EEZs in the late 1970s and the exclusion of the Community fleet from the zones of coastal States it has been essential to conclude fisheries agreements. The importance of the international dimension of the CFP may be better appreciated when it is considered that since the extension of the EEZs to 200 miles, 35% of the ocean surface and about 95% of fish stocks have been brought under the exclusive jurisdiction of coastal States. In 1999 the Community has 29 operative fisheries agreements with third countries,[75] the most important being the agreements with Morocco, Norway, Faeroe Islands, Mauritania, Latvia, Lithuania, and Estonia. In overall terms, Community fishermen have access to quotas totalling about 550 000 tonnes in third-country waters.

Article 1(3) extends the ambit of the control framework to cover the activities of the Community vessels which operate in waters of non-member countries and on the high seas, without prejudice to the special provisions contained in fisheries agreements concluded between the Community and third countries or in International Conventions to which the Community is party. The legal obligations and duties that stem from the requirement for each Member State to monitor the activities of its vessels outside the Community fisheries zone pursuant to **Article 2(2)**, and to ensure compliance with the rules applicable in those waters, are examined in some detail in Chapters 4, 6 and 10. However, at this point, it should be stressed that **Article 1(3)** when read in conjunction with **Article 2(2)** raises two fundamental issues. On the one hand, there is the seminal issue of the exclusive competence of the Community to participate in fishery organisations and conventions and to conclude international agreements which was briefly examined above;[76] on the other hand, there are the enforcement issues which arise from **Article 2(2)** and from the obligations to ensure compliance with the rules which stem from international fishery organisations (which incidentally are mostly concerned with fishing beyond the 200-miles zone) and to adhere to the terms of bilateral access agreements for Community vessels to the fishery zones of third countries.

In accordance with established practice, the coastal State will exercise enforcement jurisdiction over the activities of Community vessels which operate in waters under the sovereignty or jurisdiction of the coastal State. Typical of the requirement placed on Community vessels which operate in third-country waters are the provisions in the EU–Morocco fishery agreement which include the requirements: to hail entry/exit from the fishery zone; to embark an observer; to complete a logboook; and to make a number of landings in ports in Morocco.[77] Furthermore, there is provision for designated inspectors from Morocco to participate in the mutual observation of shore-based inspection in the port of Las Palmas in Spain. There is also a requirement for a number of vessels to participate in a satellite tracking project.

8. Co-operation and co-ordination

Co-operation and co-ordination are key elements in the legal framework underpinning fishery enforcement in the European Union.[78] Title VIIIa of the Control Regulation requires Member States to provide each other with the mutual assistance required to carry out control activities. In particular, it advocates the establishment of special inspection programmes and information exchanges to overcome the difficulties encountered in inspecting Community fishing vessels as well as in relation to third-country vessels. There is an express obligation to notify the flag Member State if there are activities carried out which are likely to infringe Community rules. In such instances special checks may be implemented. Member States are also required to take measures to permit their competent authorities and the Commission to be regularly informed of the actions taken in response to requests from the flag State for the implementation of special checks.

A major development in the European enforcement structure was the adoption in 1998 of an amendment to the Control Regulation which provided a legal basis for fisheries inspectors from Member States to accompany European Commission inspectors as observers when the latter are undertaking their verification missions in other Member State.[79] This power however is qualified in so far as the Commission requires the approval of the Member State to be visited prior to taking such an initiative. Member States are also encouraged to carry out among themselves and on their own initiative, monitoring, inspecting and surveillance programmes.[80]

This concept of mutual assistance and joint inspection programmes as well as co-operation features elsewhere in the Control Regulation, most notably in **Article 13(7)** which deals with transport documents. It is clearly evident that co-operation between different Member State authorities on the control of third-country vessels operating in the EU fishery zone is a prerequisite for a successful control scheme. Not surprisingly, there is special mention in Title VIIIa of the need for inter-Member State co-operation and exchanges of information of third-country vessels suspected of infringing Community rules. Indeed it may be argued that co-operation between Member States control authorities is one of the principal themes of the fishery enforcement regime established by the Control Regulation. However, neither the Commission nor, indeed, the Council have the power to force Member States to co-operate on given issues or specific cases which fall outside the scope of the Control Regulation. The initiative in this respect rests solely with the Member States in question. Consequently, the success of Title VIIIa and the wider concept of co-operation depend to a large extent entirely on good will and the Community spirit of the Member States. Whether the full potential of this provision is realised will only be ascertained in the fullness of time.

The idea of co-operation is not limited to inter-Member State cooperation. There is specific reference in the Control Regulation to co-operation with and between the control authorities and the industry. **Article 4(1)** clearly states that the competence for inspection is the prerogative of each Member State which may decide on the means of inspection. The need for equanimity and appropriate control is clearly identifiable in that Member States are urged to act in such a manner as to avoid undue interference with normal fishing activity and in that there is no discrimination as regards the sector and vessels chosen for inspection.[81] This 'carrot and stick' approach is balanced with the stated requirement in **Article 4(2)** (repeating the theme in the recital), of the need for persons responsible for the fishing vessels, premises or transport vehicles inspected to co-operate fully with the Member State authorities.

Commentary

Law enforcement requires effective co-ordination and co-operation between all parties involved in the enforcement process. This may involve a wide range of activities varying from the sharing of information such as surveillance and inspection data to the transfer of judicial or administrative proceedings from the jurisdiction of one Member State to another.

The international obligation to co-operate has always been an element in international maritime law. In 1974 the International Court of Justice was able to assert 'the rules of international maritime law have been the product of mutual accommodation, reasonableness, and co-operation'.[82] In view of several incidents which have occurred in international fisheries outside the European Union it is questionable whether this is still a valid observation.[83] However, the significance of co-operation and co-ordination has received a major boost as well as solid convention base in a number of major international instruments adopted in the 1990s to promote sustainable and responsible fisheries.[84] These are examined in Chapter 11.

In the Community legal order the general obligation on Member States to co-operate in enforcement stems from **Article 5** {10 and 11} of the EC Treaty. This extends as far as placing a reciprocal duty or obligation on Member States and the Commission to cooperate. The Court of Justice emphasised this obligation in the *Zwartveld Case* (referred to above as the *EC Fisheries Inspector Report Case*) and noted that:

> 'In that (the) Community (is) subject to the rule of law, relations between the Member States and the Community institutions are governed, according to **Article 5** of the EC Treaty, by a principle of sincere co-operation. The principle not only requires the Member States to take all measures necessary to guarantee the application and effectiveness of Community law, if necessary by instituting criminal proceedings, but also imposes on Member States the duty of sincere cooperation.
>
> This duty of sincere co-operation imposed on Community institutions is of particular importance *vis-à-vis* the judicial authorities of the Member States, who are responsible for ensuring that Community law is applied and respected in the national system.'[85]

Furthermore, the Court held that it was incumbent upon every Community institution to give their active assistance to national proceedings by producing documents to the national court and by authorising its officials to give evidence in proceedings.[86] The Court however reserved their jurisdiction to rule on instances which were excluded from this obligation which is to be on the basis of their assessment of 'imperative reasons relating to the need to avoid any interference with the functioning and independence of the Community.'[87] As noted by one commentator, the duty of mutual and sincere co-operation is therefore itself a legal imperative, especially in the context of collaborative enforcement which depends on the use of national criminal proceedings.[88] In instances where there is suspected fraud involving the Community financial interests the duty to co-operate has since been clearly stated in **Article 209a** {280} of the EC Treaty. To this end Member States are now obliged to organise, with the help of the Commission, close and regular cooperation between the competent departments of their administrations. The Treaty of Amsterdam saw the insertion into the Treaty on European Union of a new Title (Title VIa {VII}, **Article 5a** {11} EC) on the theme of 'closer co-operation'. This provides a legal base, once certain conditions are fulfilled, for Member States to establish greater co-operation between themselves. One of these conditions, however, is that the subject area for closer co-operation does not concern areas which fall within the exclusive competence of the Community. The application of this Title to the common fisheries policy is thus curtailed.

In reality there appears to have been little success in the efforts made or envisaged in the Control Regulation during the period 1993–1998 to foster greater co-operation and co-ordination in fishery enforcement between the respective enforcement agencies in the Member States.[89] Significantly, the municipal legal structures in the Member States provided few opportunities for co-operation. Member States rigidly adhere to their coastal state enforcement authority to conduct inspection and control operations in respect of all vessels operating in the coastal zone, and to their flag state authority in respect of their flag vessels outside that zone. In several instances the detection and sanction process for vessels suspected of non-compliance with the regulatory framework was clearly limited by national boundaries and jurisdiction. In this regard, if vessels operate in fisheries which are trans-boundary or in disputed zones and land their catches in different Member States then there is clear scope for vessels to exploit the loopholes that arise as a result of the inflexibility and normative constraints which are imposed by rigid reliance on the flag state–coastal state enforcement structure. It must be pointed out however that jurisdictional disputes between Member States are generally avoided in the fishery enforcement context by 'gentlemen's agreement' regarding issues such as surveillance, boarding and inspection of fishing vessels. A case in point is the jurisdictional clash which could have arisen between the United Kingdom and Ireland in relation to vessels which operate in the sea areas which are the subject of jurisdictional disputes ('grey areas'). However, as officially stated in the Irish Parliament (the Dáil) there are arrangements for an exchange of information between Irish and United Kingdom authorities on the movement of third States' fishing vessels in such areas.[90] Irish and British fishery enforcement vessels patrol the 'grey areas' alternately, and it has been suggested that these 'gentlemen's agreements' extend to the action to be taken in the event of the boarding and inspection of their respective fishing vessels.[91] It should also be pointed out that in a similar context the Court of Justice did not accept that the jurisdiction dispute between the United Kingdom and the Faeroes was a satisfactory reason for a Member State not to comply with Community obligation to observe quota regulations.[92]

In particular there has been little effort to harness a co-operative effort to deal with trans-jurisdictional problems such as quota hopping.[93] A pertinent example of the need for joint co-operation is the monitoring of the activities of vessels which fly flags of convenience in order to gain access to quota allocated to the flag Member State. These are colloquially referred to as 'quota hoppers', because, although beneficially owned and predominantly crewed by Spanish nationals, they benefit from the quota of the total allowable catch (TAC) attributed to other Member States. Although there are several types of quota hopping the most prevalent have been the vessels which have been registered on the United Kingdom national register which is administered in Cardiff, make occasional obligatory statutory visits to ports in the United Kingdom, fish mainly in the sea area which is mainly under Irish jurisdiction, and land their catches in the ports in Northern and North-Western Spain.[94] Other than raising interesting problems for the enforcement authorities in several Member States, the vessels which have benefited from the phenomenon of quota hopping have led to some remarkable decisions by the Court of Justice which are examined in some detail in Chapter 7. Not surprisingly, the Court has consistently upheld the obligation placed on Member States to coordinate their control activities and to 'introduce measures whereby their competent authorities and the Commission may be regularly informed of the experience gained'.[95] Indeed, as pointed out by Advocate General Darmon, the co-operative participation by Member States in control and monitoring procedures is an application of the principle of Community solidarity, which the Court has recognised as one of the foundations of the Community as well as being a Treaty obligation.[96]

On the other hand, there have been a number of joint Dutch/Belgian inspection programmes, but it appears that, for ulterior reasons, Member States were reluctant to pursue joint fishery enforcement initiatives. Since the adoption of the Control Regulation, some Member States have taken up the co-operation procedure in a bilateral or multilateral setting. For example, the French/Spanish Guidance Committee for resolving some of the issues relating to the anchovy fishery in the Bay of Biscay, and the gear conflict in the high seas albacore fishery, was set up in 1995. This provides a forum for crisis management and dialogue in instances which essentially can be resolved at a bilateral level, particularly, to avert the type of confrontation which occurred in the driftnet fishery in 1994–1995.[97] The Community did not have competence to deal with particular incidents between different Member State vessels, although it played a major role in the resolution of the gear conflict which was eventually achieved on a bilateral basis between France and Spain. Similarly, there have been agreements between Norway (third-country) and several Member States (Denmark, Ireland, Scotland, Sweden and the Netherlands) to exchange information regarding fleet movements and the reporting of vessels entering and leaving the Community fishery zone. In some instances the results of reconnaissance flights and sea surveillance are exchanged, and inspectors are allowed to observe inspections carried outside their flag State.

It is apparent that there is a need to develop and enhance co-operation between Member States. In this regard there is considerable scope within the 1998 amendment to the Control Regulation to support and formalise the exchange of national inspectors and the initiation of joint programmes. This would have numerous advantages, particularly for organisations and inspection services that have to monitor and control other Member States' fishing fleets operating in waters under their jurisdiction or sovereignty. To pursue these possibilities, financial provision has been made for inter-Member State inspector exchange and training programmes in the Community budget for the period 1996–2000.[98] It is submitted, however, that the reason why there is insufficient co-operation and co-ordination is not because there is a reluctance to share or pool information but rather because of the divergence between the rules of evidence, the legal and judicial procedures for prescribing and invoking sanctions and penalties in the Member States. The current lack of co-ordination is thus linked to the more substantive issues regarding the unique legal order of the Community and the absence of a *communautaire* system of criminal or administrative justice to operate in parallel with the Community fishery enforcement system.[99] Within the present structure it is difficult to foresee what initiatives the Community institutions can take to initiate effective co-ordination between Member States on enforcement issues or to harmonise procedures and penalties in Member States' criminal and administrative systems. Particularly as **Article K** {29–45} of the Treaty on European Union only clearly demonstrates, law enforcement and the criminal justice process are seen by the Member States in general as matters for inter-governmental, not supranational, co-operation.[100] The expanded nature of inter-governmental co-operation as a result of the Treaty of Amsterdam is mentioned in Chapter 8 and clearly supports the view that there is now a greater emphasis placed on this essential element of inter-Member State relations than was evident during the first phase of the CFP. It also needs to be stressed that while the 1998 amendment to the Control Regulation places a firm obligation on Member States to provide each other with mutual assistance, to exchange information and to establish specific monitoring programmes in fisheries which involve two or more Member States, it nevertheless clearly states that fisheries inspectors acting as observers from one Member State may not be deployed with Commission inspectors in another Member State without the approval of the Member State to be visited. The Control Regulation is silent regarding the competence, rights, obligations and interests of the personnel deployed as observers. In contrast the competence of Commission inspectors is clearly defined and is examined in Chapter 6.

The obligation placed on Member States to coordinate their control activities does not limit the concept of co-operation to inter-Member State activities. There is a responsibility placed on persons responsible for the fishing vessels, premises or transport vehicles inspected to co-operate fully with the Member State authorities.[101] This is a fundamental condition to ensure the successful implementation and operation of a fishery enforcement regime. On the first view, the concept of *full co-operation* would appear to be devoid of legislative character and purposeful in intent. However, it is evident that most Member States' national legislation prescribe punitive penalties for those that choose not to *co-operate* or in other words not to comply with the national agencies empowered to carry out inspections. Furthermore, co-operation or non co-operation may have a major implication on the outcome of subsequent judicial or administrative proceedings.

9. *Co-operation is contingent upon the identification of enforcement officers*

Co-operation between the respective parties in the enforcement process will be facilitated if those authorised to discharge the fishery enforcement function are clearly identifiable as such authorised persons. To this end, it should also be pointed out that **Article 5** of the Control Regulation provides that detailed rules may be adopted regarding, *inter alia,* the identification of officially designated inspectors and inspection vessels, aircraft and of such other means of inspection as may be used by a Member State;[102] as well as procedures to be used for the inspection and surveillance of activities in the fisheries sector.[103] The aforementioned are significant because the extent of the 'obligation to comply with the orders issued by the authorities carrying out the inspection' will be determined by, *inter alia,* the clear identification of the party vested with the powers to undertake inspection. This issue was the subject of an **Article 177** {234} referral to the Court of Justice from the *Kriminal – og Skifteret, Frederikshavn* – Denmark.[104] The *Fishery Pennant Case* arose as a result of the skipper of a fishing boat not changing the course or stopping his vessel in order to facilitate a boarding party from a Danish inspection vessel. The inflatable boarding craft did not carry or display an inspection signal pennant or symbol which was prescribed in the regulations. The skipper of the fishing boat pleaded that he was unaware that the order to stop had been given by an inspection vessel. The Danish Court referred two questions to the Court of Justice:

'(1) Is **Article 2** of Commission Regulation (EEC) No. 1382/87 of 20 May 1987 establishing detailed rules concerning the inspection of fishing vessels to be interpreted as meaning the inspection symbol or pennant described in Annex 1 to that regulation must be displayed by or be painted on the boarding boat (inflatable boat) or is it sufficient that the parent vessel *Nordjylland* displayed that pennant or had that symbol painted on it? If the answer to the first question is that the boarding boat did not have to bear the pennant or have the symbol painted on it, no further preliminary ruling is sought. If that is not the case:

(2) Has any failure to comply with the requirements in **Article 2** any legal significance regarding the question of whether the skipper of a vessel (see **Article**

3) is under *an obligation to comply* with orders from the authority carrying out the inspection and if so, what is its significance?'

On the first question, the Court ruled that the queried provision must be interpreted as meaning that every inspection vessel, regardless of type or dimension, including vessels operating independently from a main inspection vessel must display the prescribed symbol.[105] On the substantive second issue the Court held that if the symbols are not displayed by an inspection vessel, it may be presumed that the skipper of a fishing vessel to be inspected is not aware that the order to stop is being given by the competent authority of a Member State.[106] However, that presumption will be negated by proof that the skipper of the vessel to be inspected was aware of the inspecting authority.[107] It was the Court's view that the obligation to comply with the orders of a representative of the competent authority of a Member State presupposes that the skipper of the vessel to be inspected is aware of the status of that representative. Thus in the case referred to in the absence of the symbol or pennant required by the Community legislation, the skipper is presumed to be unaware of that status, unless the authorities taking action concerning the infringement prove otherwise.[108] It is thus clear from the Court's judgement that the obligation to co-operate referred to in **Article 4(2)** of the Control Regulation is contingent upon the parties subject to inspection being aware of the status of the representative of the inspection authority in the Member States.

Interestingly, the master of a vessel flying the Spanish flag which had been detained on a charge of failing to keep a proper fisheries logbook in waters under Irish Jurisdiction successfully pleaded in the Circuit Court in Cork, that the boarding craft used by the inspection party did not comply with Community regulation in that it failed to fly the stipulated pennant.[109] In this case, which is unreported, the trial Judge held that failure to fly the correct pennant constituted a basic flaw in the prosecution's case and thus instructed the jury to acquit the skipper.

Section (iii) The Control Regulation: the Management Committee procedure - application regulations

1. *Management Committee for Fisheries and Aquaculture (Article 36 - Adopting Commission legislation, the Management Committee procedure - implementation regulations)*

Background

The word *comitology* is frequently used to describe the elaborate procedures that centre around the various types of committee which oversee the implementation of Community law.[110] The legal basis of the Committee procedure is **Article 155** {211} of the EC Treaty which states that, in addition to the powers of decision it enjoys as of right, the Commission exercises 'the powers conferred on it by the Council for the implementation of the rules laid down by the latter'.

The vast bulk of fisheries legislation is in the form of regulations dealing with general principles and specific areas of the CFP. The 'basic regulations', and 'management regulations' are adopted by the Council on the basis of the Commission proposals, the procedure for fishery regulations in this regard is the same as the procedure followed in the general *corpus* of Community law. The procedure is laid down in **Article 43** {37} of the EC Treaty and entails consultation with the European Parliament. Frequently, the Council, because of its heavy work load and restrictive time schedule, delegates power to the Commission to legislate detailed application/ implementation rules. In such instances there is no requirement to consult the European Parliament. The 1983 Management Regulation provided that implementation measures were to be adopted by the Council on the basis of a Commission proposal. The legitimacy of resorting to this type of legislative procedure for adopting implementation regulations in the domain of fisheries was upheld by the Court of Justice in the first *Romkes* case.[111] There are similar enabling provisions in the 1992 Management Regulation.[112]

The choice of which type of Committee is responsible for overseeing the adoption of a legislative instrument is specified in the legislation itself. For example, as noted throughout this book, the Control Regulation stipulates in several places that further action is to be taken by the *Management Committee for Fisheries and Aquaculture* (hereinafter, referred to as the Management Committee), regarding the adoption of implementation regulations. This is similar to the general reliance on Management Committees for detailed implementation of the Common Agriculture Policy. The Management Committee for Fisheries is composed of national experts with a Commission official who acts as the chairperson. The purpose of the Management Committee is to ensure that the Commission in exercising its delegated powers takes the views of the Member States into consideration.[113] This is of the utmost importance to the CFP because in many instances the Commission relies on the scientific or expert resources in the Member States to assist in the preparation of the detailed proposals to implement the policy. Furthermore, the consultation procedure ensures that the Commission is aware of Member States' views and consequently may tailor proposals to reconcile the views of all parties. The use of the Committee procedure has provoked criticism because it is sometimes perceived as secretive,[114] but in general it has served the CFP well. By relying on the Committee procedure in the Control Regulation the Council has ensured that the Commission has power to regulate the more detailed aspects of the Community control system. This power, nonetheless, is supervised and limited.

2. *Article 36 procedure*

Article 36 provides that where the **Article 36** procedure is to be followed for the adoption of a Commission regulation, the chairperson shall refer the matter to the 'Committee' either on his own initiative or at the request of the representative of a Member State. It is the Commission which has responsibility for submitting a draft of the measure to be taken. The Committee are obliged to deliver its opinion on the said draft within a time limit, which the Chairperson may lay down in accordance with the urgency of the matter under consideration. The opinion shall be delivered by the majority as laid down in **Article 148(2)** {205(2)} of the EC Treaty which is the same procedure as the Council follow when voting on the adoption of a proposal from the Commission. The votes of the representatives of the Member States within the

Committee are weighed in the manner set out in that article.[115] In this regard the Chairperson is not entitled to vote.

The Commission adopts measures which can be applied immediately. However, if these measures are not in accordance with the opinion of the Committee, they are communicated by the Commission to the Council forthwith. In such cases, the Commission may defer application of the measures which it has decided upon for a period of not more than one month from the date of the Commission communication. In this regard the Council acting on a qualified majority, may take a different decision within the aforementioned time limit. This procedure is referred to as *filet*. In the alternative procedure, *contre filet*, the Commission is obliged to defer application of the measures for a period laid down in each Council act from the date of communication to the Council. Within this period the Council may adopt a different decision by qualified majority .

Implementation action by the Management Committee is necessary for the Control Regulation to function effectively.

3. *Implementation rules*

As noted in the introduction to this chapter, the Control Regulation is a framework regulation and thus requires many follow up measures for detailed implementation. For example, **Article 5** stipulates that detailed rules are to be adopted under the **Article 36** procedure. These include, *inter alia*, the same rules listed in **Article 3** of Council Regulation No. 2241/87[116] regarding identification of inspectors and inspection vessels, procedures for boarding fishing vessels and inspection of catches, subsequent reports, markings of fishing vessels and gear, and the certification of the characteristics of fishing vessels. Detailed rules are also required for the recording of data relating to the position of fishing vessels and the transmission of such data to Member States and the Commission, as well as provisions in relation to the movement and products held on board third-country vessels. Although no date was set for the adoption of Commission implementing regulations, it was no doubt envisaged that this would be undertaken as soon as possible in order to allow Member States to cope with their global obligations under the Control Regulation. In the interim period, the detailed rules which existed prior to the adoption of the new regime remain in force.

The procedure for adopting detailed implementing legislation is different from the procedure laid down in **Article 43(2)** {37(2)} of the EC Treaty and this has already been discussed above in the context of the function of the Committee for Fisheries and Aquaculture. Because the procedure in the Management Committee is more streamlined and participants are less subject to political sensibilities the drafting and adoption of detailed implementation rules does not generally pose a problem. However, a legal issue which arises in this context is whether the reference to implementing regulations in the Control Regulation could be used as an excuse by Member States in delaying the implementation of the general provisions. Or does the absence of detailed implementing rules preclude the enforcement of the general measures prescribed in the Control Regulation? The answer to this question is of major significance because of the many provisions which require detailed implementing regulations. The Court of Justice ruling in Case 87/82 *Rogers* v. *Darthenay*[117] may provide an answer. The case concerned the prosecution of a French vessel for using a net with illegal net attachments within British fishery limits in August 1981. The accused was charged with contravening, *inter alia,* **Article 7** of Council Regulation No. 2527/80, and **Article 8** of Fishing Nets Order (No. 2) 1980 as amended.

In *Rogers* v. *Darthenay,* Plymouth Magistrates court referred four questions to the Court of Justice, pursuant to **Article 177** {234} of the EC Treaty, for a preliminary ruling on the interpretation of Regulation No. 2527/80 laying down technical measures for the conservation of fishery resources.

The first question is analogous to the aforementioned issues:

> 'Whether Article 7 of EEC Regulation No. 2527/80 (as extended) has any effect when no detailed implementing rules have been adopted?'

In relation to this question the Court stated:

> 'The purpose of Regulation No. 2527/80 is to ensure the protection of fishing stocks and also a balanced exploitation of the resources of the sea in the interests both of fishermen and consumers. It follows that the prohibition laid down in the first sentence of **Article 7** of the Regulation constitutes an essential provision for the achievement of the objective pursued, since without that prohibition there would be no effective conservation measure at a Community level.
>
> Moreover, that first sentence of **Article 7** of Regulation No. 2527/80 is an independent and perfectly clear provision creating a prohibition with immediate effect which cannot depend upon the adoption of the detailed implementing rules provided for in the second sentence of **Article 7**.
>
> The expression 'detailed implementing rules' used in the second sentence of **Article 7** refers to the determination of certain fishing attachments the use of which appears to be compatible with the prohibition laid down in the first sentence and not to implementing measures necessary to ensure the full effect of that prohibition. Consequently, the fact that the detailed implementing rules referred to in **Article 7** have not been adopted cannot in any event prevent the prohibition laid down in the first sentence of that Article from taking full effect.
>
> The answer to the first question should, therefore, be that the prohibition in **Article 7** of Regulation No. 2527/80 takes full effect even though the detailed implementing rules provided for in the second sentence of that Article have not been adopted.'

Case 87/82 *Rogers* v. *Darthenay* clearly suggests that the absence of detailed implementing rules in certain circumstances does not preclude the enforcement by Member States of the general provisions prescribed in the Control Regulation. However, it must be added that the prolonged delay in introducing detailed implementing rules cannot be justified, particularly as Member States' administrative and computerised systems for catch registration and cross verification procedures (discussed below) are geared towards the general provisions and any change to include additional requirements would result in inordinate cost and a drain on the limited resources, not to mention further delay. There is also a precedent in the much-criticised long period of delay between the adoption of Council Regulation No. 171/83 in January 1983 and the adoption of detailed implementing measures in July 1984 and December 1984, pursuant to **Article 6** and **Article 7** of that regulation. From the enforcement perspective, procrastination reduces the opportunity to ensure that the imposition of new measures is swift and uncomplicated. It is also guaranteed to draw the reproaches of the fishing industry on to national control authorities and to lower the esteem of the European Union institutions in the eyes of all parties.

One of the questions which arises as a result of the Commission's power to lay down detailed implementing rules in accordance with the management committee procedure is: what are the limits on the Commission competence pursuant to **Article 36**? The Court has held that provisions conferring executive authority must be interpreted in the light of the scheme and objectives both of those provisions and of the rule as a whole.[118] This issue arose in the context of an **Article 173** {230} EC Treaty application by Spain and France for annulment of a Commission Regulation concerning the catch declarations of vessels flying the flag of a Member State and operating in the zone of a developing country.[119] The Commission had adopted the contested regulation on the basis of the 1987 Control Regulation. The Court pointed out that the 1987 Control Regulation was itself a regulation for the implementation of a Basic Council Regulation (the 1983 Management Regulation) and that an implementing regulation such as the 1987 Control Regulation must respect the basic elements in the Basic Regulation which in this case was to establish a Community system for the conservation and management of fishery resources.[120] The Court was thus able to conclude by examining the scheme of the Basic Regulations in question, that the Commission had no power to extend on the basis of the Control Regulation the catch recording provisions to vessels of Member States operating in the zones of developing countries. The principal reason substantiating this conclusion is that fisheries agreements with developing countries are governed by other criteria - a financial objective - as opposed to conservation of fishery resources. (However, it could be argued that in a world of depleted fishery resources this type of restrictive interpretation of international agreements is no longer valid and that a broader interpretation of the Community's global fishery conservation and sustainable management obligations, particularly to developing Countries, could now be expected from the Court of Justice.) The decision of the Court of Justice, nonetheless, provides an indication of the line of reasoning we can expect from the Court if any of the implementing regulations stemming from the Control Regulation are the subject of a legal challenge in the Court of Justice. The Court in all probability will examine the contested regulation to see that it remains within the confines prescribed in the Control Regulation and the 1992 Basic Management Regulation.

Whether the Council has the power to adopt a regulation which empowers the Commission, or extends the Commission competence, to adopt detailed rules to impose penalties or take administrative sanctions such as those discussed earlier in this study has not been ruled on by the Court in a specific fisheries case, although this was one of the pleas raised by the applicant in the *Swedish fish licence Case* but subsequently ruled inadmissible by the Court on procedural grounds.[121] The Court has held in the context of the agricultural policy that the Council has the power to impose penalties in the implementing powers of the Commission provided such power is properly delegated.[122] In any case an amendment to the Control Regulation in 1998 provides that the Council, acting on the basis of **Article 43** {37} of the EC Treaty, may draw up a list of behaviour which seriously infringes the Community rules and to which the Member States undertake to apply proportionate, dissuasive and effective sanctions.[123] It thus appears that the regulation of measures relating to sanctions and penalties will remain within the firm remit of the Council.

Detailed implementing rules have a significant impact on the execution of the enforcement function in the Member States. In this regard there has always been the requirement for rules to be drafted in a manner that balances the importance of precise and clear provisions against the hazards of over-regulation. Occasionally it may be necessary for a Member State Court or Tribunal to make a reference to the Court of Justice under **Article 177** {234} of the EC Treaty in order to obtain a ruling on the interpretation or validity of the Community provision. An example of this arose in the *Danish Fishery Pennant Case* which has already been discussed in Section (i) above.

Section (iv) National control measures

Introduction

The subject of national measures and their relationship with Community legislation is linked to the division of legislative competence between the Member States and the Community.[124] It has been the subject of voluminous literature and thus only requires brief comment here.[125] In general, Member State competence to prescribe and resort to national measures is limited to four situations: firstly, in areas where Member States have retained exclusive competence, such as the power to prescribe a particular type of judicial or administrative procedure for fisheries offences; secondly, to take measures of the kind described in *Commission* v. *United Kingdom* (discussed below) at times and places where there are no Community conservation and management rules in place;[126] thirdly, to take measures which go beyond the minimum requirement in Community regulations provided that such measures accord with Community law and are in conformity with the common fisheries policy;[127] finally, to fill lacunae in Community legislation.[128]

National measures are most frequently required when a particular problem or fishery requires specific legislation which is not provided for in Community law. The Commission, however, has noted that, in reality, political commitment is lacking in several Member States to apply national measures, which are more severe than those laid down by Community legislation, in situations where this is required for inspections to be effective.[129] The relationship between national control measures and the corpus of Community fisheries law has nonetheless resulted in a substantial amount of jurisprudence from the Court of Justice.[130] Other than being of major significance in the context of the scope of Community legislation, the case-law is important because it indicates the likely approach the Court will take in the future when considering the compatibility of national fishery enforcement measures with the Community law. It also offers a remarkable example of the manner in which economic and other arguments find reflection in the law and the difficulties that may ensue in the aftermath of judicial procedures for those parties who are left with the task of applying and enforcing the law. Ultimately, the true value of the Court of Justice jurisprudence will be to curtail Member States' temptation to take unilateral action if the CFP is knocked by further crises issues.

1. *Provisions in the Control Regulation pertaining to national control measures*

In accordance with **Article 38**, the Control Regulation shall apply without prejudice to any national control measures which go beyond the minimum requirements, provided that they comply with Community law and are in conformity with the common fisheries policy. **Article 38** also specifies the additional notification requirement that any such national control measures shall be communicated to the Commission in accordance with **Article 2(2)** of Council Regulation No. 101/76 laying down a common structural policy for the fishing industry.[131]

Commentary

The introduction of national measures is qualified by three requirements: firstly, national measures must go beyond the minimum requirements; secondly, such measures must comply with Community law and be in conformity with the CFP; thirdly, the obligation on Member States to inform the Commission of national measures. Many of the issues raised by national measures may be evaluated in the context of these requirements.

(i) National measures which are stricter than the minimum requirements in Community law.

The term *'go beyond'*, referred to in **Article 38**, is interpreted as meaning to be stricter or more onerous. Consequently, there is no scope for a Member State to introduce measures which are less rigorous than those set out in the Control Regulation unless there is express provision to that effect. Moreover, the introduction of national enforcement measures is not a substitute for Community measures. That having been said, national measures may place greater obligations on fishermen than the measures introduced on foot of the Control Regulation.[132] This latitude is not unusual when it is considered that **Article 10** of the 1992 Management Regulation expressly authorises the Member States to take measures for the conservation and management of fish stocks which are more stringent than those provided for under community rules, provided that such measures apply solely to fishermen from the Member State concerned and are compatible with the objectives pursued under **Articles 2(1)** and **2(2)** of that regulation.

An example of national measures which went beyond Community measures arose in *Wood and Cowie Case.*[133] In this instance a condition of a United Kingdom licence required the master of a vessel to report by radio his intention to cross from one ICES zone to another (this requirement is referred to as a hail requirement). This condition was upheld by the Court as being in conformity with Community law even though the same condition did not apply to vessels flying the flag of other Member States fishing for the same species in the same areas. The Court held that **Article 15** of the 1987 Control Regulation (the predecessor to **Article 38** of the 1993 Control Regulation) does not require the Member State to apply national control measures to vessels of the other Member States.[134] It was the view of the Court that community rules on conservation do not preclude a condition that is intended to ensure fishing activities subject to quotas can be maintained.[135] At first sight, the introduction of national control measures may appear to be discriminatory. However, in the context of the fishing policy, **Article 6** {12} of the EC Treaty (originally **Article 7**) does not apply to any disparities in treatment or any distortions which may result for persons and undertakings which are subject to the jurisdiction of the Community from the application to Member States of measures that are stricter than those applied in the same sphere to other Member States.[136] The application of national measures cannot be held contrary to the principle of non discrimination merely because other Member States are applying rules that are less strict.[137] Comparable situations must not be treated differently and different situations must not be treated in the same way unless such treatment is objectively justified.[138] Furthermore, in respect to a possible breach of the equal access principle, the Court interpreted this principle in the light of the national quota system as not precluding a Member State from applying to vessels flying its flag, national control measures stricter than those imposed by Community legislation, provided that such measures are intended to monitor fishing activity and prevent fraud and are not disproportionate to the objectives pursued.[139] From *Wood and Cowie,* we can surmise that it is unlikely that the introduction of stricter control measures infringe **Article 34** {29} of the EC Treaty (constitutes a measure having equivalent effect to a quantitative restriction on exports) on the grounds that it is settled case law that **Article 34** {29} applies only to national measures which have as their specific object or effect the restriction of exports in such a way as to provide advantage to the national production or domestic market.[140] It is hence doubtful that a disputed measure, if it applies to all fishing vessels without distinguishing between those that sell their products on the domestic market from those that export, would infringe Community law.

(ii) The requirement to notify and consult with the Commission

Article 2(2) of Council Regulation No. 101/76 laying down a common structural policy for the fishing industry requires that each Member State shall notify other Member States and the Commission of the existing laws and administrative rules applied by each Member State in respect of fishing in the maritime waters under its sovereignty or under its jurisdiction zones together with rules arising out of the principle of equal access.[141] Notification allows the Commission to check the conformity of the measure with Community law and to assess whether there is adequate compliance with the CFP as well as ensuring the enhancement of Community policy. The normal practice on receiving notification from the Member State is for Unit XIV.1, which is the unit responsible for legal questions in the Fisheries Directorate General (DG XIV), to review the legislation in question and see if it is compatible with Community law. There may also be consultation with the Commission's legal service. On occasion the Commission will advise the Member State to withdraw a national conservation measure and in the event of non-compliance commence **Article 169** {226} of the EC Treaty proceedings. The 1997 General Report on the activities of the European Union records that in 1997 the Commission was notified of 156 national conservation measures, 147 of which were subject to observation by the Commission. Nine measures were still under examination at the time the report was published.[142]

Ostensibly, the procedural requirement for the notification of national control measures is different from that which is necessitated by the introduction of national conservation measures. In Case 141/78 *France* v. *United Kingdom,* the Court ruled that in relation to the latter measures, notification must be given *before* a measure is brought into force.[143] It must thus be emphasised that **Article 38** of the Control Regulation and **Article 2(2)** of Regulation No. 101/76 obliges Member States to notify other Member States and the Commission of the 'existing' control provisions concerning fishing and not provisions which they 'intend' to adopt. This requirement must be distinguished from the obligation in respect to new conservation measures which by virtue of **Article 10** of the 1992 Management Regulation and **Article 14** of the 1986 Technical Regulation (and **Article 17** of the 1997 Technical Regulation) must be notified, in time for the Commission to present its observations, and prior to the introduction or amendment of national conservation and resource management measures.[144] The Court in *Cowie and Wood* held that national control measures must be notified to the Commission but not necessarily prior to their adoption.[145] **Article 38** does not mention any of the consequences of a failure to notify a national control measure, so the question arises: does non notification affect the validity of a national measure? The Court in *Cowie and Wood* points out, in relation to a similar provision in the 1987 Control Regulation, that the adoption of such a national measure is not made conditional on its prior notification to the Commission, the notification requirement having been laid down for the purpose of information only. Consequently, the absence of notification does not affect the validity of such a measure which satisfies the other criteria of compliance and conformity with Community law.[146] In drawing this conclusion, the Court pointed out that the Technical Regulation required a more detailed notification requirement which allowed the Commission to delay or prevent the entry into force of technical conservation measures notified to it.[147] Interestingly, the Advocate General arrived at the same deduction by a different route, which appears to be more logical. In his view it was necessary to go beyond the text of the provision and find that the Member States were bound to notify the text of the provision in advance on the grounds that **Articles 2** and **3** of Regulation No. 101/76 were provisions that were clearly complementary and inseparable.[148] He further considered that the measures although capable of being amenable to **Article 169** {226} proceedings, were valid in the absence of express intervention power in favour of the Commission.

In any case, it needs to be emphasised that there are express provisions in **Article 17(2)** of the 1997 Technical Regulation regarding notification of national technical conservation measures.[149]

(iii) National measures must comply with Community law and be in conformity with the CFP

The requirement for measures to conform with the CFP stems from **Article 5** {10} of the EC Treaty which stipulates that Member States shall 'abstain from any measures which could jeopardise the attainment of the objectives of this Treaty'. Although there is no statement in the treaty on the issue of Community law supremacy there is a well established body of case law establishing the doctrine that directly effective provisions of Community law have supremacy over conflicting provisions of national law which can be traced back to *Van Gend en Loos*.[150] The unilateral powers of the Member States are curtailed by Community law.[151] In *Commission* v. *United Kingdom* the Court held that since the expiration of the transitional period laid down by the Act of Accession in 1972 power to adopt as part of the CFP, measures relating to the conservation of the sea has belonged to the Community.[152] Member States are thus, after 1 January 1979, no longer entitled to exercise any power of their own which does not conform to Community law.[153] The requirement for national measures to comply with Community law and to conform with the CFP is best illustrated in the series of cases which are referred to as 'the quota hopping cases' which are reviewed in Chapter 7.

References

1 Council Regulation (EC) No. 2847/93 of 12.10.1993 establishing a control system applicable to the common fisheries policy, *OJ* L 261, 20.10.1993. As amended by Regulations (EC) No. 2870/95, *OJ* L 301, 14.12.1995; No. 2489/96 *OJ* L 338, 28.12.1996, p. 12; No. 686/97, *OJ* L 102/1, 19.04.1997, p. 1; No. 2205/97, *OJ* L 304, 07.11.97, p. 1; No. 2635/97, *OJ* L 356, 31.12.1997, p. 14; and No. 2846/98, *OJ* L 358, 31.12.1998, p. 5.
2 Council Regulation No. 2057/82, *OJ* L 220, 29.07.1982, p. 1. (Repealed.)
3 Council Regulation No. 2241/87 of 23.07.1987 establishing certain control measures for fishing activities, *OJ* L 207, 29.07.1987, *OJ* L 306, 11.11.1988, p. 2, as last amended by Council Regulation No. 343/88, *OJ* L 306, 11.11.1988, p. 2. (Repealed.)
4 Council Regulation No. 3760/92, **Article 12**, *OJ* L 389/1 31.12.1992.
5 *Ibid.*
6 The opinion of the European Parliament, *OJ* C 21, 25.01.1993, p. 55. At its 303rd Plenary Session, on 24.02.1993, the Economic and Social Committee adopted its opinion on the new control regulation by 67 votes to 5, with 5 abstentions, *OJ* C108, 19.04.1993, p. 36.
7 *Op. cit.*, fn 1.
8 *Op. cit.*, fn 1.
9 Council Regulation No. 2241/87, *op. cit.*, fn 3.
10 **Article 189(2)** {249} of the EC Treaty provides that regulations are 'directly applicable in all Member States', thus, the Control Regulation is automatically part of each national control system without the necessity of any implementing legislation. Direct applicability and direct effect are tenets of the 'new legal order', Case 26/62 *Van Gend en Loos* v. *Nederlandse Administratie der Belastingen* [1963] ECR 1, and, as Community legal principles they have resulted in a veritable library of academic comment as well as being dealt with in all Community law texts. The issue of direct effect can be traced from the seminal case of Case 26/62, *Van Gend en Loos* through a line of notable cases since including, *inter alia*, Case 9/70, *Grad* v. *Finanzamt Traunstein* [1970] ECR 825; Case 41/74, *Van Duyn* v. *Home Office* [1974] ECR 1337; Case 43/75, *Defrenne* v. *Sabena*, [1976] ECR 455; Case 96/80 *Jenkins* v. *Kingsgate*, [1981] ECR 911; Cases C-6 9/90 *Francovitch* v. *Italy* [1991] ECR I-5357. On the literature see, *inter alia*, Hartley, Winter, 'Direct Applicability and Direct Effect: Two Distinct and Different Concepts in Community Law' (1986) CML Rev. **9**, **425**; Pescatore, 'The Doctrine of "Direct Effect": An Infant Disease of Community Law', (1983) *EL Rev*, **8** 155.
11 *Ibid.*
12 See Case 24/73 *Variola* [1973] ECR 981 and Case 50/76 *Amsterdam Bulb BV* v. *Produktschap voor Siengewassen*, [1977] ECR 137.
13 The subject of criminal enforcement is examined by, *inter alios*, Berg, A., 'Enforcement of the Common Fisheries Policy, with Special Reference to the Netherlands'; Churchill, R.R., 'Enforcement of the Common Fisheries Policy, with Special Reference to the United Kingdom'; in Harding (eds.) *Enforcing European Community Rules*, (Aldershot, 1996) 62–82, 83–102. See discussion on sanctions, Chapter 6, *post*.
14 The Court has declared that treaties between the Community and third countries be 'an integral part of Community law', Case 87/75, *Bresciani* v. *Italian Finance Dept.*, [1976] ECR, 129; for a general discussion on this point see Hartley, T.C., *Foundations of European Community Law*, 187–232. See, Chapter 2, *ante*.

15 Joined Cases 21-24/72, *International Fruit Co., N.V.* v. *Produktschap voor Groenten en Fruit* [1972] ECR 1219. Case 104/81, *Hauptzollamt* v *Kupferberg* [1982] ECR 3641.
16 Case 106/77 *Simmenthal* [1978] ECR 629.
17 On this issue see Case 39/88 *Commission* v. *Ireland* [1990] ECR I-4271, discussed in Chapter 7, *post.*
18 Council Regulation No. 2241/87, *OJ* L 20, 28.01.1976, p. 19., in respect to carry over provision, see discussion on Community Logbooks in Chapter 4, *post.* Council Regulation (EC) No. 2846/98, *OJ* L 358, 31.12.1998, **Article 3**, extends the date of entry into force of some provisions.
19 *Op. cit.*, fn 6.
20 Although this issue of reconsultation did not arise in the Control Regulation it is interesting to note that in Case C-331/88, *The Queen* v. *Minister of Agriculture, Fisheries and Food and Secretary of State for Health, ex parte Fedesa and other* [1990] ECR I-4023 the Court held that there is no necessity for reconsultation if the changes were either technical or either essentially in line with the wishes of Parliament.
21 Churchill, R.R., *EC Fisheries Law*, p. 279, discusses the preference of lawyers for direct action or authoritative prescription to regulate fisheries. He suggests that when the Community is faced with any new question requiring regulation, it should consider carefully whether direct or indirect regulation, or some combination, is the most appropriate form of regulation, rather than simply assuming, as most lawyers probably would, that direct regulation should always be used in preference to indirect regulation.
22 See further discussion of sanctions in Chapter 6, *post.*
23 Chapter 2, *ante.*
24 For example in the 'quota hopping' cases, see Case 3/87 *The Queen* v. *Ministry of Agriculture, Fisheries and Food, ex parte Agegate Ltd.* [1989] I-4459; Case 216/87 *The Queen* v. *Ministry of Agriculture, Fisheries and Food, ex parte Jaderow Ltd. and others* [1989] ECR 4509; Cases C-221/89, *The Queen* v. *Secretary of State for Transport, ex parte Factortame Ltd and others* [1991] ECR I-3905; C-93/89; *Commission* v. *Ireland,* [1991] ECR I-4569; Case C-246/89, *Commission* v. *United Kingdom of Great Britain and Northern Ireland* [1989] ECR 3125. These cases are discussed in Chapter 7, *post.*
25 The role of the Commission is examined in Chapter 6, *post.*
26 The cost of enforcement is examined in Chapter 9, *post.*
27 See, *inter alia*, Freestone, D., 'Some Institutional Implications of the Establishment of Exclusive Economic Zones by EC Member States', *Ocean Development and International Law,* **23**, (1992) 97–114; Churchill, R., *EC Fisheries Law,* 56–72; and by the same author 'EC Fisheries and an EZ–Easy!', *Ocean Development and International Law,* **23**, (1992) 145–163.
28 Council Regulation No. 2847/93, **Article 2(1)**, *op. cit.*, fn 1.
29 See discussion below on coordination between Member States. The Commission had previously noted that Member States were slow to have recourse to international law rights such as that of *hot pursuit* to stop infringements of Community legislation by their own vessels, those of other Member States, or those of non-member countries, see, Report from the Commission to the Council and the European Parliament on monitoring and implementation of the CFP, SEC (92) 394 of 06.03.1992, p. 23. The reluctance of Member States to resort to international law rights can be contrasted with the frequent recourse to such provisions in other jurisdictions, and the tendency, in some cases, to exceed their international law mandate by resorting to unilateral action, the most notable example being the unilateral action by Canada in 1995 in seizing the Spanish vessel, the *Estai,* which was fishing in international waters for allegedly infringing the Canadian law, The Coastal Water Sea Fishery Conservation Act, 1994 (as amended). See, Curran P. and Long R., 'Unilateral fishery law enforcement: the case of the *Estai*', *The Irish Journal of European Law,* 5(2), (1996), 123–163.

30 In particular, see Freestone, D. and Churchill, R.R., fn 27, *supra*.

31 See, *inter alia*, Case 61/77, *Commission* v. *Ireland* [1978] ECR 417 at 446. In *Commission* v. *Ireland* the Court pointed out that the Communities' powers are exercised primarily in, and in respect of, the territories to which the Treaties apply; and institutional acts adopted on the basis of the Treaties apply in principle to the same geographical areas as the Treaties themselves. The Court set down a clear guideline regarding the criteria which it used in assessing the field of application of Community regulation. The Court stated: '... Regulation ... must therefore be understood as referring to the limits of the field of application of Community law in its entirety, as that field may at any given time be constituted', paragraphs 44–47. For a discussion of this case, see, Timmermans, C.W.A., *Sociaal-economische wetgeving*. (1978), 582–589; Churchill, R., *European Law Review* (1979), 391–396; Winkel, K. and Von Borries, R., *Common Market Law Review* (1978), 494–502.

32 This accords with the basic rule of international law regarding the interpretation of treaties: 'A treaty shall be interpreted in good faith in accordance with the ordinary meaning to be given to the terms of the treaty in their context and in the light of its object and purpose', see **Article 31**, Vienna Convention on the Law of Treaties.

33 For a discussion of the Community's competence in maritime zones and activities in these zones, see all the papers in *Ocean Development and International Law* **23** (1992), 89–259, and especially Freestone and Churchill, R.R., *op. cit.* fn 27; Birnie, P., 'An EC Exclusive Economic Zone: *Marine Environment Aspects*,' 193 and 200–206, Soons, A.H.A., 'Regulation of Marine Scientific Research by the European Community and its Member States,' 259 and 262–267. See also, *inter alios*, Fleischer, C.A., L'accès aux lieux de pêche et la traité de Rome', *RMC* **141** (1971) 148 at 151; Koers, A.W., 'The External Authority of the EEC in regard to Marine Fisheries', *CMLRev* **14** (1977), 269 at 274–5.

34 Thus, for example, the Court in *Kramer (Cornelius)* (Joined Cases 3, 4 and 6/76) [ECR] 1279, stated that it should be made clear that although **Article 5** of the original regulation laying down a common structural policy for the fishing industry (Regulation No. 2141/70) was applicable only to a geographical limited area (the territorial sea and exclusive fishery zone). It none the less follows from **Article 102** of the Act of Accession, from **Article 1** of the said regulation and moreover from the very nature of things that the rule-making authority of the Community *ratione materiae* also extends – in so far as Member States have similar authority under public international law – to fishing on the high seas, paragraphs 30–33 of the judgement.

35 Case 61/77, *Commission* v. *Ireland* [1978] ECR 417 at 446; Case 24/83, *Gewiese and Mehlich* v. *Mackenzie*, [1984] ECR 817, Case 63/83, *R.* v. *Kirk*, [1984] ECR 2689; Case 141/78, *France* v. *United Kingdom* [1979] ECR 2923.

36 Case C-286/90, *Anklagemydigheden* v. *Poulsen and Diva Navigation Corp.* [1992] ECR-I-6019 at paragraph 24.

37 Joined Cases 3/76, 4/76, and 6/76 *Kramer* [1976] ECR 127, at paragraphs 30–33; Case C-258/89 *Commission* v. *Spain* [1991] ECR I-3977.

38 Case C-286/90, *Anklagemydigheden* v. *Poulsen and Diva Navigation Corp.* [1992] ECR-6019 at paragraph 9. For the Court's application of the rules of the law of the sea, see paragraphs 25 to 27.

39 To date, with the exception of the Channel Islands, there has been no particular enforcement problems associated with these territories. Issues raised by, for example, the special status of the Isle of Man which is not part of the United Kingdom but is described under international and constitutional law as 'dependency of the British Crown' have led to some interesting case law in the Court of Justice but have not led to any significant developments or circumstances pertaining to the CFP. The rather elusive concept of the legal position of the Isle of Man is explained in some detail in Advocate General Jacobs' opinion in Case

C-355/89, *Department of Health and Social Security* v. *Christopher Stewart Barr and Montrose Holdings Ltd.* [1991] ECR I-3479.

40 Case 804/79, *Commission* v. *United Kingdom* [1981] ECR 1045. See Chapter 2, *ante*.

41 *OJ* 20 L 20, 19.01.1976, p. 19.

42 Cases 61/77, *Commission* v. *Ireland*, [1978] ECR 417. Referred to as *The Irish Fisheries Case*.

43 Case C-146/89, *Commission* v. *United Kingdom of Great Britain and Northern Ireland*, [1991] ECR I-3533. In the latter the Court was faced with a straightforward question: was the 12-mile zone, established by Regulation No. 170/83 as a derogation from the equal access principle, to be measured from the baselines in force at the time that Regulation No. 170/83 was adopted (as argued by the Commission and France), or was the zone to be measured from new baselines as were defined from time to time by Member States in accordance with international law (a concept referred to as 'ambulatory baselines'), as argued by the United Kingdom? The net effect if the answer was in the affirmative would be to move the 12-mile territorial zone several miles further seawards and to exclude certain other Member States' vessels access from some of their traditional fishing grounds. One of the arguments presented by the United Kingdom relied on the Court's judgement in Case 61/77 *The Irish Fisheries Case* [1978] ECR 417 to support the contention that any adjustment by Member States of their maritime waters automatically entails an identical adjustment in the sphere of application of Community rules (this concept that as the national maritime zones increase in size so does the scope of application of the existing body of relevant Community fisheries had been previously consistently supported by the case law of the Court, Case 141/78, *France* v. *United Kingdom*, [1979] ECR 2923. Case 24/83, *Gewiese and Mehlich* v. *Mackenzie*, [1984] ECR 817, Case 63/83, *Regina* v. *Kirk*, [1984] ECR 2689.) The Court cited the Advocate General's opinion to distinguish the grounds on which the *Irish Fisheries* case was decided, namely at issue in that case were the rules which apply without distinction to all Community fishermen (i.e. the scope of the equal access principle), as opposed to the problem of safeguarding certain fishing activity. Thus, it was the view of the Court that the reasoning of the Court in the *Irish Fisheries* Case could not be transposed to the *United Kingdom* (Baseline case). It is implicit in the Court's decision that the drawing of baselines from which the maritime zones are measured remains within the competence of Member States, even though the way in which that competence is exercised may be constrained by the provisions of Community law.

44 Case 146/89, paragraph 55, *ibid*.

45 The United Kingdom has since implemented the Court's judgement concerning the baselines to be applied under **Article 6(2)** of Council Regulation No. 170/82, since repealed by **Article 6(2)** of Council Regulation No. 3760/92. A practical point of interest, from the enforcement perspective, is that as soon as the French and Belgian fishermen objected, and when **Article 169** {226} proceedings were instituted by the Commission, the British authorities did not enforce the provisions of the Territorial Sea Act 1987 which extended the basepoints to include elevations lying between 3 and 12 nautical miles off the coast. This may have been due to lack of conviction and legal certainty which would have undermined the validity of the UK case from an international law perspective. However, in the long-term the Court took a more altruistic approach and described the UK conduct as 'exemplary' and, unusually, each party was ordered to pay its own costs. The decision of the Court in the *United Kingdom (Baseline case)* that the seaward movement of the UK baselines in light of the extension of the UK territorial sea to 12 miles did not affect the fishing zones defined by Council Regulation No. 170/83 must be regarded as an unusual case to be distinguished on exceptional circumstances (both the UK situation and the relevant regulation). The case is, however, a good example of the tension that arises from obligations derived from customary international law, as well as Member States' exclusive jurisdiction to act under public international law, and the internal requirements for

Member States to comply with Community law. Central to this case was the issue whether Member States can invoke international obligations against Community law. One authority has summarised this argument as follows: since the Community is bound by international law, Community law cannot compel Member States to violate this law if they act within their own sphere of competence. Where the Member States alone bear international responsibility, obligations imposed by public international law should take precedence over Community law. In The *United Kingdom (Baseline case)* it has been suggested that the Court has denied the very existence of this conflict. The international rules at stake did not force the United Kingdom to act as it did. They merely contained an option; it was thus possible to comply with Community law without violating international law. See further, Anderson, D., 'The Straits of Dover and the Southern North Sea: Some Recent Legal Developments', *International Journal of Estuarine and Coastal Law* **7** (1992), 85–97; Carleton, M.C., 'The evolution of the Maritime Boundary – The UK experience in the North Sea and Channel', *ibid.*, 99–112.

46 LOS Convention, **Article 73**, see Chapter 11, *post.*

47 Council Regulation (EC) No. 2846/98, *OJ* L 358/10, 31.12.1998, **Article 28a**.

48 See discussion on the subject, *inter alia,* of communication requirements, pursuant to **Article 2(2)**, (these are commonly referred to as hail requirements), Chapter 4, Section (i), *post.*

49 This table is constructed from data presented by the FAO in, 'Coastal State requirements for foreign fishing', *FAO Legislative Study,* Table D, 245–256, (Rome, 1996). It does not take account of the revised Community measures pertaining to the activities of third-country vessels which were adopted pursuant to Council Regulation (EC) No. 2846/98, *OJ* L 358, 31.12.1998, p. 5.

50 *Ibid.*

51 See Chapter 11, *post.*

52 LOS Convention, **Article 58(1)**.

53 Geneva Convention on the Territorial Sea and the Contiguous Zone of 29 April 1958, **Articles 14–23**; LOS Convention, **Articles 17–32**.

54 Case C-286/90, (1992) ECR I-6019.

55 Joined Cases 89/85, 114/85, 116–117/85, 125–128/85, *Ahlstrom* v. *Commission,* [1988] ECR 5193. This observation was made by Brandtner, B., Folz, H.P., 'A Survey of Principle Decisions of the European Court of Justice pertaining to International Law in 1991–1992', *European Journal of International Law* **4** (1993), 442–445. The case is annotated by, *inter alios,* Sanchez, R. and Luis, I., *Revista de Instituciones Europeas,* (1993), 141–153; Le Bihan, C., *Revue trimestrielle de droit européen,* (1993), 421–438; Cataldi, G., *Il Foro italiano* (1993), **IV** Col., 249–258; Slot, P.J., *Common Market Law Review* (1994), pp. 147–153.

56 *OJ* L 288/1, 1986.

57 Case C-286/90, 1992, ECR 1-6019, paragraph 9 and 10.

58 *Ibid.* paragraph 16.

59 *Ibid.* paragraph 20.

60 *Ibid.* paragraphs 24–25.

61 *Ibid.* paragraphs 28–29.

62 Case 286/90, Opinion of Advocate General Tesauro delivered on 31 March 1992, paragraphs 10–15, [1992] ECR I-4673.

63 Case C-286/90, 1992 ECR 1-6019, paragraph 31.

64 *Ibid.* paragraphs 31–33.

65 *Ibid.* paragraph 38.

66 Brandtner, B. and Folz, H.P., 'A Survey of Principal Decisions of the European Court of Justice pertaining to International Law in 1991–1992', *op. cit.* fn 55, 442–445.

67 Slot, P.J., Case C-286/90 *Anklagemyndigheden (Public Prosecutor)* v. *P.M. Poulsen and Diva Navigation* [1992] ECR I-6019, *Common Market Law Review* **31** (1994), 147–153.

68 See in particular the comments and analysis by Slot, P.J., *ibid.*, and Brandtner, B. and Folz, H.P. *op. cit.* fn 55.
69 Case C-286/90, paragraph 11.
70 International Convention for the Safety of Life at Sea, 1974; International Convention for the Prevention of Pollution from Ships (MARPOL 73/78); International Convention on Standards of Training, Certification and Watchkeeping for Seafarers, 1978; 1976 Merchant Shipping (Minimum Standards) Convention (ILO 147).
71 Discussion of port state control has been the subject of extensive analysis elsewhere, see, *inter alios*, Payoyo, P.B., 'Implementation of international conventions through port state control: an assessment', *Marine Policy*, 15(5) (1994), 379–392; Kasoulides, G., 'Port state enforcement regime through international organisations' in Soons, A. (ed.) *Implementation of the Law of the Sea Convention Through International Institutions, Law of the Sea Institute* (1990), 187.
72 See Chapter 11, *post*.
73 See Chapter 4, *post*.
74 Council Regulation No. 2846/98, *OJ* L 358, 31.12.1998, p. 5; **Article 28e**.
75 See Chapter 1, *ante*. For a list of Agreements, see pp. viii and ix, *ante*.
76 See Chapter 2, *ante*.
77 Agreement on cooperation in the sea fisheries sector between the European Community and the Kingdom of Morocco, *OJ* L 30 Vol. 40, 31.01.1997, *passim*.
78 Member State cooperation with the Commission is discussed in Chapter 6, *post*. On the obligation to cooperate pursuant to the international law regime to regulate high seas fisheries and to promote responsible fisheries, see Chapter 11, *post*.
79 Council Regulation No. 2846/98, *OJ* L 358, 31.12.1998, p. 5, **Article 34b**. See Chapters 6 and 8, *ante*.
80 *Ibid.*, **Article 34b(2)**.
81 Clear reference to the Community ideal of equality enshrined in **Article 6** {12} of the EC Treaty. This is a possible response to growing fears within certain sectors of the Community fleet that legislative measures in the Member States were targeted against particular vessels. On the wider issues raised see Churchill, R.R. and Foster, N.G., 'European Community Law and prior treaty obligations of Member States: The Spanish Fishermen's Cases', *ICLQ* **36(3)**, (1987) 504–524.
82 Judgement of the 25 July 1974, *Germany* v. *Iceland* fisheries jurisdiction case, *Recueil des arrêtes* 1974, p. 175, paragraph 45.
83 The European Union fishery dispute with Canada in 1995 is an example of an international incident which had a major impact on the drafting and subsequent ratification of the Agreement for the implementation of the provisions of the United Nations Convention on the Law of the Sea (UNCLOS) relating to the conservation and management of straddling stocks and highly migratory fish stocks.
84 Recent international developments include, *inter alia*: 'Agenda 21' a program of action adopted by UNCED; the 'Agreement to Promote Compliance with International Conservation and Management Measures by Fishing Vessels on the High Seas' (FAO Compliance Agreement); 'The Code of Conduct for Responsible Fisheries'; and the 1995 'Agreement for the implementation of the provisions of the UNCLOS relating to the conservation and management of straddling fish stocks and highly migratory fish stocks'; See Chapter 11, *post*.
85 Case 2/88, *Criminal Proceedings Against J.J. Zwartveld and Others*, [1990] ECR I-3365, paragraphs 17 and 18. The Court cited two previous decisions to support their view, Case 68/88 *Commission* v. *Greece* [1989] ECR 2965, at p. 2984, paragraph 23, and Case 230/81, *Luxembourg* v. *European Parliament* [1983] ECR 255, paragraph 37. See Chapter 6, *post*.
86 Case 2/88, paragraph 22.
87 Case 2/88, paragraphs 24–25.

88 See, Harding (ed.) *Enforcing Community Law*, (Aldershot, 1996), p.8.
89 See, Commission Report 1996, *Monitoring the Common Fisheries Policy*, COM (96), 18.03.1996.
90 Dáil Debates, vol. 298, col. 93.
91 See Symmons, C.R., *Ireland and the Law of the Sea*, (Dublin, 1993), 91–92; see also by the same author, 'Continental Shelf and Fishery Limit Boundaries: An Analysis of Overlapping Zones', *ICLQ* **28** (1979), 703, 733.
92 Case C 62/89 *Commission* v. *French Republic* [1990] ECR I-0925.
93 This issue is discussed further in Chapter 7, *post*.
94 See Chapter 1, *ante*.
95 See, *inter alia*, Case C-9/89 *Kingdom of Spain* v. *Council of European Communities* [1990] ECR I-1383.
96 Opinion of AG Darmon, Case C-9/89 *Kingdom of Spain* v. *Council of European Communities* [1990] ECR I-1383, paragraph 45. Citing the judgement in Case 77/77 *BP* v. *Commission* [1978] ECR 1513 and **Article 5** {10} of the EC Treaty to support this view.
97 See Chapter 10, *post*.
98 See Chapter 8, *post*.
99 See Chapter 12, post.
100 See, further, Schutte, J.J.E., 'The European Market of 1993: A test for a Regional Model of supranational Criminal Justice or of Inter-Regional Co-operation?', in Eser, A. and Lagodny, O, (eds), *Principles and Procedures for a New Transnational Criminal Law* (Freiburg, 1992), 387–413; Sevenster, H.G., 'Criminal Law and EC Law', *CMLRev* Vol. **29** (1992), 29–70; Van den Wyngaert, C. and Stressen, G., *European Criminal Law: A collection of international and European instruments*, (The Hague, 1996).
101 Council Regulation No. 2847/93, **Article 4(2)**.
102 Council Regulation No. 2847/93, **Article 5(a)** also states, 'and such other like means of inspection as may be used by a Member State', this possibly refers to the use of land based radar apparatus. Whether it could be as a justification for the use of submarines for fishery surveillance is unlikely. Canadian authorities have used submarines in the NAFO regulatory area for monitoring the activity of Community vessels operating in the NAFO Regulatory area. See News Release Department Fisheries and Oceans NR-N-94-113. On the diplomatic initiatives to alert the international Community to the dangers posed to fishing vessels by submersed submarines see Symmons, C.R., *Ireland and the Law of the Sea*, (Dublin, 1993), 95–101.
103 The issue of detailed implementation rules is discussed in Section (ii) *infra*. Prior to the 1998 amendment to the Control Regulation **Article 5** was more specific in that it required the establishment of detailed rules for, *inter alia*, inspectors and masters of fishing vessels if an inspector wishes to board a vessel and the procedures for inspection of the vessel, its gear or its catches'.
104 Judgement of the Court (Fourth Chamber) of 18.01.1996 in Case C-276/94. Criminal proceedings against Finn Ohrt. Reference for a preliminary ruling: *Kriminal – og Skifteret, Frederikshavn* - Denmark, [1996] ECR I-0119.
105 Case C-276/94, paragraphs 8–9, *ibid*.
106 Case C-276/94, paragraph 13, *ibid*.
107 *Ibid*.
108 Case C-276/94, paragraph 14.
109 See, *The Marine Times*, 2 June 1997, 5–6.
110 The origin of the word '*comitology*' is the anglophone translation of the French word '*comitologie*' and is alleged to have been first used by Northcote Parkinson, C. in *Parkinson Law* (1958) and defined in his *In-laws and Outlaws* (1962) as the study of Committees and how they operate, see Bainbridge, T. and Teasdale, A., *The Penguin Companion to the European Union*, (London, 1995), 46–48.

111 Case 46/86, *Romkes* v. *Officier van Justitie for the District of Zwolle* (1987) ECR 2671.
112 See Council Regulation No. 3760/92 of 20.12.1992 establishing a Community system for fisheries and aquaculture, *OJ* L 389/1, 31.12.1992, **Articles 17** and **18**.
113 On the delegated legislative powers of the Commission see, *inter alios*, Weatherill, S. and Beamont, P., *EC Law*, (London, 2nd edn, 1995), *passim*; Lasok, D. and Bridge, J.W., *Introductions to the Law and the Institutions of the European Communities*, (London, 1991), 272–278.
114 Vos, E., 'The Rise of Committees', *European Law Journal* 3(3) (1997) 210–229. In the same edition, see, Joerges C., Neyer J., 'From Intergovernmental Bargaining to Deliberate Political Processes: The Constitutionalisation of Comitology', 273–299.
115 EC Treaty, **Article 148(2)** {205(2)}.
116 Council Regulation No. 2241/87, establishing certain control measures for fishing activities, *OJ* L 207 29.07.87, as amended by Council Regulation No. 3483/88, *OJ* L 306, 11.11.1988.
117 Case 87/82, *Lt.Cdr. A.G. Rogers* v. *Darthenay*, [1983] E.C.R. 1579; [1984] 1 C.M.L.R. 135.
118 Case 25/70, *Einfuhr- und Vorratsstelle fuer Getreide und Futtermittel* v. *Koester* [1970] ECR 1161, paragraph 16.
119 Cases 6/88 and 7/88, *Kingdom of Spain and French Republic* v. *Commission*, [1989] ECR 3639.
120 Cases 6/88 and 7/88, paragraph 14, the Court cited Case 46/86 *Romkes* v. *Officier van Justitie de l'Arrondissement de Zwolle* 1987 ELR 2671.
121 Case C-135/92, *Fiskano AB* v. *Commission* [1994] ECR I-2885, paragraph 31 and 32. See discussion on the issue of penalties, Chapter 6, *ante*.
122 Case 25/70, *Einfuhr- und Vorratsstelle* v. *Köster*, [1970] ECR 1161; Case 240/90, *Germany* v. *Commission* [1992] ECR I-5381.
123 Council Regulation (EC) No. 2846/98, **Article 1(19)**, *OJ* L 358/11, 31.12.1998, amending Council Regulation (EC) No. 2847/93, **Article 31**. See Council Regulation (EC) No. 1447/1999 *OJ* L 167/5, 02.07.1999.
124 See Chapter 2 *ante*, and Section (i) *supra*.
125 All standard text books discuss the issues of national measures, in relation to fisheries, see Churchill, R.R., *EC Fisheries Law*, 87–110.
126 Case 804/79, *Commission* v. *United Kingdom*, [1981] ECR 1045. In the absence of Community rules these measures related to cases of need arising from the development of technical and biological facts. However, it may be argued that this category is mainly of academic interest in so far as the Council since 1983 has consistently managed to adopt the necessary management regulations within the prescribed time-limits.
127 There may be specific provisions in Community regulations for Member States to legislate, a pertinent example is **Article 38** of the Control Regulation, discussed *infra*.
128 In Case 87/82, *Lt.Cdr. A.G. Rogers* v. *Darthenay*, [1983] ECR 1579, the Court found that in the absence of Community rules 'it is for the competent court to fill the resulting lacuna'. In this regard national courts would appear to be allowed to uphold national measures, provided that such rules are consistent with the aims of the Community legislation in question. On a similar note, in Case 86-87/84 *Officier van Justitie* v. *Bout* [1985] ECR 0941, the Court upheld national provisions in the absence of a Community regulation.
129 Report from the Commission to the Council and the European Parliament on monitoring and implementation of the CFP, SEC (92) 394 of 06.03.1992, p. 21.
130 See in particular the discussion of quota hopping, Chapter 7, *post*.
131 *OJ* L 20, 28.01.1976, p. 19.
132 On this point see, *Opinion of Advocate General Tesauro* delivered on 10 May 1995 in Case C-44/94 *The Queen* v. *Minister of Agriculture, Fisheries and Food, ex parte National Federation of Fishermen's Organizations and Others and Federation of Highlands and Islands Fishermen and Others*, [1995] ECR I-3115.
133 Joined Cases C-251/90 and C-252/90 *Procurator fiscal, Elgin* v. *Wood and Cowie* [1992] ECR I-2873.

134 Joined Cases C-251/90 and C-252/90, paragraph 17.
135 Joined Cases C-251/90 and C-252/90, paragraph 18.
136 Joined Cases C-251/90 and C-252/90, paragraph 19; but see earlier judgments in Joined Cases 185/78 to 204/78 *Criminal proceedings against J. Van Dam et Zonen and others* [1979] ECR 2345, paragraph 10.
137 Case 379/92 *Peralta* [1994] ECR I-3453.
138 See Case 106/83, *Sermide* v. *Cassa Conguaglio Zucchero* [1984] ECR 4209, paragraph 28; cited in Case C-146/89, *United Kingdom Days at Sea Case, op. cit.*, fn 43.
139 Joined Cases C-251/90 and C-252/90, paragraphs 16–18, *op. cit.*, 133.
140 Case 15/79, *Groenveld* v. *Produktschap voor Vee en Vlees* [1979] ECR 3409, but as regards fisheries see judgement in Case C-9/89, *Spain* v. *Council* [1990] ECR 13863, paragraph 21.
141 The equal access provision in Article 2(1) of Council Regulation 101/76 have to be interpreted in light of the Court decision in Case 223/86, *Pesca Valentia Limited* v. *Ministry for Fisheries and Forestry, Ireland and the Attorney General,* [1988] ECR 0083; Case 281/81 *Anklagemyndigheden* v. *Jack Noble Kerr* [1982] ECR 4053.
142 Report on the General Activities of the European Union 1997, (Brussels, 1998).
143 Case 141/78, *French Republic* v. *United Kingdom,* [1979] ECR 2923 at 2924.
144 In Case 804/79, *Commission* v. *United Kingdom* [1981] ECR 1045 the Court stressed that, in the absence of Community measures, there was an obligation on Member States not only to undertake detailed consultation with the Commission and to seek its approval in good faith but also a duty not to lay down national conservation measures in spite of objections, reservations or conditions formulated by the Commission. In this regard the Court held that in order to meet the requirements of the Decisions of the Council and of the procedure fixed by the Hague Resolution the consultation to be engaged in by the Government of a Member State must, prior to the adoption of conservation measures, allow the Commission to weigh up all the implications of the provisions proposed and to exercise properly the duty of supervision devolving upon it in pursuance of **Article 155** {211} of the Treaty. In this instance, the United Kingdom had failed to fulfil its obligations by bringing into force conservation measures without appropriate prior consultation and in spite of the Commission's objections. Furthermore, that the United Kingdom had contravened Community law by maintaining a system of fishing licences which had not been the subject of appropriate consultation and the authorisation of the Commission, the detailed rules for the implementation of which were reserved wholly to the United Kingdom without it being possible for the other parties concerned to be legally certain how the system would actually be applied, and thus constituted a discrimination against fishermen in other Member States. The requirement to notify the Commission and other Member States is derived from the requirement of legal certainty which is obviously imperative in the fisheries sector where uncertainty may well lead to incidents and the application of serious sanctions. In Case 24/83, *Gewiese and Mehlich* v. *Mackenzie,* [1984] ECR 817, the skippers of two German vessel were convicted of fishing for herring off the west coast of Scotland in contravention of the West Coast Herring (Prohibition of Fishing) Order 1981. On an **Article 177** {234} reference to the European Court, the Court of Justice distinguished *Commission* v. *United Kingdom* on the grounds that in *Mackenzie* approval of the order by the Commission was not necessary as the order was merely the enactment of a previous defective order, which had been previously notified and approved by the Commission. In *Rogers* v. *Darthenay* (discussed above) the skipper of a French vessel argued that since it was established by the Court in *Commission* v. *United Kingdom* that Member States were no longer entitled to exercise unilateral power in the matter of conservation and thus the United Kingdom was not entitled to prescribe the impugned legislation (Fishing Nets (No. 2) Order 1980). The Court noted the exclusive powers of the Community since the end of the Accession transitional period, and thus held in the absence of Community legislation, it is for the national

courts to fill the lacuna in a manner which is consistent with protecting fish stocks and which also takes into account the fact that protecting fish stocks should be permitted. Consequently, Member States are not only obliged to consult the Commission but are also obliged not to take unilateral measures contrary to the views of the Commission.

145 Joined Cases C-251/90 and C-252/90, paragraph 24, *op. cit.*, fn 133

146 Joined Cases C-251/90 and C-252/90 , paragraph 28, *op. cit.*, fn 133.

147 Joined Cases C-251/90 and C-252/90, paragraph 29, *op. cit.*, fn 133.

148 *Opinion of Advocate General Tesauro delivered on 23 January 1992, Procurator Fiscal, Eglin* v. *Kenneth Gordon Wood and James Cowie,* References for a preliminary ruling Joined Cases 251/90 and C-251/90, C-252/90, [1992] ECR I-2873. 149 Council Regulation (EC) No 894/97, *OJ* L 132, 23.05.97, p. 1.

150 Case 26/62 [1963] ECR 1.

151 The national measures in *Kramer* and the *Irish Fisheries Case* (discussed *supra*) did not meet this requirement.

152 Case 804/79 *Commission* v. *United Kingdom,* [1981] ECR 1045.

153 On the position before and after 1979 see Churchill, R.R., *EC Fisheries Law*, 85–110.

Chapter 4
Enforcing the Community Catch Registration System

Introduction
If the Community fishery management system is to work properly it is essential that there are effective measures to ensure the proper monitoring of catches. The *Catch Registration System* is the means by which national authorities record the quantities of fish caught by their vessels and monitor their quota uptake. In the Control Regulation there is a distinct emphasis on the need for accuracy of the *Catch Registration System*. There are 13 Articles in Title II,[1] many of the (former) provisions of the 1987 Control Regulation having been carried over; some of the earlier ones are modified and there are several new provisions introduced to widen the ambit of the Member States' systems in order to provide a better mechanism to monitor the catch of the Community fleet.[2] This is necessary as the TAC/quota management concept, which is the cornerstone of the Communities conservation policy, is totally dependent on accurate catch data. The importance of the national catch registration systems cannot be overstated. Compliance with TACs and quotas is often poor and there have been considerable discrepancies between 'official' and 'actual' catches. If the new provisions introduced by Title II, as well as existing Community legislation on the monitoring and recording of catches, are applied with vigour in the Member States, then it is anticipated that the accuracy of catch data would be greatly improved.

This chapter reviews the key elements of the catch registration system. Section (i) examines the new catch registration system which has the following components; logbooks, landing declarations, sales notes, take-over declarations and transport documents. Section (ii) discusses the fishing effort scheme and Section (iii) looks at the obligations placed on Member States to record the catches by vessels flying their flags.

Section (i) Logbooks, landing declarations, sales notes and take-over declarations

1. *Fishery Logbooks*[3]

Background
Fishermen have always kept written records or logbooks of their fishing trips. Such records were kept secret, because a skipper's reputation and competence were always subject to his ability to find and catch more fish than his fellow skippers. Fishery Logbooks are now a common feature in all management systems and their completion

is the first step in an administrative or paper process to establish when a Member State's allocation of a particular quota is being fished. More specifically, accurate catch data can establish the extent to which quotas are fished or overfished, where fishing activity occurs and the variety of gear utilised, the duration of fishing trips, as well as to record the number and type of fish discarded during a fishing operation. Logbooks are thus a vital element in the management system of the CFP.

The European Community Fishery Logbook was first introduced in 1985. Its purpose is to ensure Community-wide compliance with the conservation measures and to make the monitoring of the quotas allocated to each Member State more effective. In the logbook, skippers are obliged to record the size and composition of the catch for TAC species, the date and the general location of the fishing operation, the type of gear used, the registration number of the vessel and the name and address of the skipper. The data acquired were, in most cases, used exclusively for quota management purposes.

(i) Provisions in the Control Regulation

Article 6(1) of the Control Regulation reiterates the requirement for the skippers of fishing vessels to keep a logbook recording the quantities of TAC species caught, the date and location of the fishing operation, and the type of gear in operation.[4] Up to the year 2000, **Article 6(2)** provides scope for the Council to widen the list of species other than TAC species.[5] Thereafter, any vessel retaining 50 kg live-weight equivalent of any species must record the quantity in the logbook provided that it is taken in waters other than in the Mediterranean Sea. Catches taken in the Mediterranean Sea if exceeding 50 kg live-weight equivalent must be recorded if they consist of any species on a list to be adopted. This is important because conservation measures are no longer restricted to the original list of species adopted when the CFP was established in 1983. However, until the Council adopts new measures, **Article 5** of Council Regulation No. 2241/87 remains in force.[6]

A requirement to record the number of fish discarded at sea for the purpose of evaluation is introduced by **Article 6(3)**. This is part of the effort to provide a global solution to the discard problem and has to be considered in the light of other similar measures in the control and structural policy which are aimed at rationalising the methods of fishing.

Article 6(4) provides a derogation for vessels under 10 metres (a sizeable proportion of the Community fleet) from keeping a logbook, but requests Member States to monitor on a sampling plan basis, the activities of such vessels, pursuant to **Article 6(6)**. The Council may decide by a qualified majority vote other categories of exemptions.

Article 6(8) as amended gives the Management Committee, in accordance with the **Article 36** (Control Regulation) procedure, a mandate for the adoption of detailed rules for implementation of the logbook obligation specified in **Article 6**.

In accordance with **Article 40** Community vessels operating in the Mediterranean shall keep a logbook as from 1 January 2000.

Vessels over 15 metres between the perpendiculars or over 18 metres overall must record fishing effort in their logbook when fishing in the areas defined in Annex I of Council Regulation No. 685/95.[7]

(ii) Commentary
All Community vessels which are 10 metres or greater, including those which operate in the Mediterranean Sea from 2000, are obliged to keep a logbook of their operations. The logbook provisions in the Control Regulations call for a number of comments. Firstly, there have been several cases in the Court of Justice which have clarified the extent of the obligation to complete and retain a Community logbook. For example, the obligation placed on Member States to ensure that skippers on vessels flying the flag of a Member State keep a logbook when fishing for TAC species outside the Community zone has been the subject of judicial examination. In *Commission* v. *Spain,*[8] a case which concerned competence to regulate catches made outside the Community zone, the Court held that the Community was entitled to require Member States to apply control measures (such as, *inter alia:* the keeping of a logbook by skippers; the submission of a landing declaration; the recording by the Member State of all catches and to notify the Commission of the information received from vessels; as well as the prohibition of fishing of certain stocks), to catches made outside the Community fishing zone of stocks subject to a TAC.

Specifically, the Court held that Spain had failed to fulfil its obligations under the EC Treaty by failing to require skippers of vessels flying the Spanish flag, and fishing for a stock or group of stocks subject to a TAC or quota, to keep a logbook indicating the quantities of each species caught, the date and location of such catches by reference to the smallest TAC or quota which had been fixed and administered, and the type of gear used.[9] The Court rejected Spain's argument that the restriction of fishing on the high seas would be ineffective since non-member countries allowed their vessels to operate freely, on the grounds that if the Community gave sole consideration to the stocks found in Community waters, then it would be scarcely effective if those species would not be subject to quotas once they moved outside the 200-mile zone. The Court appeared to be acutely aware of the enforcement difficulties that would arise if they were to find otherwise and justified their decision by pointing out that catches made inside the Community zone could easily be declared as having been made on the high seas.[10] It was, thus, essential to restrict fishing possibilities outside the Community zone in the light of the objectives of the CFP.[11]

Secondly, reliance on the Management Committee Procedure (discussed in Chapter 3) has also come under judicial examination. In *Kingdom of Spain and the French Republic* v. *Commission*[12] a Commission regulation which, *inter alia,* required Community fishermen to keep logbooks in the waters under the jurisdiction or sovereignty of a number of developing countries with which the Community had negotiated 'compensation agreements' (the Community gives financial aid in return for access rights for Community vessels) was held by the Court to be void on the grounds that it did not have an appropriate legal basis. The Commission was thus not entitled to impose controls in relation to fishing carried out in the waters of Countries where no Community rules for the conservation and management of resources were in force. This case was unusual and should be distinguished on its facts. As a rule, the Council clearly legislates the control aspect of external agreements and the Commission does not use the Committee procedure to extend the application of controls to waters under the sovereignty or jurisdiction of third countries. In these agreements there are generally specific provisions requiring Community fishermen to keep a logbook. Community fishing vessels operating in third-country waters may also have to comply with additional catch recording provisions. For example, in the EU–Moroccan Fishery Agreement there is the stipulation that Masters of vessels of 80 GRT or more shall keep the special logbook for fishing in Morocco's fishing zone during the periods that the vessels operate in those waters.[13] In practice EC fishermen complete an EC logbook regardless of which zone they fish in. Furthermore, fishing vessels operating on the high seas in areas regulated by a regional international fishery organisation are required to comply with additional recording provisions agreed under the international convention to which the Community is party.[14] A typical example is the requirement for Masters of vessels to keep a specific International Commission for the Conservation of Atlantic Tuna (ICCAT) Logbook if engaged in fishing for tuna and related species in areas outside Community jurisdiction in the Atlantic and Mediterranean.

A salient feature of the Community logbook provisions is the provision to record the number of fish discarded for evaluation purposes.[15] This is a practical requirement as it is estimated that, globally, between 17.9 and 39.5 million tonnes of fish are discarded each year in commercial fisheries.[16] As there are insufficient data available on the scale of the discards in the Community fleet, the logbook provisions are an effort to address one of the long-standing criticisms of the Community management system, namely that the quantities of fish discarded by vessels are not evaluated. This omission has provoked some commentators to disagree with the Community system for stock management.[17] The problem of discards arises in a variety of situations which vary with, *inter alia*, the methods of fishing and the gears employed, fishing zones and seasons, and the markets in question. The quantity of discards generated by a fishing method depends on its ability to take only fish of a marketable size of the target species but, as the Commission has noted, there are no completely selective fishing techniques at present.[18] The issue of discards has been the subject of much Community institutional comment and the focus of several questions in the European Parliament.[19] The Commission, in line with the 1991 CFP Report, has tackled the issue with a broad range of measures, including, *inter alia*, the presentation of an extensive report to the Council entitled 'Discards within EC Fisheries',[20] which examines all aspects of the problem; reviews available data on the selectivity of fishing gear in specific fisheries; and examines the introduction of economic incentives for fishermen prepared to use more selective gear to achieve substantial reductions in discards.[21] The logbook requirement is part of the effort to provide a global solution to the discard problem and has to be considered in the light of other similar measures in the control and structural policy which are aimed at rationalising fishing methods.

The logbook provisions in the Control Regulation are obviously meritorious and well intended but there may be several questions posed with regard to the validity of a sample plan as a *bona fide* means to record and assess the fishing effort of such vessels. The majority of vessels which are under 10 metres are privately-owned and their individual effort varies enormously depending on several factors including, *inter alia:* whether the owner is a part time or full time skipper; and the location of the vessels, in so far as vessels located in more favourable locations, even on a regional basis, such as the South-East Coast of England will have greater opportunity to fish than those vessels which operate in sea areas which are prone to more inclement weather such as the North and West Coasts of Scotland. Therefore the validity of each sample plan requires careful assessment. On the other hand, the inclusion of a scheme to monitor the catches of smaller size vessels is an acknowledgement that this section of the fleet also catches significant quantities of fish. In conclusion, it may be argued that the sample plan is a compromise between the adoption of a mechanism which provides some coverage of a very large number of vessels and which is capable of indicating catch trends/quantities and a system incorporating a very large number of vessels whose total catches of quota species is relatively insignificant if compared to the quota uptake of larger vessels.

Because the logbook is the basic reference for inspection at sea and in port, there have been many enforcement problems identified which are generally associated with its accurate completion. In particular, there is evidence that there has been double logbook keeping and misreporting of catches and catch area.[22] Furthermore, there has been frequent under-reporting and over-reporting of catches. It must be pointed out in this context that the issue of sanctions for non-compliance with the logbook requirements is not specifically regulated for in the Control Regulation *per se*, other than in the broad provisions prescribed in TITLE VIII concerning the transgression of fisheries law.[23] Although the completion of logbooks is at the very heart of the TAC and quota system, there appears to be evidence to support the view that the lowest fines have been imposed in the Member States for logbook infringements.[24] It should be noted however that in *Commission* v. *Spain*[25] the Court censured Spain for not taking penal or administrative sanctions in order to curb the infringement of rules in force concerning conservation and control measures which included the requirement for vessels to retain and complete a logbook.

(iii) A new logbook format

The Commission is currently drafting proposals to modify the Community logbook. There are several reasons why it is necessary to introduce a new logbook format. Principally, it is because of the legal obligation on masters of vessels, as from 1 January 1996, to record fishing effort in the logbook.[26] The obligation to record fishing effort is a new obligation for Community vessels and is examined separately in the next subsection. It will also be necessary to reconcile the new format with the new regulatory provisions regarding the obligation to record catch data such as the level of discards. In 1998 the Council agreed that any quantity of fish greater than 50 kg of live-weight equivalent of any species retained on board must be recorded in the logbook in areas other than the Mediterranean Sea. After the year 2000 this requirement will also apply to certain species taken in the Mediterranean Sea.[27] The Council is obliged to adopt the appropriate list before this date.

Interestingly, the issue of the list of species to be recorded in the Community logbook arose in the Bandon District Court in Ireland as a result of the detention of a Spanish tuna vessel by the Irish authorities on the grounds that the vessel master did comply with the requirement to retain an EC logbook on board.[28] In the case in question the Court did not impose the standard penalty for a logbook offence on the grounds that, in the absence of a definitive list of species, there was at that particular time no obligation to record tuna in a logbook (tuna has since become a TAC/quota species and there is consequently a requirement to record such catches in the logbook). This case supports the view that prior to 1998, and in the absence of the new logbook format and the relevant regulation the only legal obligation on the master of a vessel is to record TAC/quota species in the logbook.

Any new logbook would probably incorporate many of the existing provisions. This is important because the industry and fishermen in particular, as well as the control authorities in the Member States, have become familiar, over the fourteen intervening years, with the procedure for completing and processing the logbook introduced in 1985. The revised format will have to provide for the many new requirements, such as for recording fishing effort and for recording the precise catches taken during fishing operations.[29] The Commission's endeavours in this regard will be subject to the critical appraisal of Member States and their industry representatives and will have to reflect the concerns of all parties. In the long-term, it may be prudent to introduce an electronic logbook which will be compatible with the satellite surveillance system and the catch recording system ashore.[30] However, this would require extensive preparation and could entail complex legal and data confidentiality issues which may delay such an innovative move in the short-term.

2. *Advance notification if landing outside the flag Member State*

The difficulties created for the Community management and control system by vessels landing their catches outside their flag Member State are addressed by the requirement for vessels to give notice to the competent authorities in the landing location at least two hours prior to such landings and after 1 July 1999, vessels shall give four hours' notice and/or comply with any designated port scheme in force in the Member State where the landing will be made. Pursuant to **Article 7(1)**, this notice must stipulate certain details such as: the location and estimated time of landing and the quantities of each species to be landed. The purpose of notification is to counter the practice of illicit landings and fish disappearing into the so-called *'black market'* and not being recorded on the catch registration system. It may also help to monitor the activities of vessels operating under flags of convenience which have proved particularly difficult to control since the decisions of the Court of Justice in the 'quota hopping cases'.[31]

There may be difficulty in applying **Article 7(1)** in so far as fishermen will not always be in a position to notify the authorities of the exact quantities of fish to be landed, but rather 'the estimated quantities of each species to be landed'.[32] The notification requirement is supported by the provision in **Article 7(2)** that failure to make the stipulated notification will result in the master of the vessel being subject to the appropriate sanction by the competent authority. There is no provision, however, which specifies the severity of the sanction and this remains a matter to be dealt with in national legislation.

The risk of the notification requirement being too onerous for certain sectors of the fleet is diminished by allowing the Management Committee to provide exemptions for certain categories of vessels for a limited or renewable period, or make provision for another notification period taking into account *inter alia*, the distance between the fishing grounds, the landing locations and the ports where the vessels in question are registered or listed. There may be many genuine cases which merit exemption from this provision in view of the amount of additional work entailed for the limited Member State control resources. On the other hand, a valid criticism of the four-hour period is whether it is sufficient notice in some cases, or excessive notice in other instances. Particularly when one considers that, in a country such as Denmark, there are over 700 landing places, many of which are a considerable distance from the nearest inspection authority, the four-hour period may in some instances be inadequate. On the other hand, in the Baltic cod fishery which is often conducted on a daily basis, the four-hour notice would require significant resources to enforce. This may be contrasted with Belgium where there are three established fishing ports[33] and the four-hour requirement may thus be useful.[34] In 1999 the commission adopted a two-hour notification period for community vessels wishing to land catches in ports adjacent to the Baltic Sea area.

3. *Landing declarations*

Background

The second element of the paper trail, the landing declaration, is necessary because the recorded catch in the logbook is not a complete and accurate record of the quantity of fish caught and retained on board. This is because it would be unrealistic for fishermen to weigh each fish caught and then record the exact unit weight in the logbook. In addition, calculating the precise weight at sea can be notoriously difficult. There is currently a tolerance of 20% allowed between recorded catch and the quantities of fish retained on board after the fishing operation. In order to provide an accurate and definitive record of the quantity of fish landed after a fishing trip there is a requirement for a landing declaration to be completed.

(i) Provisions in the Control Regulation

In order to record accurately the quantities and location of catch of the species stipulated in **Article 6(2)**, the master or his representative of each vessel (10 metres or greater) must, within 48 hours of landing, submit a declaration to the competent authorities in each Member State.[35] In this landing declaration, the skipper or agent must record as a minimum, the quantities landed and indicate the location of such catches by reference to the smallest zone for which a TAC or quota has been fixed and administered.

In contrast to the logbook provisions, **Article 8(2)**, the Council on a proposal from the Commission may decide to extend the landing declaration to vessels under 10 metres. Similar to the logbook provisions, it allows the Council acting by way of qualified majority (on a proposal from the Commission), to extend the exemption requirement of submitting landing declarations for certain classes of vessels whose overall length is, or exceeds, 10 metres, provided that this vessel is pursuing certain specific fishing operations.[36] **Article 8(3)** requires Member States to introduce sample plans for vessels that are exempt from the landing declaration requirement and, furthermore, that **Article 36** procedure will be followed for laying down detailed rules for the implementation of **Article 8.**[37]

(ii) Commentary

The requirement to complete and submit a landing declaration is well-established in the Member States and poses few problems from the enforcement perspective. However, the 20% tolerance between the recorded catch in the logbook and the true quantities of fish retained on board after a particular fishing operation can be exploited by certain sectors of the Community fleet for fraudulent purposes. In some cases it provides an easy means to circumvent quota restrictions, particularly for vessels which operate individual vessel quotas. It may be argued that the regulatory requirements provide scope for 20% of the catch to remain undeclared in the landing declaration. This is generally an enforcement problem if there is no physical inspection of the catch by inspectors when the vessel arrives in port after a fishing trip and the catch is landed before the landing declaration is completed. Some Member States such as the Netherlands, however, have taken national measures to counter this practice by requiring landing declarations to be completed by masters as soon as their vessels arrive in port so as to ensure that the allowed difference between the recorded catch and the landed catch (i.e. the 20%) does not disappear into the '*grey market*'.

The second point to be made regarding landing declarations is that in the case of *Commission* v. *Spain*,[38] discussed above in the context of the logbook requirements, the Court held that the Community was entitled to require Member States to apply control measures, such as, *inter alia*, the submission of a landing declaration by skippers in respect of catches made outside the Community fishing zone of stocks subject to a TAC. This requirement is thus well established in Community law.

Provided a fisherman has correctly estimated his catch to within 20% of its true unprocessed weight (live weight) and duly recorded this figure and other entries in his logbook within the required period, and subsequently submitted a landing declaration with the exact quantity of fish landed within the requisite 48-hour period of his fishing trip, then he has discharged his administrative duty and legal obligation to supply accurate catch data.

4. *Sales notes*

Background

The enforcement requirement to monitor fish landed is supplemented by placing additional regulatory responsibilities on the down stream sector of the industry (i.e. auctions and marketing authorities). Within the new fishery enforcement structure, this is an essential element. Particularly as noted in other jurisdictions, collusion between fishermen and fish buyers undermines conservation objectives.[39] Furthermore, white-collar crime could become a serious problem for Community fishery law enforcement authorities. It arises from the fact that fishery products are a source of income and the more product delivered to the market the larger the profit for the fish vendor. Systematic misreporting of catches is normally a result of collusion between fishermen and processors. In New Zealand, the opportunity for this type of offence appears to be greatest amongst small and medium-size vertical companies.[40]

(i) Provisions in the Control Regulation

In order to bring all parties within the control regime in the EU, **Article 9(1)** requires auctions centres or marketing authorities to submit on first sale of fishery products, a sales note to the competent authorities in the Member State. In the event of fish not being marketed in the manner envisaged in **Article 9(1)**, then **Article 9(2)** provides that the onus is on the buyer of fishery products to submit a sales note to the competent authority or other authorised body of the Member State in whose territory this operation has been carried out. As of July 1999, catches landed but not offered for immediate sale, shall not be collected unless (a) a sales note, (b) a transport document, or (c) a take over declaration, is submitted to the competent national authorities.

Furthermore, the fishery product landed should not be removed until the sales note has been submitted and the accuracy of this note is the sole responsibility of the buyer. The responsibility for the accuracy of the take over declaration rests with the holder of this declaration. There is also a provision which exempts buyers who are consumers (i.e. the general public) from the aforementioned measures.[41]

Article 9(3) specifies the information to be provided on the sales note. As a minimum, it states that the following information will be provided for all species: where appropriate, the individual size or weight, grade, presentation and freshness; the price and quantity at first sale for each species and, where appropriate, on an individual size or weight, grade, presentation and freshness basis; the destination of products withdrawn from the market (this includes by-products, products for human consumption, and carry-over products); the name of both the seller and the buyer; the place and date of the sale; the relevant name of each species and its geographical area of origin; where appropriate, the relevant minimum fish size; and where possible the reference number of the sales contract. These requirements are elaborate, particularly those which refer to all species and not just those prescribed in **Article 6(2)** (logbook provisions).

The inclusion of the sub phrase 'where appropriate' may allow a degree of flexibility when furnishing the information stipulated in the sales note.

Information in the sales note pertaining to the external identification and the name of the fishing vessel which landed the fish, the name of the owner or the master, and the port and date of landing are also to be included pursuant to **Article 9(4)** and where appropriate reference to one of the documents provided for in **Articles 13(1)** and **9(4)(b)**. As from 1 July 1999, wherever a sales note does not correspond to the invoice or to a document replacing it, the Member State shall adopt the necessary provisions to ensure that the information on the price excluding tax for deliveries of goods to the purchaser is identical to that on the invoice.[42]

The take-over declaration mentioned above is drawn up by the owners of the landed fishery products or their agents. The declaration must contain the following information: the name of each species and its area of origin; the weight of each species and type of presentation, where appropriate the minimum fish size, the identification of the fishing vessel landing the products, identification of the master, port and date of landing, place(s) where the product are stored, and where applicable reference either to a transport document or a T2M document.

Sales note, take-over declaration, and a copy of the transport document must be sent to the competent authorities in the Member State within a period of 48 hours either by computer or by paper means and they should retain a copy for a year pursuant to **Articles 9(5) and 9(6)**.[43] If the fishery products are landed in one Member State and presented for sale in another, then the transporter of the products must transmit within 48 hours a transport document to the competent authorities of the State where the first marketing takes place. Similarly, where the first marketing of fishery products does not take place in the Member State where landed, the competent authority responsible for the first marketing shall ensure that a copy of the sales note is submitted as soon as possible to those authorities responsible for monitoring the products.

Article 9(7) provides that it is the Commission, in accordance with **Article 36** procedure may grant exemptions from the obligation to submit the sales note for vessels which are less than 10 metres in length. Such exemptions may only be granted in cases where the Member States have installed an acceptable monitoring scheme or for quantities landed of fisheries products not exceeding 50 kg of live-weight equivalent by species. As with logbooks and landing declarations, detailed rules for the implementation of sales note obligations are to be adopted by the Committee procedure pursuant to **Article 36**.

Commentary
The full implementation of **Article 9** may cause some problems in several Member States, because of the diverse nature of the different segments of Member States fishing fleets. For example the United Kingdom has a very high volume of landings from the small vessels artisan fleet, particularly those vessels based in the East and South coasts. Although this fleet is responsible for a large number of the landings its overall recorded contribution to national landings is alleged to be insignificant, in comparative terms. Numerically, in 1992 there were almost 10 000 fishing vessels in the United Kingdom, but 80% of these vessels were less than 12 metres. At the same time there were only 340 vessels over 24 metres. Therefore the administrative burden of processing a sales note every time there is a landing from the small vessel fleet, particularly considering the majority of these vessels land on a daily basis, would no doubt be a major undertaking (if not an insurmountable task) for the national enforcement authority.[44] However, in this context it must be pointed out that **Article 9(7)** provides Member States with scope to lessen the administrative burden by providing exemptions to certain categories of vessels less than 10 metres and for quantities landed of fisheries products not exceeding 50 kg of live-weight equivalent by species.

There is a strong case for the establishment of a standard format for sales notes. This would possibly simplify the administrative task of verifying the information provided, bearing in mind that the marketing of fish outside the Member State of origin is now common practice. Under the present regulation, sales notes can, theoretically, be provided in any one of the Community's 11 official languages. However, national administrative procedures for processing sales notes vary greatly, so a standard document would necessitate adjustment of these systems accordingly. **Article 9(9)** may provide a legal basis for the Commission, using the **Article 36** Management Committee procedure, to establish an EU standard sales note. Perhaps it would be sensible to provide the information in a coded standard manner which would be easily understood in all Member States. This may be a valid approach considering that **Article 9(5)** requires sales notes to be transmitted either on paper or computer to the competent authority. Indeed without having a computer system it is evident that the administrative and logistical burden required to undertake this task is formidable and may well be beyond the capability of several Member States' authorities with their limited resources.[45] For example there are over 250 fish auction houses/markets in Spain.

A legal issue which arises in this context is whether the reference to implementation regulations in **Article 9(9)** could be used as a reason by Member States for delaying the implementation of the general provisions in **Articles 9(3)** and **9(4)**. Or does the absence of detailed implementing rules preclude the enforcement of the general measures prescribed in **Article 9**? The Court of Justice ruling in Case 87/82 *Rogers* v. *Darthenay*,[46] discussed above, provides that as long as the general provisions are independent, perfectly clear, and have immediate effect, then their application does not depend upon the adoption of the detailed implementing rules. This suggests that the absence of detailed implementing rules pursuant to **Article 9(9)** does not preclude the enforcement by Member States of the general provisions prescribed in **Article 9**.

5. *Third-country fishing vessels operating in the Community fishing area*

Background

Under international law, if the Community fleet does not fish the total available/allowable catch in the Community fishery area then, in theory, third-country vessels should be allowed access to excess stocks.[47] Consequently, in principle there is limited access to the Community zone for third-country vessels. In practice, however, access to the Community zone is decided annually after the Council has considered the overall balance between fishing possibilities obtained in third-country waters and those to be allocated in Community waters. The Commission negotiates the agreements which may be for periods of one year or longer. All agreements require the ratification of the Council prior to implementation. Regulations cover the quantities to be fished, the zones where access is granted, rules governing the issuing of licences, and in some cases, certain technical and enforcement measures. The scale of this undertaking may be appreciated when it is considered that in 1995 the Council agreed that the bilateral fishery agreements in 1996 would include, *inter alia*: the arrangement with Norway for catch possibilities for each other's zones and for exploiting joint stocks; similar arrangement with the Faeroes Islands and Iceland; and the Baltic Sea arrangements with Poland, Estonia, Latvia, and Lithuania.[48] All together, the Community granted third-country access in 1996 to 700 000 tonnes of fish in Community waters whilst, in exchange for these quotas, Community fishermen have access to quotas of about 550 000 tonnes in third-country waters. In 1996 the number of authorised third-country vessels was about 500 vessels, the majority of which flew the flag of Norway. In general, the Community fleet tries to gain access to more lucrative stocks in exchanges with third countries. Since the 1970s the Community regulates access by third-country vessels to the Community zone by licences which the Commission issues and administers.[49] In addition, vessel movements (entry and exit) are communicated via coastal radio station to the Commission which informs the enforcement authorities in the Member States concerned. Responsibility for the inspection and enforcement of Community regulations pertaining to the activities of third-country vessels rests with the enforcement authorities in the coastal Member States.

(i) Provisions in the Control Regulation

As noted in Chapter 3, the term third-country vessel is given a broad definition in **Article 28a** of the Control Regulation. It includes not only fishing vessels but all vessels associated with fishing activities. With a view to ensuring that the activity of third-country vessels is closely monitored, the Control Regulation lays down strict requirements to be followed when these vessels operate in the Community fishing area. In 1998 the Council adopted additional regulatory requirements which apply to third-country vessels. In particular, pursuant to **Article 28b**, third-country vessels are obliged to have a licence as well as a fishing permit and are restricted from transhipment operations without prior authorisation from the relevant Member State. Third-country vessels are obliged to keep a logbook containing the information described in **Article 6.**[50] Significantly, all third-country vessels over 24 metres overall will be obliged to be equipped with position monitoring system from 1 January 2000. Prior to this date such vessels will have to comply with a system for reporting movement and catches retained on board as well as the rules on the marking and identification of fishing vessels and their gear. Third-country vessels are also obliged to comply with the instructions of the authorities responsible for monitoring, particularly as regards for inspections prior to leaving the Community fishing zone.

In accordance with **Article 28d**, it is the European Commission which shall determine the date from which catches of a stock or group of stocks subject to a quota taken by third-country vessels shall be deemed to have exhausted the quota. There is an obligation on the Commission to notify the third country and the Member States of this fact without delay.[51] As from this date, fishing for that stock or group of stocks by such vessels or the retention on board, the transhipment and the landing of fish taken after that date shall be provisionally prohibited. The Commission shall also determine the date up to which the transhipments and landings or final catch declarations are permitted.

Stricter rules pertaining to third-country vessels which utilise the landing facilities in Community ports were adopted in 1998. The requirement regarding notification of port and time of landing is extended to 72 hours as opposed to four hours for Community vessels, pursuant to **Article 28e.**[52] Notification must include information regarding the time of arrival at the port of landing, the catches retained on board, the zone or zones where the catch was made, whether in the Community fishing zone, in zones under the jurisdiction or sovereignty of a third-country, or on the high seas. Significantly, the master may not carry out landing operation if the competent authorities of the Member State have not confirmed the receipt of the advance notification. A major restrictive provision is the obligation on third-country vessels only to put into ports designated by the Member States whose landing facilities they wish to use. This does not apply in cases of *force majeure* or where a vessel is in distress. Neither does it apply if there are special provisions in fisheries agreements concluded between the Community and various third countries. With a view to reducing this obligation in the cases where it would be superfluous, the Commission in accordance with **Article 36**, may grant exemptions to certain categories of third-country vessels, or make provision for another notification period taking into account a number of circumstances.[53]

Pursuant to **Article 28f**, the time requirement for submitting a landing declaration remains the same as that required for Community vessels at 48 hours. Each Member State is also obliged to forward to the Commission, at its request, information on landing by third-country vessels.

Article 28g requires where the master or his representative of a third-country vessel declares that catches have been taken on the high seas, the competent authorities shall authorise landing only if it has been proved to their satisfaction by the master or his representative that:

- the species retained on board have been caught outside the regulatory areas of any competent international organisation of which the Community is a member; or
- the species retained on board have been caught in compliance with the conservation and management measures adopted by the competent regional organisation of which the Community is a member.

(ii) Commentary

In general, it may be said that the provisions introduced by the Control Regulation apply without prejudice to provisions provided for in the fisheries agreements between the Community and third countries.[54] Nevertheless, the control measures which apply to third-country vessels are important for several reasons. Firstly, from the point of view of reciprocity it is essential that third-country vessels which have access to the Community zone are subject to similar obligations placed on Member State vessels while operating outside the Community zone. Secondly, it is necessary to apply the same standards to third-country vessels that are applied to the Community fishermen who operate in the Community zone. In evaluating how Community fishermen fare in this context the control and enforcement regime in Norway can be compared with the Community fishery enforcement and control system. The Norwegian enforcement philosophy relies on inspection at sea to allow inspection staff to check compliance with technical measures and to assess the accuracy of the data recorded in the logbook. Moreover, inspectors check that vessels are not discarding unwanted fish. The enforcement task at sea is carried out by the Norwegian coast guard which is specially equipped, trained and organised to carry out fishery enforcement.[55] The Norwegian EFZ is divided up into zones in order to facilitate the inspection of fishing vessels. In particular, vessels which are fishing north of 62° North are obliged to present themselves at designated checkpoints, en route to or from the fishing grounds. The aim of this specific measure is to reduce the possibility of vessels evading inspection. The Norwegians have a wide range of sanctions and penalties at their disposal to deal with vessels which are convicted of breaching Norwegian fishery regulations. Furthermore, Community vessels landing catches in Norwegian ports are rigorously inspected, whereas in the Community up to the amendment of the Control Regulation in 1998 it was easier for third-country vessels to select their port of landing or simply make round trips to their home ports. The designation of mandatory ports of landing will thus simplify the enforcement task for the control authorities in the Member States.

There appear to be little data available regarding the level of unauthorised fishing by third-country vessels in the Community zone. Exceptions include the alleged unauthorised activity by Japanese tuna vessels in the Irish EFZ in 1996.[56] There have been other reports of unauthorised fishing by third-country vessels in areas close to the Azores, Madeira and the Straits of Gibraltar.

The monitoring of the activities of third-country vessels has been the subject of a report from the Commission to the Council in 1996.[57] The report notes that, in theory, advance notification should allow enforcement authorities to carry out timely inspections in order to verify the quantities landed or transhipped in the Community zone. However, in practice the number of inspections of third-country vessels is low in relation to their level of activity. For example, it is reported that about 500 inspections (337 at sea and 172 ashore) have been carried out in 1995, of which the major part was conducted by the United Kingdom. During these inspections, only 11 infringements were detected at sea. In comparison the report notes that the outcome of inspections conducted on Community vessels operating in Norway suggests that the applicable rules are perhaps more thoroughly enforced in the Norwegian zone and that expulsion (withdrawal of licence) is not an uncommon sanction invoked against Community vessels.[58] During 1995 the Norwegian Coast Guard inspected 899 Community vessels and issued a total of 116 warnings, detained 15 vessels for suspected violation of Norwegian fishery regulations and expelled 13 Community vessels, after judicial and administrative procedures from the Norwegian EEZ.

The report suggests a number of measures to improve the monitoring and surveillance of third-country vessels which operate in the Community zone. In particular it advocates the introduction of check-points in three geographical areas in which third-country vessels would be obliged to be present for inspection. The three zones in question are Western Waters, the North Sea (which includes the Skagerrak), and the Baltic. However, the report points out that it falls upon the Member States (Sweden, Denmark, Germany, the Netherlands, United Kingdom and Ireland) to take appropriate measures to ensure that, within their territory and waters subject to their sovereignty and jurisdiction, the number of inspections of third-country vessels is increased and that the appropriate sanctions as prescribed in **Article 31** of the Control Regulation are imposed.[59] Furthermore, the Commission restates its willingness to ensure, where appropriate, licences and special fishing permits are repealed in respect of vessels not abiding by the rules. The Commission's power to invoke this type of sanction is discussed in Chapter 6.

Finally, it ought to be pointed out that third-country vessels which do not have a licence to fish in the Community zone may still land their catches in Community ports. Indeed it may be recalled that in *Diva* v. *Poulsen Navigation*, discussed above in Chapter 3, the Court of Justice held that Community regulations will apply to such vessels once they enter the internal waters of a coastal Member State. In any case, pursuant to the Control Regulation, such vessels are obliged to observe the 72-hour advance notification requirement prior to their arrival in a Community port to land their catch. Moreover, the Community Market regulation, the Directive on veterinary checks and the Directive on health conditions of fishery products,[60] requires third-country vessels using ports in the Member States for the purpose of landing and marketing of their catches to comply with Community health and market legislation. In particular, any fishery products landed in Community ports by such vessels must be accompanied by a certificate of origin (the certificate provides detailed information regarding the origin of the catch and, where applicable, the vessel or vessels from which those products have been transhipped; the quantities broken down by species; the intended method of marketing). Individual fish sizes in the catch must also comply with the minimum size measurement stipulated in Community regulations if such products are destined for sale in the Community market. Landings by third-country vessels are only permitted at the ports designated by the Member States where health and veterinary checking facilities are established.

It should also be mentioned that if the vessels fly the flag of a country which is a contracting party of a regional organisation, there may also be a requirement for such vessels to produce the appropriate logbook pertaining to the catch for the purposes of inspection. Thus, for example, vessels which fly the flag of contracting parties of a regional organisation such as the International Commission for the Conservation of Atlantic Tunas (ICCAT) are obliged to retain the ICCAT fishery logbook on board.

6. *Transhipment (trans-shipping)*

Background
Transhipment is the operation whereby one fishing vessel transfers its catch to another vessel, which is generally a cargo ship or former fishing vessel. The receiving vessels are colloquially referred to as 'klondikers' or 'luggers'. Receiving vessels tend to remain anchored or moored in some coastal area adjacent to the fishing grounds and service the fleets of fishing vessels which harvest the shoals of pelagic fish as they migrate. Transhipment is normally associated with pelagic fleets, but not exclusively so. Major centres for these operations are the Shetland Islands in Scotland, and Lough Swilly and Beare Island Sound in Ireland. The receiving vessels or 'klondikers' traditionally bore the flag of Eastern European countries or the former USSR. They shipped their cargo predominantly to destinations in Africa or Eastern Europe, where the fish meal was used as animal feed stuff or fertiliser. Transhipment of fish can be a major problem for the enforcement authorities in the Member States.[61] The 1983 control regulation was amended, at the suggestion of the United Kingdom, to make it easier to monitor the activities of vessels engaged in transhipment, because of the difficulties encountered in this regard in UK waters.[62] These measures have been further strengthened by the 1993 Control Regulation amended in 1998.

(i) Provisions in the Control Regulation
Article 11 is a framework provision in that it only specifies general requirements on transhipment of fish from one vessel to another. Transhipments and fishing operations involving joint action by two or more vessels taking place in the maritime waters subject to the sovereignty or jurisdiction of a Member State, as well as transhipment taking place in a Member State's ports may be authorised by the port State. The masters of the vessels concerned shall comply with the rules laid down by the Management Committee regarding:

- the definition of authorised places;
- inspection and surveillance procedures; and
- the terms and conditions for recording and notifying the transhipment operation and quantities transhipped.

The provision in **Article 11** shall not apply to pair trawling activities by Community vessels.

Article 12 provides that, where landings take place more than 15 days after the catch, the information required in **Article 8** (the landing declaration) and **Article 11** (transhipment provisions in the 1993 Control Regulation) shall be submitted to the competent authorities of the Member State whose flag the vessel is flying or in which the vessel is registered not later than 15 days after the catch.

(ii) Commentary
The overall aim of the provisions in the Control Regulation relating to transhipment is to ensure that catches taken by Community fishing vessels are not privately loaded on reefers or klondykers and not registered in the catch registration system. As noted above, within the context of the overall fishing activities in the Community, the practice of transhipping is now generally limited to a small number of pelagic fisheries

(herring and mackerel). In many instances, transhipment operations are also closely controlled through national measures which prescribe licence and reporting conditions on vessels transporting pelagic species.

Council Regulation No. 2846/98 (the 1998 amendment to the Control Regulation) provides that Management Committee will adopt detailing rules which will apply to vessels which are involved in transhipment. Pending the adoption of these rules, **Article 11** of Council Regulation No. 2847/93 as well as national rules which are in conformity with Community law, will continue to apply.

7. *Transport documents*

Background

One of the principal reasons for the establishment of the Community was to ensure the elimination of quantitative restrictions upon the import and export of goods and measures of equivalent effect.[63] The jurisprudence of the Court of Justice ensured the free movement of goods in a long line of cases since its landmark decision in the *Dassonville* case.[64] Even before the advent of the single market in 1993, there existed a complex trade and distribution network for fishery products throughout the EU. Hake landed in ports in southern Ireland may be sold within 48 hours in the fish auctions of Galicia and Cantabria. Baltic cod fished in the waters of Estonia, Latvia, and Lithuania may be landed in Sweden or Germany and transported to Denmark for processing in Bornholm and finally offered for sale on a supermarket shelf in Paris. From the enforcement perspective, this practice has several implications. The traditional proximity between the landing places and the place of first sale or marketing is no longer the norm in the industry. The importation of fishery products from outside the Community requires careful monitoring because of the absence of internal border control. Because of the high level of re-export, there is scope for fraud and circumventing many of the provisions in the CFP.

(i) Provisions in the Control Regulation

To overcome some of the problems created by the movement of fish, which is facilitated by the advent of the single market, the Control Regulation introduced the transport document. **Article 13(1)** as amended requires that as from 1 July 1999, all fishery products landed in, or imported into, the Community for which neither a sales note nor a take-over declaration has been submitted, either unprocessed or having being processed on board and which are transported to a place other than that of landing or import, shall be accompanied by a document drawn up by the transporter until the first sale has taken place. The details to be provided on transport documents are the following: an indication of the origin of the consignment (the name of the fishing vessel and its external identification),[65] in case of import other than by vessel, this document shall indicate the location where the consignment was imported; the place of destination of the consignment(s) and the identification of the transport vehicle; indicate the quantities of fish (in kgs processed weight) for each species transported; the names of the consignee and the place and date of loading; as well as the relevant name for each species, its geographical area of origin and , where appropriate, the relevant minimum fish size. **Article 13(3)** reiterates **Article 13(1)**, and stipulates that the transporter must ensure that as a minimum the document must contain the information prescribed in **Article 13(2)**.

Derogations to the above requirement are provided in **Articles 13(4)** and **13(5)**. The former provides that the transporter shall be exempt from the requirements of the transport document if it is replaced by either a copy of the logbook record or the landing declaration or alternatively if it is replaced by a copy of the form T2M (which is a customs document)[66] indicating the origin and the quantities transported. Member State authorities are allowed to grant exemptions with respect to fish being transported short distances, less than 20 kilometres or in the vicinity of a port area, **Article 13(5)**. This is important because many ports in Member States are the location of processing or auction facilities, and in most instances catches only need to be transported short distances. The 1998 amendment to the Control Regulation provides that where fishery products have been declared as sold in accordance with **Article 9** (the sales note provisions) but transported to a location other than the place of landing or import, the transporter conveying the products must, at all times, be able to prove on the basis of a document that a sales transaction has actually taken place.

Article 13(6) requires each Member State to carry out control by means of sampling on its territory in order to verify that the transport obligations are respected. The level of such controls may take into account the intensity of the controls in previous stages.

Obviously, the transport of fish across international frontiers is an area where there is obvious scope for supra national coordination.[67] **Article 13(7)** provides the necessary incentive for Member States to co-ordinate their control activity to ensure that inspection is as effective and as economical as possible. In particular, Member States are required to exercise surveillance over the movement of merchandise which may have been drawn to their attention as possibly being the subject of operations contrary to Community Regulations.

In order to ensure that the transport provisions/obligations in **Article 13** are implemented in a harmonised manner, detailed rules on its application are to be adopted in accordance with the Management Committee procedure in **Article 36**.

(ii) Commentary

The proper functioning of the rules governing the transport of fishery products should ensure that the national enforcement authorities are in a position to intervene in order to prevent the carriage and eventual sale of fish which have not been comprehensively registered at the point of landing or importation or whose precise origin is unclear. The control of transport also offers several possibilities for monitoring fraudulent practices such as illegal fish imports to the Community, and it also provides a realistic deterrent to the marketing and sale of undersize fish. The measures introduced by the Control Regulation focus primarily on two issues: whether quantities of fish are accompanied by a transport document; and, if so, whether the document in question contains the proper information.

The clear intention in the Control Regulation is to allow reasonable flexibility and not to hamper the free movement of goods by overtaxing all parties with the introduction of additional and in some instances unnecessary documentation. Placing responsibility on the national authorities in this regard fits well with the concept of subsidiarity. However, the replacement of the transport document by logbooks or landing declaration does not always achieve the objective of allowing national authorities to trace the origin of the fish being transported, in so far as there may be two or more vehicles used in the transport of a specific catch. In other instances the catch may have been dispatched to different markets in different countries, thus rendering the logbook and the landing declaration less valuable for the purpose of control. Moreover, it is interesting to note that certain responsibility for the transport document is placed on the '*transporter*', a person who may have limited knowledge, if any, of the different species of fish being transported and the regulatory requirements underpinning the CFP. In such instances, however, it is unlikely that the absence of such knowledge would be an adequate defence to a charge of malpractice on the grounds that *ignorantia legis neminem excusat* (ignorance of the law is no excuse).

A further difficulty with this provision is that transport documents need to be provided only up to the point of first sale. This can pose a problem as it may be difficult for enforcement authorities to establish how and when the first sale has taken place.[68] For example, in some pelagic fisheries, the contract for sale is a verbal contract concluded prior to the fish being landed or caught, on the basis of '*no foal no fee*'. However, it must be assumed that verbal contracts do not constitute the first sale as the requirement to provide a transport document would easily be circumvented and rendered nugatory. In this regard one may argue that there appears to be a strong case for requiring a transport document before and after the first point of sale.

As with sales notes, it might be thought that there is a case for developing a standard format for transport documents. This would circumvent language problems and simplify the control task. But again, there are many different procedural difficulties at national level for dealing with transport documents which negate this suggestion. Both **Article 9** (sales note) and **Article 13** (transport document) provide a legal basis for the Commission pursuant to the Management Committee **Article 36** procedure to establish a standard format. The support of the majority of Member States would be required for this to be achieved. Member States, on the other hand, could use the coordination procedure in **Article 13(7)** to standardise or reach consensus on the use of a specialised document to serve the particular needs of fish being transported to a given market such as Irish and UK white fish being transported to the Iberian Market, or French anchovy being transported to Spanish processing plants. Such documents, of course, would need to conform with Community law.

The 1995 Control Report records that the Commission takes the view that the control on transport documents have at best been limited and that they cannot be considered to be of any significance in the overall control.[69] Consequently, the Commission concludes that the monitoring opportunities provided by this measure are not being fully exploited by Member States, particularly in the context of the problem in dealing with undersize fish.[70] This is one of the reasons why the transport document provisions were revised in 1998 to take account of fish size.

8. *The first marketing of fishery products*

Provisions in the Control Regulation

Article 14 provides that all landings by Member State vessels over 10 metres, and third-country vessels landing in the Community zone shall be recorded. This applies to all categories of landings that come within the ambit of **Articles 8, 9** and **10**. For this purpose, Member States may require the first marketing to be conducted by sale at an auction centre. If the catch is not placed on the market for the first time by sale at an auction, as provided for in **Article 9(2)** (the sales note provisions), then Member States shall ensure that the quantities involved are notified to the auction centres or other competent bodies authorised by the Member States.[71]

Because of the vast number of landing places in some Member States and the wide divergence between the marketing infrastructure throughout the European Union, it was essential that the marketing requirements be consistent with the derogations provided in other provisions of the Control Regulation.[72] **Article 14(3)** stipulates that information in respect of the vessels, whose landings are exempt from the requirements in **Articles 7** and **8** (advance notice and landing declaration requirements), or landings in ports which do not have a sufficient developed administrative structure to register landings may, on the approval of the Commission, be exempt from the aforementioned requirement for such information to be processed. Member States requiring exemptions required the approval of the Commission by 1 January 1995. The exemption criteria are applicable if such data registration would create disproportionate difficulties for the national authorities in relation to the total landings and if the respective species landed are sold locally. Lists of ports qualifying under the exemption shall be established by each Member State and notified to the Commission.

There is provision for a sampling plan to evaluate the extent of the respective landings in the qualified ports benefiting from the exemption. **Article 14(4)** requires that this sampling plan must be approved by the Commission before the derogation is granted. So as to ensure that this is not a one-off effort, there is also the requirement that Member States shall regularly transmit to the Commission the results of their evaluations from the sampling plan.

Section (ii) Fishing effort regime

Introduction

As is evident from the discussion in Chapter 2, the management of fisheries exclusively by restricting the quantities of fish exploited through TAC/quota limits for specific species, supplemented by technical conservation measures to protect immature fish, proved to be inadequate to ensure the sustainability of the fishery resources in Community waters during the first phase of the fisheries policy (1983–1992). By 1992 it was also apparent that the Community fleet size was too large and that there was insufficient progress towards its reduction.[73] This led to a pressing requirement to introduce a more comprehensive means to regulate, and manage, access to fishery resources. Specifically, it was evident that the management scheme for certain fish stocks, which in some instances were at the threshold of extinction, required the introduction of additional conservation measures to reduce the amount of time vessels spent operating in such fisheries. In other words, the spiralling trend of resource over-exploitation made it inevitable that the Community would have to introduce a fishing effort regime.

The reluctance of the Community legislators to introduce a fishing effort regime has been dealt with in the specialist literature.[74] There were many political, social and economic reasons which militated against the introduction of a new management tool which would curtail the number of days fishing vessels were allowed to operate outside their home ports. Significantly, the reluctance of the industry to accept any limitation on the time vessels spent fishing was most prevalent in the coastal regions of the United Kingdom, Ireland and France which depended almost entirely on fishing as a means of employment. It is perhaps appropriate at this point to note that a scheme for controlling fishing effort was initially discussed and introduced into the CFP as a management tool for reducing the fishing capacity and structural profile of the Community fleet only after initial attempts to reduce the size of the fleet through vessel withdrawal were unsuccessful. Consequently, the fishing effort regime was conceived as a means to link the structural and conservation components of the policy. The enforcement issues associated with the structural aspect of the policy are examined in greater detail in Chapter 5.

From the Community fishery management perspective, the introduction of a fishing effort regime was also perceived as a means to protect the principle of relative stability as well as a means to facilitate the full integration of Spain and Portugal into the CFP as from the end of the transitional period 1986–1995.[75] The debate to introduce a fishing effort scheme was also contemporaneous with the accession negotiations for new Member States (Sweden, Norway and Finland) in 1994. This resulted in a complex negotiation environment. Despite the political sensitivity of the issues associated with fishing effort, the legislation on the adjustment to the arrangements in the fisheries Chapters of the Act of Accession of Spain and Portugal was adopted in the Council of Ministers on 30 May 1994.[76] In accordance with the Council's wishes the Commission proposed a list of the fisheries which would come within the scope of the fishing scheme as well as the levels of permitted fishing effort. In addition, the regulation prescribes the terms of access to each fishery for each Member State. With a view to ensuring that the fishing effort scheme would be adequately enforced the Council subsequently amended the Control Regulation.[77] This amendment is discussed in Paragraph 7 below.

1. *How does the fishing effort scheme work in practice?*

All Member States were obliged to submit to the Commission by March 1995 the lists of vessels for each fishery in the fishing effort scheme, as well as an assessment of the necessary fishing effort for particular fisheries. For this purpose, the fisheries are grouped into seven categories (demersal, deep water, crabs, scallop, pelagic, migratory, tuna) and into 17 effort zones (see Map).[78] The entire area to which the fishing effort scheme initially applied is generally referred to as Western Waters. A fishing effort regime was adopted by the Council for the Baltic Sea in April 1997.[79] In addition, fishing vessels subject to the effort regime in Western Waters have to communicate catches taken in the various fishing zones.[80]

Practically all the responsibility for the implementation of the fishing effort scheme rests with the flag Member State. In particular, it is the flag Member States which has to provide the Commission with the appropriate details of their arrangements to regulate fishing effort and has responsibility for the day-to-day management of fishing effort. In this regard Member States may increase their allotted fishing effort in a specific fishery by exchanging quotas or by applying to the Commission and the Council to have their fishing effort ceiling increased. The Netherlands succeeded in doing this in 1995.[81] It is important, nonetheless, to emphasise that it is the Council which decides the annual fishing effort level for each Member State and for each fishery. Such decisions are published in Council Regulations which are directly effective and are directly applicable in all Member States. The level of fishing effort is published as a global figure referred to as kilowatt days. This is determined by a simple mathematical calculation which entails multiplying the number of vessels in a given fishery by their cumulative engine power (measured in kilowatts, tonnage is taken into consideration for vessels using static gear) by the number of days spent fishing.[82] Thus in 1996, for example, Ireland was allocated 123 000 kilowatt fishing effort days for demersal species in ICES area VIIa (an area which corresponds to the Irish Sea).[83] It is up to each flag Member State to regulate fishing effort by monitoring the activity of its fleet and by taking appropriate action to ensure the overall limit of fishing effort is not exceeded.

From this brief description, it is clearly apparent that the effectiveness of the management of fishing effort scheme is dependent on, *inter alia*, the measures for inspection and control to monitor the activity of vessels as well as the structural profile of the fleet. It is also evident that the scheme of fishing effort for a fleet as diverse as the Community's entails an elaborate legislative framework.

2. *The legal framework governing the fishing effort regime*

The legal framework which governs the fishing effort scheme is complex. It entails a web of eight Regulations including three amendments to the legislative framework governing the Community control system. The current fishing effort regime has its origins in Council Regulation (EC) No. 3760/92 which, *inter alia*, makes provision for the adoption of specific measures designed to achieve a balance between exploitation rates for certain stocks and the management of fishing effort. The Council, in successive years, adopted a series of specific regulations which cover all the main elements of the fishing effort management regime. Taken together these regulations require Member States to fulfil the following obligations: the management of fishing effort (Council Regulation Nos 685/95 and 779/97); the issuing of special fishing permits; the establishment and operation of a control system (**Articles 19a, b, c** and **e** of Council Regulation (EC) No. 2847/93 as amended, and Commission Regulation No. 1449/98 laying down detailed rules for effort reports); the creation of a fully computerised system in order to manage, calculate and control fishing effort; and the transmission of fishing effort data pursuant to Commission Regulation (EC) Nos 2090/98 and 2092/98. It is not proposed to discuss each of these regulations in great detail. However, with a view to ensuring that the enforcement provisions discussed in Paragraph 7 below are placed in their correct context, it is convenient to illustrate in Table 4.1 the relationship between the various elements in the legal framework. This is followed by a brief review of some of the seminal features in the regulations in question.

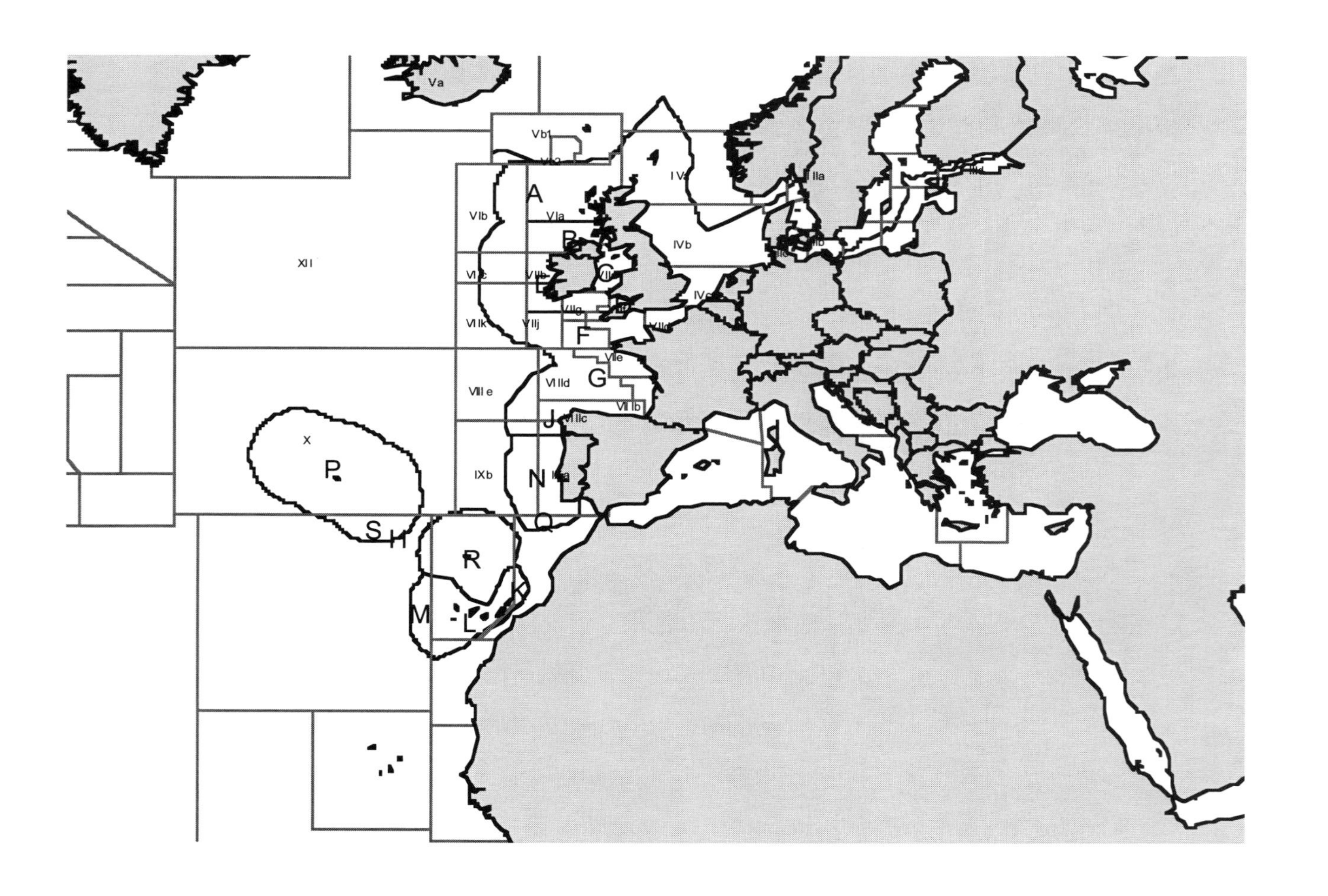
Va
Vb1
A
VIa
VIb
IVb
XII
VIIc
VIIk
VIIj
F
VIIe
G
VIIId
VIII e
VIIIb
J
N
IXb
X
P
S
H
R
M
L
Q

(a)

(b)

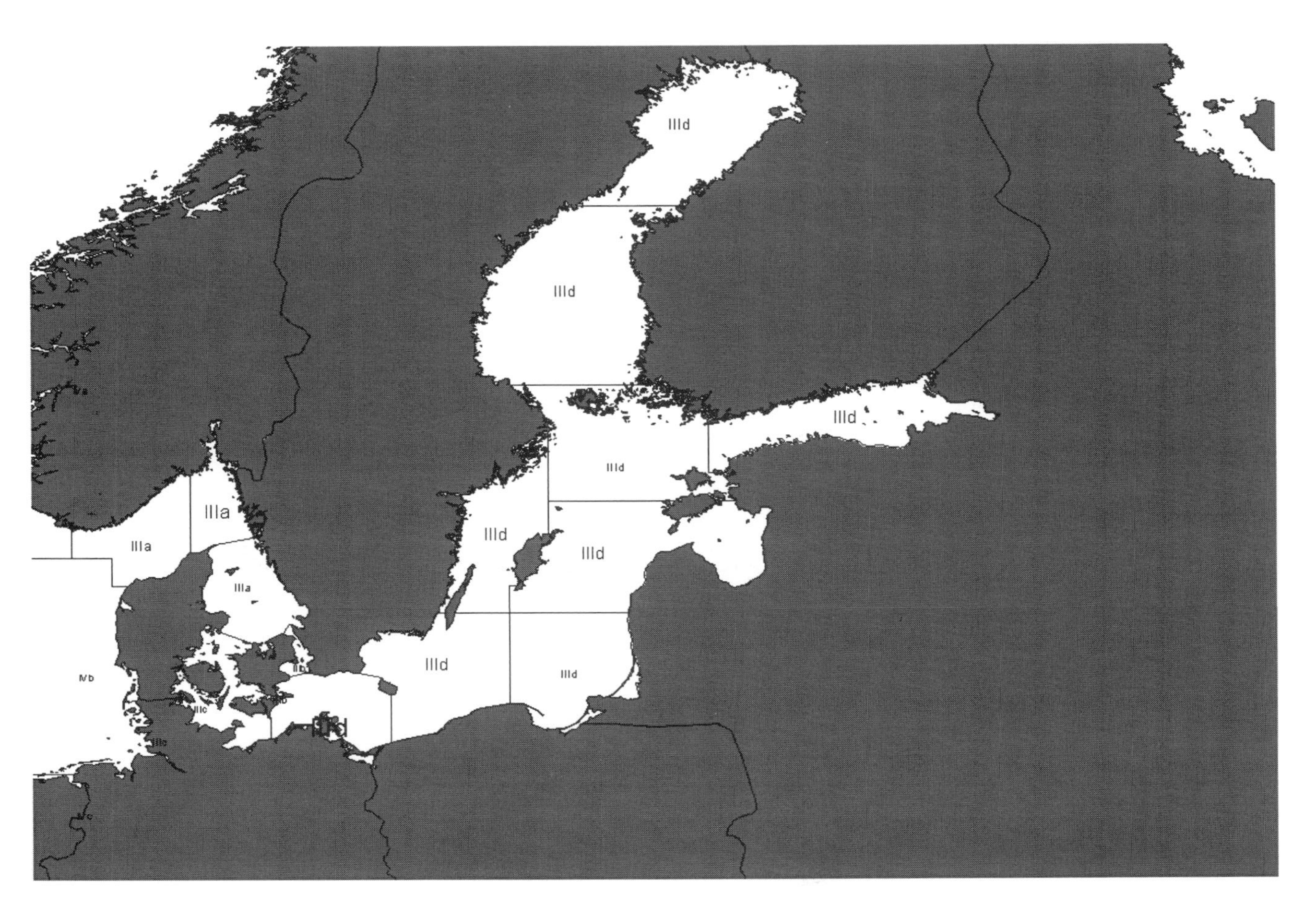

Map of the Fishing Effort Areas in the European Union. (a) Western Waters, (b) Baltic Sea.

Table 4.1 Legal framework for the fishing effort regime.

Management Regulation	Related legislation	Enforcement legislation
The 1993 Basic Regulation Council Regulation No. 3760/92 **Article 3(f)** Definition of fishing effort (capacity × activity)	Areas of application *(i) Western Waters* Council Regulation No. 685/95 *(ii) Baltic Sea* Council Regulation No. 779/97 Regulation of fishing effort by gear and type Council Regulation No. 2027/95 Fleet register and the communication of data Commission Regulation Nos 2090/98 and 2092/98. Fleet segmentation Commission Regulation No. 2091/98 Special fishing permit Council Regulation No. 1627/94	Control Regulation Council Regulation No. 2847/93 amended by Council Regulations Nos 2807/95, 2205/97, 2635/97 Application regulation on 'effort reports' Commission Regulation No. 1449/98 Logbook Regulation Council Regulation No. 2945/95 amending Council Regulation No. 2807/83.

3. *Definition of fishing effort*[84]

Article 3(f) of the 1992 Management Regulation (Council Regulation No. 3760/92) defines fishing effort:[85]

- in the case of a vessel: the product of its capacity and its activity;
- in the case of a fleet or a group of vessels: the sum of the fishing effort of each individual vessel.

The capacity of a fishing vessel is further defined as:

- in the case of a vessel using towed gear: the installed power expressed in kilowatts (kw);
- in the case of vessels using static gear: the installed power expressed in kilowatts (kw) and tonnage.

Activity of a vessel is evaluated on the time spent annually in a particular effort zone.

4. *The* ratione loci *of the fishing effort scheme*

Essentially, the fishing effort scheme applies to all waters referred to as Western Waters and certain parts of the Baltic Sea.

(i) Western Waters

The term Western Waters is not significant in the maritime legal sense. It is a term, used colloquially, to describe the areas referred to in Council Regulation No. 685/95 for the purpose of implementing the fishing effort regime.[86] This regulation defines the area of application of the effort management scheme by reference to International Council Exploration of the Sea (ICES) statistical divisions. More specifically, it refers to ICES divisions Vb, VI, VII, VIII, IX and X and CECAF areas 34.1.1, 34.1.2 and 34.2.0 as the area of application (see Map).[87] Thus, for the purpose of Community fisheries law, there is a link between ICES divisions and the 17 fishing effort zones. For illustrative purposes, the 17 fishing effort zones are alphabetically listed as A to S on the Map.[88] A detailed examination of the Map reveals, however, that in some instances the fishing effort zones bridge several ICES division areas. For example, the fishing effort zone G includes ICES divisions VIIIa, VIIIb and VIIId. Overall the effort management area includes nearly all the fishing grounds (both within and outside the sovereignty and jurisdiction of the Member States) in the north-east Atlantic. It does not apply to the North Sea or the Mediterranean Sea.

(ii) The Irish Box

The effort management area also includes the area west of Ireland and the United Kingdom commonly referred to as the *Irish Box*. This area is predominantly in Irish jurisdiction and is the area south of latitude 56°30′ North, east of longitude 12° West, and north of latitude 50°30′ North.[89] Under the adjustment of the Iberian accession arrangements 40 Spanish vessels were given access to this area in 1996.[90]

(iii) The Baltic Sea

The Baltic Sea Fishing Effort Regime applies to all areas which are within the International Baltic Sea Fisheries Council Area (subdivisions 22 to 32) falling under the sovereignty or within the jurisdiction of the Member States.[91] See Map.

5. *The fleet register*

The register contains the data on each Community fishing vessel resulting from census undertaken by each Member State on its fleet from 1989. A detailed description of the Community fleet register is provided in Chapter 5. However, it is pertinent to briefly mention the significance of the fleet register in the context of the fishing effort regime. All the information regarding the vessels which are subject to the fishing effort scheme is stored electronically on the fleet register which is located in the European Commission (DG XIV) in Brussels. There is an express obligation placed on the flag States to transmit such data to the Commission where they are made available to the relevant coastal States responsible for enforcement purposes.[92] These data are not made available to other Member States without the consent of the flag Member State.

6. *Fishing permits*

Vessels are allowed access to specific fishery for deep water species if they are included on the list of vessels which is submitted to the Commission by the flag Member State. During the period vessels operate in this fishery they are required to retain a permit for the fishery in question in addition to having a Community fishing licence. In practice, some Member States have attached the permit to the licence. The fishing permit must contain the data specified in Council Regulation No. 1627/94.[93]

7. *Provisions inserted into the Control Regulation as a result of the adoption of fishing effort regime for Community waters*

It would be pointless to introduce a fishing effort scheme without introducing appropriate rules to ensure compliance with the effort scheme and to monitor both the movement of vessels and the catches retained on board. The Control element is dealt with in several regulations:

- Council Regulation No. 2807/95 which amends Regulation No. 2847/93.
- Commission Regulation No. 1449/98 (special rules in relation to effort reports).
- Regulation No. 2945/96 which amends Council Regulation No. 2807/83 (the log-book regulation).

The fishing effort scheme initiated in 1996 for Western Waters applies to all vessels over 18 metres overall length (15 metres between the perpendiculars).[94] The control requirements are specific and vary considerably depending on the fishery, demersal or otherwise,[95] and the type of permit held by the vessel. In the interest of conciseness, it may be more appropriate to examine the provisions in the Control Regulation in the context of the obligations it places on the various parties. These obligations may be classified into three categories.

(i) The obligations placed on the master of fishing vessel
There is an obligation placed on the master of each fishing vessel over 15 metres to communicate to the control authorities an effort report before and after each entry and exit to a fishing effort area.[96] In this respect there are special provisions for vessels engaged in trans-zonal fisheries and which spend less than 72 hours at sea.[97] Interestingly, if the vessel is fishing in waters under the sovereignty or jurisdiction of a coastal Member State then effort reports have to be transmitted *simultaneously* (emphasis added) to the competent authorities in the coastal State responsible for control as well as the control or enforcement authorities in the flag State. In Annex IV of Commission Regulation No. 2945/95 there is a list of the names and addresses responsible for monitoring the maritime waters in the Member States.

Vessels which are equipped with satellite surveillance tracking systems do not have to communicate an effort report because, as explained in Chapter 11, these vessels are tracked continuously while at sea.[98]

There is also an obligation placed on the Master of the vessel to record in the logbook the movement of the vessel under the Master's command to and from port. In relation to vessels which are fishing passive gear or static gear (gill nets, tangle nets, etc.) there are specific requirements regarding the data to be communicated.[99] There are also precise requirements regarding the information to record in the fishing logbook.[100] There are also special provisions for vessels exempted from keeping a logbook.

There are restrictions regarding the type of gear retained onboard and the stowage of fishing gear.[101]

(ii) Obligations placed on the flag Member States

The flag Member State is obliged to forward data to the Commission regarding the vessels which come within the scope of the fishing effort scheme.[102] This information is registered in the Community fleet register. All Member States are obliged to account for the amount of fishing effort undertaken by the vessels flying their flag in each effort zone and in the area referred to as the Irish Box.[103] In order to allow the Commission to monitor the utilisation of fishing effort, the flag Member States must inform the Commission every month of the aggregate effort data for vessels fishing demersal species. There is a similar requirement each quarter in relation to vessels fishing pelagic species.[104] Furthermore, the flag Member State is bound to notify the Commission of the date on which all fishing effort has been utilised in a particular fishery.[105] It is the Commission, however, which notifies Member States, including the coastal Member, of the utilisation of the fishing effort allowance. In order to reduce the possibility of late notification, the flag Member State is obliged to notify the Commission when the fishing effort is 70% exhausted, as well as the legislative and other measures it is about to take to ensure that the maximum threshold of fishing effort is not exceeded.[106]

(iii) Power of the Commission

The overall responsibility for compliance with the fishing effort level rests with the Commission.[107] Consequently it is the Commission which has the power to fix the date on which fishing effort levels are deemed to have been reached.[108] Thereafter, vessels flying the flag of a Member State are prohibited from undertaking fishing activities in the relevant fisheries.[109]

8. *Commentary on the enforcement provisions*

The control and monitoring of fishing effort is a significant development in the law of fisheries in the European Community. It entails a significant administrative and enforcement commitment from the Commission, the national authorities in the Member States and the industry. In the overall context of the CFP it is essential that the fishing effort scheme works effectively in order to improve the management of fishery resources. It is also important to recall that within the planned cutbacks in the size of the Community fleet (Multi Annual Guidance Programme MAGP IV) there is a provision whereby Member States can effect part or all of their fleet reduction through means of a diminution of fishing effort.[110] It is reported that Germany, France, the Netherlands, Ireland and the United Kingdom will select this option.[111] In order to retain a strong link between the conservation and the structural policy there has to be rigid adherence to the fishing effort limitations and effective monitoring of vessel movements and activity.

The fishing effort regime depends upon effective enforcement and control measures to ensure that there is adequate compliance with Community regulations. On the one hand, it could be argued that the regime introduced in the Community supports the prerogatives of the flag Member State on the grounds that it is the flag Member State which retains responsibility for regulating fishing effort and for issuing fishing permits.[112] On the other hand, there is also an obligation on the flag Member State to notify the Commission of the relevant data relating to the utilisation and management of fishing effort. Moreover, the vessel data on the fleet register are made available to the appropriate coastal Member State responsible for ensuring Community rules are enforced. In this regard there appears to be a dual sharing of responsibility between the flag Member State and the coastal Member State. Furthermore, the fact that in some instances the masters of vessels have to report simultaneously their vessel movements to both the coastal Member State and the flag Member State supports the view that the traditional distinction inherent in the international law enforcement model between flag State and coastal State responsibilities is no longer rigidly adhered to in Community fisheries law.[113] In this respect the fishing effort regime places definitive obligations on all parties.

It is also evident that the fishing effort regime entails the communication of significant amounts of data and that the best means to achieve efficient implementation is to rely on satellite tracking for the purpose of monitoring the movements of vessels. Furthermore, for the fishing effort regime to be successful it is also essential that the flag Member State submits accurate vessel data to the Commission. The issues and problems associated with the latter obligation are examined in Chapter 5.

Section (iii) The management of national quotas

Introduction

As is evident from the discussion in Section (i) above, the monitoring of logbooks and landing declarations is a time consuming task for the national authorities. **Article 8** (landing declarations), **Article 9** (sales notes), **Article 28** (third-country vessels) and **Article 14** (marketing), provide the necessary mechanism to record the data regarding the amount of fish being landed in the Community zone. However, if this information is to be of value it must be presented and monitored so as to ascertain what is the quota uptake by each Member States. It is thus essential that all data on completed logbooks and landing declarations must be entered in the national register and used as the basis for the monthly return of catch uptake to the Commission. The Community system for catch registration is intrinsically linked to the TAC/quota mechanism, relative stability and the provisions relating to access. Therefore, from the conservation perspective, the importance of the Community catch registration system cannot be overemphasised. Although **Articles 15–19** (inclusive) establish the requirements regarding the way data are registered and transmitted to the Commission, it must be stressed that in the past catch statistics presented by some Member States were not in all cases based on logbook and landing declarations data but on other more global figures supplied by the fishing industry. Thus, one of the key objectives of the Control Regulation is to improve the methodology and accuracy in collecting and transmitting national catch data. Emphasis is placed on computerised systems after years of receiving catch data on monthly printed lists.

The aim of this section is to examine the obligations placed on the Member States to record the quantities of fish caught by national fishing sectors. Specifically it looks at catch registration, inter-Member State communication, databases, and the procedures for the suspension of fishing. An important point to emphasise at the outset is that the utilisation of quotas is a matter for the Member States. The 1993 Management Regulation states that Member States shall inform the Commission each year of the criteria and the rules they have adopted for the distribution of quotas.[114] The criteria by which Member States divide up their quota, were upheld in 1982 by the Court of Justice in *Anklagemyndigheden* v. *J. Noble Kerr.*[115] In this case *Kerr* (the skipper of a UK vessel) argued that the Danish Government establishment of a TAC for shrimp in Greenland waters and the allocation of quotas to Greenland, Danish, Faeroes and French vessels were discriminatory and contrary to **Article 7** {12} of the EC Treaty and **Article 2(1)** of Regulation No. 10/76 (national measures must be non-discriminatory). The Court in referring that such measures were for the purpose of conservation and fishery management, upheld the Danish measures on the grounds that fixing quotas was the most effective way of observing compliance with the TAC and that the criteria used by the Danish *Fiskirministry* in allocating quotas were based on a Commission proposal. One authority has asserted that *a fortiori* the Court would have upheld a similar Community measure, noting by analogy that the Court had upheld the quota system for sugar production.[116] In *Officier van Justitie* v. *Kramer*[117] (*Kramer* discussed below) the court held that fishery quotas do not constitute a quantitative restriction or a measure having equivalent effect.[118] It is also important to emphasise that the rules for the distribution of quotas needs to accord with the general principles of Community law and the CFP. Thus, for example, the Court has held that Member States cannot lay down rules for the utilisation of quotas without regard to their own obligation such as the duty to undertake the appropriate enforcement measures, particularly the provisional closure of fishing, when quotas are exceeded.[119]

1. *Monitoring of quota uptake by the Commission*

(i) Provisions in the Control Regulation

Article 15(1) provides that, before the 15th day of each month, all Member States shall notify the Commission by computer transmission of the quantities of each stock or group of stocks subject to the TACs or quotas landed during the preceding month.[120] They shall also inform the Commission of any information regarding transhipment.[121] This notification must indicate the location of the catches as laid down in vessel logbooks and landing declaration[122] and the nationality of the vessel concerned. As the Member States are compiling this information throughout the year they are obliged to inform the Commission of a quota consumption forecast which anticipates the date of exhaustion of 70% of the allocated quotas. This, of course, is only in respect of species fished by the vessels of the Member State in question. The Commission reserves the right to request more frequent information other than the monthly reports in cases where catches of stocks or groups of stocks subject to TACs or quotas are approaching TAC or quota levels.

This flow of information is two-way, as **Article 15(2)** requires the Commission to keep available for Member States, on computer, the notifications received on quota uptake.

The carrot-and-stick approach, which is a feature of the general approach in the Control Regulation, is also evident in **Article 15(3)** which states that, if Member States have not complied with the dead-line for transmitting data on monthly catches, then the Commission may set the date on which the quota allocation by that Member State shall be deemed to have been 70% exhausted and may set the estimated date on which the quota, allocation or available share may be exhausted. The Commission's estimate will be based on the historical data covering preceding years. This power of the Commission has not been tested by the Court. The vesting of this power in the Commission is a response to the reluctance of Member States, possibly as a result of inadequate national catch registration systems, to provide data to the Commission in good time.

There is also an obligation on Member States, pursuant to **Article 15(4)**, to provide notification of the quantities of stocks other than TAC and quota stocks landed during the preceding quarter. This must be done before the end of the first month in each calendar quarter.

2. *Inter-Member State transmission of landing information*

Background

Considering the size of the Community fishing area and the propensity of certain sections of the Community fleet to work away from their home ports, frequently in waters under the jurisdiction or sovereignty of other Member States, it is essential that the flow of information regarding the activities of these vessels continues on a bilateral horizontal level between Member States as well as on the vertical level to the European Commission.

(i) Provisions in the Control Regulation

Article 16(1) provides the legal basis for Member States to request from other Member States data on landings, transhipment and offers for sale in respect of a stock or a group of stocks subject to quotas by their vessels in the ports and water within the second Member State's jurisdiction. This is without prejudice to the requirement in **Article 15**. There is provision for the supply of detailed information regarding particular vessels, such as the name and the external identification of a vessel, quantities and dates, of fish landed, transhipped or offered for sale. This information must be provided within four working days unless otherwise determined between the relevant Member States. The Commission has access to this information, should it so choose, pursuant to **Article 16(2)**.

(ii) Commentary on the measures relating to the transmission of landing information
Inter-Member State transmission of landing information is particularly important with respect to flag of convenience vessels which have, since the adoption of the TAC Regulation in 1983, exploited the tolerance in Member State national legislation regarding the granting of nationality (i.e. registration) to fishing vessels. The 'quota hopping fleet' lands the greater part of its catch in ports outside the flag Member State.[123] Consequently, the verification and exchange of accurate landing data is an essential element in monitoring the activities of these vessels.

3. *Monitoring catches of Member State vessels operating in waters outside the Community zone*

(i) Provisions in the Control Regulation
Member States have to take all necessary measures to ensure monitoring of all the catches of species made by their vessels in waters subject to the sovereignty or jurisdiction of third countries and on the high seas, and to ensure verification and recording of transhipments and landings of such catches, **Article 17(1)**.

The control and verification measures envisaged are stipulated in **Article 17(2)**. Such measures shall ensure compliance with three obligations on the vessel owners and/or masters. Firstly, that a logbook will be kept on board the fishing vessels in which the masters shall record their catches. Secondly, that landing declarations shall be submitted to the authorities of the Member State of landing during landings carried out in Community ports. Finally, that the flag Member State shall be informed of the details of each transhipment of fish on to third-country fishing vessels and of landings carried out directly in third countries.

The provisions in **Article 17** apply without prejudice to the provisions of fisheries agreements concluded between the Community and third countries and international conventions to which the Community is party.[124]

In order for the information obtained from the Member States regarding the quantities of fish caught in the waters referred to in **Articles 17(1)**, and received pursuant to **Articles 17(2)**, to be registered, **Article 18(1)** specifies that each Member State is obliged to provide such data to the Commission before the end of the first month in each calendar quarter.[125] For catches made in third-country waters, such information notified shall be broken down by third country and stock by reference to the smallest statistical zone defined for the fishery in accordance with **Article 18(2)**. A similar provision exists for notification of catches made on the high seas. These shall be defined by reference to the smallest statistical zone defined by the International Convention governing the catch location and by species or group of species for all the stocks in the fishery concerned.

Article 18(3)b continues the trend of free access to information in so far as it requires the Commission to make available the information it receives to the Member States pursuant to **Article 18**. Detailed rules for the application of **Article 18** are to be adopted in accordance with the **Article 36** Management Committee procedure.

(ii) Commentary
The legal competence of the Community extends beyond the Community zone, as

noted in Chapter 2. The provisions for monitoring catches by Member States fishing vessels in areas outside the Community fishery limits are set out in **Articles 17** and **18**. These provisions assume greater importance as the Community fleet is forced to pursue fishing possibilities outside the Community zone. Community fishing vessels have access to other countries' waters on the basis of a number of arrangements which include, *inter alia*, long-term bilateral agreements. By 1998, the European Union had negotiated over 20 fisheries agreements with third countries. The most financially expensive, and arguably the most important, are those agreements with Morocco and Mauritania which give Community vessels access to stocks in return for payment. Certain Member State vessels fish stocks on the high seas, most notably the high-seas albacore tuna fishery in the North-East Atlantic, stocks of ground fish in the NAFO regulatory area in the North-West Atlantic, and pelagic stocks in the Barents Sea. In general, all Community vessels which operate in international waters have to comply with the terms of international conventions which have been ratified by the Community or their flag Member State. Moreover, Community vessels have to comply with the rules and regulations prescribed by international regional fishery organisations. The problems in relation to meeting these requirements have been under international scrutiny, particularly in the context of the EU and Member States obligations to NAFO (North Atlantic Fishery Organisation). Some of the issues raised are discussed in Chapters 10 and 11.

4. *Databases*

Background

The control regulation has created a web of administrative paper requirements in relation to obtaining accurate catch data. **Articles 3**, **6** (logbooks), **8** (landing declarations), **9** (sales notes), **10** (third-country notifications), **14** (marketing) and **17** (international waters and third-country waters), have all placed a heavy administrative burden on Member States and the Community fishing fleet. It is one of the reasons that critics of the CFP express the point of view that the policy is over-regulated. Nonetheless, modern computer systems and information technology applications provide one of the solutions to improving the efficiency and effectiveness of fishery administration in the Member States.

(i) Provisions in the Control Regulation

In order to ensure respect for catch recording obligations and the large volume of records, each Member State is required, pursuant to **Article 19(1)**, to establish a validation system comprising of particular cross-checks and verification of data resulting from the obligations created by the Control Regulation. Specifically, each Member State, pursuant to **Article 19(2)**, is obliged to create a computerised database to register all catch data collected in fulfilment of their legal obligations. In this respect Member States are allowed to create decentralised databases on condition that these and the procedures relating to the collection of data are standardised so as to ensure compatibility between them throughout the territory of the Member State.

In acknowledgement that databases cannot be conceived and implemented without some delay, there was provision for a transitional period of three years for Member

States that cannot comply with the requirements of a database for all or part of its fisheries. This transitional period expired on 1 January 1997 and the amended provision adopted by the Council in 1998 records that each Member State shall adopt the necessary measures to ensure that the information is entered into databases as soon as possible. Member States are obliged to enter data from logbooks, the landing declarations, sales notes, take-over certificates, within 15 working days of the date in which this information is received by the competent authorities. Importantly, if more than 85% of a quota has been fished, this period shall not exceed five working days.

Member States shall adopt the necessary measures to facilitate the collection, validation and cross-referencing of data. The Commission shall have remote access to duplicated computer files containing the relevant information on the basis of a specific request.

(ii) Commentary on the cross-verification procedure

The complexity of the cross-verification procedure is widely acknowledged. Experience in other jurisdictions indicates that this may lead to problems. For example, in New Zealand the fishery management system has changed to a system based on individual transferable quotas (ITQs) which requires extensive cross verification between catch volumes by individuals and reconciliation against allocated ITQs. Although this system is alleged to be sophisticated, it has transpired that subsequent investigations of suspected infringements of the system are more complex, and resulted in longer and more expensive court cases. It has been asserted by one authority that there has only been a modest increase in the number of prosecutions, and the overall cost of compliance relative to the value of production has declined.[126] In the United States it has been discovered that the final barrier to detecting illegal activity may depend on the ability of the authorities to mount a complex investigation. From a simple economic perspective, fishing is a source of income and the more product delivered, the larger the profit margin. Much of the fishing industry operates on a 'cash' basis, and, consequently, uncovering illegal activity will be extremely difficult without the help of a sophisticated unit to undertake the appropriate investigation. This unit should be capable of spotting illegal activity which is large enough, or complex enough, to have escaped detection at the lower levels of enforcement.[127]

5. *Confidentiality of data*

Background

It is obvious that if the new control system introduced by the Control Regulation is to function properly it will entail the transmission of large amounts of data regarding the activities of not only fishermen, but also many others who work in the fisheries sector. The transmission of data poses several problems for the European Community. Data-protection rules adopted by the Community pave the way for the full development of cross-border information activities and services.[128] Prior to this, different data-protection laws have impeded the single market, as some Member States had resisted exchanging essential information with other Member States whose national legislation provided a lower level of data protection. It is envisaged that the new rules adopted by the Council will mean that any person whose data are processed in the European

Union will be afforded the equivalent level of protection of their rights, in particular their right to privacy, irrespective of where the data processing is carried out.[129] Traditionally, free movement of data is particularly important for services which have a large customer base and depend on processing personal data, such as banking, insurance, and distance retailing. In this regard individuals, particularly private parties, will have the right pursuant to Community legislation to both access personal data and to know where it originated. The high-level protection is afforded through the obligations imposed on data controllers such as public authorities, enterprises and associations. The obligations on data controllers relate, for example, to the quality of the data - the processing of which must meet a specific and legitimate purpose - to security requirements and the notification of such processing to an independent supervisory authority set up in the Member States. The measures on data confidentiality in the Control Regulations have thus to be considered in the context of the Directive 96/9/EC of 11 March 1996 on the legal protection of databases;[130] Directive 95/46/EC on the protection of individuals with regard to the processing of personal data and on the free movement of such data;[131] and Commission Directive 94/46/EC of 13 October amending Directive 88/301/EC and Directive 90/388/EC in regard to satellite communications.[132] However, it may be some time before it is evident how this general data protection legislation and other specific provisions, such as **Article 37** discussed below, will inter-relate.

Provisions in the Control Regulation on data confidentiality

Other than the recent Council Directive which deals exclusively with establishing a Community standard on data protection, the Commission provides extensive guarantees in **Article 37** to the Member States to take all the necessary steps to ensure that the data received in the framework of the Control Regulation shall be treated in a confidential manner. In particular, **Article 37(2)** provides that the names of natural and legal persons shall not be communicated to the Commission or to another Member State except in the case where such communication is expressly provided for in the Control Regulation, or if it is necessary for the purposes of preventing or pursuing infringements or the verification of apparent infringements.[133] The Control Regulation data, referred to in **Article 37(1)**, shall not be transmitted unless it is aggregated with other data in a form, which does not permit the direct or the indirect identification of natural or legal persons. Pursuant to **Article 37(3)** the data exchanged between the Member States and the Commission shall not be transmitted to persons other than those in Member States or Community institutions whose function requires them to have access to it unless the Member State transmitting the data gives their expressed consent.

Professional secrecy applies to data acquired or communicated by virtue of the control regulation. It benefits from the same protection as given to similar data by Member States legislation and by corresponding provisions applicable to Community Institutions. **Article 37(5)** provides that such data are not to be used for purposes other than those provided for in the Control Regulation without the express consent of the appropriate authorities. Furthermore, it may also be subject to conditions in national legislation which prohibits the authority receiving the data from such use or communication.

The criteria for the use of data are further complicated by **Article 37(6)** which, without prejudice to international conventions concerning mutual assistance between countries, endeavours to ensure that the confidentiality requirement is not used for purposes other than those that are stated in the regulations (i.e. data confidentiality). Specifically, it provides that the aforementioned requirements **Articles 37(1)–37(5)** shall not be used as a means to negate, or in the express wording of this provision as an 'obstacle' in the framework of legal actions or proceedings subsequently undertaken for the failure to respect Community fisheries legislation. The competent authorities in the Member State are required to be informed of all instances where the said data are utilised for these purposes.

Article 37(7) provides for the notification to the Commission of the good name of a particular natural or legal person after investigation has established that this person has not been implicated in any infringement. The Commission is then obliged to pass this information to the party or parties to whom it has communicated the name of the said person, and to ensure that the data stored against this person are deleted without delay.

The data-confidentiality measures are to operate without prejudice to the Commission's ability to publish general data or studies without reference to natural or legal persons. Furthermore, data should only be stored in the form allowing identification of the person for as long as possible, and any data received further to the Control Regulation shall be available to the natural or legal person concerned.

Commentary

It must also be pointed out that there were no similar provisions on data confidentiality in the 1987 Control Regulation. The data-protection provisions in the Control Regulation are elaborate and appear to provide a guarantee to private parties that their rights will be respected. One of the key characteristics of the new measures is the balancing of this guarantee against the right of instituting legal actions in cases of suspected infringements of Community fishery law. In this regard the Court of Justice has always been acutely aware of the fundamental requirement to protect the rights of individuals. Thus, for example, the Court in *Fiskano (Swedish fish licence* case) emphasised the importance of this right to be heard as a fundamental principle of Community law which must be guaranteed even in the absence of any rules governing the procedure in question.[134] Similarly, should the need arise to present a summary of Commission inspector reports in the Court of Justice for the purpose of **Article 169** {226} of the EC Treaty proceedings, the Court accepted that it may be necessary to remove the names of the persons, the place, time and date of inspection, in order to protect the rights of third parties.[135]

6. *Closure of fishery by a Member State and/or the Commission*

Background

Once a TAC is set for a given stock then it is divided up into quotas allocated to the different Member States. The means by which this is done is referred to as the key, which is effectively the agreement achieved on quota distribution in 1982, but, as discussed in Chapter 1, has been adjusted on a number of occasions since. The measures which provide for the regulation and suspension of fishing activity are in Title IV of the Control Regulation.

Provisions in the Control Regulation

Member States are responsible for the management of their allocated quotas and are therefore obliged pursuant to **Article 21(1)** to ensure that all catches of stocks or group of stocks made by their fishing vessels shall be counted against the national or flag Member State allocated quota, irrespective of the place of landing.

Member States are responsible for determining when their national quota has been exhausted in respect of given stocks, pursuant to **Article 21(2)**. Member States are thus bound to provisionally prohibit fishing for stock as from the date of quota exhaustion, as well as the retention on-board, the transhipment and the landing of fish taken after that date. In order that this does not create too great a burden, Member States are allowed the flexibility of deciding the date up to which transhipment and landings or final declarations of catches are permitted. The closure of a particular fishery or the imposition of a fish ban for a particular species (more frequently referred to as the fish stop), must be notified to the Commission who then is obliged to inform other Member States. Ultimate responsibility for deciding when a particular quota is exhausted rests with the Commission. The reason that a Commission decision is necessary is to enable other Member States to enforce a prohibition on fishing in cases where the quota is being taken from their waters. It is interesting to note that in theory the Commission is not bound solely on the aforementioned notice by the Member State, and can decide on its own initiative, on the basis of the information available the date on which a given TAC, quota, quantitative limitation, share or allocation, to a Member State is exhausted. In the event of this happening, **Article 21(3)** further stipulates that the Commission shall advise the Member State concerned of the prospects of fishing being halted as a result of a TAC being exhausted. The Commission shall notify the Member States of the date of closure without delay.

On the date of the fish stop, the flag Member State shall provisionally prohibit fishing for that stock or group of stocks, by vessels flying its flag, as well as the retention on-board, transhipment, landing or landing of fish taken after that date and shall decide on a date up to which transhipments and landings or final declarations of catches are permitted. The Commission shall be notified forthwith of this measure and shall then inform the other Member States. In 1998 there were 40 fish stops published by the Commission. If a fish stop is issued at short notice then there is the ever present risk that vessels will be at sea and unaware that a particular stock is subject to a given order, consequently there may be an inadvertent risk of overfishing a given TAC. Furthermore, the Commission order for a closure may be retrospective, consequent on late notification by Member States. However, as Member States manage quota uptake on a regular basis after 70% of the quota is exhausted, this is not normal practice in the management of quotas.

Should it transpire that after the Commission has put a fish stop in place for the Community, and that a Member State has not exhausted its quota then **Article 21(4)** provides that if this prejudice is not removed by action in accordance with **Article 9(2)** of Council Regulation No. 3760/92, then the redress can be achieved through the procedure laid down in **Article 36**. These measures have a certain deterrent factor in so far as they may involve making deductions from the quota allocation or share of a fishing vessel of the Member State that has overfished and giving that share to the Member State that has suffered the prejudice. These deductions and consequent allocations should be made, taking into account as a matter of priority the species and zones for which the annual allocation may be made during the year in which the prejudice has occurred or in the succeeding year or years.

If it is established that vessels constantly breach fish stop rules then the flag Member State can impose additional national controls on these offenders. Furthermore, the name, external identification marks and numbers of this vessel on which additional control has been placed shall be forwarded to the Commission and the other Member States with a view to informing all parties of the supplementary control measures.

If a Member State has overfished its quota, allocation or share of a stock or group of stocks, and this has been established to the satisfaction of the Commission, then deductions to the Member States annual quota may apply. The deductions shall be decided in accordance with the **Article 36** Management Committee procedure. The functional operation of deductions is thus to be undertaken by a Commission decision on an *ad hoc* basis. However, the legislative responsibility for deciding the rules for deductions in accordance with the objectives set out in **Article 8** of Regulation No. 3760/92 rests with the Council on the basis of a proposal from the Commission. **Article 23(2)** sets out the criteria, as a matter of priority that the Council shall take into account when making such decisions, which are the degree of overfishing, any cases of overfishing in the same stock in the previous year, and the biological status of the resources concerned.

Commentary

The obligation placed on Member States to record all landings by fishing vessels flying the flag of, or registered in a Member State, fishing for a stock or group of stocks subject to a TAC or quota, and to notify the Commission of the information received, as well as the obligation placed on Member States to provisionally prohibit fishing vessels from fishing for a stock subject to a quota, have been the subject of judicial examination. In *Commission* v. *Spain*[136] the Court held that the Community was entitled to require Member States to apply control measures, such as the aforementioned, to catches made outside the Community fishing zone of stocks subject to a TAC or quota. In *Commission* v. *Netherlands*[137] the Court upheld similar measures to those in the 1993 Control Regulation which required Member States to adopt binding measures to prohibit on a provisional basis all fishing activity even before quotas were exhausted.[138] The Court had previously held in an earlier **Article 169** {226} action that the Commission, when asserting that fishing was prohibited *too late*, cannot rely on any presumption that a Member State did not fulfil in good time its obligation to prohibit fishing just because of the mere fact that quotas have been found to be exceeded.[139] The Commission was thus bound to adduce specific facts to support its submission that by a given date the Member State had failed to take the requisite action.[140] It is clearly apparent that in such cases the Commission must have sufficient evidence before it commences an **Article 169** {226} action. Indeed, in the aforementioned action taken by the Commission against the Netherlands, the Court ruled that the Commission had initiated proceedings without appropriate inquiry and consequently amended the complaints continually. This made the Netherlands Government's defence more difficult and therefore in the Court's view justified the sharing of the case costs between both parties.

The issue of provisionally prohibiting fishing in good time was also clarified in *Commission* v. *France* when the Court concluded that any exchange of quotas with another Member State for the purpose of increasing a Member State's quota must take place before the exhaustion of the initial quota or after the provisional prohibition of fishing.[141] In conclusion it is important to stress, as the Court has done in several cases,[142] that there is an obligation placed on Member States to take binding measures to prohibit provisionally all fishing activity even before the quotas are exhausted. Consequently, if a Member State is aware that a quota has been exceeded then to paraphrase the words of the Court 'it is manifestly too late to adopt the measures required by the Control Regulation'.[143]

In two different **Article 169** {226} actions against France, the Court held that a Member State cannot plead practical difficulties such as deficiencies in a statistical system, to justify the delay in closing a fishery or to justify its failure to take appropriate control measures.[144] In *Commission* v. *France* the Court found that France had failed to fulfil its obligations by failing to ensure compliance with the anchovy quotas allocated in 1991 and 1992.[145] Furthermore, that France had contravened Community law by not taking penal or administrative action against persons responsible for fishing activity after the Commission had prohibited fishing during the same period.[146] In the instance where a Member State failed to take penal or administrative action against the persons responsible for fishing and related activities after a 'fish stop' had been issued, the Court pointed out that, '*if the competent authorities of a Member State were systematically to refrain from taking action against the persons responsible for such infringements, both the conservation and management of fishery resources and the uniform application of the CFP would be jeopardised*'.[147] Interestingly, the Court did not accept the French argument that they were forced to refrain from taking action because there was a risk of major disorder likely to give rise to serious economic problems. The rejection was on the grounds that *mere apprehension of internal difficulties cannot justify a failure to apply the rules in question.*[148]

It is unlikely that vessels inadvertently fishing stocks during the period of '*retroactive closure*' would entail criminal liability as this is contrary to the general principles of law, as noted by the Court of Justice in *R.* v. *Kirk.*[149] Mr Kent Kirk, who at the time was a leading activist in the Danish fishing lobby, and subsequently became a well-known parliamentarian and Fishery Minister, took the pragmatic step of fishing with his vessel within the 12-mile territorial zone of the United Kingdom. This he hoped would bring Danish opposition to agreement on the TAC and quota regulation to a head and provide the necessary impetus for a favourable settlement of national grievances. The legal loophole which allowed this subterfuge had its origins in the principle of equal access. As noted in Chapter 1, **Article 100** of the Act of Accession of the United Kingdom had permitted the United Kingdom, in derogation from other provisions of Community law, to continue to exclude fishing boats from other Member States, including Denmark, from the United Kingdom's 12-mile zone up to 31 December 1982.[150] In theory, under **Article 2** of Council Regulation No. 101/76, if agreement had not been reached on the allocation of the resources by that date, then all Community vessels would be allowed to fish up to the beaches as from 1 January 1983.[151] The Crown Court had to decide on the legality of Mr Kirk's audacious exploitation of this interim period of opportunity which arose between the expiry of the UK derogation from the principle of equal access on 31 December 1982 and the adoption of the subsequent regime pursuant to Council Regulation No. 170/83 on 25 January 1983. Clearly under Community law there was no reason to prohibit Mr Kirk from fishing in the United Kingdom's 12-mile zone. The Court referred the case to the European Court for a preliminary ruling. The Court of Justice decided that the purported retroactive application of Council Regulation No. 170/83 could not have the effect of validating *ex post facto* the penal provisions in the United Kingdom Order in Council. The Court held that:

> 'The principle that penal provisions may not have retroactive effect is one which is common to all legal orders of the Member States and is enshrined in **Article 7** of the European Convention for the protection of Human Rights and Fundamental Freedoms of 4 November 1950 as a fundamental right; it takes its place among the general principles of law whose observance is ensured by the Court of Justice.'[152]

Thus, in the words of one leading commentator, Captain Kirk sailed into Community legal history by successfully relying on the concept of fundamental human rights which had been initially espoused by the Court for the first time in *Stauder* v. *Ulm*.[153]

7. *Final Comments on the quota management control measures*

The quota management system is at the heart of the CFP. An argument frequently advanced is that the Community management system has failed because quotas and TACs are frequently exceeded. There is clear evidence, supported by the findings of the Court in several **Article 169** {226} actions, that this is true.[154] A fundamental issue in such cases is the action taken against Member States when quotas are overfished and how effective the sanctions are in preventing recurrence. Unfortunately there appear to be no empirical data available on this matter.

With few exceptions, the TAC/quota monitoring provisions are clear and precise as to what is required of the authorities of the Member States, in particular concerning fish stops when national quotas have expired. There are no significant changes to the control and enforcement of the TAC/quota system in the 1993 Control Regulation. In this regard the forced transfer of quotas from the transgressor to the prejudiced Member State in the event of overfishing is reported as one of the provisions of the 1982 Control Regulation (as amended) which worked satisfactorily.[155]

On the other hand, the unilateral powers of the Commission to impose fish stops may prove to be fertile ground for proceedings in the Court of Justice. This in turn will depend on how many times there is recourse to this power. The existing TAC/quota monitoring mechanism must be supported by Member States and the industry to ensure that this vital component of the CFP functions as envisaged. The consequences of failure may be the reason why the Commission had sought additional scope for taking punitive action in the event of overfishing and they have been supported in this regard by the Parliament.[156] This may be partly due to the fact that conventional remedies such as **Article 169** {226} proceedings are of no conservation value with respect to overfished Member State quotas. That is to say it is not possible to restock or replace the overexploited fish stock. Indeed, there are only limited means by which the Member State could remedy the breach, the most obvious being for the Commission to reduce subsequent quotas. The use of financial sanctions as a remedy is discussed in Chapter 6, nevertheless, at this point it is interesting to note that the Court did not uphold the Commission attempts to link reimbursement from the European Agriculture Guidance and Guarantee Fund (EAGGF) with the observance of quotas when it withheld reimbursement relating to withdrawals in 1981 from Denmark, France, Ireland and the Netherlands on the grounds that these Member States had exceeded their quotas.[157] The Court, however, agreed with the Commission that export refunds granted in violation of Community conservation measures cannot be financed by EAGGF. Indeed, **Article 25(2)** of the 1992 Market Regulation expressly provides that intervention measures shall only be financed up to the limits of the quotas allocated to the Member States.[158] It should also be mentioned that the Council rejected the Commission proposal regarding the financial penalties to be imposed in cases of over-fishing in their deliberations prior to the adoption of the 1993 Control Regulation.

References

1 Title II of the Control Regulation deals exclusively with the monitoring of catches. There are however some provisions in Title VIa which deal with the monitoring of catches of third-country vessels.
2 Council Regulation No. 2241/87, *OJ* L 20, 28.01.1976, p. 19. (Repealed.)
3 For discussion on the logbook as an instrument of fishery control in Norway, see Thorvik, T., *Control and Enforcement in the Norwegian Fisheries*, (OECD, Paris, 1992), 28–29. There are several differences between the Norwegian and the Community systems: most notably in Norway the logbook is to be completed continuously, that is on a haul by haul basis, whereas in the EU the logbook is completed on a 24-hour basis. In addition in Norway the logbook is used as a primary source for fishery scientific data.
4 See discussion on **Article 40** which applies an exemption from the requirement for keeping a logbook for fishing operations in the Mediterranean until 1 January 2000.
5 Commission Regulation No. 2807/83 of 22.09.1983 laying down details for recording information on Member States' catches of fish, *OJ* L (276, 10.10.83) 306, 11.11.1988, p. 2, as amended. See Table (i) on Community secondary legislation.
6 All of **Article 5**, Council Regulation No. 2241/87 remains in force until the aforementioned lists in **Article 6(2)** of the '1993 Control Regulation' are established. This will have the effect of postponing the completion of logbooks for periods of less than 24 hours for certain categories of vessels, namely, those over 10 metres but not over 17 metres. This category was exempted from compliance with the logbook obligation under the former regulation but are obliged to comply under the 1993 Control Regulation, even when such vessels are undertaking trips of less than 24 hours.
7 See Section (ii), *infra*.
8 Case C-258/89, *Commission* v. *Spain* [1989] I-3977.
9 Case C-258/89, paragraph 2
10 Case C-258/89, paragraph 12. The opinion of Advocate General Darmon, also cites the enforcement difficulties that would arise if stocks outside the Community zone were not subject to the same limitations, and draws attention to fact that to conclude otherwise could lead to the collapse of the CFP, Opinion of Advocate General Darmon delivered on 29 May 1991. Case C-258/89, *Commission* v. *Spain* ECR [1991], I-3977, paragraphs 59–61.
11 Case C-258/89, paragraph 13.
12 Joined Cases 6/88 and 7/88, [1989] ECR 3639.
13 Agreement on cooperation in the sea fisheries sector between the European Community and the Kingdom of Morocco, *OJ* L 30 Vol. 40, 31.01.1997, p. 7.

14 In the context of international waters, there is the requirement in the NAFO Regulatory area, for masters of vessels to comply with NAFO 'hail requirements'; and other logbook provisions in the Control Regulation which deal with catch monitoring pursuant to, *inter alia*, **Article 7**, Council Regulation No. 3090/95 of 22.12.1995 laying down for 1996 certain conservation and management measures for fishery resources in the Regulatory area as defined in the Convention on Future Multilateral Cooperation in the North West Atlantic Fisheries, *OJ* L 330/108, 30.12.95.

15 Council Regulation No. 2847/93, **Article 6(3)**.

16 See Alverson, D.L., Freeberg, M.H., Pope, J.G. and Murowski, S.A., 'A Global Assessment of fisheries by-catch and discards', *FAO Fisheries Technical Paper* **339**, (Rome, 1994).

17 It would also have been a major obstacle to be overcome if Norway had joined the European Union in 1995. In this context of particular significance is that an essential feature of the Norwegian conservation philosophy is the prohibition on the practice of discarding small or juvenile fish. This issue is discussed by Thorbjorn Thorvik, in a paper entitled *Control and Enforcement in the Norwegian Fisheries*, (OECD, Paris, 1991). He points out that the Norwegian discard ban applies to all species which have an economic value. It requires that every catch of an individual species should be registered and counted against the quota for that species if a quota is established. Although this conservation philosophy is said to be shared by scientists, fishermen, administrators and politicians, there are still a number of questions to be answered as to how the prohibition is enforced. Internationally, although Russia has supported this discard policy with respect to shared stocks in the Barents Sea there has not been a similar meeting of minds with respect to stocks in the North Sea which are shared with the European Union. There are significant differences in the approach to the enforcement task between the Community and Norway. The latter focus on fishing activities as the starting point for control and enforcement of fishing effort. In Norway the discard philosophy is supported by conservation legislation which allows the Fisheries Ministry to periodically close areas in which there are juvenile fish. This in turn requires a vigilant surveillance service which is ready to undertake surveillance at any time. This 'front edge' enforcement entails high operational costs to patrol particular fisheries. In Norway closed areas are combined with the introduction and research into improved gear selectivity. It is asserted that this approach to conservation is respected by the fishing community and thus have attained a high degree of legitimacy. Interestingly, it is anticipated that the traditional measures such as those used in the EU (i.e. checking, *inter alia*, mesh size, by-catches, minimum size fish,) will be replaced in due course while measures at the aggregate level (i.e. closed areas and improved selectivity).

18 Commission reply to PQ No. 405/93, *OJ* C 207/40, 30.07.93.

19 PQs No. 1812/92, *OJ* C 86/8, 26.03.93; No. 405/93, *OJ* C 207/40, 30.07.93; No. E-2162/93, *OJ* C 219/62, 08.08.94.

20 Doc. SEC(92)423.

21 This is the plan of action as outlined by the Commission in reply to a question in the European Parliament on the reduction of discards, see, PQ. No. 1812/92, *OJ* C 86/8, 26.03.93.

22 See Chapter 2, *ante*.

23 See Chapter 6, *post*.

24 Holden, M., *The Common Fisheries Policy: Origin, Evaluation and Future*, p. 166.

25 Case C-258/89 [1989] I-3977.

26 Council Regulation No. 685/95 of 27.03.1995, *OJ* L 71, 31.03.1995, p. 5., and Council Regulation No. 2027/95 of 15.06.1995, *OJ* L 199, 24.08.1995, p. 1. As a consequence of these regulations the Council adopted a regulation No. 2870/95, *OJ* L 301 of 08.12.1995, p. 1 amending the Control Regulation. Despite the provisional arrangements contained in Regulation No. 2945/95, *OJ* L 308 of 20.12.1993, p. 18 amending Council Regulation No.

2807/83, *OJ* L 276 of 22.09.1983, the current format of the Community logbook is not entirely appropriate for the recording of fishing effort.

27 Coucil Regulation (EC) No. 2846/98, *OJ* L 358/5, 31.12.1998, **Article 6(2)**.

28 This case is unreported, however, see report *The Examiner*, 02.08.1997.

29 The revised format offers an excellent opportunity to reassess the inadequacies, from the control perspective, of the existing format. A major issue for consideration is whether to continue the 20% tolerance allowed under Council Regulation No. 2807/83, laying down detailed measures for the recording information on Member States' catches of fish, *OJ* L 276/1, 22.09.1983, as amended by Commission Regulation No. 473/89 of 24.02.1989, *OJ* L 53/54 25.02.1989. **Article 5(2)** states that the permitted margin of tolerance in estimates of the quantities, in kilograms, of fish subject to a TAC that are retained on board shall be 20%. Where such fish are held in boxes, baskets or other containers, the precise number of boxes, baskets, or other containers used shall be recorded. It ought to be pointed out that the logbook sheet itself suggests that the skipper of the vessel has the alternative of providing the number of containers or the weight in kilos. This rather confusing situation is further complicated because the relevant instructions in Annex IV of Council Regulation No. 2807/83, does not appear to be consistent in all Community languages. Significantly, the Commission advises Member States in 1988 not to institute proceedings against skippers of vessels who did not record in the pertinent logbook column the weight of catch in kilos. There is also a strong case, to omit the option of recording the catch in the traditional type of containers such as baskets as the fishing industry has become more technology driven. How the quantities of fish are recorded, live weight as opposed to processed weight and the corresponding conversion factors also needs to be considered when revising the logbook regulation.

30 See Chapter 2, *ante*.

31 See, in particular, Case 3/87 *The Queen* v. *Ministry of Agriculture, Fisheries and Food ex p. Agegate Ltd.* [1989] ECR 4509; Case 216/87, *The Queen* v. *Ministry of Agriculture, Fisheries and Food ex p. Jaderow Ltd.*, [1989] ECR 4509; Case 221/89, *the Queen* v. *Secretary of State for Transport ex p. Factortame*. These and related cases are discussed in Chapter 7, *post*.

32 *Op. cit.* fn 29.

33 Oostend, Zeebrugge, and Nieuwpoort.

34 In relation to third-country landings the notification requirement is 72 hours, **Article 28e** *infra*. Subsequent experience will no doubt dictate as to whether these notification periods are sufficient, they may be subject to observation in the Commission Reports, pursuant to **Article 35**. In 1999, the Commission adopted a two hour notification period for Community vessels. Commission Regulation (EC) No. 728/999, *OJ* L 93/10, 08.04.1999.

35 Council Regulation No. 2847/93, **Article 40** applies an exemption from the requirement for keeping landing declarations for fishing operations in the Mediterranean until 1 January 2000.

36 As with **Articles 6(2)** (other species logbook) and **6(5)** (exemption from logbook) there is no defined deadline for the adoption by the Council of the exemption from landing declarations for certain categories of vessels and pursuing specific fishing activities.

37 The Commission must approve the Member States sampling plan on evaluation of landings in certain ports, before applying any derogation, pursuant to **Article 14(3)**. This required Member States to notify the Commission before the 31.12.1994.

38 Case C-258/89, [1989] I-3977.

39 McKinney, D. *Enforcement Methodology of Federal Individual Quotal Fisheries*, (OECD, Paris, 1992).

40 McClurg, T. in a paper entitled *Two Fisheries Enforcement Paradigms: New Zealand before and after Individual TAC Quotas (ITQs)*, (OECD, Paris, 1992), 129–130.

41 Council Regulation No. 2847/93, **Article 9(8)**.

42 This is to ensure conformity with the Sixth Council Directive 77/388/EEC of 17.05.1997 on

the harmonisation of the laws of the Member States relating to turnover taxes - common system of value added tax: uniform basis of assesssment, *OJ* L 145, 13.06.1977, p. 1.

43 Council Regulation No. 2847/93, **Article 40** provides that Member States are exempted from complying with the requirement of computer transmission of sales notes until 1 January 1996.

44 *Regional Socio Economic Study in the Fisheries Sector United Kingdom* (XIV/243/93). A large segment of the Irish fleet is also small - in 1992 of some 1400 vessels, 800 boats were less than 12 metres. *Regional Socio Economic Study in the Fisheries Sector Ireland,* (XIV/243/93). Furthermore, Ireland does not have the system of fish auction houses or other bodies who are responsible for the first marketing of fishery products which are in place elsewhere in the EU. In most instances fish are sold through fishermen's cooperatives or by individual buyers whose obligations fall within the scope of **Article 9(2)**.

45 On this point, see Chapter 8, Section (ii), *post.*

46 Case 87/82, *Rogers* v. *Darthenay,* (1983) ECR 1579; (1984) 1 CMLR 135.

47 This is a based on a broad interpretation of **Article 62(2)** of the United Nations Convention on the Law of the Sea 1982. For further discussion of this issue see *inter alios*: Burke, T. *The New International Law of Fisheries*, 62–65, entitled *Harvesting Capacity and Surplus;* Churchill, R.R. and Lowe, A.V., *The Law of the Sea,* (London, 1997), *passim.*

48 *OJ* L 300, 30.12.1995, Council Regulations No. 3075/95 (Norway), No. 3081/95 (Faeroe Islands), No. 3083/95 (Latvia), No. 3085 (Lithuania) and No. 3088/95 (Poland).

49 See Case C-135/92 *Fiskano* v. *Commission* [1994] ECR I-2885, discussed in Chapter 6, *post.*

50 Council Regulation No. 2846/98, **Article 28c**.

51 On the management of quotas see Section (iii), *infra.*

52 It is intended that this requirement applies to landings by all vessels flying the flag of, or registered in, a third country, including vessels which have caught fish outside Community waters and land in Member States. Whereas the logbook requirement, **Article 28c** only applies to third-country vessels operating in the areas stated.

53 Derogations are to take into account distance between the fishing grounds, the landing locations and the ports where the vessels in question are registered or listed. The Commission adopted a regulation which provides a derogation for vessels flying the flag of Norway and Iceland in 1997. This regulation conforms with Agreement on the European Economic Area (Protocol 9, **Article 5**) in that it states that the notification period for these vessels shall be at least two hours. Furthermore, vessels flying the flag of a third country bordering the Baltic Sea who wish to use landing facilities in Denmark, Germany, Sweden, or Finland must notify at least six hours before the time of arrival in port. Commission Regulation No. 1292/97 of 03.07.1998, *OJ* L 176/21, 04.07.1997.

54 International agreements are discussed in Chapter 1, *ante.*

55 To carry out this task the Norwegian coastguard deployed 13 patrol vessels, 2 large planes and 6 helicopters.

56 Two Japanese tuna vessels were penalised almost £800,000 at Cork Circuit Criminal Court on 04.11.1996 for fishing illegally in waters under Irish jurisdiction, see report in *The Irish Times,* 05.11.1996. It is also reported that the vessels in question subsequently received additional penalties on returning to Japan which amounted to a six-week anchorage penalty, effectively depriving them of a ninth of the fishing season. The issue of double jeopardy was not commented upon.

57 *Report on Monitoring Community Conservation and Management Measures Applicable to Third Country Fishing Vessels,* COM(96) 493 final, 22.10.1996.

58 *Ibid.,* p. 4.

59 *Ibid.,* p. 7.

60 Council Regulation (EC) No. 1093/94 of 06.05.1994 setting down the terms under which fishing vessels of a third country may land directly and market their catches at Community

ports, *OJ* L 121/3, 12.05.1994. Council Directive 91/493/EEC, of 22.07.1991 laying down the health conditions for the production and the placing on the market of fishery products, *OJ* L 268/15, 24.09.1991. Amended by Council Directive 95/71/EC of 22.12.1995, *OJ* L 332/40, 30.12.1995. Council Directive 90/675/EEC of 10.12.1990 laying down the principles governing the organisation of veterinary checks on products entering the Community from third countries, *OJ* L 373/1, 31.12.1990. Amended by Regulation (EEC) No. 160/92, *OJ* L 173/13, 27.06.1992.

61 It is also a major problem for maritime safety agencies as the age of vessels tends to be old.

62 Holden, M., *The Common Fisheries Policy: Origin, Evaluation and Future*, pp. 86 and 161.

63 EC Treaty, **Articles 30** {28} and **34** {29}.

64 All Community law text books which give an account of the free movement of goods Case 52/79 *Procureur du Roi* v. *Dassonville* [1974] ECR 837. On the specific application of **Article 30** {28} EEC Treaty see, *inter alia*, Oliver, P., *Free Movement of Goods in the EEC Under the Articles 30 and 36 of the Rome Treaty* (London, 1996); Jarvis, M., *The Application of EC Law by National Courts: the Free Movement of Goods*, (Oxford, 1998).

65 It is assumed by external identification it is the vessels registration number that it is required, (most Iberian vessels have their international call sign painted on their superstructure, and these may not be particular to a given vessel). On identification requirements for Community vessels, see Commission Regulation No. 1381/87 of 20.05.1987 establishing detailed rules concerning the marking and documentation of fishing vessels, *OJ* L 132/9, 21.05.1987.

66 Prior to the ratification of the Single European Act (the SEA), and the creation of the single market, fish transported from one Member State to another had to be accompanied by a customs form, the T2M is now no longer utilised in the Member States, but remains in use for imports into the Community.

67 See Chapter 2, *ante*.

68 It is not uncommon for contracts for sale to be negotiated and agreed in the pelagic sector prior to the fish being caught and landed. In such instances, it may be ascertained that the first sale is concluded and that there is no requirement for transport documents.

69 Commission Report, 'Monitoring the Common Fisheries Policy', COM(96), Brussels, 18.03.1996, p. 12. See Chapter 8.

70 *Ibid.*

71 Council Regulation No. 2847/93, **Article 14(2)**

72 For example, Ireland has over 300 landing places and does not have an elaborate marketing structure.

73 See Chapter 1, *ante*.

74 Holden, M., *The Common Fisheries Policy: Origin, Evaluation and Future*, 180–183.

75 See Chapter 1, *ante*.

76 Council Regulation No. 1274/94, *OJ* L 140, 03.06.1994, p. 1. See Chapter 1, Section (v), *ante*.

77 Council Regulation No. 2870/95 of 8 December amending regulation No. 2847/93 establishing a control system applicable to the common fisheries policy, *OJ* L 301/1, 14.12.1995.

78 See Annex II of Commission regulation No. 2945/95 of 20.12.1995 amending Regulation No. 2807/83 laying down detailed rules for recording information on Member States' catches of fish, *OJ* L 308/18, 21.12.1995.

79 Council Regulation No. 779/97 of 24.04.1997 introducing arrangements for the managing of fishing effort in the Baltic Sea, **Article 1**, *OJ* L 113/1, 30.04.1997.

80 Council Regulation No. 2205/97 of 30.10.1997 amending Regulation No. 2847/93 establishing a control system applicable to the common fisheries policy, *OJ* L 304/1, 07.11.97.

81 Commission Regulation No. 523/96 of 26.03.1996 adjusting the maximum annual fishing effort for certain fisheries, *OJ* L 77/12, 27.03.1996.

82 A more elaborate explanation is provided in Chapter 5, *post*.

83 Council Regulation No. 2027/95 of 15.06.1995 establishing a system for fishing effort relating to certain Community fishing areas, *OJ* L 199/1, 24.08.1995, annex.
84 See Chapter 5, *post*.
85 *OJ* L 389/3, 31.12.1992.
86 *OJ* L 71/5, 31.03.95.
87 Council Regulation No. 685/95, **Article 1**.
88 There is a useful list of the effort zones appended to Commission regulation No. 2945/95 of 20.12.1995 amending regulation No. 2807/83 laying down detailed rules for recording information on Member States' catches of fish, *OJ* L 308/18, 21.12.1995.
89 *Ibid.* **Article 3(5)**, for illustrative purposes it is shaded green on the map.
90 Council Regulation No. 685/95, **Article 3(5)**.
91 Council Regulation No. 779/97 of 24.04.1997 introducing arrangements for the managing of fishing effort in the Baltic Sea, **Article 1**, *OJ* L 113/1, 30.04.1997.
92 Commission Regulation No. 493/96 of 20.03.1996 amending Regulation (EC) No. 109/94 concerning the fishing register of the Community, *OJ* L 72/12, 21.03.96, Article 8a. (Repealed). Commission Regulation (EC) No. 2090/98 of 30 .09.1998 concerning the fishing vessel register of the Community, *OJ* L 266/27, 01.10.1998.
93 *OJ* L 171/7, 06.07.94.
94 Council Regulation No. 2870/95, **Article 19a(2)**.
95 Council Regulation No. 2870/95, **Article 19a(3)**.
96 Council Regulation, **Article 19(b)**. An effort report contains the following information: the name of the vessel: the registration number: radio call sign: name of the skipper/master; geographical location; data and time of each entry and exit from a port located inside a fishing area; each entry into a fishing area; each exit from a fishing area. For further details see Commission Regulation (EC) No. 1449/98 of 07.07.1998 laying down special rules for the application of Council Regulation (EEC) No. 2847/93 as regards effort reports, *OJ* L192/4, 08.07.1998.
97 Council Regulation No. 2870/95, **Article 19c(2)**.
98 Council Regulation No. 2847/93, **Article 3**. See also Commission Regulation (EC) No. 1449/98 of 07.07.1998 laying down special rules for the application of Council Regulation (EC) No. 2847/93 as regards effort reports, *OJ* L192/4.
99 The specific requirements are prescribed in Council Regulation No. 2870/95, **Article 19c**.
100 Commission Regulation No. 2945/95 of 20.12.1995 amending regulation No. 2807/83 laying down detailed rules for recording information on Member States' catches of fish, *OJ* L 308/18, 21.12.1995.
101 Council Regulation No. 2870/95 **Article 20a**.
102 Council Regulation No. 2870/95 **Article 19f**. Commission Regulation (EC) No. 2092/98 of 30.09.1998 concerning the declaring of fishing effort relating to certain Community fishing areas and resources, *OJ* L 266/47, 01.01.1998.
103 Council Regulation No. 2870/95, **Article 19g**.
104 Council Regulation No. 2870/95, **Article 19i**.
105 Council Regulation No. 2870/95, **Article 21a**.
106 Council Regulation No. 2870/95, **Article 21b**.
107 Council Regulation No. 2870/95, **Article 21c**.
108 Council Regulation No. 2870/95, **Article 21c(2)**.
109 *Ibid.*
110 Chapter 1, *ante*.
111 *Ibid.*
112 Council Regulation No. 685/95, **Article 7**, *OJ* L 71/5, 31.03.1995.
113 See Chapters 2, *ante*, 8 and 12, *post*.

114 Council Regulation No. 3760/92 of 20.12.1992 establishing a Community system for fisheries and aquaculture, *OJ* L 389/1, 31.12.1992.
115 Case 287/81, *Anklagemyndigheden* v. *J. Noble Kerr* [1982] ECR 4053 at 4073; [1983]2 CMLR 431 at 451.
116 Churchill, R.R., *EC Fisheries Law*, p. 160. The decision of the Court of Justice on the issue of sugar production quotas, was in Case 250/84, *Eridania* v. *Ministry of Agriculture and Forestry* [1978] ECR 2749.
117 Joined Cases 3, 4, and 6/76, *Officier van Justitie* v. *Kramer* [1976] ECR 1279.
118 Discussed in the context of **Article 38**, Chapter 7, *post.*
119 Case C-52/91, *Commission* v. *Netherlands*, [1993] ECR I-3069, paragraph 21.
120 There is also a requirement, pursuant to **Article 18(1)** to notify by standard format, the catches in third-country waters and the high seas.
121 Council Regulation No. 2847/93, **Articles 11** and **12**.
122 Council Regulation No. 2847/93, **Articles 6** and **8**.
123 See Chapter 7, *post.*
124 See Table of International Conventions and Agreements, p. viii.
125 **Article 40** provides that Member States are exempt from complying with the requirement of computer transmission of sales notes and landing registrations until 1 January 1996. As noted, *supra*, Member States are obliged, pursuant to **Article 15(1)** to provide the Commission with similar data in respect of fish caught in the Community area.
126 McClurg, T., *Two Fisheries Enforcement Paradigms: New Zealand before and after ITQs*, (OECD, Paris, 1992).
127 McKinnay, D.A., *Enforcement Methodology of Federal Individual Quotal Fisheries*, (OECD, Paris, 1992).
128 For a view on some of the issues raised by Community legislation and draft proposals on Personal Data and Protection on Privacy in Telecommunications, see letter to the President of the European Parliament, from *Union des Confédérations de l'Industrie et des Employeurs d'Europe* (UNICE), Brussels, 12 March 1997, (copy on the Europlus electronic database). For a comparative examination of measures in Europe and the United States, see, Fleischmann, A., 'Personal Data Security: Divergent Standards in the European Union and the United States' *Fordham International Law Journal*, **19** (1995), 143–180; see also from the European Commission, *Developing the legislative framework for the Information society*, Background Report, No. 10, (1995) (EC office UK).
129 See, *inter alia*, Zerdick T., 'European Aspects of Data Protection: What Rights for the Citizen?' *Legal Issues of European Integration*, Deventer No. 2 (1995), 59–86; Morton, J., 'Data Protection and Privacy: R. v Brown,' *European Intellectual Property Review* 18(10) (1996), 558–561; Carlin, F., 'The Data Protection Directive: the introduction of common privacy standards', *European Law Review*, **21(1)**, 65–70.
130 *OJ* L 077, 27.03.1996, p. 20.
131 *OJ* L 365, 31.12.1994, p. 46.
132 *OJ* L 268, 19.10.1994, p. 15.
133 The issue of EC inspections and the examination of witnesses is discussed in Chapter 6, *post.*
134 Case C-135/92, *Fiskano* v. *Commission* [1994] ECR I-2885, discussed *ante.*
135 Case C-64/88, *Commission* v. *French Republic* [1991] ECR I-2727, paragraphs 7–11; discussed in Chapter 5, *ante.*
136 Case C-258/89, [1989] I-3977.
137 Case C-52/91, [1993] ECR I-3069.
138 *Ibid.*, paragraph 26.
139 Case C-290/87, *Commission* v. *Netherland*, [1989] ECR 3083, paragraphs 10–11.
140 *Ibid.*, paragraph 12.

141 *Commission* v. *France,* [1990] ECR I-925, paragraph 20.
142 Case C-62/89, *Commission* v. *France* [1990] ECR I-925, paragraphs 17 and 18; Case C-244/89, *Commission* v. *France* [1991] ECR I-163, paragraph 17; Case C-52/91, *Commission* v. *Netherlands* [1993] ECR I-3069, paragraph 26.
143 Case C-52/91, *Commission* v. *Netherlands* [1993] ECR I-3069, paragraph 27.
144 The Court cited Case 262/87, *Commission* v. *Netherlands* [1989] ECR 225, to support its finding.
145 Case C-52/95, *Commission* v. *France,* Judgement 7 December 1995; Case C-62/89, *Commission* v. *France,* [1990] ECR I-925.
146 Case C-52/95, *Commission* v. *France,* paragraph 36.
147 Case C-52/95, *Commission* v. *France,* Judgement 7 December 1995, paragraph 35.
148 *Ibid.*, paragraphs 37–38.
149 Case 63/83, *The Queen* v. *Kirk* [1984] ECR 2689; [1984]3 CMLR 522; [1985] 1 All ER 453, and the issue of retroactive criminal liability as discussed in Churchill, R.R., *EC Fisheries Law*, p. 142. This issue is also discussed in the context of fundamental human rights by Weatherill, S. and Beaumont, P., *EC Law*, (London, 1995).
150 Chapter 1, *ante.*
151 Denmark's refusal to accept the Commission's final proposal on resource allocation nearly disrupted the adoption of the TAC and quota regulation. The issue was ultimately resolved by allocating Denmark additional quantities of species to compensate for the exclusion from fishing western mackerel.
152 Case 63/83, ECR at 2718; 1 All ER at 462.
153 Case 29/69 [1969] ECR 419.
154 On EC Treaty **Article 169** {226} proceedings, see Chapter 8, *post*. The quota management system is also undermined by the Community policy on discards, discussed *infra*. See, *inter alia*, Churchill, R.R., 'Fisheries in the European Community – Sustainable Development or Sustained Mismanagement?' (paper delivered at the 25th Annual Conference of the Law of the Sea Institute, Malmo, Sweden, 1991); Biais, G. 'An evaluation of the policy of fishery resources management by TACs in European Community waters from 1983 to 1992', *Aquatic Living Resources*, 8(3) (1995) 241–251. See further Chapter 12, *post.*
155 Holden M., *The Common Fisheries Policy: Origin, Evaluation and Future*, p. 86.
156 Resolution of 13 December 1985, *OJ* 1985 C352/315.
157 Case 325/85, *Ireland* v. *Commission,* [1987] ECR 5041; Case 326/85, *Netherlands* v. *Commission* [1987] ECR 5091; Case 332/85, *Germany* v. *Commission* [1987] ECR 5143; Case 336/85 *France* v. *Commission* [1987] ECR 5173; Case 348/85 *Denmark* v. *Commission,* [1987] ECR 5225; for annotation see Churchill, R.R., *ELRev* (1988), 352–360.
158 Council Regulation No. 3759/92 of 17.12.1992 on the common organisation of the market in fishery and aquaculture products, *OJ* L 388/1, 31.12.1992.

Chapter 5

The Enforcement of Technical Conservation Measures, Structure and Market Measures

Introduction

This chapter examines three topics: the enforcement of technical conservation measures, as well as the structural and market elements of the fisheries policy.

Section (i) Enforcement and technical conservation measures

Introduction

Chapter 1 contains a brief discussion of the history of the European Union's technical measure regulations. Although the Council adopted a new Technical Regulation in 1998 which will not be implemented before 2000, the discussion here centres on the enforcement issues associated with the technical conservation measures which have been in place since 1986.[1]

'Technical conservation measures' is the term used to denote a series of conservation provisions which regulate, *inter alia,* the type of fishing gear allowed, the minimum mesh sizes, the size of fish (marine organisms) which may be landed, by-catch limits and seasonal restrictions on fishing in particular areas such as nursery areas or spawning grounds for pelagic species such as herring.[2] Perhaps the most unique technical measure is the one which prohibits the use of explosives, poisonous or stupefying substances, guns, or electric current for the purpose of fishing.[3] The majority of the technical provisions concerns the capture of demersal fish species and one of the major objectives of these measures is to reduce the possibility of capturing immature juvenile fish. Since its adoption in 1986 the Technical Regulation has been subject to frequent amendment and its ineffective implementation is believed to be one of the reasons why mortality rates of juvenile fish stocks have not been reduced. It should also be pointed out that the problems and issues associated with technical measures go beyond the subject of fishing. In particular there is concern about the environmental damage to the marine ecosystems by certain types of fishing gear (i.e. large scale driftnets) and this has resulted in a movement towards ensuring that technical measures are compatible with the broader obligation to manage the marine environment.

The reasons why the technical measures have failed to achieve their purpose are complex. It may be partially attributable to reasons such as increased fishing effort or other extraneous factors such as improved fish finding technology. It may also result from the general non-acceptance by the fishing community of rules which are frequently misunderstood. Moreover, it became increasingly apparent in recent years that the 1986 Technical Regulation was so permeated with provisions which derogate from the general rule that it could be considered to be virtually meaningless from the enforcement perspective. This was particularly true with respect to the enforcement task at sea. In accordance with the second recital of the preamble of Council Regulation (EC) No. 850/98, one of the principal aims of the new measures adopted in 1998 is to rectify the deficiencies in the 1986 Technical Regulation which resulted in problems of application and enforcement.

The broad range of technical conservation measures adopted in 1998, which will apply from the year 2000, integrates into the CFP the general principles regarding environmental protection and in particular the precautionary principle established by **Article 130r(2)** {174} of the the EC Treaty. Furthermore, with the adoption and ratification of the 'FAO Code of Conduct for Responsible Fisheries', Council Regulation No. 850/98 may be seen as a major development in finding a solution to the problems associated with the non-selectivity of fishing-gear. Within the wider international legal framework this is significant because the Code stipulates that States should require that fishing gear, methods and practices are sufficiently selective so as to minimise waste, discards and catch of non-target species, both fish and non-fish.[4] Furthermore, States should ensure that regulations are not circumvented by technical devices.

1. *Some of the enforcement difficulties associated with technical conservation measures*

One of the technical measures which have been the subject of considerable discussion in recent years is the rule which allows fishing vessels to have on board nets which have a smaller mesh size than that permitted in a particular area. Traditionally, vessels were entitled to carry gear of different sizes in order to have the capability to fish in different grounds, provided that the nets with a smaller size were lashed and stowed in such a way that they could not be readily used in areas where their use was prohibited. In many fisheries, including those outside the Community area, this restriction led to a major enforcement difficulty as enforcement officers have difficulties during inspections verifying which catch has been taken by which net.

The rules which stipulate when a given mesh size must be used in a particular fishing zone can be undermined by malpractice. Frequently, some fishermen will reduce the selectivity of the gear in use by a variety of means. These include the use of an extra smaller cod-end, or the fitting of netting or ropes around the legal cod-end to reduce the overall mesh size. These particular devices are commonly referred to as 'blinders' and by several other euphemistic terms.[5] Other illegal means to reduce the selectivity of fishing gear consist of placing heavy objects such as car tyres in the net to increase the hydrodynamic drag of the cod-end and to reduce the size of the mesh. The enforcement task is made more difficult as it is necessary to apprehend an offender *flagrante delicto.* Under both the 1986 Technical Regulation and Council Regulation No. 850/98, it is not an offence to have net attachments on board a fishing vessel, but it is an offence to use illegal attachments for fishing activity. Indeed, **Article 18** of the latter simply states that no device shall be used by means of which the mesh in any part of the fishing net is obstructed or otherwise effectively diminished. Consequently, there are simple and effective means to thwart inspection, undersize gear can be rigged in such a manner that the illegal device can be slipped out as soon as an inspection vessel arrives in the vicinity, alternatively the complete net can be slipped to the seabed to be recovered later. If time allows, the vessel will haul its gear, remove the device and resume normal fishing operations. Heavy objects recovered in the net can be validly explained as flotsam, jetsam, or other detritus commonly recovered in fishing grounds.

Interestingly, the issue of obstructing devices which have the effect of diminishing the size of a net was the subject of an **Article 177** {234} of the EC Treaty referral from Plymouth's Magistrates Court in *Lieutenant Commander A.G. Rogers* v. *H.BL. Darthenay.*[6] The Court observed that the purpose of technical conservation measures was to ensure the protection of fishing stocks and also a balanced exploitation of resources of the sea in the interest of fishermen and of consumers.[7] The prohibition on obstructing devices constituted an essential provision for the achievement of this objective. The Court noted, however, that the contested regulation did make provision for fishermen to use certain types of devices in order to protect their fishing nets.[8] (It could be suggested that the allowance for such devices is an example of rational legislation in so far as the regulators endeavoured to reconcile the practical needs of the fishermen with the requirements of scientists to have a mesh size of a particular size.) The issue of obstructing devices also arose in *Commission* v. *French Republic.* In this case, the Court declared that France had failed to carry out effective controls between 1984 and 1987 to ensure compliance with technical measures and to observe certain control measures.[9] Included in the list of omissions by the French control authorities was the failure to take measures against trawlers equipped with nets containing devices prohibited by Community rules.

With some types of gear, fishermen do not have to resort to the aforementioned illegal practice to reduce the selectivity of the gear. Manufacturers can provide nets constructed of heavy multistrand cordage which conform to the legal mesh size, but which reduce the mesh size considerably when the gear is deployed and towed. Alternatively, fishermen can extend the length of the top side chaffer to such a degree that the cod-end is nearly closed under normal fishing conditions. Prior to the application of the revised package of Technical Conservation measures adopted in 1998, such practice had a negative influence on the effectiveness of the conservation measures established in Community legislation. Problems were further accentuated because malpractice in such cases is difficult to establish. However, as a general rule, experience would indicate that if there are substantial amounts of undersize fish on board a vessel then in all probability the vessel may have been deploying illegal gear.

Several measures have been taken to combat these practices. In 1991, for example, on a proposal from the Commission, the Council adopted rules limiting the number of meshes around the circumference of certain cod-ends. The use of square mesh rules may also limit the scope for reducing the effective opening of the mesh. Another interesting possibility is the installation of a rigid screen or grate in the net to allow small fish to escape, similar to those tested and used in Norway and referred to as a Nordmann Grate. Most importantly, and as a result of the problems encountered in the application and enforcement of the 1986 Technical Regulation the Council adopted Regulation No. 850/98 which, *inter alia*, limits the range of mesh sizes that Community vessels may carry on board, aims to reduce the practice of discarding, lays down mandatory provisions of square mesh netting, specifies that certain undersized marine organisms (the species and sizes are listed in Annex XII) shall not be retained on board or be transhipped, landed, transported, stored, sold, displayed or offered for sale, but shall be returned immediately to the sea. The measurement of the size of marine organisms is set out in Annex XIII of Council Regulation No. 850/98.

2. *Provisions in the Control Regulation*

(i) Monitoring of the use of fishing gear

The 1993 Control Regulation did not introduce any new technical measures with respect to fishing gear. However, it left open the possibility of the adoption of a 'one net rule', and laid down the requirement for the Council to examine this subject. In any case, while the Council did not adopt a one net rule when it adopted Council Regulation No. 850/98, it nevertheless restricted the range of mesh sizes combinations that vessels may use from the year 2000.

All catches retained on board any Community fishing vessel shall comply with the species composition laid down in Technical Regulation (as from 2000, Council Regulation No. 850/98) for the net carried on board that vessel. **Article 20(1)** of the Control Regulation is linked with the Technical Regulation by reiterating stringent requirements in regard to the stowage of nets not in use. In particular it states that nets on board vessels which are not used must be stowed so that they may not be readily be used in accordance with the following conditions:

(a) nets, weights and similar gear shall be disconnected from their trawl boards and towing and hauling wires and ropes;
(b) nets which are on or above deck shall be securely lashed to some part of the superstructure.

In this respect there is no change in the Community Regulations regarding the carriage of nets. It is still permitted to carry nets of different mesh sizes provided that the mesh sizes are in compliance with the combination of mesh sizes listed in the relevant Annex of Council Regulation No. 850/98. (There is an exception however which pertains to vessels authorised to fish outside the Community zone, pursuant to **Article 4(3)(b)**, Council Regulation No. 850/98).

Article 20(2) is linked to **Articles 6** and **7** in so far as it stipulates the requirement to record in the logbook changes in mesh size, and the catch composition pertaining to different mesh sizes, both in the logbook and the landing declaration. Moreover, detailed rules on the keeping of a storage plan by species, of products processed, indicating where they are located in the hold, shall be adopted in due course in accordance with the procedure laid down in **Article 36.**[10]

The final decision on which nets may be carried by fishing vessels is reserved for the Council, which may on the basis of a report from the Commission, and acting on a qualified majority, decide precisely what rules are to be adopted. **Article 20(3)** envisages in the long-term either a one-net size, or more correctly a one-mesh size rule for particular fisheries, and/or specific rules for the use of nets of different mesh size for different fisheries. As noted above, the Council in 1998 did not adopt a 'one-net rule', but opted to restrict the range of combinations of mesh sizes that may be utilised by vessels.

(ii) Preventing the transport or marketing of undersize fish
In the 1998 amendment to the Control Regulation the Council inserted a new paragraph into **Article 28** to reinforce the linkage between technical conservation measures and market and transport measures. Specifically, the new paragraph states that where a minimum size has been fixed for species pursuant to **Article 4** of Council Regulation No. 3760/92, operators responsible for the selling, stocking or transporting of batches of fisheries products of that species smaller than the minimum size must be able to prove the geographical area of origin or the provenance from aquaculture of the products at all times. The Member States shall carry out the necessary controls in order to prevent occurrences that might exist in their territory because of the transport or the marketing of undersized fish.

3. *Commentary on the provisions regarding the monitoring of the use of fishing gear*[11]

Technical measures are a good example to illustrate the difficulties in prescribing comprehensive and practical regulations in fishery management. In this respect, although there may be several analytical studies completed and consultations with the industry prior to introducing new rules, there generally is some party aggrieved when the regulation is adopted and implementation commences. Significantly, the Council left the door open in 1993 on whether the European Community would adopt a 'one net' rule in due course. Obviously the Commission required time to deliberate on the subject. Subsequently, the Commission proposed the introduction of a 'two net' or more precisely a 'combination of nets' rule. After the extensive negotiations, it was agreed at the Council meeting on 30 October 1997, that as of from 1999 all Community vessels operating in specified areas will be limited to carrying a prescribed combination of nets during particular fishing trips. The range of combinations is set out in Annexes I to XI of Council Regulation No. 850/98.

From the enforcement perspective, however, several difficulties remain. It is difficult for example to enforce successfully any provision that requires the perpetrator to be apprehended *flagrante delicto* with an illegal-sized net in use. Thus, if enforcement considerations are taken seriously there was probably little real alternative but to restrict the carriage of nets of different mesh sizes. However, enforcement problems have to be balanced against the viability of fishing as an economic activity. Some vessels will always require to operate in several different fisheries targeting different species with different types of gear, and in such instances it would not be possible to comply with a one-mesh rule. In this context the 1998 Technical Regulation appears to be a pragmatic intermediate position which allows vessels to embark different gear for different fisheries and restricts the range of mesh sizes retained on board. It also expressly prohibits the use of devices by means of which the mesh in any part of the fishing net is obstructed or otherwise effectively diminished.

It must be also pointed out that within the Control Regulation there is ample scope for tightening regulatory compliance with respect to specific gear requirements for vessels operating in multi-gear fisheries. The co-operation procedure envisaged in Title VIIIa should be relied on to allow national authorities in one Member State to contact inspection authorities in the home port on a regular and informal basis and to obtain additional information regarding inspections undertaken at sea.[12] Consideration should also be given to the problem of detecting illegal practice, in this regard there is a valid argument that should a vessel slip its gear prior to inspection or during the course of inspection then there should be a presumption of malpractice, which the skipper would have to rebut. An alternative might be to assist inspection by requiring skippers not to haul their gear until the inspectors are on board the vessel and thus can observe the retrieval of the gear in use. The latter procedure is practised by several national inspection authorities in the waters under their jurisdictions or sovereignty.

In the early period of the policy, fleets in the Member States continued to catch undersize fish in flagrant violation of Community measures. Indeed, some national authorities omitted to enforce such measures. As noted above, France was censured by the Court of Justice for not observing Community rules on minimum sizes between 1984 and 1987.[13] The Court heard that the French Government by implication admitted during the pre-litigation procedure that the French Secretary of State had given instructions that only hake that were patently undersized should be seized by the authorities despite express Community regulation on the minimum size.[14] The Court, by censuring France, clearly contributed to the campaign to ensure Member States do not violate their obligations.

It should also be pointed out that the Commission reports on control, discussed in Chapter 3, specifically mention undersize fish as one of the principal enforcement issues to be addressed by the national authorities in the Member States.

(i) Use of national technical conservation measures

It must also be stressed, on the other hand, that there is the possibility that Member States may adopt technical measures which go beyond the minimum requirement in Community regulations, provided these comply with Community law and are in conformity with the CFP.[15] The Court upheld in *Officier Van Justitie* v. *L. Romkes* minimum fish size regulations which were more rigorous than those specified in Community regulation.[16] Similarly, in *Procurator Fiscal* v. *Andrew Marshall* the Court upheld a national measure which prohibited Scottish fishermen from using a particular type of net in certain waters adjacent to certain parts of the coast even though such gear was otherwise in conformity with Community law.[17] Significantly, in the latter case, the Court appears to have an excellent grasp of the enforcement difficulties in relation to this type of gear. The Court justified the national measures on the grounds that nets must be prohibited in view of their particular efficiency and because inspection checks were difficult to carry out in the area of sea in question in view of the length of coastline.[18] In this instance the gear in question (monofilament net) was used for catching salmon which had to cross the specified areas in order to reach the rivers in which they spawn. Moreover, the Court was of the opinion that it was not arbitrary that the national authorities opted to prohibit the carriage of such nets rather than to step up checks in order to show that the nets have been used for illegal purposes.[19] The findings of the Court in these two cases appears to be consistent with the reasoning followed by the Court in their judgement in *Commission* v. *United Kingdom* to the effect that even though power to adopt measures relating to the conservation and management of the resources of the sea has passed fully and definitively to the Community, it is not entirely impossible for the Member States to amend the existing conservation measures as a result of developments in biology and technology.[20] This is subject to the condition that such amendments are of limited scope and do not involve a new conservation policy on the part of a Member State.[21]

(ii) Technical conservation outside the European Union
Technical measures play an important role in conservation regimes outside the European Union. The conservation philosophy in Norway requires skippers of fishing vessels to change fishing ground if the intermixture of fish under minimum size exceeds a certain level. The focus of control and enforcement effort has gradually shifted from concentrating on technicalities like mesh sizes and attachment to nets to more general solutions like area closures. In order to achieve this there is a Special Surveillance Unit tasked with collected data on the level of undersize fish. Decisions are made after consultation with scientific advice. This policy is supported by Russia which shares joint stocks with Norway in the Barents Sea. It is also reported to be fully endorsed by the Norwegian fishing industry. In addition the Norwegian Directorate of Fisheries has expended research funds in developing improved gear selectivity and has developed a grid sorting system in shrimp and cod trawls. The Norwegian conservation policy is substantially different from that of the European Union; time will reveal which is the more successful.[22] The fishery management system in Japan relies on voluntary control from fishing communities and is based on consultation with fishery co-operatives. These organisations undertake some of the management function by shortening opening seasons and strengthening minimum size limits.[23] In Iceland there is no 'one net rule' and supervision of fishing activities at sea are carried out by the Icelandic Coast Guard whose principal task is to monitor closed areas and check fishing gear. During 1992 the Icelandic Coast Guard made 91 inspections in which catches and fishing equipment were inspected.[24] Considering the size of the Icelandic EEZ, and the diverse nature of the fishing management plan, it must be questioned whether 91 boarding operations are sufficient to expose the level of gear malpractice or to be a sufficient deterrent.

4. *Enforcement problems associated with the procedure for the measurement of the mesh size used in trawls, Danish seines and similar towed nets*

Although it is clear that from the enforcement perspective a 'combination of nets rule' will go a long way towards easing the control problems associated with some technical measures, nonetheless, it should be mentioned that there are several technical difficulties associated with the procedure for measuring nets (trawls, Danish seines and similar towed nets). In particular, the fishery inspector who has the initial decision in deciding whether the gear conforms to the regulations has to use the prescribed measurement apparatus for assessing the size of the net laid down in Commission Regulation No. 2108/84.[25] **Article 1** of this regulation describes the prescribed tool for measurement, a mesh measuring gauge. From a scientific point of view, the gauge is clearly not a precision instrument. Its principle weakness derives from the fact that it entails manual measurement which involves an element of uncertainty. For example, the harder the inspector applies pressure in inserting the gauge into the mesh the greater the diameter of the mesh will extend. This procedure has been described as akin to measuring the diameter of an elastic band. A variety of other devices have been researched, including spring loaded devices which are intended to apply constant pressure. Community legislation on the measurement of nets aims to resolve the shortcomings in manual measurement.[26] The regulations prescribe the precise sequence of the inspection procedure. In particular **Article 6** of Commission Regulation No. 2108/84 endeavours to standardise the force applied by stipulating that after a series of manual measurements have been carried out and, if the captain of the vessel contests the mesh sizes then, there remains the application of a weight or dynamometer. This is no easy task considering that the measurements may have to be made at sea in perhaps inclement and hazardous conditions.

The whole procedure for net measurement has been the subject of judicial examination. The Court of Justice in *Commission* v. *French Republic,* censured the national authority which, during 1987 and 1988, did not have the gauges provided for by Community rules to measure nets or used gauges which did not conform to those rules and for applying standards which were less strict than those in Community rules.[27] More recently, in *Criminal Proceedings Against Jelle Hakvoort* the *Amtsgericht Bremerhaven – Germany* made a referral to the Court of Justice for a preliminary ruling on the issues of mesh size, the technical rules for determining mesh size and the inspection procedure.[28] The origin of the dispute arose from the procedure adopted by the national inspectors in measuring Mr Hakvoort's net. The German inspectors did not follow the precise manual procedure set out in Commission Regulation No. 2108/84 but opted for a procedure which involved the use of the prescribed weight.[29] Mr Hakvoort based his challenge on the grounds that the procedure whereby the offence had been established was irregular. The Public Prosecutor's Office in Bremen argued that the inspectors were entitled to decide at any time to carry out a measurement using a weight. Considering the resolution of the dispute depended on the interpretation of the Community rules the *Amstgericht* (Local Court) stayed the prosecution and referred three questions to the Court of Justice. These were:

'(1) Is **Article 2(1)** of Council Regulation No. 3094/86 to be interpreted as meaning that a mesh size for the purposes of that provision is only one which has to be determined in strict compliance with the procedure for determining mesh sizes laid down in Commission Regulation No. 2108/84?

(2) Is Article 6 of Commission Regulation No. 2108/84 to be interpreted as meaning that the inspector must in every case begin by determining the mesh size of a net by manual measurement of 60 meshes and is only empowered to determine the mesh size by measurement using a weight in accordance with **Article 6(2)** of Regulation No. 2108/84 if the captain raises an objection to the first mentioned measurement or is the inspector free to undertake the measurement using a weight ... without first carrying out a manual measurement ... and in the absence of any objection on the part of the captain, and may he base his decision on whether there has been an infringement of **Article 2** of Council Regulation No. 3094/86 on that measurement alone?

(3) Are the provisions of Regulation No. 2108/84 intended also to protect the captain of the vessel undergoing the inspection?'

In reply to the first question, the Court cited its judgement in the *Kramer* case and pointed out that the Community alone is competent to adopt technical rules for determining mesh sizes and that the technical rules for determining mesh sizes referred to in Council Regulation No. 3094/86 are those contained in Commission Regulation No. 2108/84 on the determination of mesh sizes.[30] The Court then moved to the third question and noted that Regulation No. 2108/84 in laying down the procedure the inspectors are to follow (which in its view not only determines the limits of the offence itself but is also aimed at protecting the biological resources) is intended to protect the captain of the vessel inspected.[31] On the second question, the Court deduced that it was clear from the very wording of the provision that measurements by means of a weight may be carried out only *after* manual measurements have disclosed an insufficient mesh size *and* provided the captain contests the result.[32] The Court drew support for their finding by noting the Commission's observations that the two methods of measurement, manual and by applying the weight, do not always lead to the same result. In particular the former method, even if less accurate than the latter, has the advantage of covering a more representative sample of the net and may therefore reveal a larger mesh size. It was the Court's view that the procedure laid down in the Commission Regulation was thus intended to protect the captain and it followed that the complete performance of the manual inspection is the prerequisite for measurement with a weight.[33] As noted by Advocate General Van Gerven, the Commission in its observations pointed out that the weight procedure was not initially chosen to be in the regulation because it is much more complicated and time consuming.[34] That is why it was decided to use that method only if the captain of the vessel contested the results of the manual measurement. Furthermore, measurement with a weight or dynamometer is difficult to carry out on a vessel which is heaving about at sea and in those circumstances is not necessarily more accurate than manual measurement.

5. *The measurement of passive gear (gill nets, entangling nets and trammel nets)*

In 1997 the Commission adopted a new regulation which specifies the type of gauge to be used for the measurement of mesh size of nets used as passive gear as well as the procedure for measuring meshes and the calculation of the mesh size.[35] Passive gear is defined as gill nets, entangling nets, trammel nets which may consist of one or more separate nets which are rigged with top, bottom and connecting ropes, and may be equipped with anchoring, floating and navigational gear. This regulation will be significant when the new range of technical measures enters into force in the year 2000. It should also be mentioned that in cases where the master or the skipper of a vessel disputes the result of measurement in the course of inspection, there are provisions in the regulation for further and final measurement.

Section (ii) The enforcement of certain measures to improve and adjust structures in the fisheries sector

Introduction

The review of the CFP in 1992 noted that the success of the CFP requires the implementation of an effective system of control covering all aspects of the policy.[36] To achieve this it was necessary to extend the ambit of the control regime to include rules not only for the monitoring of conservation and resource management measures, but also for structural measures, and measures on the common organisation of the markets.[37] Thus, the new control framework has three distinct but interrelated features which cover the three limbs of the policy.

The significance of monitoring the structural element of the policy may be appreciated when one considers that the then President of the Commission, Mr Santer, while mapping out the broad policy framework for the period 1995–2000, identified the principal problems to be tackled in the CFP as the chronic over-capacity of European fishing fleets, the frequent threat to resources, and certain crises of essentially structural origin.[38] The identification of the structural component of the CFP as the principal priority in the Commission Work programme for 1998 and subsequent years is an acknowledgement that the structural crisis is the root of many of the problems which have undermined the CFP in recent years.

The decision to introduce a new structural package for the latter period of the CFP was made in the context of two guiding principles: first, that the Community fishing fleet was too large in proportion to the level of resources available in the Community zone; second, that there was a need to harmonise the structural policy with the objectives of the conservation policy. These principles have to be considered against the background of several factors which have conditioned the content and substance of the new policy. These in turn may be grouped into three categories: first, the failure in the initial period of the policy to tackle some fundamental weaknesses in the content and implementation of structural measures; second, the need for a more dynamic and flexible approach to structural issues; third, the challenge to link the structural policy to fishing activity and the conservation policy.

Past failures were clearly identified in a Court of Auditors' Report in 1993. While this report did not focus on the link between structural policy and fishery enforcement, it did however identify several weaknesses or oversights in the structural policy framework which are particularly germane in view of the technical checks and enforcement measures introduced by the 1993 Control Regulation. The report concluded that the Commission had not made use of all possibilities provided for in the legislation to speed up the process of standardising the measurement of tonnage of Community fishing vessels. Furthermore, that the absence of any clear position on the derating of engines led to uncertainty as to the overall engine power of Community fishing vessels (tonnage and kW power are two of the units used to measure the capacity of the Community fleet). Moreover there was, *inter alia*: the need to intensify efforts to harmonise the units of measurement of fishing capacity for the different Community fleets; the need to harmonise the Community fleet register so that it can be used for all areas of the common fisheries policy; the need to structure the Multi-Annual Guidance Programmes (MAGPs) so as to establish a stronger link between Community financial aid and the policy on the management and conservation of resources; the need to restrict capacity development aid measures to cases where the available resources may be under-fished, rather than to exclude them outright; and finally the need to ensure that aid is paid out in strict compliance with the provisions of the regulations and the conditions of granting assistance. The Court of Auditors also concluded that the Community financial aid programme to adjust the capacity of the Community fleet during the period 1987–1990 had not achieved its aim. Moreover there was poor financial management and this resulted in an aid programme which was too generous in the context of the stringency which the fleet overcapacity situation obviously required. In the context of the Commission work programme, and the Court of Auditors' Report, the measures prescribed in Title V of the Control Regulation have an increased significance.

The need for a more dynamic and flexible approach to the structural policy is reflected in the introduction of measures which the Community has adopted to deal with the excessive fish catching capacity of the Community fleet. The new measures are the '*segmentation*' of the fleet into different parts and the limitation of '*fishing effort*'.[39] These have recently become part of the Multi-Annual Guidance Programmes (MAGPs).[40] MAGPs generally run for a four-year period and are intended to fix fleet reduction targets to bring fish catching capacity into line with available fish resources. In the first series of MAGPs capacity was reduced simply by withdrawing and permanently laying up a certain number of vessels. This approach was not particularly successful and radical changes such as the segmentation of the Community fleet and fishing effort became necessary. In MAGP III, covering the period 1993–1996, up to 45% of the capacity reduction objectives imposed on Member States may be achieved by reducing the fishing effort of the national fleets. The concept of fishing effort was introduced in order to supplement 'the range of measures available to the Member States to attain the balance between the fishing capacity of their fleets and the resources available'.[41] For the purpose of the attainment of the structural policy, and in particular the MAGPs, Fishing Effort is defined as a function of Fishing Capacity and the Level of Fishing Activity. As explained in Chapter 4, Fishing Effort can thus be expressed as:

Fishing Effort = Capacity × Activity
= tonnes × days at sea

or

= power × days at sea

On the one hand, the level of fishing activity can be identified by using criteria such as 'time spent at sea', 'time spent fishing', 'time spent searching', 'soakage times', etc. On the other hand, capacity depends on such criteria as tonnage, propulsion power, auxiliary power, fish storage, gear parameters, etc.

The concept of fishing effort and the idea of limiting the number of days fishermen could fish was not accepted by the industry and its attempted introduction resulted in proceedings for judicial review brought before the High Court in the United Kingdom by a number of fishermen's organisations.[42] The High Court referred four questions to the Court of Justice for preliminary ruling under **Article 177** {234} of the EC Treaty on the interpretation of the Treaty,[43] the Council Regulation on the common organisation of the markets,[44] the Council Regulation on the Community system for fisheries and aquaculture,[45] and the Commission Decision for a Multi-Annual Guidance Programme for the fishing fleet of the United Kingdom,[46] as well as certain general principles of Community law. The Court of Justice in the *United Kingdom Days-at-Sea Case* ruled that the Council Decision No. 92/593 on a MAGP for the fishing fleet of the United Kingdom for the period 1993–1996 must be interpreted as empowering the United Kingdom to limit the number of days a year that vessels over 10 metres in length may spend at sea in so far as a maximum of 45% of the total target set in that decision may be achieved by measures other than the reductions of the capacity of the fishing fleet.[47] Subsequently, the United Kingdom did not introduce a days-at-sea regime but the principle of effort limitation is still important and it has been carried over into the MAGP IV.[48]

While most Member States support the concept of the physical withdrawal of vessels from fishing and the scrapping of vessels as the best way of eliminating over-capacity, some are particularly keen to retain the new combined approach of reducing capacity *and* fishing effort. This is particularly significant because the Community has developed rules for managing effort in specific fishing grounds such as 'Western Waters' and the 'Baltic', in parallel to the arrangements for the management of fishing effort by segment of the fleet in the MAGP.

1. *Provisions in the Control Regulation on inspection and control of certain measures to improve and adjust structures in the fisheries sector, including aquaculture*

There are four articles dealing with structural matters in Title V. Taken together they add a new dimension to the Community control regime in so far as they provide a new obligation on Member States to ensure compliance with the objectives and strategies set down by the Council to reduce the fishing capacity of the Community fleet. Each Member State is obliged, pursuant to **Article 24**, to organise regular checks, in its territory and under its sovereignty or jurisdiction, of all persons concerned by the implementation of the objectives and strategies set by the Council in accordance with Council Regulation (EEC) No. 3760/92, **Article 11**. In particular the objectives concerning the fishing capacity of the Community fleets and the adjustment of their activities.[49]

Member States are obliged to adopt provisions to verify compliance with the **Article 24** objectives. To achieve this **Article 25(1)** specifies a number of areas where there is an obligation to conduct technical controls. These controls should occur in the context of the restructuring, renewal and modernisation of the fishing fleet; the adaptation of fishing capacity by means of temporary or definitive cessation;[50] the restriction of the fishing activity of certain fishing vessels; and the restriction of design and number of fishing gear and of the method by which it is used.[51]

If the Commission has established that a Member State has not complied with the provisions of **Article 25(1)**, it may, pursuant to **Article 25(2)**, without prejudice to **Article 169** {226} of the EC Treaty, make proposals to the Council for the adoption of the appropriate general measures.

Detailed rules for the application of **Article 25** may be adopted in accordance with the Management Committee procedure for the verification of engine power, tonnage, periods of immobilisation, as well as specification of fishing gear and their number per fishing vessel. Member States are obliged to inform the Commission about the control method used and the bodies responsible for carrying out such verification.

The enforcement difficulties inherent in this approach are further complicated by the requirement of cross verification in **Article 27(1)**. This provides that Member States shall establish a system of validation which includes, in particular, the capacity to verify and compare by cross reference information on fleet fishing capacity and activity which is available from a number of sources such as the logbook (**Article 6**), the landing declaration (**Article 8**) and the register of Community fishing vessels.[52]

Furthermore, Member States are required to establish or complete existing databases containing the information about the fleet fishing capacity and activities.

2. *Enforcement of the measures relating to the structural policy*[53]

The aim of extending control of the structural measures is to enable national authorities to improve the supervision of their fishing fleets and to provide a mechanism whereby it is possible to monitor compliance with the objectives of the MAGPs.

The monitoring of compliance by the Member States with the reduction in fishing capacity or adherence to fishing effort restrictions is a separate and different type of enforcement task from that traditionally performed by national fishery inspection services.

The methodology used by national inspection services in carrying out the technical controls may vary. However, if viewed schematically it appears to fall into two distinctive schemes of inspection. For example, the first inspection method could involve an administrative or 'paper check' of vessel certificates and classification. This would entail the evaluation of all aspects regarding the technical preparation of documents and the classification of vessels within particular categories. The second inspection method could involve a physical survey of vessels to ensure that the data recorded in certificates and/or classification certificates are physically valid. The latter type of inspection requires a detailed knowledge of the principles of naval architecture and the practice of vessel construction. In this respect, it could be argued that, the most effective physical control and inspection is the one undertaken at the time the vessel is constructed or modified. Inspection at these times could be more detailed and would allow for a preventive control which would ensure that the vessel is built to the prescribed dimensions, range of engine power, propulsion efficiency and indeed that the building project corresponds to the requirement set out in the appropriate MAGP. Physical control of the structural elements of the vessel may also be undertaken when the vessel is in service. From the practical point of view, this ought to be undertaken by an authorised Marine Surveyor or Classification Inspector while the vessel is in port.

The type of checks envisaged in the Control Regulation fall into three categories which involve, first, an assessment of gross tonnage which is a basic naval architectural calculation; second, the measurement of dimensions, (i.e. the length of the vessel etc.); and third, the measurement of power at its maximum output.[54] Together these measurements allow the surveyor to establish the maximum effort during the fishing haul and at a fixed point (at zero speed) produced by the propulsion system.[55]

The monitoring of the adaptation of fishing capacity by means of temporary or definitive cessation should be easier to manage as required by **Article 25(b)**. In this regard, the entry in the register of a ship from another State ought to be checked to authenticate the history of the vessel and to provide evidence that the vessel had not been previously deprived of the right to fish in Community waters either because of the fact that it has been granted final cessation aid or because it has been offered withdrawal in support of a ship-building project.[56] Similarly, there is a need to ensure that vessels that fly the flag of a Member State which have been temporarily chartered by operators from another Member State, are not ultimately taken into consideration for the MAGP of the Member State concerned.

Even if the control authorities in the Member States adopt the approach outlined here to the inspection task and organise administrative and physical checks there are still some difficulties to overcome before attainment of efficient enforcement and control objectives. Some of these are examined below under a number of sub-headings.

3. *Difficulties with monitoring the structural policy*

(i) Fishing vessel register of the Community
The first difficulty relates to the accuracy of the fishing vessel fleet register of the Community which is one of the key components in the structural policy. The Register is located in the European Commission Office in Brussels.[57] The fleet register is not a shipping register in the maritime legal sense because all Member State vessels are registered on their respective national shipping registers in accordance with the relevant municipal legislation.[58] In this regard the French term for the fleet registry, *Fichier Flotte* (fleet database), is a more precise description from the legal point of view. Registration on the Community fleet register does not affect the rights of vessels to fly the flag of the Member States or any other entitlement or obligation arising from Convention or custom. The Community fleet register is merely an electronic database containing information on the physical characteristics of the vessels in the Community fleet. There are also data on the fishing gear used by vessels and the segment of the fleet to which the vessel belongs, as well as the activity of the vessel. The Member States forward to the Commission the information necessary to establish and manage the register. This includes information to manage the amount of fishing effort undertaken by their respective vessels.[59] The accuracy of these data is fundamental to ensure the proper implementation of the MAGP. In particular, in order to determine fishing effort, accurate data are required on the fishing effort of individual vessels or the effort of homogeneous groups of vessels. The accuracy of the Community register has been criticised.[60] In particular, inaccuracy has been caused by *inter alia*, changes in the name, owner, type of fishing, licence, structural modifications and engine or engine powers of the vessel. In practice, the registration of these changes by the national authority is slow. In some Member States a contributory factors to this tardiness is that paperwork is processed by both fishermen's representative organisations and Classification Societies and thus may involve duplication in administrative procedures. Ideally, the Community register should be sufficiently accurate to provide the basis for common parameters for measuring the capacity of vessels and so enable new parameters for the fishing effort to be introduced.

Ultimately, an accurate Community register will greatly facilitate the monitoring, development, and implementation of the MAGPs. In some Member States, however, the technical specification and certification of fishing vessels is not carried out by the same authority which is responsible for fishery control and inspection, and this in some instances may lead to incomplete or inaccurate data in the fleet register. On the other hand, it may be argued that the expertise to undertake the type of technical inspection required by the structural measures may not be available in the national authorities charged with the control and inspection function. The type of organisation varies from Member State to Member State. Thus, for example, in Belgium the national authority responsible for the verification of vessel units such as engine power or tonnage is the Ministry of Communications, whereas the *Ministerie van Verkeer een Infrastructur and Vlaamse Gemeenschap* is responsible for the verification of fishing activity. Traditionally, in the United Kingdom the Department of Transport is responsible for the verification of vessel units such as engine power or tonnage, but it is the Ministry of Agriculture, Fisheries and Food (MAFF) or the Scottish Fishery Protection Agency (SFPA) and Marine Surveyors who are responsible for the practical realisation of the inspection obligation set out in the Control Regulation. This responsibility passed to the Coast Guard in 1998.

Finally, it needs to be pointed out that if the enforcement function is properly discharged in the Member States and reveals inaccuracies or inconsistencies in relation to the national fleets then it is essential that such observations are recorded in the Community fleet register.

(ii) Units/parameters identified for monitoring fishing capacity

The second difficulty in relation to the monitoring of the structural elements relates to units/parameters identified for monitoring fishing capacity. The four units/parameters are: engine power (measured in kilowatts); gross tonnage; amount of time not fishing; and the specification of gear. Member States have not relied upon uniform criteria to measure these units. Hence, there were major discrepancies between the units of data supplied by the Member States in connection with the MAGPs and the units recorded in the Community fishing register during the period reviewed by the Court of Auditors.[61] Subsequently, considerable effort has been made to harmonise the units of measurement of capacity of the different fleets. In the relevant Community Regulations, two parameters are used to measure the fishing capacity of a vessel: the engine power in kilowatts (kW) and the tonnage. According to Council Regulation (EEC) No. 2930/86 of 22 September 1986 the tonnage of a vessel is gross tonnage as specified in Annex I to the International Convention on Tonnage Measurements of Ships ('the 1969 London Convention'), whilst the engine power is the continuous engine power determined in accordance with the requirements adopted by the International Organisation for Standardisation ('ISO standard').[62] The standardisation of units is particularly important considering that the Court of Justice had been previously asked to rule in *Officier van Justitie* v. *Bout*[63] how gross registered tonnage was to be calculated. Noting the lacunae in Community law in this regard, the Court found that 'in the absence of Community regulations it is for the Member States to determine the calculation method to be followed'.

A second problem which arises is that installed engine power is difficult to measure and, as experience in enforcement of regulations based on engine power has shown, easy to manipulate by the industry.[64] In a reply to a Parliamentary question in 1993, Commissioner Paleokrasses agreed that there was no feasible means of checking the real power of a ship's engine during inspections at sea and that the verification of engine power is done using the official documents or certificates issued by the competent authority.

(iii) Verification of the period during which a fishing vessel is immobilised[65]

The third difficulty relates to verifying fishing activity. The issue of 'Days at Sea' or restricting time spent fishing is the most controversial of the structural measures.[66] It should be stressed, however, that both the Netherlands and Denmark have national legislation which has established a restricted number of 'Days-at-Sea' regime for certain category of vessels for some time. Spain has been subject to a *de facto* 'fishing effort' regime which was imposed by the Act of Accession. These methods of restricting fishing capacity by means of temporary or definitive cessation, or by plain restriction of the activity of certain fishing vessels, has been the subject of major criticism from the industry on the grounds that it has major social and economic repercussions for many peripheral communities which have few alternative means of employment. However, from the point of view of enforcement, if the plan for a Community system of satellite monitoring of fishing vessels is fully implemented then the verification of the period during which a fishing vessel is immobilised should be more easily established.[67]

(iv) Verification of vessel category and fishing gear type
In this case, part of the inspection task would be to establish the most accurate classification of the vessel from the point of view of activity and to ensure the vessel is recorded in the correct segment for the purpose of MAGPs.[68] This type of inspection maybe undertaken at sea or in port by national inspectors and is closely linked to the monitoring of technical measures. However, there is one exception in relation to the activities and types of gear deployed by 'polyvalent vessels' (vessels which are authorised to fish with more than one type of gear). The classification and activity of this type of vessel is more difficult to assess. Obviously, vessels which used towed gear and static gear pose a particular problem (longliner/purse-seiner), particularly, as the fishing capacity of these vessels is defined as a function of tonnage (GT), total installed power (kW) and number of vessels. Thus the record of fishing effort from the use of towed gear and the fishing effort from the use of static gear need to be clearly distinguished because different parameters are utilised to regulate or limit the use of the latter from the structural perspective.

(v) Cross-checking information in order to establish the level of fishing activity
The cross verification procedure is an important feature in the new control system. The Court of Auditors had recommended that in the case of capacity withdrawn to make way for the building of a new vessel then the fishing potential should be precisely documented in terms of volume of catches, with confirmation from the authorities responsible for inspecting catches. In this regard the other aspects of the control regulation such as the requirement to keep a Community logbook, the requirement to complete a landing declaration, the requirement to submit effort reports in Western Waters and in the Baltic Sea, may be invaluable for verifying fishing activity.

(vi) Failure to carry out the prescribed technical checks
Without prejudice to **Article 169** {226} actions, **Article 25(2)** of the Control Regulation provides that if the Commission establishes that a Member State is not carrying out technical controls then it may make a proposal to the Council to take the 'appropriate general measure'. The article does not elaborate on what these measures are and it is therefore difficult to predict whether they will be non-punitive and entail the withdrawal or cancellation of Community financial aid or be punitive by imposing a financial penalty on the Member State in question. A literal interpretation of **Article 25(2)**, however, suggests that in such instances the Council could adopt the

appropriate inspection and control provisions to verify compliance with the measures to improve and adjust structures in the fisheries sector. In this context, such provisions could extend to prescribing the technical controls to be undertaken by the Member State in question.

(vii) Flag of convenience vessels
The fishing capacity of Member States' fleets has been affected by several factors including the decisions of the Court of Justice in the *Quota Hopping* cases. In particular, the capacity of the United Kingdom fleet has increased as a result of highly productive Spanish vessels on the UK register in order to fish UK quotas.[69]

Similarly, the practice in the early period of the MAGPs of exporting vessels to countries who operate flag of convenience vessels outside the Community zone, other than ensuring that these vessels were operating in direct competition with the Community fleet in international fisheries without being subject to the same obligations, posed an enforcement problem which was beyond the legislative remit and resources of Community enforcement authorities. The Council adopted a Regulation in 1993 which provides that Member States are obliged to take measures to stop vessels' fishing activity permanently or to restrict them, and if such measures include permanent transfer to a third country this must only be undertaken provided such transfer is not likely to infringe international law or affect the conservation and management of marine resources.[70] In this regard, the Community legislative initiative accords with the general thrust of several international law legal instruments which aim to curtail the activities of flag of convenience vessels. These are discussed in Chapter 11.

Section (iii) The common organisation of the market in fishery products

Introduction
The third strand of the control regime introduced by the Control Regulation consists of those provisions to monitor certain measures applicable to the common organisation of the market in fishery products.

The requirement to improve the enforcement of measures relating to the common organisation of the market came about because of several factors outside the influence of the market regulators in the Community and the Member States. These factors include: the arrival of large scale distributors with industrial requirements such as the need for regular supplies at fixed prices with quality guarantees; the increased scale of processing of fishery products because of changes in consumption patterns; reduced transport costs and improved fish conservation techniques which have eclipsed traditional micro-markets and increased competition on what has become a world market; increased competition because of the widespread availability of aquaculture products; and a Community market which has a deficit of fishery products supplied by the Community fleet. There have also been international trade agreements such as the General Agreement on Tariffs and Trade (GATT) and World Trade Organisation (WTO) Agreements which have had a major effect on the Common Custom Tariff (CCT) and lead to a number of legal constraints on the common organisation of the market in fishery products. The CCT which is sometimes referred to as the Common External Tariff (CET) is the Community mechanism whereby duties are imposed on goods coming into the Community. In principle, imports and Community products are treated equally which rules out the possibility of tariff and non tariff protection. Furthermore, the import regime is subject to the principle of trade liberalisation which prohibits quantitative restrictions or measures with an equivalent effect. It also requires that the same marketing standards apply to imports as to Community products. There is also the prohibition of any form of direct or indirect aid for the production and/or processing sector which might distort competition. Among the more important elements introduced by the Control Regulation are the measures to ensure that there is a degree of synergy between the market and conservation policies. In particular there is a need to ensure that there was coherence between the enforcement of the minimum sizes of fish as tabulated in the technical conservation measures and the minimum marketable fish size measures. Moreover, improved control would identify if the common organisation of the market intervention mechanism reinforces or detracts from the resource management policy.

In order to integrate the market policy into the Community control framework and to increase the level of monitoring, Title IV of the Control Regulation provides for the inspection and control of certain measures concerning the common organisation of the market in fishery products.

1. *Measures introduced by the Control Regulation*

Article 28(1) stipulates that in order to ensure compliance with the technical aspects of the rules regarding the measures defined in Council Regulation (EEC) No. 3759/92 of 17 December 1992,[71] on the common organisation of the market in fishery and aquaculture products, each Member State shall organise on its own territory regular checks of all persons involved in the application of the measures.

The checks are set out in **Article 28(2)** and shall concern the technical aspects of applying:

(a) the marketing standards and, in particular, minimum sizes;
(b) the price arrangements and, in particular, withdrawal of products from the market for purposes other than human consumption; storage and/or processing of products withdrawn from the market.

Furthermore, Member States shall carry out comparisons between the documents relating to the first placing on the market of the quantities referred to in **Article 9** (sales note) and the quantities landed referred to by the documents, particularly as regards their weight.[72]

As noted above, the 1998 amendment to the Control Regulation the Council inserted a new paragraph into **Article 28**. The new paragraph states that where a minimum size has been fixed for species pursuant to **Article 4** of Council Regulation No. 3760/92, operators responsible for the selling, stocking or transporting of batches of fisheries products of that species smaller than the minimum size must be able to prove the geographical area of origin or the provenance from aquaculture of the products at all times. The Member States shall carry out the necessary controls in order to prevent occurrences that might exist in their territory because of the transport or the marketing of undersized fish.

To monitor the implementation and enforcement there is a notification obligation placed on Member States pursuant to **Article 28(3)**. This requires the Commission to be notified of information relating to the Control measures adopted, the responsible competent control authorities, the type of infringements discovered and the action taken.

For commercial and fair procedure reasons this notification requirement is subject to two clauses. The first obliges the Commission, the competent authorities in the Member States, and officials and other agents, not to divulge information acquired pursuant to **Article 28** which is covered by professional secrecy. The second, pursuant to **Article 28(4)**, requires that **Article 28** shall not prejudice national provisions concerning the secrecy of legal proceedings.[73]

2. *Enforcing market measures*

Enforcing compliance with market measures is a new departure for the control regime. It poses major legal and technical problems for the enforcement authorities in the Member States. The legally-authorised market supervisory authority in Member States is frequently different from that which is responsible for fisheries control and enforcement. For example, checks on minimum sizes are undertaken by national fisheries inspectorate based in ports in the United Kingdom, Ireland and Denmark, specialised services in France and Portugal or a combination of both in other Member States. Furthermore, in some Member States the legal competence to carry out the enforcement function of market regulations is delegated on a regional basis or on an institutional basis. Moreover, the degree of integration between the competent authorities and the control/inspection authorities and how they verify and exchange information may vary considerably. This fragmentation of national authorities responsible for control is further complicated by the role producers' organisation play in operating the withdrawal of products arrangements and improving marketing on behalf of their members.[74] The areas of responsibility of national veterinary services also overlaps the bodies vested with responsibility to enforce market standards. Specifically, veterinary services are responsible for *inter alia*: the approval and inspection of production areas; monitoring the level of biotoxins in bivalve molluscs,[75] and enforcing the prescribed sanitary and hygiene conditions for fishing vessels, landing places, auctions, and transport.[76]

Article 28 provides for the technical control of marketing standards and price arrangements. The aim of the technical controls in **Article 28(2)** in relation to marketing standards is two-fold. The first is to ensure the Community market rules governing the characteristics of fishery products are applied uniformly across the Community in order to prevent the distortion of competition. In this regard it is intended that the enforcement of marketing standards will also ensure compliance with the Technical Conservation Regulation governing minimum sizes by providing that fish below such sizes shall not be marketed for human consumption. The second aim is to ensure that the market organisation price arrangements are also applied uniformly. It is in relation to the latter objective, that fishery products must be graded on the basis of size categories determined according to weight, or in some species according to size. The enforcement of common organisation of the market in fishery products is thus almost exclusively limited to the pursuance of technical controls.

There is no definition of marketing in the Control Regulation, however, one may be found in the Regulation laying down common marketing standards for certain fishery products. Marketing means the first offer for sale and/or the first sale, on Community territory, for human consumption.[77] The first point of sale is the actual sale of fish from the producer (fisherman) to the wholesaler or retailer. The market provisions in the Control Regulation would thus appear to be also limited in scope up to the point of first sale. This is important because most first-hand sales for human consumption take place through auctions thus facilitating the organisation of inspections. Even if the first marketing is not carried out through auction centres or other bodies authorised by Member States then **Article 9(2)** provides that the products cannot be removed from the landing place until a sales note has been submitted to the competent authority. This requirement should also facilitate inspection for compliance with marketing standards.

The checking of marketing standards and minimum sizes should easily be achieved within the enforcement framework. Fishery inspectors are obliged under the Technical Regulation to check minimum sizes and this is facilitated by the marketing regulation setting standards to cover classification by quality, size or weight, packing, presentation and labelling. Traditionally, grading and freshness checks (undertaken by organoleptic inspection) have been completed by the industry or by the producers' organisation in collaboration with a veterinary/sanitary expert. Fish which do not meet the prescribed classification may not be sold and must be withdrawn from the market. The producers' organisation which intends to operate a withdrawal system is obliged to ensure that the products withdrawn from the market conform to the Community marketing standards. The only authorised exception to marketing standards is fish sold by local fishermen direct to retailers or consumers where marketing standards do not apply. Furthermore, one of the major issues regarding fish which are harvested and sold without being recorded in the catch registration system (frequently referred to as '*black fish*'), or fish recorded in the catch registration system as a different species (frequently referred to as '*grey fish*') is that they do not comply with the prescribed marketing and veterinary standards.

The importance of undertaking frequent checks to ensure compliance with marketing standards in respect of fish put up for sale and not withdrawn, belonging to the same species as fish made subject to withdrawal, is particularly significant in view of the decision the Court of Justice in the *Queen* v. *Intervention Board for Agriculture Products ex parte Fish Producers' Organisation Ltd and Grimsby Fish Producers' Organisation Ltd*.[78] The Court held that the Community rules on the common organisation of the market and those laying down rules for the granting of financial compensation for fishery products must be interpreted as meaning that no financial compensation is to be granted to a producers' organisation for fish withdrawn at the Community withdrawal prices, if such fish having been graded and marketed in accordance with the community marketing standards, and if that producers' organisation had failed to a significant extent to comply with the common marketing standards laid down by the regulation in respect of other fish of the withdrawn species put up for sale but not withdrawn during the same period.[79] Furthermore, the Court ruled that the offering for sale of a product which does not conform with the marketing standards is only to be regarded of limited importance to the rules of financial compensation, if the infringement is occasional and relates to minimal quantities of the product in question, and secondly is not of such a nature as to disturb the market. The burden of proving the limited importance of the infringement rests on the producers' organisation concerned.[80] The importance of compliance with marketing standards is self-evident from the decision in the *Grimsby Fish Producers' Organisation* case. In particular the decision of the Court clearly emphasised the need for rigorous inspection of marketing standards and provided a major fiscal incentive to producers' organisations to ensure adherence in all cases to the applicable Community regulations.

The checking of the technical aspects of the price arrangements may create certain legal difficulties because price arrangements are normally administered by the 'Producers' Organisations', which in administrative law terms is the exercise of a public function by a private body.[81]

The technical check of the withdrawal of products from the market for purposes other than human consumption and storage and/or processing of products withdrawn from the market is an area which should be closely monitored to prevent fraud. In particular the producers' organisation must dispose of the fish for purposes other than human consumption or in such a way that does not interfere with the marketing of the fish in question. In this instance the national inspectors could check that the fish really is converted into oils or animal foodstuffs or distributed to schools or hospitals. Other than the regulatory requirements there may be other valid reasons, such as political reasons, for verifying the withdrawal of products from the market which generally come about because of the poor image such withdrawal creates for the CFP and the industry. It is thus important to assuage the opinion of the Community taxpayers who dislike the process of butter mountains or wine lakes being financed at their expense.

The requirement to check price arrangements in relation to storage and/or processing is necessitated because the Market Regulation (Chapter 2, **Articles 16–18**) provides for a system of private storage aid for 'high value products' which may be granted to the producers by Member States provided the products conform to the marketing standards when placed on the market again. Storage is thus a specialised form of withdrawal. Producers' organisations have no role to play in storage arrangements which usually are undertaken by a duly authorised party in the Member States. There is scope for fraud because producers must conclude a standard type of contract relating to the storage with the appropriate authorities in the Member State. The aim of the contract is to ensure proper preservation of the product in storage and effective supervision of the storage arrangement. The frequent monitoring of contracts and the price arrangements regarding storage provides a realistic deterrent to would-be offenders. It is also necessary in order to ascertain the preservation of the product and conformity with marketing standards.

In general it may be said that the provisions introduced by Title VI of the Control Regulation are timely and appropriate, particularly as the Commission had experienced difficulties in the early 1990s in endeavouring to monitor Member States' transmission of market data in order to ensure that the mechanisms of the market organisation for fishery products were working satisfactorily and that Member States were discharging their obligations.[82] On three occasions the Commission had to open

Article 169 {226} infringement procedures against Member States for failures to respect the rules in force. On all three occasions the Court of Justice held for the Commission. In *Commission* v. *Ireland* the Court held that Ireland had failed to fulfil its obligations under the Regulation on the common organisation of the market on fishery products and on the Commission Regulation which required notification of prices and fixing the list of representative wholesale markets and ports for fishery products.[83] The Court rejected Ireland's argument that the provision of information on an annual basis would be adequate for the Community to fix guide prices.[84] The Court also rejected the argument that it was not possible to deploy its limited number of fishery inspectors at the representative ports for the purpose of gathering the pricing information in question on the grounds that those inspectors are required to monitor some 900 ports and landing places in addition to the representative ports.[85] The Court noted its judgement in *Commission* v. *Italy* to support its view that a Member State may not plead internal circumstances in order to justify a failure to comply with obligations and time-limits resulting from Community law.[86] Moreover the Court had held on several occasions that practical difficulties which appear at the stage when a Community measure is put into effect cannot permit a Member State to opt out of fulfilling its obligations.[87] In a similar **Article 169** {226} action the Court held that Greece had also failed to fulfil the same obligations as Ireland.[88] Again the Court rejected arguments based on internal circumstances such as technical shortcomings, failures on the part of producers and structural difficulties caused by the organisation of administrative services.[89] Interestingly, the Court pointed out that, under the scheme established by **Article 169** {226} of the EC Treaty, the Commission enjoys discretionary powers in deciding whether to institute proceedings for a declaration that a Member State has failed to fulfil obligations and that it is not for the Court to decide whether or not that discretion was widely exercised.[90] The third successful action taken by the Commission against Italy was based on the same grounds as the actions against Ireland and Greece.[91] From the three **Article 169** {226} actions against Ireland, Greece, and Italy it is clear that the Court has upheld and promoted the operation of the common organisation of the market in fishery products. Furthermore, the Court has sent a clear signal to the Member States that nebulous practical or administrative reasons are unacceptable grounds on which to base non-compliance with Community market obligations.

References

1 For a detailed account of the evolution of technical conservation measures in Community law, see Holden, M., *The Common Fisheries Policy: Origin, Evaluation and Future*, 71–87. For a more recent document which deals with enforcement, see 'Communication from the Commission, Implementation of Technical Measures in the Common Fisheries Policy', COM(96) 669 final, Brussels, 15.12.1995. The technical measures which come into force in the year 2000 are prescribed in Council Regulation (EC) No. 850/98 of 30.03.1998 for the conservation of fishery resources through technical measures for the protection of juveniles of marine organisms, *OJ* L 125/1, 27.04.98.

2 This definition is paraphrased from the preamble of Council Regulation No. 3094/86, *OJ* L 288, 11.10.1986, p. 1. as amended. For an elaborate definition, see recital 10 in the preamble to Council Regulation (EC) No. 850/98.

3 Council Regulation No. 850/98, **Article 31**.

4 FAO Code of Conduct for Responsible Fisheries, (Rome, 1995), paragraph 8.5.1. See Chapter 11, *post*.

5 For example, Spanish fishermen occasionally refer to such devices as '*condoms*' (a particularly illustrative euphemism). The most infamous example being the alleged illegal device in the net recovered by the Canadian authorities from the seabed on the Grand Banks of Newfoundland after their unilateral enforcement action against the Spanish fishing vessel the *Estai* in March 1994. The particular device in question was subsequently exhibited to the world press by Mr Brian Tobin, Canadian Minister for Fisheries, on a barge in New York harbour on the occasion of the United Nations Conference on Straddling Stocks and Migratory Species as alleged evidence to support the Canadian enforcement action on the high seas.

6 Case 87/82, [1983] ECR 1579-1594. See Chapter 3, *ante*.

7 Case 87/82, paragraph 10, *ibid*.

8 Case 87/82, paragraph 18 and 23, *ibid*. See Commission Regulation No. 3440/84 of 06.12.1984 on the attachment of devices to trawls, Danish seines and similar nets, *OJ* L 318/23, 07.12.1984.

9 Case C-64/88, [1991] ECR I-2727, paragraphs 15–17.

10 As of January 1998 these measures have not been adopted. However, see Commission Regulation (EEC) No. 1381/87 of 20.05.1987 establishing detailed rules concerning the marking and documentation of fishing vessels, *OJ* L 132/9, 21.05.1987, which pursuant to **Article 3(2)** requires that official documents be kept on board fishing vessels indicating the capacity of fishing rooms in cubic metres and chilled or refrigerated sea-water tanks. Documents are required to be certified by a competent authority. The regulation does not elaborate on which authorities may discharge this function and thus it must be assumed that this is an issue to be addressed in the national legislation of the flag Member States.

11 The enforcement difficulties associated with the use of driftnets is examined in Chapter 10, *post*.

12 See Chapter 3, *ante*.

13 Case C-64/88, *Commission* v. *French Republic*, [1991] ECR I-2727, paragraphs 20–24.

14 *Ibid*.

15 On Community legislative competence in fisheries, see Chapter 3, *ante*. Council Regulation (EEC) No. 850/98, **Article 46**, prescribes precise requirements regarding the obligation on Member States to notify the Commission of any plans to introduce or amend national technical measures.

16 Case 53/86, [1987] ECR 2691. See discussion of this case by Churchill, R.R., *European Law Review* (1988) 352–360.

17 Case C-370/88, [1990] ECR I-4071. See Chapter 3, *ante*.

18 *Ibid*., paragraph 26.

19 *Ibid*.

20 Case 804/79, [1981] ECR 1045, paragraphs 17 and 22.

21 *Ibid*.

22 See Thorvik, T., *Control and Enforcement in the Norwegian Fisheries*, (OECD, Paris, 1992), 21–24.

23 See Ohnishi, M., *Review of the Fishery Management and Enforcement System in Japan*, (OECD, Paris, 1992), 107–119.

24 See Pálmaston, R., *Supervision of the Utilisation of Fishery Resources Off Iceland*, (OECD, Paris, 1992), 175–192.

25 Commission Regulation (EEC) No. 2108/84 of 23.07.1984 laying down detailed rules for determining the mesh size of fishing nets, *OJ* L 194/22, 24.07.1984. As amended by Commission Regulation (EC) No. 2550/97 of 16.12.1997, *OJ* L 349/1, 19.12.1997.

26 See discussion of this problem and related issues by Derham, P.J., 'The Implementation and Enforcement of Fisheries Legislation', in *The Regulation of Fisheries: Legal Economic and Social Aspects*, (Strasbourg, 1987), p. 77.
27 Case C-64/88, [1991] ECR I-2727, paragraph 14.
28 Case 348/88, *Judgement of the Court (First Chamber) of 2 May 1990. Criminal Proceedings against Jelle Hakvoort. Reference for a Preliminary Ruling: Amtsgericht Bremerhaven – Germany*, [1990] ECR I-1647.
29 Case C-348/88, paragraph 6.
30 Case C-348/88, paragraphs 12–16.
31 Case C-348/88, paragraphs 17–20.
32 Case C-348/88, paragraph 21.
33 Case C-348/88, paragraph 22..
34 *Opinion of Mr. Advocate General Van Gerven*, delivered on 20 February 1990, [1990] ECR I-1647, paragraph 8.
35 Commission Regulation (EC) No. 2550/97 of 16.12.1997, *OJ* L 349/1, 19.12.1997.
36 Chapter 2, *ante*. A major criticism of the control regime which existed between 1983 and 1993, was that the ambit of the inspection regulation was limited to monitoring the conservation and resources policy. In effect that meant that the Community inspectors were unable to verify compliance with market organisation rules and national measures for implementing the structural policy. Indeed, there is little evidence of any national effort being made in the Member States to monitor these aspects of the policy during the same period. This omission, as the Commission ironically commented, led to a paradoxical situation where fishermen who systematically failed to respect certain conservation measures, nevertheless continued to receive grants under the market regulations or the structural policy. See, Report from the Commission to the Council and the European Parliament on monitoring and implementation of the CFP, SEC (92) 394 of 06.03.1992, p.23.
37 Council Regulation No. 2847/93, **Article 1(1)**, discussed Chapter 3, *ante*.
38 Address by J. Santer, President of the Commission, to the European Parliament on the occasion of the investiture debate of the new Commission, Strasbourg, 17.01.1995. See the Commission's work programme for 1995, entitled – Common Fisheries Policy, COM(95) 26 final, p. 37. The precise quote is as follows:

> 'The main feature of 1995 will be the implementation of the reforms begun in 1992 relating to the conditions of access to Community waters and resources. The problems to be tackled are the chronic over capacity of European fishing fleets, the frequent threat to resources, and certain crises of essentially structural origin.
>
> Steps must therefore be taken to resolve the socio-economic problems arising from the regulation of fishing effort and the restructuring of fishing activities and to overcome the most serious obstacles to the rational and responsible use of fish stocks (static gear, driftnets, discards).
>
> All interested parties will have to be involved (fishermen's associations, national and regional authorities, operators) to rebuild the necessary confidence for structural change, which is the sole guarantee of the continuation of fishing activities and the safeguarding of the socio-economic fabric of the coastal communities and islands dependent on the industry.
>
> All these measures ... should make for closer supervision of fishing activities, markets and structural aid.'

39 The new practice of expressing the objectives of the structural policy in terms of fishing effort in the MAGP began in the transitional MAGP of 1992, where 25% of the requisite capacity reductions could be achieved through cutting fishing effort. Council Regulation No. 3946/92 of 19.12.1992 amending for the third time Regulation No. 4028/86 replaces the

phrase 'adjustment of fishing capacity' in **Article 1(1)(d)** with the phrase of 'adjustment of fishing effort', *OJ* L 401, 1992. On the subject of fishing effort see Chapter 4, *ante*.

40 The MAGPs are examined in Chapter 1, *ante*. Since 1983 there has been three MAGPs and the fourth, MAGP IV, is scheduled to run from 1997–2002.

41 See second recital, Council Regulation No. 394/92, *OJ* L 401, 1992.

42 High Court of Justice (England), Queen's Bench Division, Divisional Court Order of 02/12/93, (CO/2062/93), [1994] *CMLR* Vol.1, 907–920.

43 **Articles 6** {12}, **34** {29}, **39** {33}, and **40(3)** {34(3)}.

44 Council Regulation No. 3759/92 of 17.12.1992, *OJ* 1992 L 388, p. 1.

45 Council Regulation No. 3760/92 of 20.12.1992, *OJ* 1992 L 389, p. 1.

46 Commission Decision 92/593/EEC of 21.12.1992, which was made pursuant to Council Regulation No. 4028/86, *OJ* 1992 L 401, p. 33.

47 Case C-44/94, *The Queen* v. *Minister of Agriculture, Fisheries and Food, ex Parte National Federation of Fisherman's Organisations and Others and Federation of Highlands and Islands Fishermen and Others*, [1995] ECR I-3115.

48 See *Communication from the Commission relating to management guidelines for the 4th generation of multi-annual guidance programmes (MAGPs)*, COM(96) 203 final, Brussels, 30.05.1996; Proposal for a Council Decision concerning the objectives and detailed rules for restructuring the Community fisheries Sector for the period from 1 January 1997 to 31 December 2002 with a view to achieving a sustainable balance between resources and their exploitation, COM(96) 237 final.

49 These objectives concern the detailed rules for restructuring the Community fisheries sector with a view to achieving a balance on a sustainable basis between resources and their exploitation. These are discussed in Chapter 1, *ante*.

50 Title VII of Council Regulation (EEC) No. 4028/86 lays down the rules whereby Community support is given to measures to eliminate excess fishing capacity. In the event of a temporary fall in fish stocks, a temporary withdrawal premium is granted until the stocks are replenished. If there is a permanent imbalance, ie. the fishing capacity in the long term exceeds in terms of activity and mortality the level which the stock could sustain without jeopardising its survival, a final cessation is granted in order to adjust fishing capacity to stocks. In accordance with Community regulations, the two types of aid may be granted by Member States, who are then entitled to reimbursement of 50% of their expenses in accordance with the scales annexed to the regulation. The distribution of both forms of aid were criticised in the Court of Auditors' Report, *op. cit.* fn 175, Chapter 1 *ante*, pp. C 2/40–2/42. From the enforcement perspective the Court of Auditors observed that in the case of vessels withdrawn for non-fishing purposes, there is a need to organise checks to ensure that the intended purpose is abided by and not altered. Particularly, vessels intended for recreational fishing may in particular constitute an appreciable risk. Furthermore, exportation of vessels to countries which operated flag of convenience registers, meant that these vessels were operating in direct competition with the Community fleet, without being subject to the same obligations, although the amendment to Council Regulation No. 3944/90 should preclude this subterfuge.

In addition, one of the eligibility criteria for the final cessation premium, was the length between the perpendiculars. This was not assessed or certified for virtually any of the older vessels in the fleet, and consequently was assessed differently from one Member State to another. Another criterion was that vessels had to be engaged in *fishing activity* for at least 100 days during the calendar year prior to the grant application. In this case in the absence of a definition of *fishing activity*, there was no requirement to consult or cross verify primary evidence sources such as fishing logbooks or landing declarations.

The eligibility criteria have since been addressed by the Commission. The new control measure should enable Member States to improve their supervision and close of some of the loopholes in the structural policy, see Commission Replies to Court of Auditors' Report, *op. cit.*, fn 45, p. C2/54.

51 **Article 25(1)(e)** also refers to the development of the aquaculture industry and coastal areas. These are major areas of responsibility which are outside the remit of this study.

52 See paragraph 3, *infra*.

53 There has always been some control of the financial aspects of the structural measures such as that described above in the *Court of Auditors' Report* or of the type of financial control which leads to the recovery of aid improperly paid, or the investigation of fraud in the system of Communities subsidies. See, Joined Cases T-231/94 R, T-232/94 R and T-234/94 R, *Transsacciones Maritimas SA, Recursos Marinos SA and Makuspesca SA* v. *Commission of the European Communities*, [1994] ECR II-0885. The application to the Court of Justice stemmed from events subsequent to an inspection visit by Commission Officers to two Spanish shipbuilding companies and the ensuing suspension and reimbursement of Community financial aid as a result of irregularities revealed in the companies' accounts.

54 This is usually undertaken by measuring the torque-couple produced on the shaft by the engine.

55 In this regard, the propulsion line is composed by the engine, shaft, propeller, and rudder and the efficiency of these produce the total efficiency of the propulsion system.

56 The Court of Auditors' Report in 1993 revealed that in the case of Ireland and the United Kingdom, fishing boats from other Member States had been registered without the body responsible for the register being aware of or able to disclose the name and the registration number of the ship in its country of origin. See fn 175, Chapter 1 *ante* pp. C 2/19 and C 2/20.

57 Commission Regulation No. 109/94 of 19.01.1994 concerning the fishing vessel register of the Community, *OJ* L 19/5, 22.01.1994. (Repealed). Commission Regulation (EC) No. 2090/98 of 30.09.1998 concerning the fishing register of the Community, *OJ* L 266/27, 01.10.1998.

58 Such as the legislation which covers the registration of vessels flying the flag of the United Kingdom, Part 1 of the Merchant Shipping Act 1894.

59 Commission Regulation (EC) No. 2092 of 30.09.1998 concerning the declaration of fishing effort relating to certain Community fishing areas and resources, *OJ* L 266/36, 01.10.1998.

60 Court of Auditors' Report *op. cit.*, fn 175, Chapter 1 *ante*.

61 *Ibid.*

62 Harmonisation has been long overdue, as past disparities in the tonnage measurements criteria in the Member States resulted in the Court of Auditors being able to cite the example of two ships of almost identical physical characteristics being built by the same shipyard for ship owners in two different Member States being 60% different in tonnage registered, even though in both cases, it was expressed in gross registered tonnes (grt). *Op. cit.* fn. 175, Chapter 1 *ante.*, paragraph 2.6, C2/12, and Annex I.

As far as tonnage is concerned it should be noted that in international law, as well as in practice, several systems of tonnage measurement exist side by side. For example, gross tonnage as defined by the London Convention is only obligatory as a unit of measurement for all vessels over 24 metres since 18 July 1994. Until then the system of tonnage defined by the Oslo Convention (Unit of Measurement: *le tonneau de jauge brute Tjb*; Gross Registered Ton (GRT); *Brutto Register Ton* (BRT) will continue to be valid, whilst measuring units defined at national level (sometimes using a simplified formula) may be used to determine the tonnage of any vessel which can operate without an international tonnage certificate.

The Community rules which apply to fishing vessels under Regulation (EEC) No. 2930/86 provide:

(a) that the Community definition of tonnage is to apply to any vessel which entered service or was modified on or after 1 January 1987, but will be applicable to other vessels as from 18 June 1994;

(b) that the tonnage of a vessel is to be determined as provided in Annex I to the London Convention, which lays down rules for calculating the gross and net tonnage of ships.

Whereas under **Article 4(1)(b)** of the London Convention the scope of the Convention is limited to ships over 24 metres long (79 feet long), the effect of the reference to Annex I of the Convention is that the tonnage of all the fishing vessels in the Community fleet is established according to the calculation rules laid down by the London Convention irrespective of length. This approach which goes further than the general provisions in international law, provided that the tonnage of the vessels in question have been established or recalculated prior to that date in accordance with the rules laid down in Annex I to the London Convention.

63 Case 86-87/84, [1985] ECR 0941.

64 The power of fishing vessels engines has been the subject of much comment. According to **Article 5** of Regulation No. 2930/86, which defines the characteristics of fishing vessels, the engine power is the total of the maximum continuous power determined in accordance with the requirements adopted by the International Organisation for Standardisation, taking into account any gearbox incorporated into the engine. The importance of engine power varies according to the type of fishing activity and the gear in use. Recent trends are towards more powerful fishing vessels. This has been to the detriment of the structural policy objective which has endeavoured to reduce the overall power of the fleet. To reconcile these two conflicting aims, it has become common practice to fit a governor or derating device to engines. At the time of writing this book, the Commission guidelines allow for this practice with respect to certain vessels applying for Community aid for the construction or modernisation of vessels, but only subject to strict conditions and well defined limits. To put it in simplified terms an engine can be derated either when it is being manufactured or during installation in the shipyard. In the latter case the usual method of derating is to modify the injectors by fitting seals which can be removed at any time, especially if there is an emergency at sea. If the derating has been done by the manufacturer, reversing the process is a more difficult practice, and is only done on the manufacturer's premises.

65 The issues which arise as a result of the link between fishing capacity and fishing effort are complex and are examined separately in Chapter 4, *ante*.

66 See Case C-44/94, *The Queen* v. *Minister of Agriculture, Fisheries and Food, ex Parte National Federation of Fishermen's Organisations and Others and Federation of Highlands and Islands Fishermen and Others*, [1995] ECR I-3115.

67 See Chapter 11, *post*.

68 The type of data and the requirement to communicate to the Commission are dealt with in Commission Regulation (EC) No. 2091/98 of 30.09.1998 concerning the segmentation of the Community fishing fleet and fishing effort in relation to the multiannual guidance programmes, *OJ* L 266/36, 01.10.1998.

69 See Chapter 7, *post*.

70 Council Regulation No. 3699/93 of 21.12.1993 laying down the criteria and arrangement regarding Community structural assistance in the fisheries and aquaculture sector and processing and marketing of its products, *OJ* L 346/1, 31.12.93, **Article 8**.

71 *OJ* L 388, 31.12.192, p. 1, as last amended by Council Regulation No. 1891/93, *OJ* L 172, 15.07.1993, p. 1.

72 This second paragraph of **Article 28(2)** is difficult to comprehend in the English language version of the Regulation. The French text is clearer and suggests that the intention is to place an obligation on Member States to compare the documents recording the quantities and the weight of fish when first placed on the market and quantities and weight of fish recorded in the sales note.

73 The non disclosure clause with respect to professional secrecy, and the without prejudice

clauses in regard to legal proceedings, **Articles 28(3)** and **28(4)**, are standard considering the type of information acquired by those who are concerned with the practical implementation and enforcement of these provisions.

74 For the role of producers' organisations, see Title II Chapter 1 of Council Regulation No. 3759/92 of 17.12.1992 on the common organisation of the market in fishery and aquaculture products, *OJ* L 388/1, 31.12.1992. See, *inter alia*, Council Regulation No. 105/76 of 19.01.1976 on the recognition of producers' organisations in the fishing industry, *OJ* L 20/39, 28.01.1976; Commission Regulation No. 2939/94 of 02.12.1994, *OJ* L 310/12, 03.12.1994. In many Member States POs play a central role in the management of fisheries. For example, in the United Kingdom regulatory responsibility for the distribution of quotas amongst its members rests with POs. See Goodlad, J. 'Fisheries management by sectoral quotas, an assessment of the UK system', *EAFA, Proceedings of the IV Annual Conference, April 22nd–24th, 1992 University of Salerno, Italy*. For a general examination of the role of POs see, *inter alia*, Svein, J. 'Fisheries co-management: delegating government responsibility to fishermen's organisations', *Marine Policy*, **13(2)**, (1989), 137–154. The limited role of POs in Denmark is examined by Nielsen, J.R., 'Participation in fishery management policy making', *Marine Policy*, **18(1)** (1994), 29–40. Hatcher, A.C., 'Producers' organisations and devolved fisheries management in the United Kingdom: collective and individual quota systems', *Marine Policy*, **21**, (6), 519–533.

75 Council Directive 91/492/EEC of 15.07.1991 laying down the health conditions for the production and the placing on the market of live bivalve molluscs, *OJ* L 286, 24.09.1991, 1–14. On failure to transpose this Directive and Directives 91/67/EEC, 91/493/EEC and 92/48/EEC within the prescribed period, see C-325/95, *Commission* v. *Ireland*, [1996] ECR I-5615.

76 Council Directive of 22.07.1991 laying down the health conditions for the production and the placing on the market of fishery products (91/493/EEC), *OJ* L 268/15, 24.09.91, as amended by Council Directive 95/71/EC of 22.12.1995, *OJ* L 332/40, 30.12.1995. See also Council Directive 92/48/EEC.

77 Council Regulation No. 2406/96 of 26.11.1996 laying down common market standards for certain fishery products, *OJ* L 334/1, 23.12.96, **Article 1**.

78 Case C-301/88, [1990] ECR I-3803.

79 Case C-301/88, paragraph 15, *ibid.*

80 Case C-301/88, paragraph 22, *ibid.*

81 See Churchill, R.R., *EC Fisheries Law*, 231–254.

82 Joint Answers to Written PQs Nos 900/93 and 926/93, *OJ* C 301/8, 08.11.93.

83 Case C-39/88, [1990] ECR I-4271.

84 Case C-39/88, paragraphs 8–9.

85 Case C-39/88, paragraph 10.

86 Case 254/83, [1984] ECR 3395, paragraph 11.

87 The Court cited the judgements in Case 39/72 *Commission* v. *Italy* [1973] ECR 101 and in Case 128/78 *Commission* v. *United Kingdom* [1979] ECR 419.

88 Case C-200/88, *Commission of the European Communities* v. *Hellenic Republic*, [1990] ECR I-4299.

89 Case C-200/88, paragraphs 10–11.

90 Case C-200/88, paragraph 9.

91 Case C-209/88 *Commission* v. *Italian Republic* [1990] ECR I-4313, paragraph 13.

Chapter 6

The Rôle of the European Commission and Some of the Issues Pertaining to the use of Sanctions in the Enforcement Process

Introduction

This chapter reviews the verification procedures which are in place to ensure that Member States comply with Community rules. More specifically, the topics examined in Section (i) include a review of the rôle of the Commission Inspectorate in the enforcement process, the procedure for initiating administrative inquiries into non-compliance with the CFP, and the annual report completed by the Commission on the implementation of the Community fishery enforcement policy in the Member States. In Section (ii), on the other hand, there is a discussion of some of the issues pertaining to the use of sanctions for non-compliance with Community fishery regulations. These are the core issues at the heart of the Community fishery enforcement system and are provided for in Titles VII and VIIIa of the Control Regulation. These Titles introduce and expand the Commission's power to verify the application of the obligations placed on the Member States, and specify the action to be taken in the event of non compliance with Community regulations. These measures are important, principally because fishermen often dispute the justification for conservation measures, and their recalcitrant attitude extends to the authorities responsible for enforcement. Furthermore, fishermen are extremely sensitive to the question of the uniform enforcement of Community measures throughout the Community. Such uniformity is extremely difficult to achieve if one considers that the operational and legal competence for inspections remains firmly in the hands of the national authorities.[1] Thus the Commission's competence to monitor Member State control authorities is an essential component in establishing the integrity of the Community control regime.

While in legal and jurisdictional terms the power to undertake specific acts of enforcement is vested in national enforcement authorities in the Member States, the responsibility for verifying and monitoring the application of Community fishery regulations, nonetheless, rests with the Commission inspectorate.

Section (i) The rôle of the Commission

1. *The Commission's inspectorate – (the Community fishery inspectors)*[2]

Background

The Directorate of Fisheries in the Commission has an inspectorate which was established in 1983 as a means to monitor the enforcement and implementation of the

CFP in the then seven north-European 'fisheries' Member States.[3] Initially the Inspectorate consisted of 13 inspectors recruited from the national services in the Member States who were employed by the Commission on a temporary contract. The original concept was that inspectors would return to their Member States after their period of employment in the Commission and apply the international experience gained as a Community inspector in national enforcement organisations. By 1999, however, the inspectorate has increased in staff to over 20, with representatives from all Member States other than Luxembourg and Austria. Inspectors have been recruited from all sectors of the industry, enabling the Commission to draw on a pool of experience which is not limited to the national enforcement organisations. This is particularly important as the ambit of the Control Regulation extends to all aspects of the CFP.

2. *The* modus operandi *during the period 1983–1993*

The rôle and powers of the Commission inspectors were originally set out in **Article 12** of Council Regulation No. 2057/82 (the 1982 Control Regulation), which, if read in conjunction with **Article 155** {211} of the EC Treaty, obliged the Commission to ensure that the provisions of the Treaty and the measures taken by Community institutions were applied. As the Commission inspectors did not have legislative powers of enforcement to investigate the actions of private citizens or legal bodies, inspection missions in the Member States were limited to observing the activities of national inspectors. Moreover, since the Commission inspectors operated by accompanying national inspectors, any national enforcement limitations placed on the latter also circumscribe the Commission inspectors monitoring rôle. Problems were further exacerbated, by national authorities limiting the inspection programme to contrived situations specially prepared for the Commission inspectors.[4] Although the Commission had the right to initiate an administrative inquiry under **Article 12** of the 1982 Control Regulation, and ultimately to initiate **Article 169** {226} of the EC Treaty proceedings if a Member State 'failed to fulfil an obligation', experience had shown that successful actions in the Court of Justice would largely depend on hard evidence of irregularities being tendered by the Commission.[5] Indeed without such evidence the Commission Legal Service would be unlikely to proceed beyond the initial procedural stages of **Article 169** {226} of the EC Treaty. Thus, during the first period of the policy the necessity of a broad mandate to allow Commission inspectors to gather unequivocal evidence of irregularities became clearly apparent. Unfortunately, the Council did not adopt the broad proposals to enhance the powers of the inspectorate until September 1986,[6] despite the contribution the Commission inspectors made in exposing the 'grey market' operated by the Dutch fishing industry,[7] and to collecting sufficient evidence for the Commission to commence **Article 169** {226} proceedings against Belgium, Denmark, Ireland and the Netherlands for failing to correctly apply and observe the 1982 Control Regulation.[8] A typical example of the work carried out by Commission inspectors in this period was their establishment that in 1986 and 1987 the Spanish authorities had failed to record catches subject to a TAC or quota fished outside the Community fishing zone.[9] In this instance because the Spanish authorities had not observed their enforcement obligation nor had they taken penal or administrative action in respect of those vessels, the Commission was able to successfully conclude **Article 169** {226} proceedings.

By 1992 the Commission inspectorate had been in operation for nine years, and as noted in the 1992 Control Report, implementation of Community legislation was at that point in time still subject to serious shortcomings. In the lead-up to the revision of the policy in 1993 the Commission sought autonomous inspection powers as well as authorisation to increase the range of inspections to allow its services to investigate all activities associated with fisheries.[10] However, despite a favourable view by some Member States and the European Parliament, not all these powers were included in the new measures introduced as a result of the adoption of the 1993 Control Regulation.

3. *The powers of the Commission inspectors (EC inspectors)*

Provisions in the Control Regulation

(i) Verification of Member States enforcement activities

Article 29 as amended sets out the Commission's competence to verify the application of the Control Regulation by the Member States by means of the examination of documents and by conducting on-the-spot visits to ports and at sea. This refers to the previously established procedure of frequent inspections by Community Inspectors to check on the way national authorities are implementing Community Regulations in the Member States. However, **Article 29(1)** introduces a new approach which allows the Commission to decide if it considers it necessary to carry out verification in a Member State without providing national enforcement authorities with prior notice or advance notification of the visit by Commission inspectors. This power of unannounced inspection visits to Member States has been sought by the Commission for several years and had been identified, in several Commission Reports, as one of the principal deficiencies in the Community control system. To facilitate the practical operational aspect of this concept, the Commission is obliged to issue written instructions to inspectors proceeding on such missions indicating their authority and the objectives of their investigations.

The criteria for inspection missions in the Member States are set out in **Article 29(2)**. The decision making powers are vested in the Commission and it alone decides whenever it is deemed necessary for the Commission inspectors to be present at control and inspection activities carried out by national control services. Having decided to conduct missions, the Commission services contact the appropriate authorities in the Member States concerned with a view, whenever possible, to establishing a mutually acceptable inspection programme. Further specific guarantees and clarification of procedure are outlined, pursuant to **Article 29(2)(a)–(c)**. These require Member States to cooperate with the Commission in order to facilitate it in the accomplishment of its tasks. The duty to cooperate with the Commission or its representatives follows from **Article 5** {10} of the EC Treaty. In particular, Member States are to take all necessary steps to ensure that the inspection missions are not subject to publicity injurious to the inspection and control operations.[11] Should inspectors experience problems or encounter difficulties in the execution of their duties, the onus is on the Member State to provide the Commission with the means to accomplish its task and give the inspectors the opportunity to evaluate the specific control operations. Another important provision which introduces flexibility into the Community monitoring procedure is that if circumstances *in situ* do not allow the inspection and control operations envisaged in the framework of the initial inspection programme, then the Commission inspectors, in liaison and agreement with the competent national control service, can modify the initial inspection and control operation. This is important because fishing and its related activities are frequently affected by adverse weather and sea conditions and the enforcement programme has frequently to be adjusted to take account of such factors.

(ii) Autonomous inspection missions

Article 29(3) introduces the concept of the autonomous inspection missions. These arise wherever the Commission has reason to believe that specific control tasks are not being fully or properly carried out by Member States. In order to ensure better control, the Commission may request the Member State to draw up a control programme designed to address particular enforcement weaknesses. The Commission may then subsequently verify the implementation of such programmes through inspection missions. Examples of these types of missions include the enforcement of driftnets and controls on landings of vessels authorised to fish within the framework of European fisheries agreements. So as to maximise the full benefits of autonomous verification, this article also requires the Commission to transmit to the relevant Member State an evaluation report on the programme and, if appropriate, recommend further control measures.

In December 1998, the Council agreed that Commission inspectors conducting inspections without prior notice should be accompanied by national inspectors and should have access to documents, places, premises and means of transport in order to collect data for the accomplishment of their tasks. It was also agreed that the Commission should provide a report to the Member States concerning the conclusion of this category of inspection. These two provisions are included in **Article 29(3a)**.

Article 29(4) clarifies competence in the context of inspection missions conducted by aircraft at sea or ashore. It provides that the authorised inspectors may not carry out controls with respect to natural persons.

Article 29(5) sets out the administrative power required in inspection visits. In particular it provides that in the framework of their visits mentioned in **Articles 29(2)** and **29(3)**, inspectors authorised by the Commission may have on-the-spot access, in the presence of the services responsible, to the information in specified databases either in aggregate or individual form and may also examine all documents pertinent to the application of the Control Regulation.

If national provisions provide for the confidentiality of investigations, communication of this information shall be subject to the authorisation of the national court.

(iii) Verification with prior notice with inspectors from another Member State acting as observers

As a result of the amendment to the Control Regulation in 1998, there is a provision for the Commission to arrange for its inspectors visiting a Member State to be accompanied by one or more fisheries inspectors from another Member State acting as observers. This provision, **Article 34b**, may only be invoked in the case of missions undertaken where the host Member State has prior notice of the verifications to be undertaken by Commission inspectors. It also requires the approval of the Member State to be visited. Furthermore, upon request from the Commission, the sending Member State can nominate at short notice the national fisheries inspectors selected as observers. In order to deal with the practical aspects of organising such observer missions there is a requirement for Member States to draw up a list of the national inspectors whom the Commission might invite to be present at such verifications. The Commission shall, where appropriate, place the list at the disposal of all the Member States.

Significantly, the Control Regulation is silent on the powers of inspectors deployed as observers. Traditionally, in marine fisheries, observers have little if any powers of enforcement. The precise powers of inspectors, as noted above, falls within the scope of national legislation in the Member States, or may be prescribed in bilateral agreements or arrangements between Member States.

4. *The administrative inquiry*

Provisions in the Control Regulation

In order for the Commission to have unfettered scope to follow up on the reports of their Inspectors, **Article 30** provides that wherever a Member State is considered to have not fulfilled their obligations with respect to the Control Regulations or that existing monitoring provisions and methods are not effective, an administrative inquiry may be requested. Such inquiries are conducted by the Member States concerned and focus on specific shortcomings or omissions identified and reported by the Commission. Commission personnel, usually inspectors, participate in the investigations mounted by the national authorities.

Article 30(3) stipulates that where Commission officials participate in an inquiry, that inquiry shall at all times be conducted by the officials of the Member State. Commission officials may not on their own initiative use the powers of inspection conferred on national officials. On the other hand, Commission officials have access to the same premises and to the same documents as national officials. However, if there are national legislative provisions regarding criminal proceedings which reserve certain acts to officials specifically designated by national law, then in such instances, Commission officials are prohibited from taking part in such actions.

In particular they shall not participate in searches of premises or in the formal questioning of persons under national criminal law. They shall, however, have access to the information thus obtained. Because of the type and quality of information that inspectors may have access to, **Article 30(4)** contains a without prejudice clause which states that **Article 30** shall not prejudice national provisions concerning the secrecy of legal proceedings.

At the end of the inquiry, a report is submitted to the Commission. In accordance with the amendment to the Control Regulation adopted in 1998, this report shall be submitted within three months after the Commission's request to conduct an inquiry. This period may be extended if the Member State concerned submits a motivated request for an extension which should be for a reasonable period only. Such reports normally contain recommendations designed to address the problems identified prior to the inquiry.

5. *The enforcement rôle of the Commission inspectorate*

The Control Regulation imposes several obligations on Member States. The first obligation, which is of a preventive nature, requires each Member State to enforce Community fisheries law. The second obligation, which is of a punitive nature, requires Member States to take penal and administrative action against parties which violate Community law. In practice, for the enforcement system to work, it is essential that the Commission observe Member States' adherence to these obligations. This task is the responsibility of Commission inspectors. The preamble of the Control Regulation states that Commission inspectors are to ensure the uniform application of Community rules and to verify control carried out by the competent authorities of Member States. To achieve this task inspectors undertake visits to the Member States to observe the work of national enforcement authorities responsible for enforcement. Visits by Commission inspectors are generally referred to as missions (derived from the French word *mission* - meaning to go on an assignment). Missions may be of a few days' duration or run to several weeks depending on the inspectors' brief. During the course of the mission, it is important to distinguish the enforcement competence of national inspectors from that of Commission inspectors. In this context, competence refers to the legal power to undertake a particular act. On the one hand, national inspectors are generally empowered with a broad panoply of inspection and enforcement powers which may be invoked when carrying out their day-to-day tasks.[12] On the other hand, Commission inspectors have no rights with respect to third parties, nor have they any power to conduct independent inspection of fishing vessels or premises. As a consequence of this delimitation of authority, the ability of the Commission inspectorate to fulfil its mandate is largely dependent on the cooperation of national authorities. It is thus not surprising that the Control Regulation places an express obligation on Member States to co-operate with the wishes of the Commission in this regard. Fortunately, poor co-operation is generally not an issue that impedes the progress of inspections and in the period prior to the adoption of the Control Regulation in 1993 there is only one recorded instance of proceedings being commenced against a Member State, Spain, for refusing to co-operate with Commission inspectors.[13]

6. *Inspection methodology*

It is clearly evident that a key element in the Community enforcement policy is the deployment of Commission inspectors on mission. The organisation and purpose of missions have changed considerably as a result of the adoption of the 1993 Control Regulation. In general missions may be categorised into different types, such as the standard/routine mission, the specialist mission, and the rapid response mission. It is proposed to briefly look at each of these categories.

The object of the routine mission is to gather information for the Commission. With the introduction of new provisions in the 1993 Control Regulations such missions may be one of the following:

(1) Announced visits to assess the performance of the national inspection services, both in port and on the fishing grounds;

(2) Missions without prior notice which are the same in substance as announced missions but are initiated without informing Member State authorities (specialised missions discussed below may also be undertaken without prior notice);
(3) Independent missions.[14]

The routine inspection mission format offers several advantages if the overall objective is to obtain information on enforcement in the Member States. It guarantees the opportunity of monitoring or the uniform implementation of Community rules in all Member States. This is particularly important if Member States are disinclined to rigorously perform their enforcement obligations without the frequent presence of Commission inspectors. From the Commission viewpoint, routine missions provide a broad perspective with which a comparative assessment of developments and compliance with Community law in the Member States can be made. This in turn allows the Commission to draw various strands of the resource management policy and the control policy together. It also provides more transparency as all Member States receive the appropriate number and type of mission. Thus, for example, the 1995 Report on Monitoring the Common Fisheries Policy records that all Member States received at least one visit from Commission inspectors, and some Member States, namely Spain, France, the Netherlands, and the United Kingdom were host to several inspections.[15] The type of evidence gathered in routine missions may be accumulated over a number of years thus forming the basis of **Article 169** {226} proceedings such as those taken against France in 1988.[16] In that particular case, the Commission inspectors' reports established that the French enforcement authorities, during the period 1984–1987, had conducted 73 inspections in 26 ports, and could have been deemed to have failed to comply with their obligations to inspect and take action in order to ensure compliance with the conservation measures.[17]

On the other hand, the purpose of the specific or specialist missions is to investigate an area of concern or one of the new areas of enforcement. Such missions may be directed at monitoring the level of compliance with, for example:

- market measures (**Article 28** Control Regulation);
- structural measures such as verification of the data on the fleet register;
- the introduction of new technology for control purposes such as computerised cross verification of catch data, satellite vessel monitoring systems etc.;
- the fishing effort management measures;
- the control requirements pursuant to Third-Country Fishery Agreement with the European Union; and
- specific weaknesses in the enforcement of technical measures and/or catch registration.

In 1995 Commission inspectors spent a total of 128 inspector days in the Member States monitoring compliance with the legislative measures which regulate these areas.[18]

The decision to deploy inspectors on a routine as opposed to a specific mission depends on the issue under investigation and whether there is a requirement for a qualitative rather than a quantitative approach. The advantages of specific mission format is that it allows greater focus and facilitates a more detailed analysis of the enforcement activities in the Member States. In this regard it is more likely to yield tangible results for use in subsequent administrative inquiries or ultimately in Court of

Justice proceedings. Another advantage is that the specialised mission makes it more difficult for national inspectors to present a misleading picture of their enforcement endeavours. It also allows for the efficient use of the previously acquired knowledge and for the compilation of detailed reports.

Periodically, the Commission requires detailed information regarding specific and critical enforcement problems and this leads to the deployment of inspectors on rapid response missions at short notice. The objectives of these missions have been to ensure the presence of inspectors during periods of conflicts between fleets at sea, such as occurred in the albacore driftnet fishery in 1995. Inspectors have also been deployed to cover new temporary enforcement tasks as in the case of their deployment in the North West Atlantic following the conflict with Canada over the ground-fishery for Greenland halibut in international waters in March/April 1995. On that occasion, the Commission inspectors acted as observers on the European Union fleet pending the full implementation of an observer scheme. In addition to the missions described above, the Commission inspectors participate in direct enforcement activities in the context of the European Union's membership of NAFO – the North Atlantic Fisheries Organisation. NAFO operates a joint scheme of international inspection in international waters which applies to all members of the organisation. Commission inspectors (deploying from a charter vessel), acting on behalf of the European Union, conduct inspections of all fishing vessels from all contracting parties to NAFO. Enforcement is directed towards a range of technical measures (gear type, mesh and fish sizes) and quotas. This type of enforcement by Commission inspectors is the only exception to the general rule that they may not conduct direct inspection and issue reports which can be used in legal or administrative proceedings initiated for non-compliance. This task presents a heavy burden for the Commission's inspectorate as it must be undertaken throughout the fishing season which is in effect all year long. As noted in Chapter 2, Commission inspectors are also legally competent to undertake inspection in Antarctic Convention Area pursuant to the Convention on the Conservation of Antarctic Marine Living Resources but have not to date undertaken inspection in that zone.

7. *Use of Commission inspectors' reports and Commission inspectors in judicial and administrative proceedings in the Member States and the European Court of Justice*

At the end of their mission, inspectors will provide informal observations to the national authority, frequently at a senior level, and on return to Brussels, inspectors present a full report to the Commission. Inspector's reports are used as the basis for the completion of the annual control report on monitoring the Common Fisheries Policy. Because Commission inspections focus on the activities of national enforcement agencies and not directly at the activities of fishermen, their reports or observations are generally not used as evidence for the purpose of securing fishery prosecutions in the Member States. The one exception to this rule arises in NAFO, where the inspectors issue inspector reports which if there is evidence of infringements may be used in subsequent prosecutions initiated by the flag Member State. In these circumstances such reports are treated by national authorities as having the same evidential value as reports submitted by national inspectors.

Commission inspector reports may be used in judicial proceedings in the Member States whenever such reports are considered pertinent. Application for access to inspector reports must be made to the Commission. If the Commission does not wish to acquiesce to such a request, then it must provide the appropriate information to the Court of Justice to allow it to decide whether such a refusal is justified. The Court of Justice ruled on this issue in the *EC Fisheries Inspector* Case which arose out of criminal proceedings in the Netherlands relating to alleged forgery committed by the management of a fish auction regarding the quantities of fish landed and sold (grey fish).[19] The Dutch authority – *Rechter-Commissaris* – investigating the fraud deemed it essential to the successful completion of the investigation to have access to reports drawn up by EC inspectors during the period 1983–1986 and if necessary to examine the said inspectors on their findings. The Commission refused to acquiesce to this request on the grounds that the documents formed part of a file on legal matters pending in the Commission. The *Rechter-Commissaris* submitted an application to the Court of Justice for a 'request for judicial co-operation' in order to pursue his request.[20] Other than contesting the issue of admissibility of the application, the Commission argued that the inspectors' reports are purely for internal use and furthermore that the release of the documents might affect the relations between the Commission and the Member States in the delicate field of 'control'. The Court, pointing to the principle of loyal co-operation contained in **Article 5** {10} of the EC Treaty, and noting the functional character of immunity in the Protocol on the Privileges and Immunities of the European Communities, ruled that it was up to the Commission to either provide the documents requested and to permit its officials to be examined as witnesses, or else to state to the Court the imperative reasons relating to the functioning and independence of the Community which may justify its refusal to co-operate.

The first ruling necessitated a second because the Commission contested that their continued refusal was founded on imperative reasons based on the need to maintain the division of enforcement competence between the Commission and the Member States, as well as the need to protect the rights of third parties from exposure to judicial or administrative measures. The Court in the second order reiterated that the Commission is obliged to co-operate loyally with national judicial authorities responsible for enforcing Community law within the national legal order. On the reasons advanced by the Commission, the Court held that it did not exclude refusing the request where this could compromise fulfilment of the Commission's own task. In this instance the Commission was unable to prove that such was the case and the Court ordered communication of the relevant documents and authorised the said witnesses.[21]

The *EC Fisheries Inspector* Case is of interest for several reasons. At first sight it is remarkable that the Commission, which has always sought to increase the mandate of Commission inspectors, was reluctant to participate in national criminal law enforcement proceedings and based one of the arguments supporting its view on the traditional division of competence between the Community and the Member States. Secondly, although it has since been suggested by one commentator that the Court's decision now allows the Commission to play a part in the enforcement of Community law through the national courts,[22] the Court's decision has had little impact in reality on the enforcement of Community law. This is because the Court's decision has not been followed by similar types of requests from other Member State judicial and administrative authorities. Indeed the procedure developed in EC Fisheries Inspector Case (*Zwartveld*) has not been relied upon in the domain of fisheries since this case. It appears that there are few fishery cases which depend on the involvement of the Commission to deal successfully with violation of Community law in national criminal or administrative proceedings. This case has not resulted in a new and more inclusive approach to the enforcement of Community fishery law. The Court's Order nonetheless qualifies the Commission's right to invoke immunity which has positive consequences for improving the transparency of Community law and institutional accountability. On the other hand, the positivist approach in the judgement to the issue of enforcement as a natural extension of the Commission task to monitor compliance with Community law under **Article 155** {211} of the EC Treaty supports the expectation that the Court may adopt a similarly broad interpretation should the provisions in the Control Regulation be subject to judicial review.

8. *Annual 'Member State Reports', and the Commission 'Consolidated Report'*

Provisions in the Control Regulation

As a means to guarantee that there is uniformity in the application of the measures introduced by the Control Regulation, and to keep all parties informed of the general implementation of Community fisheries law, **Article 35** requires that, before 30 April each year, Member States shall transmit to the Commission a report on the application of the Control Regulation over the previous calendar year. In 1998 the Council amended **Article 35**, and as from July 1999, the Commission shall prepare an annual factual report based on national reports and an assessment report every three years. The latter shall be presented to the Council and European Parliament and Member States shall have the right to present their observations on it prior to publication. The Commission shall publish this assessment report along with the Member States' replies and, where appropriate, measures and proposals to mitigate the shortcomings discovered.

In order that Member States provide comprehensive information on enforcement including the imposition of penalties, the Commission will adopt rules setting those areas where information is required. An indication of the requirements is provided in **Article 35(3)** which lists four broad headings:

(1) the technical and human resources for fisheries monitoring, and the time effectively devoted thereto,
(2) the laws, regulations and administrative provisions that the Member State adopt to prevent and prosecute irregularities,
(3) the results of inspections or checks carried out pursuant to this regulation, including the number and type of infringements discovered and the action taken, in particular with respect to the types of behaviour which seriously infringe Community rules (those referred to in **Article 31(2a)**),
(4) the application measures and actions pursuant to **Article 19**, particularly with regard to the assessment of the reliability of the data.

Detailed rules for the provision of this information shall be adopted in accordance with the procedure laid down in **Article 36**.

Commentary
The first Annual Report was completed in 1996 and examined the monitoring and control of the CFP for the year 1995. This particular report is examined in Chapter 8. It is sufficient to point out here that the requirement of an Annual Report is derived from the broader principles of transparency and accountability which underlie the Community regulatory system. From the enforcement perspective, the report is also an effective and flexible instrument to apply pressure and to cajole reluctant Member States into fulfilling their enforcement obligations. Although it may be premature to draw conclusions about the effectiveness of the Annual Report as a catalyst for improvement, the concept should nevertheless be judged in the light of the impact the 1992 Control Report and similar previous reports had on the establishment of a new enforcement regime adopted in 1993.[23] The annual control report may also prove to be a useful document from which to evaluate the evolution and harmonisation of national enforcement strategies and legislation.

Section (ii) Transgression of fisheries law - the issue of sanctions

Background
The issue of proceedings and sanctions lies at the very core of most schemes of law enforcement.[24] Under international law a corollary of the enforcement authority of the coastal state in the EEZ or EFZ is the power to invoke such measures as may be necessary to ensure compliance with its fishery laws and regulations.[25] Coastal state legislation authorises, *inter alia*, the detection, arrest or detention on account of: illegal fishing; failure to have the prescribed fishing licence; unauthorised fishing in a sensitive area or for restricted species; failure to observe a hail requirement; operating improper gear; and the violation of many other measures. Coastal State legislation specifies the penalties to be imposed and the procedure to be followed in a particular offence. However, under international law in cases of the arrest or detention of foreign vessels, such vessels and their crews must be released upon the posting of a reasonable bond.[26] It should also be pointed out that penalties imposed for the violation of fishery law and regulation may not under international law include imprisonment in the absence of agreements to the contrary by the states concerned, or any other form of corporal punishment.[27] Furthermore, the flag State has to be notified promptly of the action taken by the coastal State and of any penalties subsequently imposed.[28]

The division of competence between the EU institutions and Member States conforms with the international law model.[29] In so far as the onus of responsibility is on Member States to verify, within the territory or waters subject to their jurisdiction or sovereignty, or vessels sailing under their flag on the high seas, compliance with the CFP. This includes compliance with the landing, selling and transport of fishery products, as well as structural and market provisions of the revised CFP. Moreover, it is only the Member State in its capacity as either the flag State or the coastal State that can take penal or administrative action for non compliance with Community law. There is one exception relating to the withdrawal and suspension of fishing licences granted to vessels flying the flags of third countries which have access to the Community zone, the latter power is vested in the Commission.[30] The range and severity of the sanctions prescribed in national legislation in the Member States varies considerably. While it is not possible to tabulate the complete range of sanctions, however, national discrepancies are clearly evident if the range of penalties for unauthorised fishing by non Member State vessels in the coastal waters of the Member States is examined. Table 6.1 summarises the provisions in national legislation.

It is generally acknowledged that there is a lack of uniformity in the range of penalties and the diversity is compounded by the way different Member States institute proceedings or invoke sanctions in the context of fisheries law enforcement in the Community.[31] In some instances the proceeding pursued and the sanctions applied by Member States are similar to those relied upon in other jurisdictions outside the European Union such as New Zealand, Japan, and the USA.[32] In other Member States proceedings and sanctions are entirely different because they are part of the continental penal code or administrative law system (Germany, Spain, Portugal) which differs considerably from the common law model in the United Kingdom and Ireland. The wide divergence, both in procedure and the scale of sanction, between the criminal law system in some Member States such as Ireland or the United Kingdom and the administrative law system in other Member States, is considered one of the principal weaknesses in the Community enforcement regime.[33] With regard to the case law of the Court of Justice, the Court has held that criminal law and rules of criminal procedure as well as inter-state co-operation in the field of criminal law are matters for which, in principle, the Member States are responsible.[34] Indeed, the discretion of the Member States in this area was guaranteed by Title VI (provisions on police and judicial co-operation in criminal matters) of the Treaty on European Union and this position is not changed in the Treaty of Amsterdam.[35] Legislative developments to protect the Community's financial interests as well as the inter-relationship between Community law, criminal law and administrative law in the Member States have been evaluated by several commentators.[36] In particular the jurisprudence of the Court of Justice regarding the obligation on Member States to invoke penalties and sanctions for non-compliance with Community law,[37] and the extent of the requirement that any penalty provided for in national law must be effective, proportionate and dissuasive,[38] are dealt with in the literature.[39]

1. *Provisions in the Control Regulation*

There are four articles (**Articles 31** to **34**) in Title VIII, which prescribe the measures and procedure in the event of the transgression of the CFP. **Articles 31** and **32** set out Member States' obligations in respect to procedure and sanctions. **Articles 33** and **34** specify particular notification requirements in the event of infringements being detected. It is proposed to examine these provisions in turn.

(i) Proceedings and sanctions

EU fisheries law is enforced at two levels. On the supra-national level or Community level, the Commission initiates **Article 169** {226} of the EC Treaty proceedings against

Table 6.1 Penalties for unauthorised foreign fishing.[40]

Member State	Maximum fine	Imprisonment	Forfeiture gear, catch
Belgium	40,000 BF (€9808)	No	Court may order
Denmark	Unspecified	No	Court may order
Finland	As Court may order	2 years	Court may order
France	500,000 FF (double for second offence) (€75,550)	No	Court may order
Germany	150,000 DM (€75,900)	No	Court may order
Greece		Up to 6 months	
Ireland	100,000 (€126,600)	No	Court may order
Italy	1 million lira (€510)	Up to 2 years	Court may order
Netherlands	Depends on criminal legislation applied		
Portugal	PTE 1 million to 5 million (€24,750)	No	Court may order
Spain	PES 4 million (fine not to exceed 35% value of the boat) €23,880	No	
Sweden	SEK 1000 per horse power of the vessel (€1,148 per horse power of the vessel)	6 months	Court may order
United Kingdom	Up to £50,000 on summary conviction (€75,200)	No	Court may order

Member States for failure to ensure compliance with Community fishery obligations.[41] On the national level of enforcement, the Control Regulation stipulates that Member States should take direct action against private parties for transgressing Community fishery law. This obligation is derived from **Article 5** {10} of the EC Treaty which provides that Member States shall take appropriate measures to ensure fulfilment of the obligations arising out of the Treaty or resulting from action taken by the Community institutions.[42] Specifically in the Control Regulation, **Article 31(1)** stipulates that Member States must ensure that the appropriate measures be taken, including administrative action or criminal proceedings in conformity with their national law, against the natural or legal persons responsible for infringements.

Article 31(2) states that the proceedings initiated (including the sanctions imposed), pursuant to **Article 31(1)** shall be capable, in accordance with the relevant provisions of national law, of effectively depriving those responsible of the economic benefit of the infringements, or of producing results proportionate to the seriousness of such infringements. This type of sanction is referred to as a 'reparative sanction' by one authority,[43] that is to say, by taking away the benefits and other advantages unduly received, the sanction is aimed at restoring the situation before the Community law was infringed. An ancillary objective is to effectively discourage further offences of the same kind.

There has been considerable debate on the most suitable form of sanction for fisheries offences. Lawyers and fishery administrators have traditionally favoured the direct penalty and this preference is reflected in Community fishery law. Therefore, it is no surprise that the use of traditional penalties is advocated in the Control Regulation. The sanctions arising from the proceedings mentioned in **Article 31(2)** may, depending on the gravity of offence, include:

- fines
- seizure of prohibited fishing gear and catches
- sequestration of the vessel
- temporary immobilisation of the vessel
- suspension of the licence[44]
- withdrawal of the licence.[45]

As the range of penalties and perceptions of the gravity of offences differ between Member States, similar offences such as logbook violations often attract widely varying penalties. In order to address this issue, **Article 31(2a)** states that the Council on the basis of **Article 43** {37} of the EC Treaty is invited to draw up a list of types of behaviour which are regarded as serious infringements of Community rules on the common fisheries policy (i.e. the rules listed in **Article 1** of the Control Regulation). Member States will then undertake to apply proportionate, dissuasive and effective sanctions. In effect this means that the Commission and Member States will try to agree upon a range of types of behaviour which seriously undermine the proper application of the CFP. This list was published in June 1999.

Transfer of Proceedings

Provision is made in **Article 34** for the transfer of evidence and responsibility for eventual legal proceedings from one Member States to another. This transfer would occur when the port Member State where the landing or transhipment takes place decides that it is in the best interest of enforcement to have the transgressor prosecuted under the authority of the flag Member State. The latter is frequently in a better position to impose appropriate sanctions (i.e. in many instances the port Member State has no jurisdiction to apply withdrawal/suspension of fishing licence which is one of the penalties envisaged in **Article 31(2)**). Such transfers are subject to a number of conditions:

- the competent authorities of both Member States shall agree to the transfer;
- the results as set out in **Article 31(2)** are more likely to be achieved through such transfers;
- the Commission shall be informed of such transfers.

Article 32 provides for the type of proceedings to be instituted against vessels which are non-flag vessels. In particular, **Article 32(1)** provides that where an infringement of the provisions of the Control Regulation is discovered by the competent authorities of the Member State of landing or transhipment, the said authorities shall take appropriate action against the master of the vessel involved or against any other person responsible for the infringement in accordance with **Article 31**.

If the Member State of landing or transhipment is not the flag Member State and its competent authorities do not undertake, in conformity with their national law, appropriate measures including the initiation of administrative action or criminal proceedings against the natural or legal persons responsible, or do not transfer prosecution in accordance with **Article 31(4)**, the quantities illegally landed or transhipped may be set against the quota allocated to the former Member State. This provision is an incentive for Member States to ensure the enforcement of Community law against non flag vessels and in particular against vessels engaged in the practice of quota hopping. **Article 32(2)** incorporates a consultation procedure in so far as the quantities of fish to be set against that Member State's quota are concerned, these shall be fixed in accordance with the procedure laid down in **Article 36** after the Commission has consulted the Member States concerned. Moreover, if the Member State of landing or transhipment no longer has a corresponding quota at its disposal, **Article 21(4)** (closure by the Commission of a fishery) shall apply *mutatis mutandis*. In such cases the quantities of fish illegally landed or transhipped being deemed to be equivalent to the amount of the prejudice suffered by the Member State of registration. This is similar to the procedure introduced by amendment to the 1983 Control Regulation.

Commentary

There are many obligations placed on Member States by the Control Regulation. One of these is punitive in nature and requires Member States to take penal or administrative action against the skipper of a vessel infringing Community fishery laws. Title VIII (**Articles 31–34**) discussed above clearly relates to this obligation. One commentator has argued that **Article 31** is a harmonisation provision on the grounds that it goes further than merely laying down conditions regarding national sanctions and provides for the type of sanctions to be imposed by all Member States.[46] On a literal interpretation of **Article 31(3)** there would appear to be little evidence to sustain this argument, on the grounds that **Article 31(3)** does not impose on Member States a duty or obligation to invoke one of the sanctions listed in that article. Indeed the prescribed list is prefixed by the caveat, *may include*, which would appear to indicate to that the list of sanctions is illustrative and non-binding. Furthermore, there would appear to be little support for the conclusion by the same authority that the sanctions in **Article 31(3)** are classified in accordance with the seriousness of offence and that lesser offences are dealt with by fines and the more serious breaches by revocation of licences. One only need point to one Member State, Ireland, where there is automatic forfeiture of catch and gear for some fishery offences under the Sea Fisheries Protection Act 1959 (as amended). Thus in the case of a Japanese bluefin tuna vessel which was detained and prosecuted for illegal entry and fishing in the Irish fishing zone in 1995, the fine for the offence amounted to £100,000 whereas the value of the catch and gear amounted to £700,000. Furthermore, it should be noted that the most serious offences are dealt with in Ireland by sequestration of the vessel (No. 3 on the list) and not withdrawal of the licence.[47] Sequestration of the vessel, in the majority of cases, will also entail the mandatory forfeiture of the catch and gear, and in most instances this will entail the mandatory suspension of the fishing licence.

On the other hand, it may be argued that the 1998 amendment to the Control Regulation which places the onus on the Council to draw up a list of types of behaviour which seriously infringe the rules of the CFP and to which the Member States undertake to apply proportionate, dissuasive and effective sanctions is a clear vector which indicates a move towards the position where similar infringements attract similar penalties throughout the Community.

The right to impose sanctions or penalties falls primarily within the exclusive competence of Member States with only one exception. This relates to the withdrawal and suspension of the fishing licence and special permits issued and the refusal to issue a new fishing licence and special permits for the period of one calendar year to vessels flying the flag of third countries and operating under the terms of EU/third-country agreements. Competence to withdraw/suspend such licences lies with the Commission.[48] The exercise of this power by the Commission is subject to a number of conditions which includes notification from Member States of infringement committed by the vessel and a detailed statement of the facts of the case including sanctions imposed in respect of the infringement.[49] The Commission is further obliged to assess the seriousness of the infringement in the light of the judicial and administrative decision in the Member States and in particular the commercial benefit which the vessel owner may have gained and the impact of the infringement on fishery resources. Moreover, the Commission is obliged to give the owner of the vessel the opportunity to express his views on the alleged infringement.[50] The Court of Justice has upheld the right of the impugned party to contest the Commission's decision by an action for annulment under the second paragraph of **Article 173** {230} of the EC Treaty.[51] Interestingly, for procedural reasons, the Court did not have to rule on whether the Council had the power to adopt a regulation which granted the competence to the Commission to impose penalties such as those outlined above.[52]

The measures prescribed in Title VII must be viewed in the context of the Court of Justice decisions on the obligation placed on Member States to inspect fishing vessels and to take penal and administrative action against skippers who do not observe the regulations in force concerning conservation and control measures.[53]

In this regard, in *Commission* v. *Netherlands,* the Court of Justice required the

Commission to adduce specific and concrete evidence if it wished to prove that a Member State had failed to fulfil such an obligation.[54] In the absence of such evidence the Court rejected the Commission assertion that the efforts made by the Netherlands authorities were inadequate as regards the prosecutions of infringements and that the action taken against illegal fishing could have been more effective.[55] More recently the Court has pointed out that if the competent authorities of a Member State systematically refrained from taking action, such as ensuring compliance with quota regulations, both the conservation and management of fishery resources and the uniform application of the CFP would be jeopardised.[56] Moreover, the Court has rejected arguments that national authorities were unable to conduct appropriate enforcement because of an unfavourable economic climate for fisheries, or the threat of public disorder.[57]

The imposition of sanctions can only take place through the appropriate judicial or administrative process. The European Court in several cases including, *Hoffman La Roche* v. *Commission*[58] has held that the general principles of Community law require that when an administrative body adopts an administrative measure which is likely to prejudice the interests of an individual, then that party must be allowed to express its point of view.[59] The Court applied this principle in *Fiskano (Swedish fish licence* case) and stressed that the observance of the right to be heard is, in all proceedings initiated against a person which are liable to culminate in a measure adversely affecting that person, a fundamental principle of Community law which must be guaranteed even in the absence of any rules governing the procedure in question.[60] Drawing support from its previous judgements,[61] the Court ruled that observance of the right to be heard requires that any persons on whom a penalty may be imposed must be placed in a position in which they can effectively make known their view of the matters on the basis of which the Commission imposed a penalty.[62] In this instance, the Court held that the decision of the Commission contained in a letter informing the Swedish authorities of a penalty imposed on a Swedish vessel under the Agreement on fisheries between the Community and Sweden must be annulled if the owner of the vessel has not been given the opportunity to submit observations before the adoption of the decision.[63]

It ought to be pointed out that the Court of Justice has held that there is an obligation on Member States to penalise infringements under Community law in the same manner as infringements under national law.[64] Nonetheless the list of penalties in **Article 31** is not exhaustive and consequently there is still scope for Member States to invoke indirect penalties which on occasion may be more effective than the aforementioned prescribed sanctions. Indirect penalties include the withdrawal of grants, the non payment of withdrawal prices for fishery products withdrawn from the market, or linking the payment of structural funds to effective enforcement procedures.[65] Alternatively Member States may encourage fishermen or producers' organisations towards greater compliance with the Community management framework. The latter approach has the advantage of providing an incentive for fishermen to set an example, it may also redress the psychological impact which affects the majority in the fishing community who uphold the law but frequently witness offenders escape punishment. The possibility of combining *direct action* and *indirect action* has also arisen in respect of the issue of economic incentives and disincentives to induce fishermen to use more selective fishing gear.

(ii) Notification requirements

Article 33(1) provides that the competent authorities of Member States shall, without delay and in compliance with their procedures under national law, notify the flag Member State or the Member State of registration, of any infringement of the Community rules referred to in **Article 1** (which establishes the Community control system to ensure compliance with the CFP), indicating the names of the master and the owner, the circumstances of the infringement, any criminal or administrative proceedings or other measures taken and any definitive ruling related to such infringement. Upon request Member States shall notify the Commission of this information in specific cases. To ensure that Member States deal in the appropriate manner with violators, **Article 33(2)** states that following a transfer of prosecution pursuant to **Article 31(4)**, the flag Member State or the Member State of registration shall take all appropriate measures as set out in **Article 31**. The notification requirement extends to the Commission pursuant to **Article 33(3)**, in that the flag Member State or the Member State of registration shall notify the Commission without delay of any measures taken in accordance with **Article 33(2)**, along with the name and the external identification of the vessel concerned.

As noted above, detailed rules for the provision of information for the Annual Report are to be adopted in accordance with the **Article 36** procedure. There will thus be a requirement to provide information regarding laws, regulations or administrative provisions adopted by Member States in order to prevent and prosecute irregularities, pursuant to **Article 35(3)**. Notification of inspections or monitoring carried out by Member States to fulfil their obligation to the Control Regulation, pursuant to **Article 35(3)**. This must include the type and number of infringement discovered and the action taken by the Member State.

Commentary

As pointed out in the introduction to this section the obligation to notify the flag state of the alleged fishery offences committed by a vessel flying its flag in the coastal zone of another State stems from **Article 73(4)** of the LOS Convention. The latter states that in cases of arrest or detention of foreign vessels the coastal State shall promptly notify the flag State, through appropriate channels, of the action taken and of any penalties subsequently imposed. The obligation to notify the Commission of pending prosecutions and the level and number of penalties imposed stems from the requirement, pursuant to **Article 35** of the Control Regulation, for the Commission to have sufficient information to complete the annual reports. In this regard such notification does not prejudice Member State competence to prescribe or invoke penalties or appropriate sanctions.

Interestingly, in a regulation adopted by the Commission in 1985 to harmonise the form in which Member States communicate information about inspections of fishing activities carried out by national control authorities, there are several definitions which are germane to the notification requirement.[66] In accordance with that regulation:

- 'Official written warnings given' shall mean written notification by the authorities to a captain or other person responsible for an infringement, who is not brought to Court, of infringements committed by him and of penalties which may be imposed if the infringements are repeated;

- 'administrative penalties imposed' shall mean financial or other penalties imposed by the authorities as a result of an infringement or administrative decision infringement taken as a result of that adversely affecting the activities of a captain or other person responsible for an infringement;
- 'infringement brought to Court' means infringements brought to Court, whatever the action of the Court.

References

1 See, Chapter 2, *ante*.
2 The rôle of the Communities fisheries inspectors in the monitoring and verification of the implementation of the Community rules in relation to driftnets is discussed in Chapter 10, *post*.
3 France, Germany, Belgium, the Netherlands, Denmark, Ireland, and the United Kingdom.
4 See Commission Reports discussed in Chapter 2, *ante*, particularly, COM (85) 490 final 17.09.1986.
5 See Chapter 8, *post*.
6 *OJ* 1986 C245/5.
7 See EP Doc A2-34/85, p.11.
8 See Chapter 8, *post*.
9 Case C-258/89R, *Commission* v. *Kingdom of Spain*, [1991] ECR I-3977, paragraph 4.
10 See Chapter 2, *ante*.
11 While Council Regulation No. 2847/93, **Article 29(2)(a)** places an express obligation on the Member States to ensure that inspection missions are not subject to publicity injurious to inspection and control operations, the discretion on how to discharge this obligation also rests with the Member States as is evident from the broad non-contextual nature of the provision.
12 See, for instance, Sea Fisheries Act 1959 (Ireland), Chapters II and III, as amended, concerning the special powers of Sea Fisheries Protection Officers while undertaking their enforcement duties in Ireland.
13 See, Ninth Annual Report on monitoring and application of Community law - 1991, *OJ* C 250/34, 28.09.1992. The 'Twelfth annual report on monitoring and application of Community law - 1994' records that these proceedings were in 1994 still very much alive, *OJ* C 254, 29.09.1995.
14 Council Regulation No. 2847/93, **Article 29(3)**.
15 Report from the Commission, *Monitoring the Common Fisheries Policy 1995*, COM(97) 226 final, 67–68.
16 Case C-64/88 *Commission* v. *French Republic* [1991] ECR I-2727.
17 *Ibid.*, paragraph 9.
18 *Op. cit.* fn 15.
19 Case 2/88, Order of the Court of 13 July 1990, *Criminal Proceedings Against J.J. Zwartveld and Others*, [1990] ECR I-3365. For annotation, see, *inter alia*: Watson, J.S., *Common Market Law Review*, (1991) 428–443; Simon, D., *Journal du droit international* (1991) 445–447; Clemente, G., *Rivista della Corte dei Conti* (1991), 342–355.
20 On the significance of the procedural route pursued for the referral, and the significance of the decision in the context of judicial cooperation and the 1959 Convention on mutual Assistance in Criminal Matters, see, Chapter 2, *ante*.
21 Case 2/88, Order of the Court of 6 December 1990, *Criminal Proceedings Against J.J. Zwartveld and Others*, [1990] ECR I-4405.

22 Watson, S., *op. cit.* fn 19.
23 See Chapter 2, *ante.*
24 On the issue of sanctions in Community law, see, European Communities, *The System of Administrative and Penal Sanctions in the Member States of the European Communities,* 2 vols., (1992, Brussels); Harding, C., *European Community Investigations and Sanctions,* (Leicester, 1993); Grasso, G., *A New Approach to Community Administrative Penalties,* 30–42; Vervaele, J.E., 'Administrative Sanctioning Powers of and in the Community: Towards a System of European Administrative Sanctions,' in Vervaele (ed.), *Administrative Law Application and Enforcement of Community Law in the Netherlands,* (Deventer, 1994) 161–201; Guldenmund, R. and Westeroun Van Meerteren, L., 'Towards an Administrative Sanctioning System in the Common Agriculture Policy', in Harding (ed.), *Enforcing European Community Rules* (Aldershot, 1996), 103–123. Vervaele, J.E. (ed.) Compliance and Enforcement of European Community Law (The Hague, 1999).
25 LOS Convention, **Article 73(1).**
26 LOS Convention, **Article 73(2).**
27 LOS Convention, **Article 73(3).**
28 LOS Convention, **Article 73(4).**
29 State practice outside the Community also appears to have developed along the lines of the United Nations LOS Convention model, however some countries such as Tanzania have made legislative provision for imprisonment for unauthorised exploitation of EEZ resources, see Mlimuka, A.K.L.J., 'The influence of the 1982 United Nations Convention on Law of the Sea on state practice: the case of Tanzanian legislation establishing the Exclusive Economic Zone', *ODIL* (1995), **26**, 57–73.
30 See Case C-135/92, *Fiskano* v. *Commission* [1994] ECR I-2885.
31 This difference in national procedures is also evident in many areas of Community law outside the domain of fisheries.
32 On national legislation see Moore, *Coastal State Requirements for Foreign Fishing,* FAO Legis. St. 21 Rev. 3 (1988). The prosecution and penalty structure for fisheries offences in New Zealand, is discussed by McClurg, T. in a paper entitled *Two Fisheries Enforcement Paradigms: New Zealand before and after ITQs,* (OECD, Paris, 1992), which notes the high costs of mounting a successful prosecution case and the reluctance of Judges to impose severe penalties. The author points out that in New Zealand fisheries offences are criminal offences and that the prosecution of cases of minor infringements must meet the same standard of evidence as that for serious offences such as in a murder trial.

Penalties for fishery offences in Japan are discussed by Masatake Ohnishi in a paper delivered at the same OECD conference, entitled, *Review of the Fishery Management and Enforcement system in Japan,* (OECD, Paris, 1992), 115–117. Penalties fall into two categories, judicial or administrative. There are provisions which allow the punishment of both the violators and the responsible party such as the fishing manager. Penalties may be as severe as imprisonment of three years and a fine of 2 million yen and include the confiscation of vessels, fishing gear and catches involved in violations. On this basis it would appear to be a costly business to ignore Japanese fisheries regulations. Penalties and sanctions for violating 'The Lacey Act', in the USA, are discussed by Kuruc, M. in a paper entitled *The Lacey Act* delivered at the same conference, (OECD, Paris, 1992) 95–97. The Act is one of the USA's primary laws directed at illicit interstate or foreign trade in illegally taken species. The Act is used in conjunction with smuggling laws, money laundering statutes, theft statutes, conspiracy laws and other federal statutes. It is Kuruc's opinion that the fines and sanctions contained in the Act do not serve as a deterrent, offer adequate punishment, or reflect the collective loss as a result of participating in such illegal enterprises. One leading American authority has expressed a similar view regarding the inadequate penalty structure for fisheries offences in the United States, see prepared statement of J. Gauvin to the

House of Representatives Judicial Committee Hearing on the Enforcement of the Magnuson Fishery Conservation and Management Act and related Laws and Regulations, 1 October 1992, *US Government Publication*, Serial 90, (Washington, 1993), 3–9, (copy Dalhousie Law Library). An interesting view from the Canadian perspective on the Lacey Act is presented by Kraniotis, P. in a paper entitled *The Fisheries enforcement agreement between the United States and Canada*, (OECD, Paris, 1992), p. 59. It is his opinion that the $10,000 penalties in the Lacey Act did not deter vessels intent on poaching in Canadian waters and the Magnuson Act was a better deterrent as Congress had increased penalties from $25,000 to $100,000. Canada and the USA have a special agreement, which was signed in Ottawa on 26.09.1990, to ensure that each country take appropriate measures consistent with international law to ensure that its nationals, residents and vessels do not violate, within the waters or zones of the other Party, the national fisheries law and regulations of the other party (**Article 1**). By 1992, the US attorney in the Region in the 36 cases filed under the agreement, and in 28 of the cases listed issued penalties: totalling $2,903,500, 41 years of permit sanctions, and $400,000 in catch seizures. Although, the high penalties had resulted in protracted proceedings, it was hoped that the penalties would send a strong message to the fishing industry, *op. cit.* p. 59. The bilateral agreement between Canada and the USA is outside the remit of this study, but it offers an excellent comparative subject for assessing not only the enforcement aspect of EU international agreements with third countries, but also for assessing the deterrent effect of the prescribed penalties and sanctions for violating Community fisheries law.

33 See, *inter alios*, Churchill, R.R., 'Enforcement of the Common Fisheries Policy, with special reference to the United Kingdom', 83–102; Berg, A., 'Enforcement of the Common Fisheries Policy, with special reference to the Netherlands', in Harding and Swart (eds), *Enforcing European Community Rules*, (Aldershot, 1996), 62–82.

34 Case 1/78, *Draft Convention of International Atomic Energy Agency on the Physical Protection of Nuclear Materials, Facilities and Transports*, [1978] ECR 2151; Case 203/80, *Casati*, [1981] ECR 2595.

35 See Chapter 1, *ante*.

36 See, for example, the discussion of the PIF Regulation and Convention on the protection of the European Communities' financial interests of 26 July 1995, *OJ* C 316, 27.11.1995, p. 48, by Swart, B., 'From Rome to Maastricht and Beyond: The Problem of Enforcing Community Law' in Harding and Swart (eds), *Enforcing European Community Rules*, (Aldershot, 1996), 1–21, 66–67. European Community Rules, (Aldershot, 1996), pp. 1–21.

37 Case 14/83 *Von Colson and Kamann* v. *Land Nordrhein-Westfalen*, [1984] ECR 1891.

38 Case C-68/88 *Commission* v. *Greece*, [1989]ECR 2965.

39 *Op. cit.* fn 24.

40 'Coastal State requirements for foreign fishing', *FAO Legislative Study* **57**, Table E, (FAO Rome, 1996), 259- 273.

41 The reports, examined in Chapter 2, *ante*, noted that the Commission was reluctant to initiate Article 169 {226} EC Treaty proceedings against Member States in all but the most flagrant violations of fisheries law. On the issue of enforcement actions, see Dashwood and White 'Enforcement Actions under Articles 169 and 170 EEC', *ELR* (1989) **14**, 388–413. See also authorities cited in Chapter 8, *post*.

42 On the Courts' interpretation of **Article 5** {10}, see, *inter alia*, Case 50/76, *Amsterdam Bulb* v. *Productschap voor Sierewassen*, [1977] ECR 137; Case 68/88 *Commission* v. *Greece*, [1989] ECR 2965.

43 This term was used by the *Ad Hoc* Group of Government experts in their report on Community law and criminal law, Vervaele, J., *Fraud against the Community*, (1992), p.316.

44 In relation to the Commission suspending fishing licences, see, Order of the President of the Court of 22 April 1986, Case 55/86 *Asociacion Provincial De Armadores de Buques De Pesca de*

Gran Sol De Pontevedra (Arposol) v. *Council of European Communities,* [1988] ECR 0013; Annotation Churchill R.R., *ELRev.* [1988] pp. 352-360; Case C-135/92 *Fiskano AB* v. *Commission* [1994] ECR I-2885, paragraph 39.

45 The 1992 Management Regulation, Council Regulation (EC) No. 3760/92, *OJ* L 389/1, 31.12.1992, **Article 5(1)**, required the establishment of a community system laying down rules for the minimum information to be contained in fishing licences. In 1993, the Council adopted Regulation No. 3690/93 establishing a Community system laying down rules for the minimum information contained in fishing licences, *OJ* L 341/93, 31.12.1993. The Council meeting in Luxembourg in June 1994, adopted regulation pertaining to fishing permits, Council Regulation No. 1627/94 of 27.06.1994 laying down general provisions concerning special fishing permits, *OJ* L 171/7, 06.07.94. The provisions on licences and permits which includes the control aspects and sanctions required further implementing regulations and because the issues were sensitive politically the Commission proposal in this regard was revised several times. See, Commission Regulation No. 2943/95 of 20.12.1995 setting out detailed rules for applying Council Regulation No. 1627/94 laying down general provisions concerning special fishing permits, *OJ* L 308/15, 21.12.1995.

46 See Berg, A., 'Enforcement of the Common Fisheries Policy, with special reference to the United Kingdom', in Harding and Swart (eds), *Enforcing European Community Rules,* (Aldershot, 1996), 66–67.

47 On this point as regards practice in the United Kingdom, see Churchill, R.R., 'Enforcement of the Common Fisheries Policy, with special reference to the United Kingdom', in Harding and Swart (eds), *Enforcing European Community Rules,* (Aldershot, 1996), 83–102.

48 Council Regulation No. 1627/94 of 27.06.1994 laying down general provisions concerning special fishing permits, *OJ* L 171/7, 06.07.94, **Articles 9** and **10**. See also discussion of the legal nature of sanctions which arose in Case C-135/92 *Fiskano AB* v. *Commission* [1994] ECR I-2885, and Case 55/86 *Asociacion Provincial De Armadores de Buques De Pesca de Gran Sol De Pontevedra (Arposol)* v. *Council of European Communities,* [1988] ECR 0013, by Berg, A., 'Enforcement of the Common Fisheries Policy, with special reference to the United Kingdom', in Harding and Swart (eds), *Enforcing European Community Rules,* (Aldershot, 1996), pp. 66–67.

49 Commission Regulation No. 2943/95 of 20.12.1995 setting out detailed rules for applying Council Regulation No. 1627/94 laying down general provisions concerning special fishing permits, *OJ* L 308/15, 21.12.1995, **Article 5**.

50 Case C-135/92 *Fiskano AB* v. *Commission* [1994] ECR I-2885, paragraph 39. See discussion on the right to be heard, *infra.*

51 *Ibid.*

52 Paragraphs 31 and 32 *ibid.*

53 Case C-290/87, *Commission* v. *Netherlands,* [1989] ECR 3083, paragraphs 18–21.

54 *Ibid.*

55 *Ibid.*

56 Case C-52/95, *Commission* v. *French Republic,* [1995] ECR 1-4443, paragraph 35.

57 paragraph 37, *ibid.*

58 Case 85/76, *Hoffmann-La Roche & Co. AG* v. *Commission,* [1979] ECR 461.

59 This is examined by Churchill, R.R. and Foster, N.G. 'European Community law and prior treaty obligations of Member States: The Spanish Fishermen's Cases', *ICLQ* **36(3)** (1987), 504–524.

60 Case C-135/92 *Fiskano AB* v. *Commission* [1994] ECR I-2885, paragraph 39.

61 Case 85/76 *Hoffmann-La Roche* v. *Commission,* [1979] ECR 461; Case 234/84 *Belgium* v. *Commission* [1986] ECR 2263 (*'Meura'*); Case 40/85 *Belgium* v. *Commission* [1986] (*'Boch'*), Joined Cases 46/87 and 227/88 *Hoechst* v. *Commission* [1989] ECR 2859; Case 301/87 *France* v. *Commission* [1990] ECR I-307; Case C-142/87 *Belgium* v. *Commission* [1990] ECR I-959; Joined Cases C-48/90 and C-66/90, *Netherlands and Others* v. *Commission* [1992] ECR I-565.

62 Case C-135/92, paragraph 40.
63 Case C-135/92, paragraph 44.
64 Case 68/88, *Commission* v. *Greece*, [1989] ECR 2965.
65 The option of linking the amount of structural aid to how well public authorities manage and conserve fish stocks was suggested by the Court of Auditors in their Report, see Chapter 5, *ante*. By analogy and in the context of the CAP, the Court has upheld the total loss of the right to aid for a producer of peas and beans as a result of the non observance of the stipulated application periods in the applicable Council Regulations, see, Case 358/88, *Oberhausener Kraftfutterwerk Wilhelm Hopermann* v. *Bundesanstaltfuer Landwirtschaftliche Marktordnung ('Aid scheme peas and field beans case')*, [1990] ECR I-1687.
66 Commission Regulation (EEC) No. 3561/85 of 17.12.1985 concerning information about inspections of fishing activities carried out by national authorities, *OJ* L 339/29, 18.12.1985, **Article 1(2)**.

Chapter 7

Quota Hopping: An Insoluble Enforcement Problem?

Introduction

Quota hopping is the practice whereby persons (legal or otherwise) and vessels of one nationality, obtain the right to catch a portion of the available catch of another country's national quota under the common fisheries policy. In effect, quota hopping is a form of the well-established practice in merchant shipping of vessels sailing under a flag of convenience, or being registered in a country, when in fact the vessel is beneficially owned and controlled by persons who live elsewhere. In general this category of vessel seldom, if ever, enters the jurisdiction or ports of their country of registry. In the fisheries sector quota hopping has largely been associated, but not exclusively so, with the efforts by Spanish interests to obtain access to Irish and British quotas.

Initially, Member States endeavoured to deal with the phenomenon of quota hopping through national legislative measures which restricted and qualified the conditions for the registration of fishing vessels and the granting of licences. The compatibility of such measures with Community law has been examined by the Court of Justice in a series of cases commonly referred to as the 'quota hopping cases'. The Court's deliberations and judgements in these cases reaffirm the primacy of basic principles of the Treaty as regards the granting of flag rights and licences to fishing vessels and were of major political significance in many sectors outside fisheries. The case law helps to define the limits of Member State legislative autonomy and discloses its relation with some of the fundamental norms of Community law. In practice the Court's decisions have had major implications for the enforcement of Community fisheries law. In particular, the quota hopping cases illustrate that, despite the almost exclusive prescriptive jurisdiction of the Community in the domain of fisheries and the development of a number of broad principles by the Court of Justice in respect of Member State obligations to deal with non-compliance of Community law as well as the supervisory role of Community institutions, responsibility for the practical enforcement of Community fisheries rests almost exclusively with the Member States. In this chapter it is proposed to briefly review the phenomenon of quota hopping and to examine some of the associated enforcement issues. It is also proposed to canvas a number of solutions, from the enforcement perspective, to the problems associated with monitoring the activities of quota hopping vessels. This chapter concludes by mentioning some of the licence conditions which came into force in the United Kingdom in 1999 and which aim to maintain a real economic link between coastal Communities in the United Kingdom and vessels which fly the flag of the United Kingdom and have access to national quotas.

Table 7.1 The number and category of vessels engaged in quota hopping in 1998.

Former flag	Convenience flag	Number of vessels	Vessel type	Target species	ICES area
Spain	United Kingdom	110[2]	Side/stern trawlers	Demersal	VI and VII Western Waters
Netherlands	United Kingdom	30–40[3]	Beam trawlers	Flat fish Sole and plaice	IVA (North Sea) VII
Spain	France	25–30	Side/stern trawlers	Demersal	Western Waters
Spain	Germany	9	Side/stern trawlers	Demersal	Western Waters
Spain	Ireland	5	Side/stern trawlers	Demersal	Western Waters

1. *The number and type of vessels which are involved in quota hopping*

Data available on the type and number of vessels which engage in the practice of quota hopping suggest that the number of vessels appear to have increased significantly since the early 1980s. Between 1980 and 1983, approximately 60 vessels registered in the United Kingdom were owned or controlled by Spanish interests. In 1998, the number and category of vessels engaged in quota hopping would appear to be as in Table 7.1.

Vessels engaged in quota hopping in the European Union are registered and sail under the flags of Member Sates other than their Member State of origin. Such vessels are generally nominally owned by companies registered in the United Kingdom, France, Germany and Ireland although the true beneficial ownership may rest with Spanish or Dutch commercial interests. Vessels operate out of ports in North and North-Western Spain and fish mainly for white fish (hake, monk, sole etc. - mainly TAC/quota species) in waters which are predominantly under Irish jurisdiction or elsewhere in Western Waters. The former Dutch vessels generally operate in the North Sea. Quota hopping vessels rarely visit their ports of registry unless statutory obliged to do so under national legislation.[4] Consequently, quota hopping vessels present a unique and inter-jurisdictional enforcement problem for the control services in the flag Member State, the coastal Member State, and to a lesser extent the port State of landing.

2. *Origin of the quota hopping phenomenon*

The phenomenon of quota hopping has increased dramatically since the early 1980s and it is now possible to identify some of the contributory factors to the evolution of this practice. These include, *inter alia*:

- the loss of traditional fishing grounds for several European fleets with the extension of coastal state jurisdiction in 1976;[5]
- the overcapacity of Member States' fishing fleets in relation to the resources available in the Community fishing zones;
- the large size of the Spanish fleet prior to Spain's accession to the Community in 1986;
- the limited access regime for Spanish vessels to the Community zone, and in particular for the fleet which traditionally fished waters under Irish and British jurisdiction;[6]
- inadequate arrangements during the 1980s for monitoring the withdrawal of surplus or old capacity in some Member States and in particular the Netherlands;
- inadequate restrictions in some Member States on the uptake of fishing licences, and in particular the free availability of non-pressure stock licences in the United Kingdom;
- absence of arrangements in some Member Sates to ensure that there was a valid economic link between vessels and their flag State.
- the free-market ideology in the United Kingdom which favoured growth in the number of vessels (increased tonnage) on the national register and the freedom of exchange of fishing licences;

- the absence of Government subvention for the development and restructuring of the United Kingdom fleet during the same period which lead to the decline of the indigenous fishing industry.

3. *The jurisprudence of the Court of Justice in the quota hopping cases*

For the purpose of reviewing and summarising the case law, the phenomenon of quota hopping may be seen to have occurred in three phases.

(i) First phase

The first phase of quota hopping commenced before Spain's accession to the Community in 1986 and concerned Member States' efforts to curtail the possibility of third-country interests (mainly from Spain) from acquiring fishing possibilities under the Irish and the United Kingdom quota allocation. Both the United Kingdom and Ireland adopted national legislation stipulating that 75% of the crew on board vessels flying their respective flags and fishing within exclusive fishery limits had to be nationals or Community nationals. These measures were contested by an Irish–Spanish joint venture company *Pesca Valentia* which ultimately led to a Court of Justice decision upholding the Irish legislation which required a minimum proportion of a vessel's crew to be Community nationals.[7] In *Pesca Valentia* the Court of Justice did not refer to discrimination on the grounds of nationality *vis-à-vis* Spanish nationals (Spain was not a Member State at the time the reference was made) and ruled that under Community law the conditions for flying a flag of a Member State or registered in that State were left to be defined in the legislation of each Member State, subject to the condition that such legislation accorded with the principles of Community law.[8] As a result of the Court's decision, Ireland and the United Kingdom ostensibly succeeded in limiting Spanish quota hopping vessels from operating within their respective jurisdictions. However, this proved to be a somewhat pyrrhic victory because the vessels in question recruited crews with the required nationality in order to comply with the crewing requirement.

(ii) Second phase

As explained in Chapter 1, the Community agreed a common policy for fisheries in 1983 which allocated quotas on a national basis. The second phase of quota hopping jurisprudence commenced in 1986 with the impending accession of Spain and Portugal to the Community. With a view to curtailing Spanish vessels from circumventing the quota allocation to Member States, the United Kingdom enacted legislation which attached special conditions to fishing licences. These measures provided special operating conditions for vessels, introduced a crewing requirement as well as specific social security obligations which applied to all crews of United Kingdom fishing vessels. These provisions were subsequently challenged in the High Court in the United Kingdom which made two **Article 177** {234} of the EC Treaty referrals to the Court of Justice. The Court delivered two judgements (*Agegate* and *Jaderow*) pertaining to the United Kingdom legislation.[9] The judgement in *Agegate* upheld the Community nationality requirement and a second requirement whereby the crew of a vessel are obliged to contribute to a Member State social security scheme. However, the Court rejected the requirement that the crew should be resident in the Member State on the grounds that this requirement was not justified by the national allocation system for fish quotas. In *Jaderow* the Court ruled that conditions that are designed to ensure that there is a real economic link between the Member States are justified if their purpose is to ensure that populations that are dependent on fisheries and related industries should benefit from the quotas. In particular the obligation on vessels to operate from a national port conforms with the aim of the quota system if it merely involves the obligation to operate habitually [and not for each fishing trip] from a national port. Most significantly, in both *Agegate* and *Jaderow,* Advocate General Mischo came to a different conclusion to the Court, his view being that the system of quotas and their utilisation presented sufficient justification to limit the application of the Treaty in the situations under examination.[10]

The second phase of quota hopping continued after Spain acceded to the Community in 1986 and previous national legislative measures to contain the problem of 'quota hopping' proved to be ineffective. The United Kingdom Parliament enacted the Merchant Shipping Act 1988 and the Merchant Shipping Regulations 1988 which, *inter alia,* sought to thwart Spanish shareholders in British companies from owning and operating vessels flying the flag of the United Kingdom. The intention was to preclude access to United Kingdom quotas and to stop the circumvention of the limited access under the Iberian accession regime for the Spanish fleet to Community resources.

In December 1988, *Factortame Ltd and other* (who represented the owners of 95 fishing vessels registered in the United Kingdom) applied for a judicial review and challenged that the 1988 Act contravened several Treaty provisions including **Articles 7** {12} (prohibition of discrimination on the grounds of nationality), **34** {29} (prohibition of quantitative restrictions on exports), **40(2)** {34(2)} (prohibition of discrimination between producers or consumers within the Community), **48–51** {39–42} (freedom of movement of workers), **52–58** {43–48} (right of establishment), **59–66** {49–55} (freedom to provide services within the Community), and **221** {294} (nationals of other Member States should be subject to the same treatment as own nationals as regards participation in capital of companies and firms) of the EC Treaty. The applicants also applied for interim relief. The Commission initiated **Article 169** {226} proceedings against Ireland and the United Kingdom on the grounds that national legislation contravened Community law.

As a result of requests for a preliminary ruling in *Commission* v. *United Kingdom* the Court, by way of an interim order, instructed the United Kingdom to suspend application of the nationality requirement with regard to the registration requirement of vessels fishing under the United Kingdom flag.[11] In its ruling of 19 June 1990 the Court in *Factortame Ltd (No. 1),* confirmed that national courts were obliged in such cases to ensure full compliance with Community law.[12] Referring to the decision in *Simmenthal*[13] the Court affirmed that directly effective provisions of Community law, *'render automatically inapplicable any conflicting provision of . . . national law'*.[14] On the issue of whether national courts should disapply national law in favour of Community law by granting interim protection based on that law when it had not yet been established that the Community law created rights for the Applicants, the Court relying on the policy argument of effectiveness, found:

> 'The full effectiveness of Community law would be . . . impaired if a rule of national law could prevent a court seised of a dispute governed by Community law from granting interim relief in order to ensure the full effectiveness of the judgement to be given on the existence of the rights claimed under Community law. It follows that a court which in those circumstances would grant interim relief, if it were not for a rule of national law, is obliged to set aside that rule.'[15]

The House of Lords subsequently granted the applicants, who were contesting the restriction introduced by the British legislation in question, the necessary legal protection while the main action was pending.[16] On the issue of supremacy, the House of Lords also overruled the impugned provisions on the grounds that they were in conflict with Community law. The applicants were thus successful in their application for interim relief.

On 25 July 1991 the Court of Justice gave its eagerly awaited ruling in Case 221/91 *Factortame Ltd (No. 2)*. The Court held that it was for the Member States to determine the conditions governing the right to fly their flag but in exercising that power, Member States had to respect the primacy of Community law and in particular **Articles 7** {12}, **52** {43} and **221** {294} of the EC Treaty whereby the right to fly a particular flag cannot be refused on the grounds that the owners or operators of the vessels concerned are nationals of another Member State or that they reside or have their principal place of business in another Member State.[17]

The Court gave its judgement in Case 93/89 *Commission* v. *Ireland* on the same day as the *Factortame Ltd (No. 2)* decision and ruled that **Article 52** {43} of the Treaty was infringed by Irish legislation requiring nationals of other Member States, without a similar requirement being imposed on its own nationals, to set up an Irish company before obtaining a licence for sea fishing.[18] Furthermore, the national legislation could not be justified by the existence of a Community system of national fishing quotas since the licences were not intended as detailed rules for the utilisation of those quotas.[19] Ireland subsequently amended the Fisheries Amendment Act 1983. In a similar **Article 169** {226} proceedings taken by the Commission against the United Kingdom on the grounds that the Merchant Shipping Act 1988 contravened Community law,[20] the Court held that the United Kingdom had failed to fulfil the obligations under **Articles 7** {12}, **52** {43} and **221** {294} of the EC Treaty. The facts in this case were the same as in *Factortame (No. 2)* and in all material respects the Court's judgements are identical in both cases.

As noted by one leading authority,[21] the Court's judgements in this phase of quota hopping cases would make it difficult for Member States to limit the practice of quota hopping. In effect all Member States could do is to require fishing vessels to be managed and operated from within its territory and to operate some of the time from one of the flag State ports. However, the same authority suggests that measures which accorded with the general provisions of Community law could be introduced in a regional basis rather than on a national basis to control the activities of these vessels.

(iii) Third phase

The third phase of the Court of Justice jurisprudence relating to quota hoppers concerns two enforcement actions taken by the Commission against Ireland and the United Kingdom.

In Case 279/89, the Commission brought an action concerning UK legislation which incorporated in fishing licenses nationality and residence requirements and to conditions whereby vessels must return periodically to British ports.[22] It also commenced proceedings against Ireland in Case 280/89 *Commission* v. *Ireland* because of provisions to exclude UK vessels from Irish waters unless they obtained a fishing licence in accordance with the conditions challenged in *Commission* v. *United Kingdom*.[23] On 17 November 1992 the Court held that the United Kingdom had failed to meet its obligations under the Treaty by laying down these conditions relating to the composition of the crews. These findings were in line with the previous ruling in *Agegate* which dealt with the same national legislation. The Court, however, did not decide on the condition requiring periodic return to British ports, since this condition had been examined by the Court in *Jaedrow* and held to be incompatible with Community law in certain circumstances but was not contrary to the provisions of Community law in every circumstance. On 2 December 1992, the Court delivered its judgement in Case C-280/89 *Commission* v. *Ireland* and ruled that the Irish legislation breached the principle of equal access if it enacts a fishing ban on certain vessels flying the flag of another Member State within its exclusive fishery limits. That breach cannot be justified by the fact that the exclusion by other Member States of those same vessels from fishing against its national quota is not contrary to Community law. Moreover, a ban by a Member State on the transhipment of fish within its fishery limits and landing of fish in its territory imposed on certain vessels flying the flag of another Member State constitutes a barrier to free movement of goods which is prohibited by **Article 30** {28} of the EC Treaty.

During the course of initiating proceedings against the United Kingdom and Ireland, in relation to granting of flag rights and licences, the Commission also reviewed the legislation in other Member States. It emerged that nearly all the Member States had similar legislation which in the Commission's view did not conform with Community law. Reasoned opinions were issued in 1993 and infringement proceedings were initiated against all the relevant Member States.[24] In 1994 the Commission was able to terminate infringement proceedings against Germany following the entry into force of new legislation which remedied the previous shortcomings. More recently, the Court in *Commission* v. *France* ruled that France had failed to fulfil its obligation under **Articles 6** {12}, **48** {39}, **52** {43} and **221** {294} of the EC Treaty, as well as Council Directive 75/34/EEC and Commission Regulation No. 1251/70, by retaining in the French Customs Code legislation which restricted the right to a register a vessel in the national register to certain categories of natural and legal persons linked to France.[25] Furthermore, the Court found that France had failed to fulfil its obligations under **Article 171** {228} of the EC Treaty by not taking appropriate legislative measures to comply with an earlier judgement of the Court which had found that the French Maritime Labour Code impugned Community law.[26] Similarly, the Court held that Ireland had also failed to comply with **Articles 6** {12}, **48** {39}, **52** {43} and **58** {48} of the EC Treaty, as well as Council Directive 75/34/EEC and Commission Regulation No. 1251/70, by maintaining national legislation which limited the right to register a vessel (other than a fishing vessel) in the Irish shipping register to a vessel which is owned in whole or in part by the Government, a Minister of State, an Irish citizen or an Irish body corporate.[27] Interestingly, in both judgements the Court stated that access to leisure activities is a corollary to the freedom of movement under **Article 48** {39} and **52**

{43} of the EC Treaty and thus national legislation which restricted the registering of pleasure craft was also contrary to Community law.[28]

The Court's jurisprudence in the aforementioned cases resolved the major points of Community law in respect of national measures which were introduced by Member States to regulate quota hopping. The Court has since ruled in a historic judgement that Member States are obliged to make good the loss or damage caused to individuals for breaches of Community law attributable to the State and this is applicable where the national legislature was responsible for the breach in question.[29] There were also indications that the issues relating to quota hopping were to be reviewed at the intergovernmental conference in 1994–1996. At one stage during the negotiations leading to the revision of the Treaties, the United Kingdom advocated that the Treaty of Amsterdam be amended to allow Member States to reserve quotas for national fishermen.[30] However, after the Amsterdam summit meeting in June 1997, it was revealed that there had been an exchange of letters between the Commission and the United Kingdom in order to clarify the measures to be taken against quota hopping vessels.[31] Subsequently, it was disclosed in a House of Commons debate that the measures in question represented substantial progress towards resolving the problems associated with quota hopping.[32] In particular, the United Kingdom was reported as considering the insertion of provisions in fishing licences which consolidate the economic link between vessels and the ports of their flag State.[33] These measures which take effect from 1999 are discussed below.

4. *Enforcement problems which have resulted from the phenomenon of quota hopping*

The Court's adjudication of the compatibility with Community law of the national measures dealing with the phenomenon of quota hopping has elicited much academic comment,[34] and is generally considered remarkable on the grounds of principle. Several practical issues, however, remained unaddressed.[35] One relates to what extent, subject to compliance with the Treaty and given the underlying principles and specific features of the CFP, a relationship must exist between a fishing vessel and the Member State whose flag it flies and whose quotas it fishes? The second is: to what extent the constraints imposed by the Community's structural policy may be reconciled with the fundamental principles of the Treaty? Other questions relate to a number of enforcement and practical application issues which were left unresolved in the wake of the judgements. It is the latter which are addressed here.

The Court in the 'quota hopping cases', by adhering rigidly to the fundamental principles established by the Treaty, may have undermined some core elements in the resource management policy which is constructed with a framework of Community secondary legislation and depends almost exclusively on national enforcement measures for its effectiveness. In effect, on the substantive issues, the Court's jurisprudence undermined a system of resource allocation which hinged on a system of national quotas.[36] In this approach the Court exposed the absence of a solid Treaty base for the management tools of the CFP and in particular the established criteria for resource allocation and distribution. It also demonstrated that a common conservation policy and the equality of Community fishermen in the eyes of the law do not readily co-exist with an enforcement structure which is closely linked to the sovereignty of the Nation State and divides duties and responsibilities on the basis of the flag State–coastal State international law of the sea enforcement model.[37] The decisions have to be viewed in the light of the previous decision in *Romkes* which upheld the allocation method of quotas among the Member States as being compatible with Community law.[38] The decision in *Factortame Ltd (No. 2)* undermined Member State ability to take effective action to curtail the use of their flag as a means of obtaining access to national fish quotas. It has made the monitoring and supervision of the uptake of national fish quotas more difficult and has placed national enforcement agencies in the invidious position whereby national efforts to monitor their flag fleet can be easily rendered nugatory. The Court's decisions upheld the equal treatment of fishermen in principle. However, because Member States have not ceded a suitable range of inter-jurisdictional inspection and surveillance powers to national inspection services (or to a Community inspection service) the Court's decision may have ensured the disparate application of the law in reality.[39] Indeed, one commentator has suggested that the Court in *Factortame Ltd (No. 2)* may have been endeavouring to promote a *communuataire* means of conservation.[40] Whatever the motive, it is obvious that a resource management method which involves a delicate balance in the distribution or allocation of quotas to Member States requires legislative safeguards to ensure that there is rigorous implementation and compliance with the overall plan.

(i) Community legislative action to control the activities of vessels engaged in quota hopping

During the period of the Court's deliberation, the Council amended the 1987 Control Regulation to facilitate the exercise of monitoring the activities of vessels engaged in quota hopping.[41] The 1987 Control Regulation was amended with a view to strengthen the application of the fisheries conservation rules by improving co-operation between Member States to prevent overfishing. In accordance to the recitals in the preamble, Community rules must include measures for the inspection and monitoring by the authorities of the Member States of all fishing vessels, including the vessels of non-member countries, at sea and in port. In particular, the 1987 Control Regulation as amended allowed the flag Member State to obtain more rapid and detailed information on catches landed by their vessels in other Member States.[42] It also introduced several onerous provisions on the Member State of landing including the possibility of deduction of catches from national quotas in instances where there was failure to transfer proceedings to the flag State. Some of these measures, introduced by the 1987 Control Regulation Amendment, have been carried over into the 1993 Control Regulation.[43]

(ii) Legal challenge to Community measures by Spain

Spain challenged the 1987 amendment to the Control Regulation in the Court of Justice on four grounds.[44] First, it was alleged the new measures burdened the Member State where landings are made with fishery enforcement measures and that it thereby transfers responsibility from the flag Member State to the port Member State (i.e. the Member State where the vessel lands the catch). In relation to this argument, the Court pointed out that the control measures prior to the disputed regulation required all Member States to ensure that the rules limiting catches were complied with and thus the new regulation was only adding further details to this obligation.[45] Secondly, Spain argued that the task of checking whether a particular vessel holds a licence may fall on the Member State of landing and that the licence system was thus in the process of being turned into a Community system. The Court noting that a licence system was one of the methods to ensure compliance with catch limitations imposed by the Community conservation system and that the Council was thus entitled to strengthen co-operation without infringing any rule under the Treaty.[46] Thirdly, Spain contended that the new regulation required vessels which failed to comply with the rules to carry on board a specific document certified by the Member State of registration stating that the latter had inspected the vessel within the previous two months. Spain alleged that this amounted to a measure having an effect equivalent to a quantitative restriction on exports contrary to **Article 34** {29} of the EC Treaty, on the grounds that it restricted, or removed the option for, a vessel from landing the catch in the port of its choice. The Court drawing the distinction between an additional control measure (the extra document) and the placing of restrictions on the exports of fish, noted that it had previously held that **Article 34** {29} concerns only those measures which have as their specific object or effect the restrictions of patterns of exports, and thereby the establishment of a different treatment between the domestic trade of a Member State and its export trade in such a manner as to provide a particular advantage for national production or for its domestic market at the expense of the production or trade of other Member States.[47] As Spain had failed to show that the measures complained of restricted specifically the patterns of exports, the Court rejected this argument.[48] Fourthly, the Court upheld the provision which provided that Member States infringed the conservation rules if during landing or transhipment in their ports they failed to initiate proceedings against parties responsible for breaching Community rules (in such instances Member States may transfer judicial or administrative proceedings to the flag State). Furthermore, the Court also upheld the punitive provision which allowed for the deduction of the quantities of catches unlawfully landed or transhipped from the quotas of the Member State which has failed to comply with the obligation to take penal or administrative action or failed to transfer proceedings. In arriving at this conclusion the Court found the argument based on the 'territoriality' inapplicable on the grounds that the contested regulation did not require the Member State to extend their jurisdiction beyond the generally accepted principles governing the distribution of criminal jurisdiction between States.[49] Moreover, in relation to the quota deduction the Court accepted the opinion of the Commission (which had intervened with the United Kingdom) that it was reasonable to deduct the quantities from the Member State of landing on the grounds that their control system was inadequate.[50] In this regard the Court found no breach of the principle of proportionality and upheld the contested measures.

(iii) Monitoring compliance by flag of convenience vessels with quotas

In 1989 the Commission published a document entitled *Commission Communication on a Community Framework for Access to Fishing Quotas*,[51] outlining a solution which one authority has argued, if adopted, would not have been effective in preventing quota hopping.[52] Furthermore, the same authority suggests that this may indicate a Commission desire to move away from a system of national quotas to a system of regional quotas.[53] The merits of such arguments are inconclusive. It is evident, however, that Case C9/89, *Spain* v. *Council*, supports the view that the Court will view favourably measures which are intended to improve the implementation of, and compliance with, the Community fishery management system. Despite this unequivocal endorsement, under **Article 16** of the Control Regulation the transmission of landing information on a bilateral level is dependent on the flag State requesting this information from the Member State where the landing is made. The question thus arises, 'if the flag State is not aware of the landing, then will it be in a position to initiate a request for such information in the first instance?'. It could be argued that there should be a mandatory obligation on the port State to report all individual vessel landings to the appropriate flag State. However, it ought to be stressed that there is a penalty structure in place to ensure that such landings are monitored. Specifically, **Article 32(2)** of the Control Regulation is an incentive to improve the monitoring of activities of quota hopping vessels. It provides that if the Member State of landing or transhipment is not the flag Member State and its competent authorities do not undertake, in conformity with their national law, appropriate measures including the initiation of administrative action or criminal proceedings against the natural or legal person responsible for the landings, or do not transfer prosecution proceedings, then the catches in question may be counted against the national quotas of the Member State of landing.[54] There are two inherent weaknesses in **Article 32(2)**. The first one relates to the issue of evidence. That is to say the Commission will not always be in the position to verify that the Member State where the landing occurred did not invoke the appropriate national measures. Secondly, to exercise the quota deduction power, the Commission requires the support of the majority of Member States in the Management Committee Fisheries in order to adopt the appropriate implementation regulation pursuant to the **Article 36** procedure. In practice, this might be difficult to obtain if such action was considered by other Member States to offend political sensibilities and contrary to the harmonious and constructive working atmosphere required by the Management Committee to complete their work on an ongoing basis. On the other hand, there may be scope for the two Member States in consultation with the Commission to exchange the quotas on a bilateral basis without resorting to a vote in the Management Committee.

In conclusion, the issue of controlling the activities of quota hopping vessels deserves special attention. The judgements in *Pesca Valentia, Agegate, Jaderow, Factortame Ltd* and the various Commission **Article 169** {226} cases clearly limit Member States legislative scope in preventing nationals from other EC Member States fishing against national quotas. As one authority has succinctly noted, the phenomenon of quota hopping is a reflection of the tension between the system of national quotas and general Community law, particularly the non-discrimination principles in **Articles 7** {12}, **40** {34}, **52** {43} , and **221** {294} of the EC Treaty.[55] From the enforcement perspective there are still difficulties regarding the monitoring of flag of convenience vessels landing their catches outside the flag Member State. The more stringent requirements regarding, *inter alia*, notice of landing (**Article 7**), submitting landing declarations (**Article 8**), quota uptake (**Article 15**), notification requirements (**Article 16**) and sanctions (**Article 21**), may have the appearance of improving the control system but may prove to be ineffective, principally because the power and resources to enforce these measures are not in the hands of the flag Member States but rest with the authorities in the ports of landing. Indeed, the power in **Article 32(2)** to deduct quotas for failure by the Member

State of landing or transhipment to undertake inspection obligations will improve enforcement in the Member States only if it can be invoked on a systematic and regular basis. In this context, the 1998 amendment to the Control Regulation which allows for non-flag Member State inspectors to act as observers (**Article 34b)** in another Member State is a major step which will facilitate the exchange of information and introduce greater transparency into the monitoring inspection and surveillance programmes undertaken by the port Member State (i.e. Member State of landing).

5. *Confronting the enforcement dilemma posed by quota hopping vessels*

Although the issue of quota hopping is essentially one related to the surplus fishing capacity of national fleets and will ultimately have to be addressed as a structural issue, quota hopping has led, nonetheless, to enforcement problems. These arise because the quota hopping fleet fish predominantly outside the jurisdiction of the flag State. This is most evident in the case of the *Anglo-Spanish* fleet generally (i.e. United Kingdom/Irish registered quota hoppers) which has benefited from access to United Kingdom and Irish quotas although obliged to make statutory visits to the United Kingdom, fish principally for stocks (mainly hake) which are in the Irish jurisdiction, and land their catches directly into ports in North-Western Spain.

Admittedly, the fact that Community law has limited the number of legislative options available to Member States to deal with the problem and in particular the decision of the Court in *Jaderow*, means that it is not possible to place restrictions on vessels with regard to specifying mandatory ports for the landing of all quota species after each and every fishing trip.[56] It is thus essential that the enforcement strategy in the Member States should focus on at-sea inspections. Furthermore, there exists within the Control Regulation several possibilities for improving compliance by the quota hopping fleet with their regulatory obligations. These include, *inter alia*:

- the implementation of an inter-Member State enforcement strategy to deal with flag of convenience vessels which entails a joint inspection programme and the frequent exchange of inspectors (including inspectors to act as observers);
- improved Member State co-operation in the exchange of data relating to logbook and landing declarations;
- the imposition of strict national measures in relation to vessels landing outside flag States, such as the non-landing of catches prior to the completion of landing declarations and the mandatory verification of catches by national inspectors in the port State;
- transfer of proceedings. If there is evidence that an offence has been committed in the waters under the sovereignty or jurisdiction of a coastal state then consideration should be given to transferring proceedings to the flag State or to the coastal State where the alleged infringement was committed; and
- mandatory satellite tracking of all flag of convenience vessels.[57]

Essentially, the ability of quota hopping vessels to land their catches outside the flag Member State is the source of the enforcement dilemma for national inspection services. Therefore, a second possible solution to reconcile the enforcement problems associated with quota hopping vessels would be to allow inspectors from the flag

Member State to inspect national vessels landing in other Member States. In response to a Parliamentary question on this issue in 1992,[58] Commissioner Marin, recalling the exclusive competence of the coastal Member States for the monitoring and inspection of fisheries activities, pointed out that Member States are free to take appropriate action to optimise the effectiveness of their inspections. In this regard, the Commissioner expressed the view that the Commission considered the exchange of inspectors as a valuable practice in the common interest of standardising inspection procedures through the sharing of experiences. However, to this he added the caveat, which could be construed as a statement of policy, that, theoretically, the idea of fisheries inspectors from Member States carrying out inspections in other Member States within a legal framework established at Community level would have a certain attraction if this would lead to a common confidence that the rules were applied equitably and efficiently across the Community. The idea would also seem to fit well with the subsidiarity principle. In practice, however, it would be extremely difficult to ensure that all Member States actually participated in such a scheme. Moreover, there would be consequences for Member State sovereignty, if inspectors of other Member States were granted rights of enforcement against private citizens of that Member State. Finally, Commissioner Marin suggests that the Commission's own inspectors are the most practical and effective means of monitoring Member States' compliance with conservation and control measures.

It is also evident that the 1998 amendment to the Control Regulation clearly establishes a legal base for inspectors to act as observers outside their flag State. It also expressly encourages the need for effective bilateral agreements in so far as **Article 34(1b)** states that Member States may also carry out, among themselves and on their initiative, monitoring, inspecting and surveillance programmes. Both these measures may prove to be a useful means to enforce Community measures.

However, adopting such a course of action, non flag State inspections in the Member State of landing or transhipment, is constrained by Title VI {VI} of the TEU which clearly establishes that the future developments in the areas of Justice and Home Affairs are largely a matter for inter-governmental co-operation and not on the basis of supranational structures and procedures. Moreover, in accordance with the Amsterdam Treaty, mutual co-operation in criminal cases remains a matter for inter-governmental co-operation.[59] It is also evident that, although the powers of Commission inspectors have been expanded by the 1993 Control Regulation, there is little scope for a small unit to closely monitor the activities of the 200 flag of convenience vessels (approximately). The solution to this problem may lie in a three-strand approach. One based on the inter-Member State co-operation procedure and joint inspection programmes envisaged in **Title VIIIa** of the Control Regulation.[60] The second strand based on the development of the criminal justice agreements, conventions, protocols to tackle inter-jurisdictional fisheries crime at an inter-governmental level.[61] The third solution is for the Community in conjunction with the Member States to introduce a broader range of enforcement tools. This could include an observer scheme on board vessels suspected of infringing Community regulations and the mandatory application of satellite tracking to all vessels engaged in quota hopping regardless of their size, as well as the setting up of enforcement check points, similar to those relied upon by the Norwegian authorities in the Norwegian zone, through which vessels have to pass when en route to and from the fishing grounds.

Furthermore, Member States ought to consider reviewing the level and severity of their sanctions. It ought to be borne in mind that, in 1992, 15 foreign flag vessels were found to have hidden compartments for holding fish in the Irish exclusive fishery zone. This offence attracted a minimum fine prior to 1992. Significantly, after the legislation had been amended to increase the penalty tenfold to a maximum penalty of £200,000, there has been only one reported incidence of this offence recurring in the period 1993–1996.[62] Furthermore, there is also evidence which suggests that vessels which exceed their individual sector or non-sector quotas occasionally are penalised severely by Courts in the Member States.[63] Again however there is little empirical data available regarding the deterrent value of such penalties. The frequency of such offences, on the other hand, underlines the need for enhanced Member State co-operation in the exchange of data relating to logbook and landing declarations.

In conclusion, there are many factors which have given rise to quota hopping, a phenomenon which however is not unique to the European fishing industry. As noted in Chapter 2, it was apparent at the end of the first period of the CFP that the success and thus the effectiveness of the CFP depends in large measure on the degree of compliance with Community regulations. Essentially, the legacy of the quota hopping jurisprudence is that it has made the task of the State authorities entrusted with the enforcement function more difficult and thus, arguably, facilitated non-compliance with Community rules. Moreover, the absence of a readily apparent antidote to non-compliance by the quota hopping fleet is the most conclusive confirmation of the thesis that the flag Sate–coastal State law enforcement model is not a totally suitable means to attain the objectives of the CFP. Ultimately, the resolution of the dilemma posed by quota hopping will depend on how the competent authorities in the flag State, the coastal State, and the port States reconcile their diverging approaches to the enforcement of Community fishery law.

Finally, it ought also to be pointed out that there are several international obligations stemming from the Code of Conduct for Responsible Fisheries, the Compliance Agreement on High Seas Fisheries, and the United Nations Straddling Stocks and Highly Migratory Species Agreement which require the Community and the Member States to monitor closely the activities of vessels sailing under their flags.[64] In particular there is an onus on the Community to ensure that vessels flying their flags do not undermine international fishery management schemes. In addition there is an express requirement pursuant to the Code of Responsible Fisheries for the port State to provide assistance to the flag State in cases where there is non-compliance with regional or global conservation and management measures.[65] It may be argued that the preponderance of international instruments clearly support the view that a *laissez-faire* approach to enforcement is no longer acceptable.

6. *The economic link between vessels flying the flag of the United Kingdom and the United Kingdom*

Following consultations between the Commission and the Ministry of Agriculture, Fisheries and Food (MAFF), the United Kingom authorities finalised in 1997 their proposals for ensuring an economic link between vessels flying the flag of the United Kingdom and populations dependent upon fisheries and related industries. These proposals included a new licence condition to be introduced with effect from 1 January 1999, requiring all vessels over 10 metres to demonstrate an economic link with the United Kingdom. In accordance with the legislation, an economic link may be established by:

- landing at least 50% of the vessel's catch of quota stocks into the United Kingdom; or
- employing a crew of whom at least 50% are normally resident in a United Kingdom coastal area; or
- demonstrating an economic link by other means (including combinations of the above) providing sufficient benefit to populations on fisheries and related industries.

Vessels over 10 metres which land in total less than 2 tonnes of quota species in any quota year are exempt from the above conditions as are vessels less than 10 metres in length which, because of their size, are assumed to meet landing requirements.

Vessels failing to meet these requirements in any licence year will be liable to prosecution for breach of their licence conditions and/or withdrawal of the authority to fish for quota stocks. Furthermore, United Kingdom registered fishing vessels which do not establish an acceptable economic link with the United Kingdom but are otherwise eligible for a licence may be issued with a licence entitling them to fish for non-quota stocks in domestic waters only. However, they may not transfer or exchange the quota based on the track record to other vessels. Where applicable, the quota will be retained by MAFF.

Commentary

The United Kingom measures are clearly intended to strengthen the economic link between vessels which fly the British flag and access to the United Kingdom quota allocation. Ultimately, the purpose of the licence conditions is to ensure that the populations dependent on fisheries and related industries benefit from national quotas. In this regard the measures are required to conform with the principle of proportionality, that is to say they do not go beyond what is necessary to secure a real economic link between the vessel and the United Kingdom.

If vessels opt for the landing requirement in order to establish the requisite economic link, this should provide national enforcement authorities with sufficient opportunities to conduct inspections in port. Vessels which choose any other alternative means of establishing a link are required to make one visit to a United Kingdom port every six months.

References

1 See more detailed definition and for a discussion of the case law resulting from the first phase and quota hopping, see Churchill, R.R., 'Quota hopping: The Common Fisheries Policy wrongfooted?', *CMLRev* **27** (1990), pp. 209–247. See also, *inter alia*, Munir, A.E., *Fisheries after Factortame*, Current EC Legal Development Series, (London, 1991); O'Reilly, J., 'Judicial Review and the Common Fisheries Policy in Community Law', in *Constitutional Adjudication in European Community and National Law*, (Dublin, 1992), 51–65; Noirfalisse, C., 'The Community System of Fisheries Management', (University of Chicago, 1992), 325–351;

Magliveras, K. D., 'Fishing in troubled waters: the Merchant Shipping Act 1988 and the European Community', *ICLQ* **39** (1980) 899, at 903.

2 These figures are approximate and have been deduced from a number of sources in the industry. The Anglo/Spanish convenience flag phenomenon has occurred in three phases. Initially, a pre-CFP phase, when 40 vessels which belonged to the Spanish 'Gran Sol' fleet which were not included in the list of vessels granted licences under the terms of the bilateral fishery agreement in 1979–1981, were registered in the United Kingdom during this period. Interestingly, the catches of these vessels augmented the United Kingdom share of the TAC when it was agreed in 1983. The second phase occurred during the period 1979–1998, 40 former United Kingdom licensed vessels were purchased by companies with Spanish shareholders. The third phase, during the period, 1990–1997, 30 vessels which were formerly registered in central America (under convenience flag arrangements) or Latin America were registered on the United Kingdom register.

3 Former Dutch beam trawlers acquired by United Kingdom fishing interests and registered as United Kingdom vessels. Several old Dutch vessels re-registered in the United Kingdom by their Dutch owners in order to obtain non-pressure stock fishing licences and access to United Kingdom quotas.

4 The issue of statutory visits are discussed below.

5 Chapter 1, *ante*.

6 For example, only 300 vessels were placed on the basic list of licences pursuant to Article 158 of the Spanish Treaty of Accession, despite the fact that the indigenous fleet for the demersal fishing grounds West and South-West of Ireland the United Kingdom was comprised of 430 vessels.

7 Case 223/86, *Pesca Valentia Limited* v. *Minister for Fisheries and Forestry, Ireland and the Attorney General* [1988] ECR 83.

8 Case 223/86, paragraph 13; on the significance of this point see O'Reilly, J., 'Judicial Review and the Common Fisheries Policy in Community Law', in *Constitutional Adjudication in European Community and National Law*, (Dublin, 1992), p. 56.

9 Case C 3/87, *the Queen* v. *Ministry of Agriculture, Fisheries Food, ex parte Agegate Ltd* [1989] ECR 4459. Case C-216/87, *The Queen* v. *Ministry of Agriculture, Fisheries Food, ex parte Jaderow* [1989] ECR 4509.

10 The Advocate General, in both cases, reached a different conclusion to the Court on the grounds that the residency condition included in the UK licence constituted the corollary, so to speak, of the derogation from certain rules of Community law entailed in the quota system itself.

11 Case C-246/89, Order of the President of the Court of 10 October 1989, *Commission* v. *United Kingdom*, [1989] ECR 3125.

12 Case C-213/89, *The Queen* v. *Secretary of State for Transport, ex parte Factortame and others*, [1990] ECR 2433. On this aspect of the case, see *inter alios*: Gravelles, 'Disapplying an Act of Parliament pending a preliminary ruling: the *Factortame* case' (1989) *PL* 568; Barav, 'Enforcement of Community rights in national courts: The case for jurisdiction to grant interim relief', *CMLRev.* **26** (1989), p. 391. Greenwood, C., 'Relationship of Community law and English law: supremacy and direct effect: the *Factortame* saga', *All ER Annual Review* (1990), 102. Lewis, 'Statutes and the EEC: interim relief and the Crown', CLJ **48**, 347 (1989). For general annotation, see Alott, P., *The Cambridge Law Journal*, (1990), 377–380; Lord Donaldson of Lymington, *The Law Teacher*, **25** (1991), 4–10; Drexl, J., *The American Journal of Comparative Law*, (1993), 551–571; Papadis, L., *Legal Issues of European Integration* (1994/1995), 153–193; Aragonés, J., *Fordham International Law Journal*, (1990–1991) 778–818; Craig P., *Yearbook of European Law*, (1991), 221–255.

13 Case 106/77 *Amministrazione delle Finanze dello Stato* v. *Simmenthal SpA* [1978] ECR 629, [1990] 3 CMLR 589.

14 Case C-213/89, paragraph 18.
15 Case C-213/89, paragraph 21. It is the opinion of several authorities, that the principal concern of the Court in their decision in *Factortame (No. 1)* seems to be the establishment and enhancement the effectiveness of Community law, rather than being guided by textual or contextual indications. Whatever may be said about the Court's reasoning, it is clearly apparent that *Factortame* removed any vestige of doubt regarding the supremacy of Community legislation over conflicting national measures.
16 Merchant Shipping Act Amendment Order 1989, SI 1989/2006. This order only affected the nationality requirements and not the other restrictions in the principal Act.
17 Case C-221/89, *R* v. *Secretary of State for Transport, ex parte Factortame Ltd (No. 2) and others,* [1991] ECR 3905. The decision had been predicted, see Churchill, R.R., 'Factortame No. 2', *CMLRev* (1992) 405.
18 Case 93/89, *Commission* v. *Ireland,* [1991] ECR I-4569. In Ireland, the Fisheries (Amendment) Act 1983, which predated Spanish Accession, linked the use of fishing boats to licences which were only granted if the boat was owned by an Irish citizen or a body corporate established and subject to Irish law and having its principal place of business in the State. In 1989 the Commission commenced enforcement proceedings against Ireland regarding this legislation on the grounds that it made the granting of a fishing licence subject to a nationality requirement thus restricting the freedom of establishment. For annotation of this case, see, *inter alios,* Cataldi, G., *Foro Itiliano,* **IV** (1992) Col. 321–340; Boutard Labarde, Marie-Chantal, *Journal du Droit International,* 1992, 453–454; Costa, V., *Revista Juridica de Catalunya,* (1992) 405–414; Churchill, R.R., *CMLRev* (1992), 405–414.
19 Case C-93/89, paragraph 14.
20 Case C-246/89, *Commission* v. *United Kingdom* [1991] ECR I-4585.
21 See Churchill, R.R., *CMLRev, op. cit.,* fn 17.
22 Case C-279/89, *Commission* v. *United Kingdom.*
23 Case C-280/89, [1992] ECR 6185.
24 See, '11th Annual Report on the Monitoring and Application of Community Law-1993', *OJ* C 254, Vol. 38, 29.09.1995.
25 Case C-334/94, [1996] ECR I-1307. For a commentary see *inter alia* Chaumette P., *Le droit maritime français* (1996) 752–758; Luby, M., *Journal du droit international* (1997), 560–561.
26 Case 167/73 *Commission* v. *France* [1974] ECR 359.
27 Case C-151/96 *Commission* v. *Ireland* [1997] ECR I-3327.
28 Case C-334/94, [1996] ECR I-1307, paragraph 23, and Case C-151/96, [1997] ECR I-3327, paragraph 23.
29 Joined Cases C-46/93 and C-48/93 *Brasserie du Pêcheur Sa* v. *Bundesrepublik Deutschland and the Queen* v. *Secretary of State for Transport, ex parte: Factortame Ltd and Others,* Judgment of the Court of 5 March 1996, [1996] ECR I-1029. On this case, see fn 80, Chapter 8, *post.* The principle of State liability for breach of Community law was established in *Factortame 3* in relation to the period when the nationality provisions in the Merchant Shipping Act 1988 were in force. The period in question was between 1 April 1989 and 2 November 1989. As noted above, the various provisions of the Act were suspended after the President of the Court of Justice made an order for the interim suspension pending final judgement on their compatibility with **Articles 7, 52, 58** and **221** of the EC Treaty, see Chapter 8, *post.*
30 See report *Financial Times,* 20.07.1996.
31 *Financial Times,* 24.06.1997; *The Independent,* 19.06.1997.
32 *Ibid.*
33 *Ibid.*
34 See the authorities cited, *supra;* and Munir, A.E. 'Fisheries after Factortame', *op. cit.,* fn 1; see also the *Commission Communication on a Community Framework for Access to Fishing Quotas, OJ* 1989, C 224/3.

35 See, '9th Annual Report on the Monitoring and Application of Community Law-1991', *OJ* C 250/33, 28.09.1992.
36 See, *inter alia*, Churchill, R.R., *CMLRev* (1992), 405–414; Noirfalisse, C. 'The Community system of Fisheries Management and the *Factortame Case*', *Yearbook of European Law*, (1993), pp. 325–351.
37 See Chapter 2, *ante*.
38 In Case 46/86, *Albert Romkes* v. *Officier van Justitie for the District of Zwolle* [1987] ECR 2671 the Court justified the quota system on the grounds that it apportioned on an objective basis the sacrifices to be made by the fishermen in each Member State.
39 See discussion on co-operation and co-ordination in Chapter 2, *ante*. The disparate application of the law is examined further in Chapter 12, *post*.
40 Noirfalisse, C., 'The Community System of Fisheries Management', *op. cit.*, fn 36, p. 350.
41 Council Regulation No. 2241/87.
42 Council Regulation No. 3483/88 amending Council Regulation No. 2241/87 establishing certain control measures for fishing activities, *OJ* 1988, L 306, p. 2.
43 The obligation on Member States in which the vessel landed their catches to take penal or administrative action is carried over, see discussion of **Article 32**, Chapters 4 and 6, *ante*. Some of the 1987 sanction measures have not been carried over into the 1993 Control Regulations, particularly those relating to the action to be taken against vessels which had failed to comply with rules concerning conservation or control measures. Neither is there a requirement in the 1993 Control Regulation for vessels to have certificates authorising them to use ports outside flag Member States.
44 Case C-9/89 *Kingdom of Spain* v. *Council of European Communities* [1990] ECR I-1383.
45 *Ibid.*, paragraph 10.
46 *Ibid.*, paragraph 16.
47 Paragraph 21, the Court cited Case 15/79 *Groenveld* v. *Productschap voor Vee en Vlees* [1979] ECR 3409.
48 *Ibid.*, paragraph 22.
49 *Ibid.*, paragraph 27.
50 *Ibid.*, paragraphs 30–33.
51 *OJ* 1989, C 224/3.
52 Churchill, R.R., *op. cit.*, fn 1, 230–236.
53 The issues raised are examined in the context of sanctions and national measures, Chapter 6, *ante*.
54 As previously noted, in such instances the Commission is obliged to consult the two Member States concerned. Furthermore, there are provisions to deal with situations where the Member State in question has no quota available. See Chapter 4, *ante*.
55 Churchill *op. cit.*, fn 1, p. 245, discussion on the broader implications of the quota hopping issue.
56 However, see, exchange of letters between the United Kingdom and the Commission which suggests otherwise, as reported in the *Financial Times*, *op. cit.*, fn. 31. See also paragraph 6 *infra* regarding the licence requirement for vessels flying the flag of the United Kingdom as from 1999.
57 Satellite surveillance (VMS) will apply to all Community vessels over 24 metres as from the year 2000. See Chapter 11, *post*.
58 PQ No. 2859/92, 07.04.93, *OJ* C 99/24.
59 See Chapter 1, *ante*.
60 See Chapter 3, *ante*.
61 For discussion on a Community justice system, see, Sevenster, H.G. 'Criminal Law and EC Law', *CMLRev* 29(1), (1992) 29–70, and other authorities cited in Chapter 12, *post*.
62 See *Irish Times*, 28.11.1996.

63 For example fines totalling £92,500 were imposed on to two Anglo-Spanish vessels in Truro Magistrates Court, 31 October 1996, for cheating on their allocated boat quotas.
64 Some of the issues raised in these instruments are examined in Chapter 11, *post*.
65 FAO Code of Conduct for Responsible Fisheries, paragraph 8.3.2.

Chapter 8
Ensuring Enforcement by Member States

Introduction

In every legal system there is a difference between law in theory and law in practice: Community fisheries law, *strictu sensu,* is no different. Determining how the law operates in practice is no simple task. Indeed, our knowledge of the implementation and the enforcement of Community law in general has been described by one legal scholar as a 'black hole'.[1] With respect to Community fisheries law it is proposed to examine the dichotomy between theory and practice and to assess whether the 'black hole' metaphor is an appropriate description of the implementation and enforcement of the rules underpinning the CFP. The two strands in this discussion concern, on the one hand, the effectiveness of the Control Regulation, and, on the other hand, the limitations of the Community and international legal order from the point of view of enforcing Community fisheries regulations. This chapter focuses primarily on the first strand, the wider debate emanating from the second strand is introduced and is further elaborated upon in the case study on driftnets examined in Chapter 10.

The purpose of this chapter is to review the implementation of the enforcement regime created by the Control Regulation and to assess the effectiveness of the regulatory structure during the period 1994–1997. To simplify this task this chapter is divided into three subsections and while the discussion is primarily of a technical character, emphasis will be placed on the problems associated with implementation and the overall thrust of change as a result of the new enforcement framework established in 1993. A subsidiary objective of this chapter is to assess compliance by the Member States and the industry with their regulatory obligations. Compliance is important not least because the effectiveness of Community fishery law is easily rendered nugatory if non-compliance with Community rules is widespread. The obligation to ensure the effectiveness of Community law rests with the Community institutions. In this regard, the Commission, acting upon the legal base set out in **Articles 5** {10} and **155** {211} of the EC Treaty, has relied upon three distinctive approaches to ensure the effectiveness of the regulations which underpin the CFP. These are: litigation, in particular infringement proceedings under **Article 169** {226}, **Article 170** {227} and **Article 171** {228} of the EC Treaty; so called 'soft-law'; and structural reform involving national administrations and Community institutions.[2]

The initiation of **Article 169** {226}, **Article 170** {227} and **Article 171** {228} of the EC Treaty proceedings represents the formal approach to law enforcement and is evaluated in Section (i). Experience with the CFP, however, suggests that there is frequently insufficient evidence to initiate infringement actions in all cases where the Commission suspects that Community law is violated.[3] It is thus essential that there are other less formal mechanisms to monitor and to assess, on a regular basis, compliance by Member States with their enforcement obligations and to ensure that particular sectors of the fishing industry do not derive any advantage from any failure to implement and enforce Community law. One of the means selected by the Community to evaluate compliance with the regulatory framework is the annual performance report which is examined in Section (ii). The report acts as a barometer of Member States' performance, and can be a useful instrument with which to cajole recalcitrant parties into fulfilling their legal duties.[4] Reliance on similar means to ensure the effectiveness of Community law has been referred to by one commentator as 'soft law', because in principle it has no binding force, but in effect it may have many practical consequences.[5] The third approach, 'structural reform', is examined in Section (iii). Structural reform refers to the modification or reshaping of legal, economic and political structures in the Community and the Member States. (It must not be confused with the structural reform of the fishing industry examined in Chapter 5 above.) In the context of this chapter structural reform may be undertaken by the Community institutions, the Member States, or it may result from judicial activism. Structural reform is facilitated by the allocation of Community finances to improve the resources and structures in the Member State available to undertake the enforcement task, a subject which is examined separately below.

Section (i) Enforcement litigation - Article 169 {226}, Article 170 {227} and Article 171 {228} of the EC Treaty

Introduction

A brief overview is presented in this section of the rôle of the Community institutions and the Member States in fishery law enforcement proceedings, more commonly referred to as **Article 169** {226}, **Article 170** {227} and **Article 171** {228} of the EC Treaty proceedings.[6] It needs to be emphasised at the outset, however, that the rôle of the Commission as guardian of the Treaties pursuant to **Article 155** {211} is not confined to **Article 169** {226} proceedings, but extends to other articles of the EC Treaty. Thus, for example, the Commission is also enforcing Community law when it commences litigation against other institutions (**Articles 173** {230} and **175** {232}), when it scrutinises State aid for legality (**Article 93** {88}), when it acts to combat fraud against the Community budget,[7] and when it secures compliance against restrictive practices and abuse of dominance in the market place (**Articles 85** {81} and **86** {82}). Furthermore, the most common and direct means by which Community law is enforced is through the medium of private action by individuals against national governments in Member State courts. This is possible through the application of the doctrine of direct effect which provides a very effective means of enforcing Community law.

One situation in which national proceedings are appropriate is when a national government purports to restrict the freedom of action of a citizen in a way which is contrary to Community law. An affected party may plead Community law as an argument in a national court or tribunal and a preliminary reference may be made to the European Court pursuant to **Article 177** {234} of the EC Treaty to determine the way the provision should be interpreted. The Court of Justice may, on the one hand, simply provide a general indication of the contested provisions and state that it is a question of fact to be decided by the national court whether a particular case comes within its scope: or, on the other hand, it may provide a precise and specific meaning and the case is sent back to the national Court for a formal judgement.[8] In fisheries cases, preliminary references are made fairly frequently and are an effective means of enforcing Community law. In the period 1976–1998, there have been 26 such references resulting from fisheries cases in the Member States.[9] In some of these cases the Court has established important principles of Community law. Some of the most significant jurisprudence resulted from the Court's decision to uphold the rights of Spanish fishermen who invoked Community law to overturn the restrictions imposed on foreign vessels being registered in the United Kingdom and Ireland.[10]

It is thus apparent that there are several distinct and separate procedures by which Community law may be enforced. The ultimate method, however, remains **Article 169** {226} of the EC Treaty proceedings which are initiated by the Commission against a Member State which is alleged to have failed to fulfil a Community obligation. In the period 1977–1998, there have been 22 judgements handed down by the Court of Justice as a result of **Article 169** {226} proceedings pertaining to the CFP.[11]

The President of the Court may also grant interim measures while an **Article 169** {226} of the EC Treaty action is pending.[12] It is proposed to draw attention to two decisions of the Court in relation to fisheries cases which were important in the development of the procedures in seeking interim relief from the Court of Justice while the main action is pending. This is followed by a discussion of the rôle of the Commission and the Court in **Article 169** {226} proceedings with special emphasis on actions in the domain of the CFP.

1. *Interim measures*

The Court of Justice may exercise wide powers in interim proceedings and the procedure is now well established for the granting of interim relief.[13] Indeed, one of the landmark decisions establishing the procedure resulted from unilateral fisheries conservation measures introduced by Ireland in 1977 which were challenged by the Commission on the grounds that they were contrary to the EC Treaty.[14] In this instance the Commission commenced **Article 169** {226} proceedings and concurrently applied to the Court for an interim order for Ireland to suspend the measures. The Court upheld the view of the Commission and ruled that there were serious doubts as to whether the Irish conservation measures were compatible with Community law and that in such a case deemed that the suspension of the measures was justified. The Court, however, qualified this ruling by noting that unless the measures were replaced by other provisions such a suspension would have had a harmful effect on conservation and thus ruled that the most appropriate course of action was to defer a judicial decision for one month in order to allow both parties to reach a settlement on the alternative provisions to replace the contested legislation. After prolonged negotiations and several applications for adjournment, the Court eventually ruled that Ireland was obliged to suspend the contested national legislation within five days, a time period considered adequate to allow for the adoption of alternative measures which had to be approved by the Commission. In this instance the Court not only clarified the procedure in relation to interim measures but also effectively curtailed Member States' power to adopt unilateral measures.

Mention should also be made of the leading decision in *Factortame Ltd (No. 2)* which resulted from proceedings regarding the granting of flag rights to fishing vessels.[15] In this instance, the House of Lords made an application seeking clarification on the obligation placed on national courts to grant interim relief while a case was pending in the Court of Justice. The Court of Justice held that where only a rule of domestic law prevented the Court from providing interim relief, it should put aside that rule of domestic law and in the appropriate circumstances grant interim relief. The House of Lords subsequently granted the applicants the necessary legal protection while the main action was pending.

2. *Article 169 {226} of the EC Treaty proceedings – the rôle of the Commission*

The ultimate legal option available to the Commission to enforce Community law is **Article 169** {226} of the EC Treaty proceedings. This type of action entails a two-stage process which is extensively documented and commented upon in the literature.[16] The first stage is administrative and includes an invitation by the Commission to the Member State to submit their observations on the alleged breach of Community law. Member States are obliged to co-operate with the Commission during this stage.[17] On the basis of observations submitted this invitation may be followed by the delivery of a reasoned opinion by the Commission. The second stage concludes when the Commission brings the contested matter before the European Court of Justice for adjudication. Significantly, in **Article 169** {226} proceedings, the majority of actions are settled prior to the judicial phase. Moreover, despite the deceptive appearance of simplicity which one might assume from the brevity of **Article 169** {226}, actions in fishery cases may result in complex and lengthy administrative and judicial stages which in some instances may take up to five years to conclude.[18]

The procedural requirements to institute proceedings are well established as is the discretion to litigate which rests with the Commission.[19] In fact a strict interpretation of **Article 169** {226} enables the Commission to institute proceedings every time it forms the opinion that a Member State has not fulfilled an obligation under Community law, without being drawn into distinctions regarding the nature and gravity of the infringement.[20] In this regard the use of **Article 169** {226} proceedings is not limited to situations which arise from a disagreement on the interpretation to be given to the Community rules between the Commission and the authorities of the Member State whose conduct is in question. Moreover a Member States' non-compliance with an obligation imposed by a rule of Community law itself constitutes a failure to fulfil obligations. The fact that it had no adverse impact is irrelevant for the purpose of deciding whether an action brought under **Article 169** {226} is well founded.[21] The Court has also held that the subject of an action under **Article 169** {226} is determined by the Commission's reasoned opinion, and even when the default has been remedied after the time-limit prescribed by the second paragraph of that article has expired, there may still be an interest in pursuing the action in order to establish the basis of liability which a Member State may incur as a result of its default towards other Member States, the Community, or other parties.[22] Finally, it ought also be pointed out that the Commission's discretion whether or not to open an investigation, or to issue a letter of formal notice, or to issue a reasoned opinion, or to commence proceedings, is not susceptible to judicial review.[23] The burden of proof that a Member State has failed to comply with an obligation rests with the Commission despite the fact that the obligation to undertake inspection and take direct enforcement action rests with the Member State.[24] As previously explained in Chapter 3, this division of responsibilities poses its own problems which in some instances may lead to several practical difficulties in collecting the appropriate evidence with which to initiate **Article 169** {226} proceedings in the first place.

A review of **Article 169** {226} of the EC Treaty proceedings pertaining to fisheries in the period since 1983 reveals that the pattern of cases conforms to the general pattern of Community law in so far as cases are generally terminated in the administrative procedural stage and before the judicial stage leading to a judgement by the Court of Justice.[25] Furthermore, it is also clearly evident that, *pro rata*, sea fisheries is not as litigious as other sectors of Community law such as the agriculture or environment policies. For example in 1994 there were 359 suspected infringements of Community law in the environmental, nuclear safety and civil protection sector, 317 suspected infringements in the agriculture sector, and 9 suspected infringements in fisheries.[26] It is also obvious that the number of suspected infringements is quite insignificant if compared to the number of suspected violations in the internal market and financial service sector which in 1995 was recorded as 512.

In 1990 the Commission took the unprecedented step of releasing in the Annual Report figures on a country by country basis the number of **Article 169** {226} letters that had been issued in the various sectors. Ostensibly, this was a deliberate political move in order to highlight the issue of implementation and one that caused considerable disquiet among some Member States.[27]

It is also useful to evaluate, on a comparative basis, the number of alleged infringements by Member States in the domain of sea fisheries. The following data (Table 8.1) are extrapolated from the Commission Reports covering the period 1990–1996.[28]

A number of comments may made about the data presented in Table 8.1. The first concerns the total number of infringements which at first sight appears to be insignificant. It must be borne in mind, however, that in the case of Community fisheries law that **Article 169** {226} proceedings are generally only initiated after detection by

Table 8.1 Article 169 {226} of the EC Treaty infringement proceedings during the period 1990–1996.

Year	B	DK	D	Gr	E	F	IRL	I	L	NL	P	UK	Total
1990	0	2	0	0	1	2	0	1	0	0	1	4	11
1991	0	0	0	0	0	1	3	0	0	0	4	0	9
1992	1	1	1	1	5	2	2	2	0	0	0	2	17
1993	0	1	1	0	0	2	0	1	0	1	0	1	7
1994	0	1	0	1	2	3	1	0	0	0	0	1	9
1995	0	0	2	1	1	1	0	0	0	0	0	0	5
1996[29]	1	1	0	0	1	1	0	0	0	0	0	0	4
Total	2	6	4	3	10	12	6	4	0	1	5	8	62

the Commission's own resources of an alleged infringement by a Member State or as a result of an initiative in the European Parliament. The limitations of these sources have already been highlighted.[30] It should also be noted that fishing and associated industries are undertaken outside public view and consequently are not subject to the same critical appraisal as some of the other more visible Community policies. In contrast to fisheries, a significant number of proceedings are initiated as a result of complaints lodged by the public regarding alleged breaches of Community environmental law. In 1996, for example, there were 955 complaints lodged by Community citizens regarding suspected infringements of Community law and over a third of the infringement cases registered by the Commission in the Court of Justice concerned violations of environmental legislation (chiefly industrial projects launched without the appropriate environmental impact assessment).[31] Another third of the complaints registered concerned the internal market which included all manner of problems such as barriers to the freedom of establishment for dentists.[32] Furthermore, in other sectors such as agriculture, one of the principal reasons prompting the initiation of infringement proceedings is the non-transposition of Community directives into national legislation, this problem, as explained in Chapter 1, does not beset the CFP as the latter is based almost exclusively on regulations. Nonetheless, even when the latter two reasons are taken into consideration, there still appears to be a significant shortfall in the number of **Article 169** {226} actions. This is partly explained by the reliance on informal procedures by the Commission to deal with recalcitrant Member States when there is evidence of non-compliance with Community obligations. In this regard it is quite common for the Commission to seek informal information regarding breaches of Community rules from the Committee of Permanent Representatives of the Member States (COREPER) or directly from the appropriate national authority.[33] In general, these informal requests for information are conducted with discretion, on the basis of observations made by Commission inspectors during their visits to Member States, or on the basis of the administrative inquiries conducted into irregularities resulting from inadequate implementation of the Control Regulation.[34] The details of the Commission's investigations are not published and generally lead to substantial discussions between the Commission and the Member State authorities to ascertain facts and to reach a settlement regarding the alleged dereliction, at the informal stage. It is only when it is clearly evident that there are significant matters to be addressed and that there is no possibility to pursue alternative options that the formal phase of **Article 169** {226} proceedings is commenced. Indeed, there may also be an amicable conclusion to the proceedings at various stages during the formal phase. In conclusion, the Commission thus makes full use of the pre-judicial or pre-litigation stage of infringement proceedings to persuade the offending Member State to remedy the deficiency or to negotiate a settlement. It is only after the Commission has failed to achieve a negotiated solution, and the administrative phase has been completed, that the judicial phase commences. As the Court has held, referral of an action to its jurisdiction is the last resort, 'the *ultimo ratio*' enabling the Community interests enshrined in the Treaty to prevail over the inertia and resistance of the Member States.[35]

The second aspect of the data presented in the table which calls for comment, is the relative uniform distribution of **Article 169** {226} actions among Member States (except, of course, Luxembourg), which suggests that the application of Community fisheries law is uniform throughout the Community. Almost all Member States have been the subject of at least one infringement proceeding, and the majority of Member States have had to contest four such actions. Interestingly, Belgium and the Netherlands have the most laudable record in so far as their respective failures have only led to one **Article 169** {226} action prior to 1996. Surprisingly, in the period in question, the Commission had taken enforcement proceedings against the United Kingdom and Spain on eight and ten occasions respectively, which suggests that the common perception that there are substantial differences in the manner Community law is applied in both Member States is unfounded.[36] Equally, Ireland, with a small fleet and modest fishing industry in Community terms, has been the subject of six enforcement proceedings. On the other end of the scale, France, with 12 infringement proceedings during the said period, has, *prima facie*, the poorest record. However, a word of caution: too much significance should not be given to the data presented in Table 8.1, on the grounds that it is not possible to give a complete assessment of the effectiveness of the enforcement procedure because there is no precise information available on the number of cases which are terminated in the informal phase of proceedings.

3. *Rôle of the Court of Justice in Article 169 {226} proceedings*

On the basis of the aforementioned discussion, it is easy to deduce that the rôle of the Court in **Article 169** {226} proceedings is clearly restricted by the nature and number of proceedings instituted by the Commission. In fisheries, typical cases which reach the adjudication phase are those based on the failures by Member States to comply with Community fishery management measures such as the duty to prohibit fishing after the exhaustion of a quota.[37] In this regard, the Court has held that a Member State cannot rely in practical difficulties related, for example, to deficiencies in the national statistical system for collecting fisheries data in order to justify failure to adopt appropriate control measures with regard to compliance with fisheries quotas.[38] Furthermore, when a Member State observes non-compliance with Community legislation they are required to take penal or administrative action. In several cases the Court has pointed out that Member States cannot plead internal circumstances in order to justify a failure to comply with obligations and time limits resulting from Community law.[39] Moreover, a Member State cannot justify its failure to fulfil obligations under the Treaty by pointing to the fact that other Member States have also failed, and continue to fail, to fulfil their own obligations.[40] In this regard, it is the Court's view that under the current legal order established by the Treaty, the implementation of Community law cannot be made subject to reciprocity, and that in such instances, **Articles 169** [226} and **170** {227} provide suitable means of redress for dealing with the failure by Member States to observe their obligations.[41] Interestingly, the Court rewarded the exemplary conduct of a Member State (the United Kingdom) which voluntarily suspended the application of contested measures while an **Article 169** {226} action was pending (thereby rendering unnecessary an application to the Court for interim measures), by ruling that this magnanimity on the part of the Member State constituted an exceptional circumstance within the rules of procedure so as to justify in this instance the order for each party to bear its own costs.[42] As the Court subsequently held that the contested national measures constituted a failure to fulfil obligations, the United Kingdom were thus not liable for the costs of the Commission.

One of the distinctive characteristics of the fisheries jurisprudence is the way the Court has sought to reconcile the aims of the CFP and general principles of Community law.[43] In the fishery cases, one may identify two approaches. In the first phase, one can trace how the Court has been motivated by the desire to strengthen the Community;[44] increase the power and effectiveness of Community law;[45] and to enlarge and protect the powers of the Community institutions,[46] which may collectively be referred to as the promotion of European integration. There is some evidence from the fisheries jurisprudence, on the other hand, to suggest that the Court has not, on occasion, taken into full consideration the policy implications of its decisions and has been overtly concerned with the advancement of general legal principles.[47] In particular in the quota hopping cases the Court ruled on important issues of principle but did not give appropriate weight in their judgements to the underlying axioms and specific features of the CFP such as relative stability in resource allocation.

Overall, from the enforcement perspective, the Court has exercised its enforcement jurisdiction with a view to ensuring the effective implementation of Community fisheries law. The Court has censured several Member States for inadequate enforcement of Community conservation and management measures.[48] Interestingly, of the twenty-two **Article 169** {226} decisions reported, the Court has ruled in favour of the Commission in all actions which have reached the judicial phase. Indeed, one can only surmise if there would have been a greater improvement in Member State adherence to their Community obligations if the Commission had initiated more **Article 169** {226} proceedings. On the other hand, there appear to be several reasons why the Commission is not more systematic in the pursuit of errant Member States. Indeed, one authority has identified three categories of cases in which the Commission would not rigorously pursue **Article 169** {226} infringement proceedings.[49] The first category concerns isolated acts by national officials. This category is particularly pertinent to the CFP because the myriad of secondary legislation which underpins the policy depends in large measure on national enforcement agencies enforcing Community law. In this case the Commission generally considers that such isolated acts do not warrant the rigorous application of the law. Furthermore, there is little doubt that both the informal and formal phases of **Article 169** {226} proceedings would be unworkable and an overburden if the Commission sought to pursue every dereliction by national officials. The second category are failures by national courts to uphold Community obligations. This equally applies to sea fisheries, the Commission has not so far taken a Member State to task for an alleged breach by a national court of a Community obligation in respect to fisheries. The third category concerns situations where the Commission is hesitant to take action on the grounds that it will inflame politically-sensitive situations. With respect to this category, it may be recalled that there are several instances where France has failed successfully to defend **Article 169** {226} infringement proceedings by merely pleading political and economically sensitive reasons as a justification for failure to honour a Community obligation. Nevertheless, this category may apply to the CFP if the particular issue in question is overtly politically-sensitive and the Commission considers it counter-productive to commence litigation. In reality, however, this problem does not arise because generally litigation is commenced several months after the services of the Commission have prepared the file on the contested issue. Conveniently, this acts as a deflationary period and allows time for parties to reflect on their respective positions.

4. *Article 170 {227} proceedings*

In Community law an action may be taken by a Member State against another Member State if the former believes that the latter has not fulfilled a Community obligation. The procedure for such an action is laid down in **Article 170** {227} of the EC Treaty.[50] In reality this provision is not frequently relied upon as a means to enforce Community law, although significantly the first case to reach the judicial phase was an action by France against the United Kingdom on the grounds that certain fishery conservation measures adopted by the United Kingdom were contrary to certain Treaty provisions.[51] In this instance the Commission was requested by France to give an opinion on the subject and after both parties had submitted their views in writing, the Commission held a hearing in which both parties were allowed express their views in oral pleadings. On the basis of the various submissions, the Commission held that the United Kingdom had infringed the Treaty. Subsequently, France then took the matter before the Court of Justice and the Commission applied for leave to intervene in support of France. The latter application was granted and the hearing proceeded in the normal way with the Court finding that the United Kingdom had in fact infringed Community law.

It may be argued that **Article 170** {227} of the EC Treaty actions provide a unique mechanism for the enforcement of Community fishery law in so far as it offers Member States a ready made mechanism to challenge any other Member State that is suspected of failing to comply with Community obligations. However, because only one fishing case has been brought before the Court of Justice under **Article 170** {227}, in the period up to 1999, suggests that national governments, if left to their own devices, will rarely initiate enforcement proceedings. In reality, the hard law option regarding litigation remains a matter for the Commission.

5. *Article 171 {228} proceedings*

Prior to the adoption of the Treaty on European Union, among the criticisms of enforcement proceedings under **Article 169** {226} and **Article 170** {227}, was that Community institutions were largely impotent in relation to ensuring that the judgements of the Court were complied with by Member States which had failed to fulfil an obligation under the Treaty and that no effective sanction was applied. However, pursuant to **Article 171** {228} of the Treaty of European Union, the Commission may, if it considers that a Member State has not complied with a judgement of the Court, issue a reasoned opinion after giving the Member State the opportunity to submit its observations. The opinion must specify the points of non-compliance by the Member State and must prescribe a time limit for compliance. Moreover, if the Member State does not comply with the Commission's request within the specified time limit, the Commission may bring the case before the Court and seek the payment of a lump-sum fine or penalty payment in order to penalise the Member State in question. In general a penalty payment is calculated on the basis of a fine for a specified amount *per diem,* or other time, until the Member State complies with the judgement of the Court. In theory the latter type of fine should provide a major incentive to the Member State to comply with the judgement.[52]

Article 171 {228} proceedings have the potential to improve the enforcement of Community fishery law as there has been a marked tardiness by Member States in the past in complying with the judgements of the Court of Justice. However, because the procedure has not been relied upon in many cases there are some issues in relation to **Article 171** {228} actions which have yet to be resolved. These include the means to enforce the fines and the issue of successive actions. Indeed in considering the effectiveness of **Article 171** {228} proceedings in the domain of enforcing Community fisheries regulations a number of issues have to be placed in context. Firstly, in common with the competition policy, the Commission has had power to penalise Member States for failure to uphold Community law. Specifically, whenever a quota is exceeded, the Commission may with the prior approval of the Management Committee, deduct the quantity overfished from a Member State's quota allocation in subsequent years.[53] Similarly, the Commission may make deductions from national quotas if a Member State fails to take appropriate measures in cases of non-compliance with Community law.[54] This is a major weapon in the armoury of the Commission and it is not clear if **Article 171** {228} proceedings have the same deterrent and symbolic effect as the reduction of fishing possibilities. However, the Commission does not have power to withhold Community subvention of financial aid to Member States if there is non-compliance with Community law. The latter was a power that the Commission had included in its original draft proposal for the 1993 Control Regulation.[55] Secondly, while the Commission has initiated **Article 171** {228} infringement proceedings in one fishery related case,[56] the procedure under **Article 171** {228} suffers from the same defects regarding expediency as those under **Article 169** {226} proceedings. In this regard the Court has recently pointed out that although **Article 171** {228} does not specify the period within which a judgement must be complied with, the importance of immediate and uniform application requires that the process of compliance must be initiated at once and must be completed as soon as possible.[57] Thirdly, it must also be recalled that it is open to claimants who have also suffered a loss as a result of a Member State's violation of Community law to bring proceedings for damages in national courts.[58] More recently, the Court has upheld an application for compensation as a result of loss suffered during the period while the quota hopping cases were pending.[59] Ultimately, this form of action, or the threat of this action, may prove to be equally effective and persuasive means to improve compliance with Community law.

6. *Conclusions regarding the hard law option to law enforcement*

In conclusion, it may be said that **Article 169** {226} proceedings, despite shortcomings, continues to remain the ultimate means available to the Commission to ensure the enforcement of Community law. While **Article 170** {227} is rarely relied upon as a means to ensure that Member States fulfil their obligations in Community fisheries law, nevertheless, such proceedings do offer the possibility for Member States to play a more active part in the enforcement process. The potential of **Article 171** {228} proceedings as an effective instrument to ensure that recalcitrant Member States comply with their obligations has not to date been realised and it will be a number of years before the utility of such proceedings can be assessed. Moreover the development of Member State liability by the Courts is a relevant development in the law in this area, which now provides a major incentive to Member States to comply with their obligations and should thus have a greater impact on the enforcement of Community law.

It may be suggested at this point that **Article 169** {226} of the EC Treaty proceedings are an ill-suited means by which to enforce Community fisheries law. This conclusion is not surprising, nor unique to the fisheries policy, and stems from the fact that the procedure is a general one for all areas of Community law and as such predates the CFP. In particular, the procedure is too attenuated to be effective in fisheries conservation cases which by their nature often require prompt remedial action to arrest malpractice or to swiftly correct a Member State's dereliction to observe a Community obligation. This weakness is further exacerbated by the fact that the Commission is not in a position to order immediate interim protective measures without making an application to the Court of Justice for such measures.

In conclusion, it may be said that in the fisheries sector failures to fulfil Community obligations are more likely to be resolved by negotiation and consultation than through **Article 169** {226} proceedings in the Court of Justice.[60] The use of **Article 169** {226} litigation as such follows a carrot-and-stick approach in so far as all persuasive means are exhausted before **Article 169** {226} proceedings are initiated and this in the majority of cases is preceded by the threat of litigation and the concomitant threat of penalty payments or fines under **Article 171** {228} of the EC Treaty. An example of this approach is evident in respect to the 1993 Control Regulation in that the Twelfth Annual Report on the Implementation of Community Law records that in order to give the competent national authorities time to familiarise themselves with the new system of control and enforcement introduced by the regulation and to facilitate steps to put the regulation into practice, the Commission decided to postpone or suspend, depending on the case, for one year, outstanding **Article 169** {226} actions in 1994 relating to over-fishing.[61] In contrast, in the Thirteenth Annual Report which surveys the year 1995 the Commission records that while there was a substantial improvement over previous years as regards over-fishing in several Member States and that the Commission was able to terminate several infringement proceedings, it nonetheless decided to go ahead with infringement proceedings that were pending against four Member States (France, the United Kingdom, Spain, and Denmark).[62]

Section (ii) The soft law approach to fisheries law enforcement and compliance with the Control Regulation during the period 1993–1996

Introduction

The adoption of the 1993 Control Regulation coincided with a greater general appreciation, within the European Union, of the need for transparency and accountability regarding the enforcement of Community law. Against this background, specific provisions (**Articles 34** and **35**) were inserted in the Control Regulation to ensure, *inter alia*, that more details and information would be available on enforcement. This would facilitate comparisons between Member States in terms of control tasks and resources devoted to enforcement functions. As noted above in Chapter 6, the requirements regarding the notification of information to the Commission were amended in 1998 and are now contained in **Article 35** of the Control Regulation.

In order to fulfil the legal requirement in **Article 35** of the Control Regulation, annual reports have been completed every year since 1994. Other than acting as a barometer of Member States' enforcement performance, they have also proved to be a useful instrument with which to cajole recalcitrant Member States into discharging their legal duties.[63] Reliance on such means to ensure the effectiveness of Community law is often referred to as 'soft law', because in principle it has no binding force, but in effect it may have many practical consequences.[64] The use of annual reports is similar to the use of the quasi-legal form of the *communication* and has several advantages over the 'hard law' option such as **Article 169** {226} proceedings.[65] Some of these include: the comprehensive examination of the entire sector on an annual basis; and an independent evaluation in so far as the report is produced by the Commission, thus in effect bypassing the Council.[66] This approach allows for a proactive approach to enforcement issues and it enables the Commission, without waiting for a decision from the Council or the Court of Justice (as is the case in **Article 169** {226} actions), to present their interpretation on enforcement issues as well as identifying and proposing solutions for the enforcement problems encountered in the entire sector. In the report the Commission may also provide guidelines for ameliorating the effectiveness of Community fisheries law. The main disadvantage with the 'soft law' approach is that it lacks legal exactness and it is difficult to forecast at what point it will develop into, or be followed by hard law or **Article 169** {226} of the EC Treaty proceedings which may also entail **Article 171** {228} of the EC Treaty penalties.

1. *Compliance by Member States with their enforcement obligations*

The Annual Reports on Monitoring the CFP are the principal means used by the Commission to report on the level of conformity by Member States with their fishery enforcement obligations. The 1994 Report on Monitoring the CFP, published in 1995, was made in response to the formal obligation in **Article 35** of the Control Regulation and was the first exercise of its kind to survey monitoring activities in the Member States on an annual basis.[67] The 1994 Control Report is nevertheless an incomplete assessment of the enforcement efforts undertaken in 1994 by the Member States. For example, in relation to the inspection and monitoring of fishing vessels and their activities, the report records that although most Member States submitted figures corresponding to the inspection activities there appeared to be a large difference in the number of inspections of fishing vessels undertaken by various Member States. One reason for the differences can be attributed to the definition of an inspection from a simple sighting to a detailed inspection operation at sea.[68] Another reason may stem from the differences in the preliminary investigative procedures relied upon in Member States (Ireland, the United Kingdom and the Netherlands) which deal with fishery offences through the criminal law process from those Member States (Spain, Portugal) which use administrative law procedures. In the report there is also a dearth of information regarding the extension of the enforcement regime to other aspects of the common fisheries policy. There is, however, a detailed overview of enforcement in relation to the conservation policy and an assessment of the performance of each Member State.

The 1995 Report, published in June 1996, is the second annual report and contains an assessment of enforcement in Sweden and Finland in addition to the other 11 Member States.[69] Interestingly, the Commission, in the introduction to this report, acknowledge the valuable contribution made by Commission inspectors in assessing the enforcement effort in the Member States. The 1995 Report also points out that the conclusions in the 1994 Report remain valid and thus underscore the importance of the first report.

Despite some shortcomings, the annual report is an essential document to promote synergy between Member States and to dissipate suspicion that the enforcement and control structures are not being implemented in a uniform manner. Because of its importance in supporting and underpinning the scheme of enforcement introduced by the 1993 Control Regulation it is intended to examine here some of the issues raised in both the 1994 and 1995 Reports under a number of sub-headings, which include, *inter alia*: resources available for enforcement; co-ordination and co-operation between the different authorities in the Member States empowered to discharge the enforcement function; catch registration; misreporting of catches; transport documents; technical measures; market measures; structural measures; sanctions; and the future content of the annual reports on the monitoring of the CFP. Furthermore, it is proposed to examine the issues raised in the reports in the light of some recent developments in the Community legal order including agreement on a new Treaty for Europe which was signed in Amsterdam on 17 June 1997.

(i) Resources available for enforcement[70]

Both the 1994 and 1995 Reports note that the organisation of monitoring, control and surveillance differs considerably between Member States. Some Member States have a single competent authority for fisheries enforcement while other Member States rely upon an amalgam of several government departments which may undertake additional tasks not related to fisheries enforcement.[71] Overall, the 1994 Report concludes that there is a broad range of organisation types which vary from a comparatively well-organised service using qualified staff to a poorly co-ordinated set of national and regional departments with non-specialised personnel.[72] This is a theme which is repeated in the 1995 Report. Both Reports record that there are insufficient human resources available in some Member States to discharge the enforcement obligation and this raises a doubt as to whether the enforcement regime applicable to the common fisheries policy is being properly implemented. In conclusion, there appears to be no major difference in the quantity and quality of resources available in the Member States to discharge the enforcement function from the position which existed in the pre-1993 phase of the policy.[73]

(ii) Co-ordination of enforcement strategies and inter-Member State co-operation in the enforcement of Community fishery law

The need for co-ordination and co-operation in the enforcement of Community law is not unique to the domain of sea fisheries. From a general perspective outside the domain of sea fisheries, there have been several developments since 1983 which have led to greater judicial and administrative co-operation and co-ordination between the various authorities in the Member States vested with responsibility to apply and enforce Community law. The agreement by Member States, on a new Treaty for Europe, the Treaty of Amsterdam, has several implications for the improvement and development of the law in the area of justice and home affairs. Indeed, overall, one of the predominant features in the Amsterdam Treaty is the *leitmotif* 'closer co-operation'. Thus, as one authority has pointed out, the Amsterdam Treaty inserts into the Common Provisions of the TEU a *general* flexibility clause, complemented by similar clauses in the EC Treaty and Title VI of the Treaty on European Union.[74]

It is against this background that co-ordination and co-operation in fishery enforcement must be considered, particularly as there is an obligation on Member States to co-ordinate enforcement tasks and strategies. However, according to both the 1994 and 1995 reports there appears to be, in practice, little effort by Member States to co-ordinate the day to day activities of the national control services. In particular there is an absence of bilateral agreements in relation to issues such as data exchange and the monitoring of vessels landing catches outside their flag Member State. The only exceptions appear to be the United Kingdom and Belgium who have a reciprocal enforcement agreement. Some Member States (Denmark, Ireland and the United Kingdom), however, have bilateral enforcement agreements with Norway. There has also been an element of co-ordination between enforcement authorities of Spain, France, Ireland and the United Kingdom in the driftnet fishery in the North-East Atlantic and between Italy, Greece and Spain in the Mediterranean.

Given the absence of adequate structures to ensure effective and frequent co-ordination of enforcement tasks between Member States, the issue of inter-Member State co-operation in the enforcement of Community fishery law is paramount. Despite the major emphasis on inter-Member State co-operation in the 1993 Control Regulation,[75] the 1994 report notes that the legal framework applicable in the Member States provides very few opportunities for co-operation between Member States in the domain of enforcing Community regulations. Moreover, rigid adherence to flag Member State jurisdiction has resulted in many practical difficulties in cases where fishing vessels operate in the waters under the jurisdiction or sovereignty of several Member States and land their catch outside their flag Member State. This enforcement problem is further complicated if vessels are fishing under individual vessel quota allocation and these quotas are fished in waters outside the jurisdiction or sovereignty of the flag State. There is general consensus that in such cases, in the absence of co-operation between Member States, there is little prospect of detecting and sanctioning malpractice such as vessels exceeding their quota allocation.

The 1994 Control Report points out that co-operation between Member States envisages procedures for the exchange of information and evidence as well as taking action against vessels which have escaped inspection in other Member States.[76] While Member States have widely diverging rules of evidence, the 1994 Control Report suggests that the lack of co-operation may reflect the permissive attitude adopted by national authorities in respect of their national fishing industries.[77] Indeed, this problem is not unique to the enforcement of fisheries law and is linked to the broader issue of whether the use of punitive criminal sanctions or administrative penalties is the most appropriate means to ensure compliance with Community rules.[78] The problem stems from the different structures for the investigation and prosecution of offences in Member States and also from differences in judicial procedure. In particular, the separation of investigative, prosecutorial and judicial functions in the Member States with a legal system based on the common law is not shared with those Member States which have a legal system which is based on a *code civil*. Furthermore, the inability of Courts and Tribunals which have jurisdiction to deal with offences to compel foreign based witnesses to give *viva voce* evidence presents a particular obstacle to co-operation in fishery enforcement.

Measures to improve co-operation in fishery enforcement Because it is clearly apparent that the level of co-operation and co-ordination in enforcement is modest several practical measures towards improving enforcement were identified by senior enforcement officials responsible for control in the Member State at a meeting in Dublin in 1996 organised under the auspices of the Irish Presidency of the European Council.[79] These measures include, the convening of frequent information seminars on control issues for fishery enforcement personnel; the development of joint-training programmes; the frequent exchange of inspectors; the evaluation of a standard infringement form for use throughout the Community; co-operation on landing inspections in Member States other than the flag Member State; and the convening of regional meetings to assist in the improvement of control. Overall, it may be said that all the measures proposed are practical and politically non-contentious and may thus prove to be the most appropriate catalyst to improve co-operation between Member States.

The steps proposed at the Dublin meeting accord with international developments in the enforcement of international fishery law such as the Code of Responsible Fisheries, and the United Nations Agreement on Straddling Stocks and Highly Migratory Species.[80] Indeed the proposed measures exceed the international framework in many respects and demonstrate that the uniqueness of the legal order in the European Community as a suitable framework within which to foster and to develop suitable co-operative structures to be used by Member States in the domain of enforcing fishery regulations. However, the size of this task must not be underestimated and it may be some time before major obstacles such as the principle in international law of territoriality and the differences which stem from the diverse criminal and administrative structures in Member States are overcome.[81]

In the 1995 Control Report, the Commission places a significant emphasis on the requirement to improve the operational co-operation between Member States as well as the need for better understanding of the enforcement structures in other Member States among the national enforcement authorities, the scientific Community and the fishing industries.[82] The 1995 Control Report concludes by pointing to the availability of financial assistance for improving inspection and enforcement structures and forewarns Member States that the Commission will examine in due course whether the present Community enforcement provisions require reinforcement.[83]

As noted in Chapter 3, in response to the shortcomings in co-operation among the authorities responsible for monitoring in the Member States, the Control Regulation was amended in 1998 to strengthen and facilitate greater cooperation. In particular a new Title was inserted into the Regulation, Title VIIIa, which provides a legal basis for the establishment of a general framework under which all the authorities concerned can request mutual assistance and the exchange of relevant information. It also introduces a legal basis for inspectors to act as observers outside their flag State.

(iii) Catch registration in the Member States[84]

The 1994 Report records that the number of returned documents (logbooks, landing declarations, and sales notes) seems in general to be satisfactory. However in relation to the collection of catch and other data the report notes that there are large differences in the approach to, and the effectiveness of data collection and verification.[85] It appears that only Belgium, Denmark, Germany and the Netherlands regularly cross-check different information sources to verify the quality of catch information as recorded by their respective fishing fleets. The report draws attention to two specific problems which are cause for concern: the landing of catches outside auctions and the difficulty in estimating the catches made by vessels which are not obliged to use a logbook (i.e. vessels less than 10 metres).

(iv) Misreporting of catches

The misreporting of catches which was identified as a major weakness in the CFP during the period 1983–1993 is a major theme of the 1994 Control Report.[86] The 1994 Report records that the proportion of catches not recorded by fishermen in their log-books is considerable.[87] The report does not quantify in more precise terms the scale of the problem as this information is obviously difficult to obtain.[88] Furthermore, in the 1994 Report the Commission expresses the view that the situation has not shown any tendency to improve since the introduction of the CFP in 1983.[89] In order to ensure compliance with the regulations it is suggested by the Commission that national inspections should be concentrated on a relatively small number of ports and on the activities of processors as well as fishermen.[90] Moreover, a close analysis of logbooks ought to be able to uncover inconsistencies if compared with sighting information from inspections undertaken at sea, or with sighting information provided by satellite surveillance systems.[91] Indeed, while the report does not mention the advent of modern surveillance technology it is inevitable that the implementation of the satellite surveillance system in the European Community should prove a useful tool in monitoring vessels making illegal and unregistered landings of catches.[92] The 1994 Report records that the problem of discard continues to be a major problem for specific fisheries, such as those in the East Atlantic and the North Sea.[93]

(v) Transport documents

As is evident from the discussion in Chapter 4, the transport of fish on land must be accompanied by transport documents which describe the origin of the consignment, the content of the transport, the destination of the transport vehicle, as well as the consignee and the place and date of landing. On the basis of the reports from the Member States and evidence collected by the Commission Fisheries Inspectorate, the Commission took the view that enforcement of the measures pertaining to transport documents has been at best very limited in the majority of Member States and that they cannot be considered to be of significance in terms of overall enforcement of fishery measures.[94] Consequently, the Commission concludes that this element of the enforcement framework does not work satisfactorily and remains underutilised.

(vi) Technical measures[95]

The 1994 Control Report records that the observance of some technical measures (such as area restrictions) are the easiest regulatory provisions to enforce, particularly in fisheries which are close to the coast and patrolled by aircraft or patrol vessels. The report notes that no Member State appears to have an inspection strategy to cross-check the results of technical inspection at sea, with the results of inspections on landing, or after first sale. Interestingly, and possibly indicative of a more robust response to the requirement to fulfil Community obligations, is the fact that the Commission initiated **Article 171** {228} proceedings in 1995 because France did not give prompt effect to the judgement of the Court of Justice in Case 64/88 when the Court held that France had failed to discharge the obligation to enforce technical conservation measures.[96]

(vii) Progress on the enforcement of measures relating to the marketing of fishery products[97]

From the discussion in Chapter 5, it is evident that the rigorous application of the relevant body of Community rules should ensure the proper functioning of the common organisation of the market of fishery products is not destabilised by undeclared imports. Moreover it should also ensure that undersized fish are not presented for sale through publicly or privately owned auction centres that benefit directly or indirectly from the price withdrawal system. As regards the actual implementation of the enforcement measures required by the Control Regulation, the Commission noted in the 1994 Control Report that it could not conclude that the new provisions had been satisfactorily implemented. Furthermore in relation to the procedures governing the withdrawal of fish from the market, Member States did not report on compliance and therefore this important aspect of the common organisation of the market is not commented upon in the Report.

(viii) Structural measures[98]

In the Community enforcement system introduced by the Control Regulation there is a major emphasis on effective measures to monitor the application and implementation of the structural policy.[99] Furthermore, it may be recalled from the discussion in Chapter 5 above that the structural enforcement measures presented in the Control Regulation are an abridged version of what the Commission sought, in so far as the Council did not approve the Commission suggestion that it should be given powers to restrict the number of days at sea authorised for certain categories of vessels of a Member State which did not comply with the aim of the MAGP, and that these restrictions should at least be equivalent to the amount by which the aim of MAGP had been exceeded. Indeed from the data in both the 1994 and 1995 Control Reports it is evident that the implementation and the enforcement of Community provisions has been unsuccessful and will remain the focus of particular attention for Commission action in the future.[100]

(ix) Procedure for prosecuting suspected infringements in the Member States and the imposition of penalties and sanctions[101]

From a general perspective the 1994 Report records that fisheries enforcement in the European Union may be divided into three different phases, that is to say, the disclosure of an alleged infringement, the initiation of infringement procedures, and the imposition of sanctions. In relation to the latter, the 1994 Control Report records that the differences between the national systems are also reflected in the way in which sanctions are imposed for fishery offences.[102] In this respect sanctions may be generally classified into three main categories:

- penalties imposed in criminal court proceedings (financial fines, imprisonment, forfeiture of catch or gear and licence withdrawal);

- civil fines imposed by non-criminal courts;
- administrative penalties imposed by administrative authorities in the countries which have an administrative system of fisheries enforcement.

The 1994 Report records that the Control Regulation contains certain provisions relating to the effectiveness of sanctions to be applied in cases where the Community measures are not observed in order to safeguard the objectivity and integrity of the action taken following infringements.[103] However, most Member States in 1994 did not provide information on the type or level of penalties in their reports. Moreover, in cases where the Member States did notify the Commission of the number of infringements prosecuted, little or no information about the eventual penalty imposed, if any, was made available. The divergence of Community rules may be gauged from the statistics presented in the Annex to the Report. For example in 1994, Portugal is recorded as prosecuting 642 fishery offences in Court, compared to 162 in the United Kingdom and 67 in Ireland. Throughout the Community, 257 Spanish-flagged vessels were prosecuted for violating fishery regulations in other Member State fisheries zones. The latter ratio of inspection *vis-à-vis* infringements suggests, on the one hand, that there is a high incidence of non-compliance with Community regulations by these particular vessels. On the other hand, it may be argued that the figures only reflect the level and intensity of inspections of vessels flying the flag of Spain.[104]

Moreover, the overall picture of the level of sanctions imposed by Member States in cases of fishery offences is further complicated by the practice adopted by some Member States of systematically reporting all infringements committed by vessels from other Member States and by third-country vessels. In this regard, there are no figures available in relation to fishery offences prosecuted in Spanish Courts or administrative tribunals (although the figure of 11 812 is noted as a total number of infringements detected in Spain. It needs to be stressed, however, that the prosecution of fishery offences in Spain falls into distinctive phases. The first is essentially an administrative phase which normally entails the citing of a vessel for an infringement of Community law by a fishing inspector. The subsequent imposition of a sanction, generally a fine, under the administrative authority of the Director General of Fisheries. The second phase commences if the infringement or the sanction is contested by the impugned party. This phase is essentially a judicial phase and may involve oral and written pleadings in Court. The Court may uphold or overturn the sanction imposed by the administrative authority. It is believed that the majority of fishery offences in Spain are dealt with during the administrative phase.

The 1994 Report concludes that as long as the Commission does not receive systematic information on the amount and type of penalties imposed by national courts or administrative authorities, it cannot properly assess the efficiency of fisheries enforcement in the Member States.[105]

(x) Future reports

Future reports are likely to focus on issues such as the application of technology to the enforcement task as well as the resources in place to ensure adherence, structural measures, particularly the adjustment of fleet capacity and the monitoring of fishing effort under MAGP IV. Other areas of interest will include the monitoring of market measures, checks on transport documents, and the development of systematic cross-checks between different sources of information. There may also be emphasis on the problems in each fishery that lie behind major infringements of Community regulations to the possible detriment of other Member States such as over-exploitation of Member State quotas in the North and Baltic Seas.

2. *Conclusion regarding the soft law option to fisheries law enforcement*

The informal approach to law enforcement, characterised as soft law at the beginning of this section and symbolised by the annual reports of 1994 and 1995 on monitoring the CFP, is now firmly established as a means to achieve greater transparency in the application of the Control Regulation. While the level of implementation attained by the Member States in the initial phase of application of the Control Regulation is less than satisfactory, the annual report on the other hand is an important development in the enforcement of Community fishery law for several reasons. Firstly, it establishes the soft law approach as a prompt and direct means to redress the major shortcomings in the Member States. In particular, it is a means which is free of the debilitating tardiness of **Article 169** {226} proceedings. This conclusion is supported by the fact that the 1994 Report was succeeded by a Heads of Inspection meeting in Dublin which provided a forum to discuss and redress the shortcomings revealed by the report. Secondly, the soft law approach offers Member States an opportunity to assess what is taking place in the Community, which should in turn influence their respective positions and input into the future direction of the policy. Thirdly, the soft law approach has emphasised that many fishery offences are inter-jurisdictional in character and thus may not be addressed effectively and dealt with in isolated structures in the Member States or through the medium of municipal law in the Member States.

While the Amsterdam Treaty has clarified the obligation on Member States to co-operate in police and criminal justice matters it is contended that it will be difficult in the short term to develop through the medium of 'administrative decisions' precise enforcement processes and procedures to ensure compliance with Community law in the Member States. Consequently, increased use may be of codes of practice, recommendations, guidelines, resolutions, declaration of principles, standards.[106] It may thus be argued that these instruments are particularly suitable means of achieving greater harmonisation and assimilation of Member States' enforcement practices and standards. Indeed, the principal attraction of the soft law approach is that Member States are able to undertake obligations and simultaneously retain a degree of autonomy over their actions and resources.[107] Furthermore, the soft law approach blurs the traditional distinction between flag State and coastal State competence in the domain of enforcing Community fishery law. In this regard it is clearly apparent from the Monitoring Reports that this particular 'international law enforcement model' is probably incapable of ensuring that the objectives of the Control Regulation are achieved. It is also evident from the Reports that without effective implementation of the broader enforcement approach envisaged in the Control Regulation there is little chance of attaining improved compliance with the substantial body of rules which underpin the CFP.

Section (iii) Structural reform and fishery law enforcement

1. *Structural reform*

Structural reform entails the harmonisation of administrative and enforcement organisations (not to be confused with structural reform of the fishing industry). It is undertaken on the legal basis of **Article 5** {10} of the EC Treaty. To date there have been several examples of structural or bureaucratic co-operation in relation to the CFP, the most significant example of this concept is evident in the emphasis the Court of Justice placed on the principle of 'sincere co-operation' in its decision in the *EC Fisheries Inspector Case.*[108] In this decision attention was drawn to the obligation on the Commission to assist national courts with the enforcement of Community law unless the Court of Justice ordered otherwise.

The obligations imposed by **Article 5** {10} of the EC Treaty affect parties in different ways. These have been grouped into four different categories by one commentator.[109] If these categories are examined in the general context of the CFP and in the specific context of the Control Regulation then the concept of structural reform becomes more tangible. The first obligation is the duty on Member States to consult with the Commission when there is a doubt as to whether a national measure is contrary to Community law. This is the case with the Control Regulation which expressly provides that Member States are obliged to consult the Commission prior to adopting national measures which exceed Community obligations.[110] Secondly, Member States have an obligation to provide information to the Commission regarding practice and procedure in their respective Member States. Again the Control Regulation requires Member States to furnish the relevant information to allow the Commission, for example to complete the Annual Report on the monitoring of the CFP.[111] Thirdly, pursuant to **Article 5** {10} of the EC Treaty, the Commission and the Member States have a reciprocal duty of co-operation in the Community sphere, that is to say, when Member States are implementing Community measures or policies, are acting on behalf of the Community, or are using powers which are regulated by the Community. In the context of the Control Regulation there is an express requirement placed on Member States and the Community to co-operate and to co-ordinate a common enforcement policy.[112] Fourthly, it has been suggested that **Article 5** {10} of the EC Treaty may be invoked in order to prevent a Member State seeking to link unrelated policy or legislative measures in Council discussion. This issue is more difficult to analyse in the context of the fisheries enforcement framework in so far as structural reform is concerned. Although in general the CFP is now well established as an intrinsically distinct policy, linkage rarely arises outside the confines of the policy. Linkage, however, does occur at Council meetings as a means to achieve agreement on contentious proposals, although to what degree in the context of fisheries has not been commented upon. Finally, the most obvious example of the inter-organisational exchange in the context of enforcing the CFP is the agreement by the Heads of Inspection Service Meeting in Dublin in 1996 to organise an inter-Member State exchange scheme for inspectors and fishery enforcement officers.[113] The amendment to the Control Regulation in 1998 which provides a legal basis for inspectors to act as observers is also further proof of the evolution of structural reform. Furthermore, the fact that consideration has been given to the introduction of a standard inspection form to be used on a Community-wide basis is a significant step towards achieving uniformity in the inspection effort undertaken by national authorities.

2. *Conclusion regarding structural reform*

As noted above, the ability of the Members States to enforce Community law is restricted by the international law model which divides powers and responsibilities between the flag Sate and the coastal States. In the European Union, however, structural reform and organisational arrangements are two of the keys which may open the door leading to improved fishery law enforcement. Unfortunately, the door has been double-locked. On the one hand there is the locking device put in place by the Member States in Titles VI {VI} and VII {VIII} of the Treaty of European Union and similar provisions in the Amsterdam Treaty which firmly anchor law enforcement and co-operation in criminal justice in the domain of inter-governmental co-operation. On the other hand, there is the security mechanism which has always existed as a result of the different structures in the Member States for the investigation and prosecution of fishery offences. Indeed, in some Member States such as Ireland and the United Kingdom there are strict lines of division between the legal powers of the different bodies charged with responsibility to discharge the investigative, prosecutorial and judicial functions, which may also inhibit Member States from inter-State co-operation. This is specifically relevant to the procedures and processes for collecting evidence, an area in which common and civil law systems differ fundamentally. So although we may see evidence outside the inter-governmental framework which signifies an integrated pattern of enforcement behaviour and policy making, such progress is trammelled by guiding axioms which remain outside the exclusive domain of fisheries law. Indeed, it may be argued that this defect is not unique to fisheries as is evident from the dilatory progress that has been made in developing the appropriate structures and procedures for EUROPOL.[114] Furthermore, in the context of the fight against fraud of the Community budget, it is also evident that the only effective way to combat transnational fraud is to create a judicial space providing for improved co-operation between investigators, prosecutors, and judges.[115] Thus the upgrading of Title VI of the TEU in the Amsterdam Treaty, is a progressive move which within the prescribed areas will entail, *inter alia*: the improvement of co-operation between police forces, customs authorities and other executive authorities in the Member States, either directly or via EUROPOL; extend the co-operative process to national authorities and the national courts; and may entail the approximation of certain elements of criminal law and administrative provisions in the Member States. It may thus be concluded that structural reform is an on-going process which will be effected by the aforementioned provision in the Treaty of Amsterdam.

3. *Conclusions on enforcement and compliance during the period 1994–1997*

This chapter has been concerned with some of the critical issues pertaining to the enforcement of Community fisheries law and has focused on three different options with which to enforce the rules of the CFP. Firstly, there is the hard law option namely enforcement proceedings under **Articles 169–171** {226-228} of the Treaties; secondly, the soft law option such as the annual enforcement reports which presents a systematic over-view of progress since the adoption of the 1993 Control Regulation; and thirdly, the seeds of structural reform which may germinate from the provisions in the Control Regulation. At the beginning of this chapter, it was pointed out that one eminent commentator had referred to our knowledge of the enforcement of Community law as a black hole.[116] In view of the aforementioned discussion, it is suggested that this metaphor is not entirely appropriate in describing our knowledge of the enforcement of Community fishery law.

Overall, it may be said that the Community institutions have displayed a reluctance to pursue the formal approach to law enforcement which entails **Article 169** {226} proceedings.[117] The soft law option and structural reform to law enforcement have consequently increased in significance as a means to ensure that all parties comply with their obligations.[118] Moreover, because the CFP has a complex regulatory framework, it is inevitable that the effectiveness of the Control Regulation will depend on the Community institutions, Member States, the industry, and individuals. There appears to be little evidence available during the period 1994–1998 to support the view that the present legal order in the Community is entirely adequate to ensure an effective system of fishery law enforcement and to sustain a reasonable expectation that the preponderance of deviant behaviour will be detected and subjected to the appropriate sanction. One of the main reasons for this conclusion is that the approach to enforcement, including the structural reform achieved to date, is insufficient to overcome obstacles such as the lack of cohesion and incompatibility between the systems of law and procedure in the Member States. While some progress has been made in improving judicial co-operation in criminal matters pursuant to Title VI of the TEU and the amending provisions in the Amsterdam Treaty, it is increasingly apparent that there is a major deficit in appropriate inter-Member State co-operative structures to deal with fishery offences which by their very nature are frequently inter-jurisdictional. The existing structures and proposed improvements do not facilitate the pursuit of inter-jurisdictional fishery offences in an expeditious and effective manner. From a policy perspective, what is important in this context is that the widespread concern regarding the effectiveness of the CFP is not mitigated by the approaches adopted to law enforcement in the Member States, irrespective of which of the present enforcement options (hard law, soft law, structural reform), are pursued.

In conclusion, several obstacles, not unique to the fisheries sector, must be overcome before there is an improvement in the enforcement of Community fishery law. In particular, the restrictions which stem from the principles of flag and coastal state jurisdiction, the principle of territoriality and the differences in criminal and administrative procedures in the Member States, all detract from an effective enforcement system.[119] Bearing in mind, however, that there are exceptions in international law to the principle of territoriality on the basis of nationality and mutual consent to extra-territorial jurisdiction, and also taking into consideration that flag State jurisdiction allows a Member State to exert a 'long arm' jurisdiction over its vessels, there are few substantive reasons why Member States cannot agree to the appropriate structures and procedures to deal with fishery offences. In this regard it appears that one of the principal difficulties may be the possible unsuitability of criminal law as the means to achieve greater policy compliance. Indeed, it may be argued that criminal proceedings are unsuitable for dealing with fisheries infringements. This arises because criminal law requires, *inter alia,* the Courts to adopt a rigid interpretation of regulatory and other legislative provisions; a high degree of proof in the prosecution of offences; and grants certain defences and privileges to the accused, such as the privilege against self-incrimination. All of these pose difficulties in the context of enforcing fisheries law. It is thus important to examine how the Community enforcement system has worked in practice since 1993, a task which is undertaken in Chapter 10 in the context of a case study on the driftnet fishery.

References

1 Weiler, 'The Transformation of Europe', *Yale Law Journal* **100** (1991), 2463.

2 See, Snyder, F., 'The Effectiveness of European Community Law: Institutions, Processes, Tools and Techniques', in Daintith (ed.), *Implementing EC Law in the UK,* (London, 1995), 51–87.

3 On the rôle of the Commission inspectors and **Article 169** {226} proceedings see Chapter 6, *ante.*

4 This evaluation is undertaken in the light of several reports completed since 1994. These include, *inter alia,* the Commission Report on Monitoring the Common Fisheries Policy, COM(96) 100 final, 18.03.1996; the European Parliament Committee on Fisheries' Report on the Commission report on Monitoring the Common Fisheries Policy, (COM (96)0100 - C4-0213/96); and the statement of conclusions of a special meeting to discuss ways of improving enforcement, held in Dublin under the auspices of the Irish Presidency and attended by the Heads of Member State inspection services, September 1996, 9768/96. See also, *inter alios,* Berg, A. and Vervaele, J.A.E., *Fisheries Legislation in the Member States: Issues of Enforcement and Co-operation in the Member States,* a study prepared for the European Commission by the University of Utrecht, 1994; Audretsch, *Supervision in European Community Law,* (2nd ed., 1986), especially at 250–400.

5 See, Snyder, F., *op. cit.* fn 2 , *supra,* and the authorities cited therein, in particular, Wellens and Borchardt, 'Soft Law in European Community Law', *ELR* **14**(5) (Florence, 1989), 267–321; Thurer, 'The Role of Soft Law in the Process of European Integration', in Jacot-Guillarmod (ed.), *L'avenir du libre-échange en Europe: vers un Espace économique européen?;* Snyder, F., 'Soft Law and Institutional Practice in the European Community', European University Institute (EUI) Working Paper, (Florence, 1993); Delhousse and Weiler, 'EPC and the Single Act: From Soft Law to Hard Law?' in Holland (ed.), *The Future of European Political Co-operation,* (1991).

6 The jurisprudence emanating from a series of fisheries cases upon which the Court of Justice adjudicated during the period 1976–1996 is examined above in the context of the many provisions of the Control Regulation, Chapters 3 to 6, *ante.*

7 Council Regulation No. 2988/95 on the protection of the financial interests of the Community, *OJ* L 312/1, 23.12.1995; Convention on the Protection of the European Community's financial interests, *OJ* C 316/48, 27.11.1995; see White, S., 'A Variable Geometry of Enforcement? Aspects of European Community Budget Fraud', in *Crime, Law and Social Change* **23**, p.325.

8 See, Hartley, C., *The Foundations of European Community Law,* 4ed., (Oxford, 1998), Chapter 9, *passim.*

9 Case C-4/96, *Northern Ireland Fish Producers' Organisation Ltd. (NIFPO) and Northern Ireland Fishermen's Federation* v. *Department of Agriculture Northern Ireland,* [1998] ECR I-0681; Case

C-38/95, *Ministero delle Finanze* v. *Foods Import Srl.* [1996] ECR I-6543; Case C-311/94, *IJssel-Vliet Combinatie BV* v. *Minister van Economische Zaken,* [1996] ECR I-5023; Case C-276/94, *Criminal Proceedings against Finn Ohrt,* [1996] ECR I-0199; Case C-44/94, *The Queen* v. *MAFF, ex parte National Federation of Fishermen's Organization and Others and Federation of Highlands and Islands Fishermen and Others,* [1995] ECR I-3115; Case C-405/92, *Establissements Armand Mondiet SA* v. *Armement Islais SARL.,* [1993] ECR I-6133; Joined Cases C-251/90 and C-252/90, *Procurator Fiscal, Elgin* v. *Kenneth Gordon Wood and James Cowie,* [1992] ECR I-2873; Case C-221/89, *The Queen* v. *Secretary of State for Transport, ex parte Factortame Ltd and others,* [1991] ECR I-3905; Case C-348/88, *Criminal Proceedings against Jelle Hakvoort,* [1990] ECR I-1647; Case C-216/87, *The Queen* v. *MAFF, ex parte Jaderow Ltd.,* [1989] ECR 4509; Case C-3/87 *The Queen* v. *MAFF, ex parte Agegate,* [1989] ECR 4459; Case 223/86, *Pesca Valentia Ltd* v. *Ministry for Fisheries and Forestry, Ireland, and the Attorney General,* [1988] ECR 0083; Case 63/83, *Regina* v. *Kent Kirk,* [1984] ECR 2689; Case 24/83, *Wolfgang Gewiesse and Manfed Mehlich* v. *Colin Scott Mackenzie,* [1984] ECR 0817; Case 87/82, *Lt.Cdr. John Rogers* v. *H.B.L. Darthenay,* [1983] ECR 1579-1594; Joined Cases 50 to 58/82, *Administrateur des Affairs Maritimes, Bayonne and Procureur de la République* v. *José Dorca Marina and others* [1982] ECR 3949-3960; Joined Cases 13 to 28/82, *José Arantzamendi-Osa and Others* v. *Procureur de la République and Procureur Général,* [1982] ECR 3927-3938; Joined Cases 138 and 139/81, *Directeur des Affaires Maritimes du Littoral du Sud* v. *Otazo and Manuel Prego Parada* [1982] ECR 3819-3836; Joined Cases 137 and 140/81, *Directeur des Affaires Maritimes du Littoral du Sud-Oest and Procureur de la République* v. *Alfonso Campandeguy Sagarzazu,* [1982] ECR 3847-3864; Case 21/81, *Criminal Proceeedings against Daniël Bout and BV I. Bout en Zonen,* [1982] ECR 0381-0391; Case 269/80, *Regina* v. *Robert Tymen,* [1981] ECR 3079-3095; Case 181/80, *Procureur Général près la Cour d'Appel de Pau and Others* v. *José Arbelaiz-Emazabel,* [1981] ECR 2961-2983; Joined Cases 180 and 266/80, *José Crujeiras Tome* v. *Procureur de la République and Procureur de la République* v. *Anton Yurrita,* [1981] ECR 2997-3018; Case 124/80, *Officier van Justitie* v. *J. van Dam & Zonen,* [1981] ECR 1447-1461; Case 812/79 *Attorney General* v. *Juan C. Burgoa,* [1980] ECR 2787-2809; Case 88/77, *Minister for Fisheries* v. *C.A. Schonenberg and Others.,* [1978] ECR 0473-0493; Joined Cases 3, 4 and 6-76, *Officier van Justitie* v. *Cornelius Kramer and Others* [1976] ECR 1279-1316.

10 See Chapter 7, *ante.*

11 Case C-316/96 *Commission* v. *Italian Republic* [1997] ECR I-7231. Case 325/95, *Commission* v. *Ireland* [1996], ECR I-5615; Case 52/95, *Commission* v. *French Republic* [1995] ECR I-4443; Case 334/94, *Commission* v. *French Republic* [1996] ECR I-1307; Case C-131/93, *Commission* v. *Federal Republic of Germany,* [1994] ECR I-3303;Case C-228/91, *Commission* v. *Italian Republic* [1993] ECR I-2701; Case C-52/91, *Commission* v. *Netherlands,* [1993] ECR I-3069; Case C-280/89, *Commission* v. *Ireland,* [1992] ECR I-6185; Case C-258/89, *Commission* v. *Spain,* [1991] I-3977; Case C-246/89, *Commission* v. *United Kingdom* [1991] ECR I-4585; Case C-146/89, *Commission* v. *United Kingdom* [1991] ECR I-3533; Case C-93/89, *Commission* v. *Ireland* [1991] ECR I-4569; Case C-62/89, *Commision* v. *French Republic,* [1990] ECR I-10925; Case 209/88, *Commission* v. *Italian Republic,* [1990] ECR I-4313; Case 200/88, *Commission* v. *Hellenic Republic,* [1990] ECR I-4299; Case C-64/88, *Commission* v. *French Republic,* [1991] ECR I-2727; Case 39/88, *Commission* v. *Ireland,* [1990] ECR I-4271; Case 290/87, *Commission* v. *Netherlands,* [1989] ECR 3083; Case 100/84, *Commission* v. *United Kingdom,* [1985] ECR 1169-1184; Case 32/79, *Commission* v. *United Kingdom* [1980] ECR 2403-2452; Case 804/79, *Commission* v. *United Kingdom,* [1981] ECR 1045-1080; Case 61/77 *Commission* v. *Ireland* [1978] ECR 0417-0454.

12 See, for example, Case 246/89, Order of the President of the Court of 10 October 1989, *Commission* v. *United Kingdom,* [1989] ECR 3125.

13 Cases 31/77R and 53/77R, *Commission* v. *United Kingdom (Pig Producers)* [1977] ECR 921; for a discussion of this case and others see Gray, 'Interim Measures of Protection in the European Court', *ELRev* **80** (1979), pp. 96–8; see also Case C-195/90R, *Commission* v. *Germany* (road tax case), [1990] ECR I-3351.

14 Case 61/77R, *Commission* v. *Ireland,* [1977] ECR 937; for further details see Chapter 7, *ante.*
15 Case C-213/89, *The Queen* v. *Secretary of State for Transport, ex parte Factortame Ltd. and others,* [1990] ECR I-2433. See, Chapter 7, *ante.*
16 See, *inter alia,* Kunzlik, P., 'The enforcement of EU Environmental Law: Article 169, the Ombudsman and the Parliament,' *European Environmental Law Review,* 6(2) (Feb. 1997), 46–52; MacRory, R., 'The Enforcement of Community Environmental Laws, some critical issues', *CMLRev* **29** (1992), 347–369; European Commission, Implementing Community Environmental Law, COM(96), 500, 22.10.1996; Dashwood, White, 'Enforcement actions under Articles 169 and 170 EEC Treaty', *ELRev* **14** (1989) 388–413; Hartley, T.C., *The Foundations of European Community Law,* (Oxford, 1994), Chapter 10; Barav, 'Failure of Member States to fulfil their obligations under Community Law', *CMLRev* **12** (1975), 385; Ebke, 'Enforcement Techniques within the European Communities' *Journal of Air Law and Commerce* **50** (1985), 685; Oliver, P., 'Enforcing Community Rights in the English Courts', *MLR* **50** (1987), 881.
17 This follows from **Article 5** {10} EC Treaty, on this point see Case 240/86, *Commission* v. *Greece* [1989] ECR 1853.
18 The working of **Article 169** {226} proceedings has been subject to a review by the European Ombudsman in 1996. On the function and rôle of the latter see Magliveras, 'Best Intentions but Empty Words: The European Ombudsman', *ELR* **20** (1995), 401.
19 Case C-209/88, *Commission* v. *Italy* [1990] I-4313. For a discussion of the Commission's discretion, see Hartley, *The Foundations of European Community Law,* 311–313.
20 Case C-209/88, *Commission* v. *Italy* [1990] I-4313, paragraph 13.
21 To support their view the Court cited the judgment in Case 95/77, *Commission* v. *Netherlands* [1978] ECR 863, paragraphs 13–14.
22 *Ibid.*
23 The Court has held that it is clear from the scheme of **Article 169** {226} of the EC Treaty that the Commission is not obliged to commence proceedings under that provision but has discretionary power in this regard which excludes the right of private individuals to require that institutions adopt a specific position. Case 247/87, *Star Fruit Company SA* v. *Commission,* [1989] ECR 291; Case 48/65, *Lütticke* v. *Commission,* [1966] ECR 0027.
24 Case C-64/88, *Commission* v. *French Republic,* [1991] ECR I-2727; Case 290/87, *Commission* v. *Netherlands,* [1989] ECR 3083.
25 See Annual Reports on the monitoring and the application of Community Law drawn up by the Commission in response to a request by the European Parliament and the Member States (point 2), Declaration No. 19, Treaty European Union 1992, the most recent edition at the time of writing is the Fifteenth Annual Report on monitoring the application of Community Law, 1997, *OJ* C 250/01, 10.08.1998.
26 *Ibid.*
27 See, MacRory R., 'The Enforcement of Community Environmental Laws: Some Critical Issues', *CMLRev,* **29** (1992), 347–369, at p.365.
28 *Op. cit.,* fn 25.
29 The 1996 Report (Fourteenth Annual Report on monitoring the application of Community Law, 1996, *OJ* C 33/01, Vol 40, 03.11.1997) erroneously records that the Commission commenced infringement proceedings against Sweden and Finland.
30 Chapter 6, *ante.*
31 Fourteenth Annual Report on monitoring the application of Community Law, 1996, *OJ* C 33/01, Vol 40, 03.11.1997.
32 *Ibid.*
33 Discussed above in Chapter 6, *ante.*
34 Article 28, Council Regulation No. 2847/93, discussed in Chapter 6, *ante.*
35 Case 25/59, *Netherlands* v. *High Authority,* [1960] ECR 0723.
36 On this point see, for example, the view expressed by Churchill, R.R., 'Enforcement of the

Common Fisheries Policy, with Special Reference to the United Kingdom,' in Harding (ed.), *Enforcing European Community Rules*, (Aldershot, 1996), p.96.

37 Case C-290/87, *Commission* v. *Netherlands*, [1989] ECR 3083.

38 Case C-52/95, *Commission* v. *France*, [1995] ECR I-4443.

39 Case 209/88, *Commission* v. *Italy*, [1990] I-4313; Case 39/88, *Commission* v. *Ireland*, [1990] ECR I-4271.

40 Case C-146/89, *Commission* v. *United Kingdom of Great Britain and Northern Ireland*, [1991] ECR I-3533.

41 *Ibid.*

42 *Ibid.*

43 On a more academic level this issue of supporting Community policy, as defined as the values and attitudes of the judges and the objectives which they wish to promote, has been examined by a number of distinguished scholars in the context of several areas of Community law. See, *inter alia*, Pescatore, P., *L'Ordre Juridique des Communautés Européennes* (1975); Brown, L., Neville and Jacobs, F.G., *The Courts of Justice of the European Communities*, 3ed (1989).

44 Case C-221/89, *The Queen* v. *Secretary of State for Transport, ex parte Factortame Ltd and Others*, [1991] ECR I-3905.

45 Joined Cases 3, 4 and 6–76, *Cornelius Kramer and Others* [1976] ECR 1279-1316.

46 Case, C-25/94, *Commission of the European Communities* v. *Council of the European Union supported by the United Kingdom of Great Britain and Northern Ireland*, [1996] ECR I-1469.

47 See, Churchill, R.R. 'Quota Hopping, the CFP Wrong Footed', *CMLRev* **27**(2), 209–247, and discussed in Chapter 5, *ante*.

48 Case C-290/87, *Commission* v. *Netherlands*, [1989] ECR 3083; Case C-64/88, *Commission* v. *France*, [1991] ECR I-2727; Case C-62/89 *Commission* v. *France*, [1990] ECR 925; Case C-258/89, *Commission* v. *Spain*, [1991] ECR I-3977; Case C-52/91, *Commission* v. *Netherlands*, [1993] ECR I-3069.

49 Evans, 'The Enforcement of Article 169 EEC: Commission Discretion', *ELRev* **4** (1979), pp. 449–455.

50 See Hartley, T.C., *The Foundations of European Community Law*, (Oxford, 1998), 322–324.

51 Case 141/78, *France* v. *United Kingdom* [1979] ECR 2923. Annotation, Churchill, R.R., *ELRev*, (1980), 71–73; Scovazzi, T., *Diritto comunitario e degli scambi internazionali*, (1981), 53–58.

52 For a different view of the potency of **Article 171** {228} to improve compliance with Community law, see, Curtin, D., 'The Constitutional Structure of the Union: A Europe of Bits and Pieces', *CMLRev* **30** (1993), 17–69.

53 This power was granted to the Commission under Council Regulation No. 2241/87 of 23.07.1987 establishing certain control measures for fishing activities, *OJ* L 207/1, 29.07.87, (Repealed); Commission Regulation No. 493/87 of 18.02.1987 establishing detailed rules remedying the prejudice caused on the halting of certain fisheries, *OJ* L 50/13, 19.02.87. The Commission, pursuant to the Management Committee procedure, may redress overfishing through the means outlined in **Article 21(4)** of the Control Regulation, discussed in Chapter 4, *ante*.

54 The Commission, pursuant to the Management Committee procedure, may redress non compliance through the means outlined in **Article 32** of the Control Regulation. The issue raised regarding failure to initiate criminal and administrative proceedings is examined in Chapter 5, ante.

55 See Chapter 3, *ante*.

56 The Fourteenth annual report on monitoring the application of Community law - 1996, *OJ* C 332, Vol. 40, 03.11.1997, p. 55, records that the Commission sent France a reasoned opinion for failure to implement aspects of the judgement of the Court in Case C-64/88 for failure to discharge the obligation to enforce technical conservation measures the

Commission had initiated **Article 171** {228} proceedings. Specifically, the Commission noted the differences between the French gauges and those used by the Community for measuring net size, absence of control measures in relation to by-catches, non-compliance with rules on minimum sizes of fish.

57 In Case 334/94, *Commission* v. *France*, I-1307, paragraph 30, the Court cited Case 169/87 *Commission* v. *France* [1988] ECR 4093, para 14, to support this view. In Case 334/94 the Court ruled that under **Article 171** {228} of the EC Treaty the French Government had failed to comply with a judgement relating to the French Customs Code delivered by the Court delivered more than 20 years ago, Case 167/73, *Commission* v. *France* [1974] ECR 359, discussed in Chapter 7, *ante*.

58 This was established in Case C-6/90, 9/90, *Francovich and Others* v. *Italy*, [1991] ECR I-5357, a case which followed **Article 169** {226} proceedings taken by the Commission against Italy.

59 Cases C-46/93 and C-48/93, *Brasserie du Pêcheur* v. *Germany, and the Queen* v. *Secretary of State for Transport ex parte Factortame Ltd. and Others*, [1996] ECR I-1029. See, *inter alia*, Convery, State liability in the United Kingdom after *Brasserie du Pêcheur, CMLRev* **34**(3), 603–634; Van Gerven, 'Bridging the Unbridgeable: Community and National Tort Law after *Francovich* and *Brasserie*', ICLQ (1996) 507; for a comprehensive casenote, see Oliver, P., *CMLRev* **34**(3) (1997), 635–680.

60 On this point see Snyder, F., *op. cit.*, fn 2, 61–63; Harding, 'Who Goes to Court in Europe? An analysis of Litigation against the European Community', *ELR* **17** (1992), 105–125.

61 Twelfth Annual Report on Monitoring the Application of Community Law - 1994, *OJ* C 254, Vol. 38, 29.09.1995, p.254/42.

62 Thirteenth Annual Report on Monitoring the Application of Community Law - 1995, *OJ* C 303, 14.10.96, p. 303/47.

63 *Op. cit.*, fn 4, *supra.*

64 See, Snyder, F., *op. cit.* fn 2, *supra.*

65 This analysis is taken from Snyder, F., *op. cit.*, fn 2, *infra.* See also Ganz, *Quasi-Legislation*, (1987).

66 The European Parliament nevertheless presents a report on the Commission report, see for example the McKenna Report which was produced by an Irish MEP in 1996 referred to above in fn 4, *infra.*

67 On this obligation, see Chapter 6, *ante.*

68 See Chapter 2, *ante.*

69 Report from the Commission, *Monitoring the Common Fisheries Policy 1995*, Com(97)226 final.

70 For a general description of resources available in the Member States for enforcement see, Chapter 2, ante.

71 See Chapter 2, *ante.*

72 1994 Report, p. 4, *op. cit.* fn 4.

73 See Chapter 2, *ante.*

74 See, Editorial Comments, 'The Treaty of Amsterdam: Neither a Bang Nor a Whimper', *CMLRev* **34** (1997), 767–772.

75 See Chapter 3, *ante.*

76 See the Commission Report on Monitoring the Common Fisheries Policy, COM(96) 100 final, 18.03.1996, p. 18.

77 *Ibid.*

78 This issue is further discussed in Chapter 12, *post.*

79 See, footnote 1, *infra.*

80 These instruments are both examined below in Chapter 11, *post.*

81 This issue is further examined in Chapter 12, *post.*

82 Commission Report, *Monitoring the CFP 1995, op. cit.* fn 4, *supra.*

83 *Ibid.*, p.8.

84 See Chapter 4, *ante.*

85 *The 1994 Report on Monitoring the CFP*, pp. 6–8, *op. cit.*, fn 4.

86 The working groups charged with monitoring the various stocks have reported on the size and trend of the misreporting problem. The overall seriousness of the situation has also been stressed by ICES experts, see, *Report of the Statistics Committee Liaison Working Group*, (ICES CM 1995/D:1). An earlier Commission report summarises estimates of the amount of misreporting for key stocks, Annex 2 to COM(95) 243 final. The issues raised are also examined by Biais, G. who evaluates the misreporting problem in the context of the problem of discards, 'An evaluation of the policy of fishery resource management by TACs in European Community waters from 1983 to 1992', *Aquatic Living Resources*, **8**(3) (1995), 241–251.

87 *The 1994 Report on Monitoring the CFP*, *op. cit.*, fn 4.

88 The figure of 40% had been previously suggested by a European Parliament Report, the Guilland Report, discussed in Chapter 2, *ante.*

89 The Commission Report on Monitoring the Common Fisheries Policy, COM(96) 100 final, 18.03.1996.

90 *Ibid.*

91 *Ibid.*

92 See Chapter 11, *post.*

93 In the demersal fisheries, in the Community's central and southern Atlantic waters, misreporting appears to be limited because the TACs in this zone are precautionary and are set at levels imposing no real quota constraints on Member States (exceptions, however, are the French anchovy fishery, northern hake and monkfish fishery for the Spanish fleet, megrim or southern hake for the Portuguese fleet). Interestingly, the report suggests that the lower level of discards may be partly explained by the fact that there is a competitive market in southern Europe for undersize fish. Further north, on the Atlantic coast, catch misreporting problems are also prevalent but do not appear to have any distinct trend. However, in the North Sea the under-reporting problem is particularly acute. This stems from the fact that fishing effort has not been reduced in parallel with reduced TACs and quotas. Indeed, the Commission express a belief that the twin phenomena of fraud and discards illustrate the urgency for a restriction of fishing effort in the North Sea. This finding could also be supported by the significant black market for fish which has developed in the United Kingdom since 1991, see Memorandum submitted by the United Kingdom National Federation of Fishermen's Organisation, House of Lords Select Committee, Report on the Review of the CFP, Session 1992, (2nd report), 23 June 1991.

94 In their reports to the Commission, Member States did not indicate the extent to which random checks have been conducted to check on the distribution of fish catches, nor do they report that their particular checks/controls yielded significant results. The reports from UK, Ireland, and Denmark, however, record that transport documents are randomly checked. In contrast in Denmark the checks may involve the police. In Belgium, imports appear to be monitored closely. There is no information in the Portuguese and French reports regarding this form of control.

95 See, Chapter 5, *ante.*

96 Case 64/88, 11.06.1991, see Chapter 5, *ante.*

97 See, Chapter 5, Section (ii) *ante.*

98 *Ibid.*

99 The specific rules are contained in **Articles 25** to **27** of Regulation (EEC) No. 2847/93.

100 *The 1994 Report on Monitoring the CFP*, *op. cit.*, fn 4.

101 For a discussion of sanctions in Community fisheries law see, Chapter 6 and the sources cited therein, *ante.* In order to get a better understanding of the legal systems applicable in

the Member States concerning fisheries enforcement, the Commission ordered a study which examined the majority of Member States. Some of the issues recorded in the 1993 Control Report were identified in this study. The way Community law is applied and the remedies that are invoked varies greatly from Member State to Member State. See, *inter alia*, Berg, A.J., Vervaele, J.A.E., *Fisheries Legislation of the EC Member States: Issues of Enforcement and Co-operation*, (Centre for the Enforcement of European Community Law, University of Utrecht, 1994). For other comment on this point, if not somewhat out of date, see, Bridge, 'Procedural Aspects of the Enforcement of EC Law through the Legal Systems of Member States' *ELRev* **9** (1984), 28.

102 The report also notes that even if a fine deprives a fisherman of the gains of the fishing trip, it will at the most only marginally increase annual costs. In addition, national courts have been known to impose very low fines because they consider Community legislation complex.

103 **Article 30**, Council Regulation No. (EEC) 2847/93.

104 This point is developed further in Chapter 9, *post*.

105 *Ibid.*

106 *Ibid.*

107 This is a theme which is examined in further detail in Chapter 12, *post*.

108 Case C-2/88, *Criminal Proceedings Against J.J. Zwartveld and Others*, [1990] ECR I-4405.

109 *Op. cit.*, fn 2.

110 **Article 38**, Council Regulation No. 2847/93, *OJ* L 261/16, see Chapter 3, *ante*.

111 **Article 35**, *Ibid.*, see Chapter 6, *ante*, and Section (ii), *supra*.

112 See Chapter 3, *ante*, and Section (ii), *supra*.

113 See, Section (ii), *ante*.

114 See, *inter alia*, Monace, F.R., 'Europol: The Culmination of the European Union's International Police Co-operation Efforts,' *Fordham International Journal* **19**(1) (1995), 247–308; Den Boer, M., 'Police Co-operation in the TEU: Tiger in a Trojan Horse,' *CMLRev* (1995), 555–578.

115 See Council Documents 5331/96 JUST 4, which outline a draft action plan to improve judicial co-operation between Member States.

116 *Op. cit.*, fn 1.

117 Section (i), *infra*.

118 Sections (ii) and (iii), *infra*.

119 These issues are explored further in the case study in Chapter 10, *post*.

Chapter 9
Financing the Enforcement of the CFP

Introduction

From the discussion in Chapters 1–2, it is evident that the Community fishery control and enforcement system has evolved considerably since the pre-1993 phase of the policy. It now entails an elaborate legislative framework with an intricate web of enforcement officers, administrators and law officers to oversee the day-to-day application of the policy. Contemporaneous with internal developments, the onus on the Community to negotiate successfully and to resolve enforcement issues on the external plane has increased dramatically in recent years.[1] Furthermore, major developments in the international law of fisheries, which stem from the conclusion of, *inter alia*, the Agreement on the Implementation of the United Nations Provisions on Straddling Stocks and Highly Migratory Species (the New York Agreement) and the Code of Responsible Fisheries, continue to have a major impact on the evolution of appropriate enforcement structures in several international fisheries in which the Community fleet participates.[2] In parallel with the obligation to improve enforcement came the corresponding requirement of providing adequate financial investment to establish the appropriate structures and to equip the various agencies in order to discharge this obligation.

The justification for the Community contribution to expenditure in the Member States is that fishery enforcement is no longer considered exclusively as a national interest and that the investment necessary to acquire suitable enforcement resources exceeds the funding available from national budgets. Moreover, it is universally recognised that improvements need to be brought about in the implementation of management plans for fisheries. Failure to do so will lead to losses in revenue in excess of 10% to 30% on the short-term, with the possibility of the destruction of the resource in the long-term. Furthermore, non-compliance and fraud undermine the credibility of the CFP as a policy which aims to deliver sustainable fisheries and successfully engender prosperity to the sector.

This chapter examines the financial implications of fishery law enforcement and its impact on the Community budget. This chapter is divided into four sections, the first of which presents a brief overview of the Community budget with an explanation of how the CFP is financed. Section (ii) examines the cost of fishery control in the European Union and endeavours to explain the distribution of the control budget between Member States. Section (iii) looks at the type of projects that are co-financed by Community funds and the procedure for processing Member State applications for budgetary subvention. Finally, Section (iv) analyses the cost effectiveness of fishery enforcement.

As with the many other areas of fishery enforcement law there is a dearth of published material on the financing of the CFP. This chapter relies heavily on information provided in several reports and communications submitted by the Commission to the Council and to the European Parliament.[3] It also relies on Commission replies to several parliamentary questions posed in the European Parliament and a diffuse number of secondary legislative instruments adopted at Commission and at Council level since 1978.

In accordance with Community legislative provisions for the adoption of the euro every reference to the ecu (the European currency unit prior to the introduction of the euro) in a legal instrument shall be replaced by reference to the euro at a rate of one euro to one ecu after 1 January 1999. The unit of currency used in this chapter is therefore the euro. The irrevocable conversion rates between the euro and the currencies of the Member States are recorded in Council Regulation (EC) No. 2866/98.[4]

Section (i) Overview of the Community budget

1. *The budget and the CFP*

The Council of Ministers and the European Parliament constitute the joint budget authority.[5] The Court of Auditors conducts the annual audit. The procedure for adopting the budget, the rôle of the European Parliament, and the many inter-institutional agreements, have led to a three-way tension between the supranational, the intergovernmental, and the bureaucratic.[6]

The proportion of the budget which is spent on the CFP is relatively modest. In 1996 it amounted to €890.7 million, or a little over 1% of the total EU budget.[7] In recent years the amount of Community funds expended on the CFP has grown considerably. This is particularly noticeable from 1985 to date. Budget appropriations have increased four fold between 1985 and 1996 (at constant prices), and tripled in real terms. Since 1985 the funding of the structural elements of the policy has increased, with the view to reconciling the over-capacity in the Community fleet with the available fishery resources.[8] In the same period, the financing of third-country agreements has commanded a greater proportion of funds. In 1995 these two groupings of expenditure account for an astounding 90% of the fishery budget. The funding by the Community of the market support structure has fallen, in percentage terms from 13.8% in 1985, to 4.8% in 1996. Table 9.1 illustrates the distribution of the budget in 1996.

Table 9.1 Budget of the Common Fisheries Policy in 1996 (in € million).

CFP Element	Amount in € million
Structures	520.4
External Fisheries Agreements	295.7
Markets	39
Control	35.7
Research	33.7
Total	924.5

The information in Table 9.1 may be illustrated schematically, and so for comparative purposes a similar graph for 1985 is reproduced (Figure 9.1).[9] The graphs illustrate the changing pattern of the CFP Budget. The external fishery policy has been particularly expensive, costing on average €68 million a year in the period 1983–1990, which represented 29% of the total budget for the CFP. This had increased in 1994 to €284 million, which represents 36.7% of the budget. Further increases were necessary in 1996 (to €295.7 million) with the conclusion of the Moroccan agreement in late 1995,[10] and the additional costs entailed by the fishery dispute with Canada. As already noted in Chapter 1, one commentator has expressed serious doubts as to whether the value of the fish caught, pursuant to the international agreements, exceeds the cost of the agreements *per se* to the Community budget.[11]

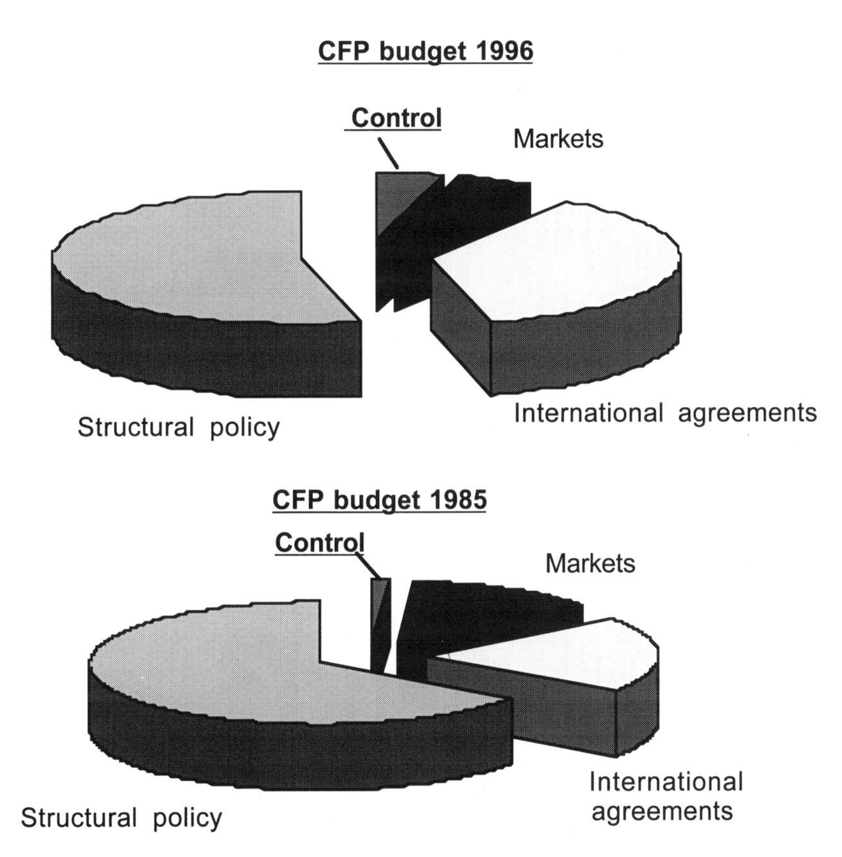

Fig. 9.1 The distribution of the CFP budget in 1996 and 1985.

From this brief overview of the budget it can be concluded that the CFP is not a major drain on Community funds. The expenditure on fisheries is only some one sixteenth of the Community's market support for butter and less than half the Community's market support for wine and tobacco. It must also be noted that the fishing industry is valued in the region of €7 billion to the Community economy per annum and supports approximately 300 000 jobs in the catching sector. The precise number of jobs in the downstream and upstream sectors is not known and difficult to quantify but is thought to support a similar number of jobs.[12]

Finally, the cost of the enforcement aspect of the policy is modest when compared with the cost of the external fishery agreements and the implementation of the structural elements of the CFP. In percentage terms the control element of the budget cost, in 1996, approximately 4% of the fishery budget, while the structural policy 58.5% and international agreements 33.5%.

It is now proposed to examine how much, in real terms, the Community spends on the control policy.

Section (ii) Funding enforcement

1. Funding of enforcement and surveillance cost by the European Union

Enforcement of fisheries regulations is one of the most costly and problematic features of fishery management programmes. In some programmes, for example those outside the European Union, enforcement can account for upwards of half all expenditure on fisheries management.

Costs arise from the need to equip services which operate both at sea and in port and in some cases services which monitor the activities of specific sectors of the fishing industry, for example the marketing and the processing sectors. Although comprehensive comparative costs are not readily available, the overall cost of enforcement to major maritime States are considerable. For example,[13] the annual costs of monitoring in Norway is estimated at about €60 million, while in Canada these can be in the region of €55–60 million. In the United States, the fisheries budget of the national Coast Guard which is the main offshore enforcement service is estimated about €390 million. These figures relate to annual operational costs and do not take account of the on-going need for capital investment such as the replacement of patrol vessels or surveillance aircraft.

Within the European Union, the Member States bear the majority of costs associated with enforcement. Available information on operational costs suggests that the United Kingdom has an annual expenditure of approximately €35 million,[14] while Ireland would spend upwards of €10 million. The level of investment required to maintain and upgrade these services is equally high. As most Member States have an operational service, investment is usually geared towards improving existing organisations rather than the creation of a new service.

Investments usually centre on the acquisition and improvement of:

- seagoing equipment, aircraft, helicopters and land vehicles;
- systems for recording fishing activities, and the processing of catch data; and
- new technology for enforcement, information technology systems and vessel monitoring through satellite tracking.

Prior to the adoption of the CFP, national control services enforced national rules and regulations. However, as the CFP was broadened and deepened to embrace a wide range of new measures, national services had to make qualitative and quantitative adjustments to their enforcement rôles. Within the European Union, the burden of actual enforcement is unevenly distributed in the sense that smaller coastal states such as the Netherlands and Belgium have a considerable more modest inspection rôle at sea compared to the United Kingdom and Ireland.

As national services began to devote more control resources to the enforcement of Community rules, there was a growing realisation at Community level, that some financial assistance from the Community budget would be both necessary and appropriate. The *raison d'être* for such financial assistance is based on the principle that surveillance and enforcement of a common (Community) resource goes beyond purely national interests and responsibilities. Initially, financial assistance was granted to two Member States in 1978 when both Ireland and Denmark received contributions to cover their respective increased enforcement responsibilities following the declaration of a 200-mile exclusive zone by the Member States of the European Union.[15]

The provision of financial aid was placed on a firmer basis in the late 1980s when the Council of Ministers adopted two decisions. The first in 1987 made provision for the granting of €5.5 million.[16] The bulk of the aid was granted to Denmark (€1.6 million), Ireland (€1.3 million) and Portugal (€1 million). This decision was supplemented by a second on 27 November 1989.[17] This covered the period 1991–1995 and extended the scope of the community aid to monitoring operations in the Mediterranean pending adoption of Community fishery rules for that area. The principal objective of the 1989 Decision was to contribute to the development of monitoring and inspection facilities in the Member States between 1 January 1991 and 31 December 1995. Operational costs entailed by national services were deemed ineligible for financial aid. To cover anticipated costs over the five-year period, the Council fixed an annual budget of €22 million per year and a total ceiling of €110 million for the reference period. Within the framework of the decision, Member States had to submit proposals for financial assistance and put up matching investments funds. Proposals submitted by Member States, on this basis, could be eligible for 50% co-finance by the Community.

Between 1991 and 1995, the Commission on behalf of the Community made a total financial contribution of €118 million. The additional sum of €8 million was made available from a supplementary budget in 1994 to fund the two-year pilot project on satellite monitoring of fishing vessels.[18] The project was funded in its entirety by the Community.

2. *Main recipients of Community financial aid*

Details on payments to Member States during the lifetime of the 1989 Decision are published in the Commission Report to the Council and the European Parliament,[19] and these indicate that Ireland and Portugal received about 50% of the available funds. The justification for Ireland's share would be the disproportionate cost borne by the Irish authorities in monitoring the fishing activities of fleets flying the flag of other Member States in the extensive Irish fishing zone. Indeed Annex VII of the Hague Resolution expressly states that the Council recognises that the protection and the control of the fishing zone off Ireland must not result in a charge to the Irish authorities which is disproportionate to the volume of Community fish resources which can be exploited by Irish fishermen in the Irish zone.[20]

Portugal, a relatively new member of the Community, had to construct and organise its national control services and consequently was one of the major recipients. The Commission Report of 1995 provides details of the type of projects supported over the five-year period. Member States with extensive coastlines and/or exclusive zones to monitor tended to invest in seagoing craft, navigational radar and communications equipment and aircraft for fisheries surveillance. Other projects included investments in information technology for data processing, as well as land vehicles.

Section (iii) Community aid

1. *Review of co-finance by the Community*

The adoption of the 1992 Control Regulation and the expiry date of the 1989 Council Decision both stimulated the Commission and the Member States to review and plan the future of Community aid for enforcement purposes. In the Commission Report (1995) a partial analysis of the results obtained from the 1991–1995 co-financing programme is attempted. It notes that the increase in investment has enabled fisheries control to be intensified. For example, the report cites data on inspections at sea which indicate that there was a 30% increase in at-sea inspections and a 20% rise in inspection days at sea. It also notes that the number of infringements detected rose from 2393 in 1990 to 5092 in 1993. Nevertheless, the report concludes that increased inspections have not resulted in any large reductions in fraudulent practices which the report ascribes to commercial difficulties facing fleets. The overall conclusion reached in the Commission Report is that further developments are necessary and that these can only be brought about through ongoing and increased investment in this segment of the fisheries management. Such investment was also necessary due to the expanded regulatory framework within which fishing activities take place.

The new elements included, *inter alia*: the direct management of fishing effort, the link between the conservation policy and limiting catching capacity; the use of satellites and computer technology, licences and special fishing permits; and the adoption of measures applicable to the Mediterranean and as a result of enlargement of the Community to include Sweden and Finland. The review of co-financing which took place primarily in 1995 also provided an opportunity to review the enforcement cost burden between Member States.

2. *Identifying control priorities for financing*

The main task of the Commission is to ensure that the monitoring by Member States is effective. However, one must also point out that the function of the Commission is not limited solely to attaining this objective. The Commission also acts as an initiator of progress and policy and in this capacity it endeavours to ensure that Community financing for monitoring is clearly focused and defined. In its 1995 Report the Commission identified areas which should be the main focus of future investment:

- the acquisition of heavy equipment for air and sea patrols;
- developing integrated projects, i.e. satellite tracking, information and communication networks;

- training and the development of new control methods to deal with, for example, the enforcement of the structural aspects of the CFP; and
- special support for specific control measures whose implementation is beyond the scope and capacity of one single Member State, i.e. enforcement difficulties encountered in the tuna fishery.

The acquisition of heavy equipment could be justified on the basis of necessity and by virtue of the high costs involved in the purchase of aircraft and surface patrol vessels. Support for integrated projects would facilitate better co-operation and communication between national control services and harmonise the introduction of new technology. In its Report, the Commission also pointed out the need for the optimum utilisation of available resources which should be concentrated in areas of most concern such as quota uptake, fishing effort, adherence to structural reduction targets. Moreover, the Commission recommended that national control authorities develop appropriate operational strategies to facilitate the targeting of areas where non-compliance with Community rules is most widespread. In addition to identifying a set of broad control priorities, the Commission Report laid particular emphasis on the need to incorporate the notion of evaluation and planning into any future co-finance programme. More particularly, the Commission stressed the need to achieve a greater degree of cost effectiveness in enforcement notwithstanding the difficulties in attempting to assess efficiency.

3. *The Community Scheme to finance control*

On 8 December 1995, the Council adopted a Decision which provides the framework for co-financing the cost of enforcement,[21] the budget allocated amounted to €205 million over a five-year period (1996–31 December 2000). The decision identified the broad categories of expenditure:

- the acquisition or modernisation of inspection and control equipment;
- specific measures intended to improve the quality and effectiveness of monitoring of fishing and related activities.

Inspection equipment includes vessels, aircraft and land vehicles, systems to detect and record fishing activities and systems for recording and managing and transmitting data relating to control. Specific measures cover joint inspection programmes, the introduction of new technologies, the implementation of specific control programmes and the computerisation of catch and fishing effort data.

The Community contribution available to Member States towards the cost of enforcement now varies between 35 and 50% of the cost of the programmes and the projects submitted. In addition, Ireland is eligible for an annual subvention of €3 million to take account of specific operational control costs. The Control Decision lays down detailed rules on how applications for aid are presented and assessed by the Commission. Member States are obliged to submit a five-year inspection programme which identifies expected annual expenditure. In addition, Member States must submit annual reports on progress achieved with respect to targets and forecasts of the previous year. Further detailed rules on the application of the Council Decision are laid down in subsequent rules and are concerned with two aspects in particular: training of enforcement personnel, and the exchange of national enforcement personnel.[22]

4. *Implementing co-finance during the period 1996–1997*

Since the adoption of the 1995 Council Decision, Member States have submitted their overall expenditure programmes for the five-year period (1995–2000) and estimates of their expenditure in 1996 and 1997 (Table 9.2). With respect to the former, Member States estimate the total expenditure at approximately €373 million, or €75 million per year. Greece and Ireland anticipate expenditure of approximately €60–64 million per year, while Spain and Portugal both plan investments of approximately €53 million.[23] These figures exclude costs associated with the installation of satellite tracking systems on board Community vessels. The global figures above probably indicate the upper amount limits of anticipated expenditure as the applications for financial programmes submitted for both years 1996 and 1997 demonstrate.

In recognition of its particular enforcement responsibilities, Ireland received two rebates on operational costs in 1996 and 1997 of €2 457 911 and €2 996 381 respectively. Community contributions in 1996 and 1997 were €24 684 793 and €33 561 911 respectively.

The planned expenditure for the two-year period amounted to €187 648 000:[24]

Table 9.2 Anticipated expenditure on control by the Member States by the Member States on control in 1996–1997.

Member State	Expenditure 1996	Expenditure 1997
Belgium	13 545	336 000
Denmark	4 614 523	5 839 993
Germany	364 291	4 621 941
Greece	13 273 926	17 043 380
Spain	2 873 150	6 317 293
France	4 584 243	2 597 540
Ireland	1 136 784	2 153 465
Italy	4 071 894	6 675 175
Netherlands	638 832	705 306
Portugal	11 160 521	11 620 373
Finland	380 499	430 067
Sweden	2 732 568	916 971
United Kingdom	3 224 568	916 971
Total	49 069 344	60 174 475

Table 9.3 Total expenditure on control by the Member States and the Community in 1996–1997.

Expenditure by Member States	
1996	49 069 516
1997	74 877 687
Total Member State expenditure	123 947 203
Expenditure by the Commission	
1996	24 684 793
1997	33 561 911
Rebate to Ireland	
1996	2 457 911
1997	2 996 381
Total Commission expenditure	63 700 996
Total expenditure	187 648 199

2. *Procedure for obtaining financial aid for fishery control and inspection projects*

Member States wishing to benefit from a Community financial contribution during the period 1996–2000 were obliged to forward to the Commission an application for funding by 15 November 1995, and by 30 June each year thereafter.[25] Applications are required to be detailed and must specify, *inter alia*, the timetable for expenditure, the technical specification of the equipment, the planned use of the equipment and how it will improve control of fishing and related activities. Each of these elements needs to be supported by evidence which justifies the proposed expenditure and by comprehensive information on the national control effort. Interestingly, if the Commission considers that the facilities are not being used for the intended purpose or in accordance with the stipulated conditions then the Member State shall be obliged to conduct an administrative inquiry in which Commission officials shall participate.[26]

The Commission decides, on a year by year basis, the eligibility of Member States' applications for Community co-finance.[27] In effect, Member States' applications are closely evaluated and scrutinised by both Directorates A and C in the Directorate General For Fisheries (DG XIV), which is responsible for the general co-ordination of budgetary and financial aspects of the CFP, and Directorate General For Budgets (DG XIX). Applications must conform with the objectives of the control policy and the CFP in general. After the inter-service consultation and the views of the Management Committee Fisheries and Agriculture have been appraised, the Commission then decides on the amount of the Community contribution. As previously stated,

Community subvention has to fall within the ceiling of the Council decision, and must be in proportion to the agreed guidelines for the five-year period 1996–2000.

Payment is normally made after the Member State submits the appropriate invoices and on submission of a completed 'public tender questionnaire' which must make reference to the notices of public contracts published in the *Official Journal*. In the case of non-publication of the notices, then the beneficiary is obliged to certify that the contract has been awarded in conformity with Community legislation. Reimbursement, like all Community expenditure, is linked to compliance with the provisions of the Directive co-ordinating procedures for the award of public works and supply contracts.

To demonstrate transparency and to ensure that the best use is made of Community funds the Member State must also give assurances regarding the effectiveness of their inspection services and the use that is to be made of the particular equipment. In recent years the Commission has also given a 'Financial Monitoring Inspection team' a mandate to visit Member States and check the conformity of the equipment with the information disclosed by national administrations. This allows the Financial Auditor to evaluate and verify the correct application of the Community subvention. To date the findings of the Financial Audit Team have not been published.

In common with Community financial assistance in many other policy areas, there is considerable disquiet if it is perceived that certain Member States have received disproportionate financial aid from the Community budget.[28] It would appear, for this reason, that the issue of Community financial contributions to national control authorities, and indeed to national fishing industries has been the subject of numerous questions in the European Parliament.[29] This practice will no doubt continue and remain one of the inherent democratic safeguards in the financial aid process.

Section (iv) The cost-effectiveness of fishery enforcement

1. *Evaluation of the Community financial aid to fishery control and some comments on its cost effectiveness*[30]

The financial resources invested by the European Union and the Member States since 1978 in national surveillance and inspection services are considerable. It is difficult to analyse or to identify what is the direct return from such investment. In any case, expenditure must be balanced against not only the value of fish stocks protected, but conversely against the damage caused by over exploitation and poor resource management. With respect to the latter two, there is a conservative estimate that a least 10% of catches landed go undeclared,[31] and that mismanagement results in the annual loss of €3000 million in revenue to the Community. The Community is not unique when it comes to improper exploitation of marine resources, overfishing on a global level is estimated by the FAO to cost between €15 000 million to €30 000 million.

The net benefactors of sound fishery control and surveillance are not only fishermen but also the downstream and upstream sectors of the catching sector. As noted in the introduction to Chapter 2, these sectors are very often located in regions of the Community where there are no other viable alternatives for employment. Moreover, the evaluation of Community financial aid has to take account of the ecological impact, and wider implications for other policy areas, if the control and enforcement element of the

CFP is poorly implemented. This could have a detrimental effect on the political unity of the Community and further disintegration of the regional areas. It would tarnish the European Union image and its role in international fishery fora, which are responsible for the management of joint and straddling stocks, and high sea fisheries.[32] Moreover, the European Union's ability to successfully conclude agreements that allow Community vessels access to third-country fishery zones would be impaired if the Community fleet was perceived to be non law-abiding. Consequently, it is submitted that cost analysis cannot be limited to the readily apparent benefits to the Community fishery sector but has to be extended to take account of many other global issues.

The CFP has since its inception required a three-pronged approach in order to achieve the objectives of the policy. This entails political action, a regulatory framework and financial subvention. None of these elements, in isolation, is sufficient to achieve a satisfactory policy. In essence, it requires political agreement to achieve the regulatory framework which in turn depends on financial support from the Community for effective implementation. The size of the budget to achieve effective enforcement of the rules which underpin the CFP is insignificant if compared with the losses which are incurred by fraud in national fishery sectors or the waste in the mismanagement of the resource.

The Community only partially finances Member State projects and the greater part of expense is borne by national authorities. However, without the Community incentive few Member States would be prepared to bear the full cost of enforcement. Furthermore, as fisheries are a common resource which in theory is established on the basis of equal access to all resources for the fishing fleets of all Member States, then the cost of fishery enforcement must likewise be considered as a Community burden to be borne by all Member States. During the period 1996 to 1997, anticipated expenditure on the development of control was approximately €187 million. This compares favourably with expenditure by other maritime States.[33]

The issue of whether the potential cost to the Community budget would be much greater if the fishery enforcement and control function was exclusively performed by a centralised body or agency under the auspices of the Community institutions is not examined in great detail here. However, any such reorganisation might well entail major cost advantages and savings on the enforcement budget.

It should be borne in mind that the co-financing by the Community, of equipment, is no guarantee that the equipment is going to be used for the stated purpose of fishery control. The question of deployment will always remain within the sole competence of the Member State. Moreover, in some instances adequately equipped inspection services continue to be ineffective because they are deployed on other priority tasks. In other cases the Community financed equipment is only partly deployed for fishery control as Member States attempt to off-set start up and operational costs by deploying the equipment in a multi-purpose rôle. This is particularly true in respect of patrol ships and surveillance aircraft which are managed as part of a Member State's armed forces. In such cases, cost evaluation is very difficult to undertake because even if the units are deployed in a fishery protection rôle they may at the same time be undertaking other contingency tasks such as environmental control, search and rescue etc. The greatest abuse is when equipment is never, or rarely, deployed in the rôle for which co-finance was sought and granted. A typical example being the use of fishery surveillance aircraft for military or governmental transport purposes.

The penultimate issue to be discussed here is, even if the available resources are fully deployed in a fishery inspection and surveillance role, how can you measure their effectiveness? Firstly, it must be assumed that the mere presence of patrol vessels, surveillance aircraft, and fisheries inspectors will have a deterrent effect, and will thus reduce the number of irregularities committed by fishing vessels. This deterrent effect cannot be evaluated for cost-effectiveness. The standard indicators of effectiveness, relied on by the Commission from their control reports, are statistics which include *inter alia*: days at sea for patrol vessels; flying hours for surveillance aircraft and helicopters; number of fishing vessel inspected at sea or on shore etc.[34] These statistics constitute quantum indicators of activity and are not qualitative measures of effectiveness or efficiency. However, they form a base or data which when supplemented by other information, will provide a clearer picture of the overall effectiveness of the enforcement effort. The second element is the number of fishery offences or irregularities pursued by the judicial or administrative process. The third element is the number and level of sanctions invoked by the courts. These indicators when combined and presented together will give a reasonably clear picture of the rigour and standard of fishery control. This particular type of analysis is, however, flawed because the level of infringements observed will not only depend on the effectiveness of the control and inspection effort, but will also be in proportion to, or remain a function of, the number of offences being committed. The greatest value in this type of *activity versus efficiency* analysis is that it may provide a statistical base on which to plan the deployment of the limited inspection resources, thus increasing cost effectiveness and allowing the concentration of enforcement effort in the areas where it is most required.

In considering a Member State's enforcement performance regard should be had to the following considerations:[35]

- the prevention, discovery and pursuit of infringements against the conservation and control rules;
- the presence in national legislation and the application in practice of penalties that are commensurate with the seriousness of infringements and effectively discourage further infringements of the same kind;
- the reliability of catch figures forwarded by the Member State to the Commission and the Member State's ability to prevent overfishing of national quotas;
- the amount and the effectiveness of the human and material resources devoted by the Member States concerned to fisheries enforcement;
- the diversity of the fishing activity in the fishery zone of the said State;
- the degree of co-operation in fisheries enforcement between the Member State and other Member States and the Commission; and
- in appropriate cases, that Member State's contribution to fisheries enforcement in areas governed by international conventions to which the Community is a Contracting Party and the scale and effectiveness of this enforcement.

Conclusions

The Community scheme for financing the control component of the CFP is well established. Enforcement, despite being within the exclusive legal competence of

Member States, is now considered a Community matter. Financial aid is of increased importance since the scope of monitoring has been extended by the Control Regulation. The Council has emphasised this importance by giving a clear commitment to the financing of additional measures and projects which are required to be introduced forthwith. One of the themes advanced in this book is that enforcement issues and regulatory compliance are now the linchpins on which the success of the CFP rests. Indeed, if viewed from the law and economics viewpoint it is evident that the investment in fishery enforcement and the rules which regulate internal and external fisheries are essentially a means to maximise wealth. In this regard it may be argued that the cost of fishery enforcement is relatively modest and (is largely disproportionate to the other elements of the CFP budget) that increased investment in law enforcement is the best way to achieve the efficient allocation of resources. It is interesting to note that recent evidence presented to the House of Commons select Committee on Agriculture indicates that while the United Kingdom spent some €35 million on enforcement each year, the value of fish landed in the United Kingdom in 1997 was valued at €682 million and a further €226 million of fish was landed abroad by vessels flying the flag of the United Kingdom. As the European Union pursues an integrated approach to fishery enforcement there could be a substantial increase on the financial demands in the budget and a greater need for a proportionate share to be directed into this aspect of the CFP. Additional investment should be seen as money well spent, which may in the long term provide a safeguard for the future of the Community fishing industry.

References

1 This is evident from the discussion regarding the enforcement of Community provisions relating to the use of driftnets in Chapter 10, post. It is also the case with respect to the fishery agreements which the Community has negotiated with third countries, discussed in Chapter 1, *ante*.

2 See Chapter 11, *post*.

3 For further background information on the Community budget, see, *inter alios*, Shackleton, M., *Financing the European Community*, (1990); Kolte, 'The Community Budget: New principles for Finance, Expenditure, Planning and Budgetary Disciple' *CMLRev* **25** (1988), 487; Zangl, 'The Inter-institutional Agreement on the improvement of the Budgetary Procedure' *CMLRev* **26** (1989), 675. See also, *Report from the Commission to the Council and the European Parliament*, COM(95) 243 final, Brussels, 09.06.1995. The Commission was obliged to submit this report to the Council and the European Parliament before June 1995 on the application of Council Decision No. 89/631. The object of this report was to evaluate Community financial contribution towards expenditure incurred by Member States for the purpose of ensuring compliance with the common fisheries policy.

4 *OJ* L 359, 31.12.1998.

5 The Parliament has some influence over the Community budget, most notably it can reject the budget and did so in 1979, pursuant to **Article 203(8)** {272 (8)} of the EEC Treaty, see also Case 34/86, *Council* v. *Parliament* [1986] ECR 2155; [1986] 3 *CMLRev* 94.

6 Some of the issues raised were subject to evaluation in the course of the 1996–1997 Inter-Governmental Conference. The Conference was obliged to review, *inter alia*, 'arrangements relating to compulsory and non-compulsory expenditure ... in order to achieve inter-institutional cooperation on a partnership basis', see statement annexed to the Inter-institutional Agreement dated 29.10.1993 which deals with budgetary discipline and the

improvement of budgetary procedure, *OJ*, 07.12.1993. The October 1993 Agreement followed on from the previous Agreement which expired in 1992. It confirms the financial perspective framework approved by the Edinburgh European Council for the period 1993–1999. The Agreement also records that within the ceiling of Community expenditure, the 'own resources' contribution will only be allowed to rise to a maximum of 1.27% of overall GNP by 1999, thus ensuring a managed structured approach to Community finances over the said period.

7 European Parliament, Final adoption of the general budget for the European Union for the financial year 1994 (*OJ* L 34, 07.02.1994).

8 Discussed in Chapter 1, *ante*.

9 Source DG XIV, fact sheet, July 1994.

10 The EU–Moroccan agreement is, in monetary terms, the most expensive fishery agreement ever concluded by the European Union. In return for fishing opportunities for the Community fleet in waters under the jurisdiction and sovereignty of Morocco, the EU provides financial compensation of €505 million over a period of four years. See Protocol setting out fishing opportunities and the financial contribution attached to Council Regulation No. 150/97 of 12.12.1996 on the conclusion of an Agreement on co-operation in the sea fisheries sector between the European Community and the Kingdom of Morocco and laying down provisions for its implementation, *OJ* L 30, Vol. 40, 31.01.1997.

11 Holden, M., *The Common Fisheries Policy: Origin, Evaluation and Future*, p. 38.

12 *Ibid.*

13 See, Suthien, J.G., *Summary and Conclusions of the Workshop on Enforcement*, (OECD, Paris, 1994), p. 7.

14 Recent evidence to the House of Commons Select Committee on Agriculture indicates that the United Kingdom Fisheries Departments spend some €35 million on fisheries enforcement which broadly falls into the following categories: enforcement on land €9 million, surface surveillance €20 million; aerial surveillance €6 million.

15 Ireland, Denmark, (Greenland), Council Decision 78/640/EEC.

16 All Member States in 1987 (€22 million), Council Decision 87/178/EEC and 87/279/EEC; 1989 (€22 million), Report from the Commission to the Council and the European Parliament on monitoring and implementation of the CFP, SEC (92) 394 of 06.03.1992.

17 Council Decision of 27 November 1989, (89/631/EEC), *OJ* L 364/64, 14.12.1989.

18 See Chapter 11, *post*.

19 COM (95) 243, 09.06.1995.

20 Annex VII of the Hague Resolution is cited by the Court of Justice in Case C-4/96, *Northern Ireland Fish Producers' Organisation Ltd (NIFPO) and Northern Ireland Fishermen's Federation* v. *Department of Agriculture for Northern Ireland*, [1989] ECR I-0681, paragraph 4. See Chapter 1, *ante*.

21 Council Decision 95/527/EC, *OJ* L 301, 30, 14.12.1995.

22 Commission Decision 96/286/EC, *OJ* L 106/37, 11.04.1996.

23 Commission Report (1995), p. 65, *op. cit.*, fn 3.

24 Data based on Commission Decisions 96/229/EC of 17.04.1996, *OJ* L 114/35, 08.05.1996; Commission Decision 97/297/EC of 28.04.1997, *OJ* L 122/24, 14.05.1997.

25 Commission Decision 96/286/EC, **Article 4**, *op. cit.*, fn 22.

26 *Ibid.*, **Article 10**, see discussion on administrative inquiries in Chapter 6, *ante*.

27 *Ibid.*, **Article 6**.

28 Churchill, R.R., *EC Fisheries Law*, p. 140.

29 For example, see *inter alia*, written parliamentary questions (PQ): E-1672/93, *OJ* C219/29, 08.08.94, on Community funding for the Irish control authorities.

30 This subject is related to the 1992 Control Report, Chapter 2, *ante*, and a special report from

the Commission to Council on the Memorandum submitted by Ireland on the fisheries sector, DOC. SEC(93) 882 final.

31 See Chapter 2, *ante*.

32 See Chapter 6, *post*.

33 In the smaller States such as New Zealand the cost of enforcement is significantly less. In New Zealand the cost of fishery enforcement in 1992 was $14 million which was totally funded by the government see McClurg, T., *Two Fisheries Enforcement Paradigms, New Zealand before and after ITQs*, (OECD, Paris, 1992), 121–141.

34 Report from the Commission to the Council and the European Parliament on monitoring and implementation of the CFP, Annex III, SEC (92) 394 of 06.03.1992.

35 See Annex I, Council Decision of 27 November 1989, (89/631/EEC), *op. cit.*, fn 17.

Chapter 10
Enforcing Community Rules on the Use of Large-Scale Driftnets

Introduction

A driftnet is a wall of vertical net which is designed to enmesh, entrap or entangle fish by drifting on the surface of the sea or below the surface of the sea.[1] Driftnet fishing as a rule is targeted at species which are pelagic in behaviour, that is, fish that spend part of their adult life in the upper layers of the sea, and more often than not are highly migratory. The use of driftnets expanded in the 1960s and the 1970s with the availability of synthetic fibres for net manufacture along with other developments in fishing technology. By the 1980s there were large fleets of driftnet vessels operating in the North Pacific, South Pacific, Indian and North Atlantic Oceans and in the Mediterranean sea.[2] Driftnets are used to fish salmon in the Baltic Sea,[3] and there is also an artisan salmon fishery with driftnets in Ireland and the United Kingdom which has traditionally been limited to fishing inside the territorial waters. In the Mediterranean, driftnetting is chiefly targeted at swordfish and, to a lesser extent, albacore and other small types of tuna. The North-East Atlantic deep water driftnet fishery is predominantly for albacore which is a highly migratory species moving from the high seas through the exclusive zones of several Member States of the European Community. The significance of the European driftnet fishery has increased rapidly in recent years. For example, the number of vessels using driftnets involved in the North-East Atlantic fishery increased from zero vessels in 1986 to 90 in 1993.[4] During the same period the quantity of albacore landed by these vessels had risen to 4900 tonnes by 1993.[5]

Driftnet fishing is often considered to be a marine environment issue because it has been linked to the biological state of marine mammals.[6] Driftnet fishing has been challenged on the grounds that this type of gear entraps a substantial non-targeted catch, including marine mammals and sea birds. Entrapment occurs during actual fishing operations and by 'ghost fishing', that is by lost and discarded nets. Furthermore, driftnet fishing is perceived to be an indiscriminate and wasteful fishing method which threatens the effective conservation of living marine resources. Driftnets are also believed to constitute a danger to safe navigation.

This chapter examines the enforcement of the Community policy on the use of large-scale pelagic driftnets in the North-East Atlantic and Mediterranean high seas fisheries. The North-East Atlantic and Mediterranean high seas driftnet fisheries are selected as a case study because they offer a cogent example of the recent evolution in fishery enforcement policy in the European Community. In the North-East Atlantic and the Mediterranean high seas driftnet fisheries, one can identify a transition from the traditional *laissez faire* approach to enforcement issues to a more dynamic proactive approach. The issues raised in the driftnet fishery also provide a good illustration of the close inter-relationship between Community law and international law, albeit as it applies to the high seas area.[7] The problems that have been encountered in these fisheries have arisen relatively recently and particularly since 1992. It has, thus, been within the framework of the Control Regulation that the Community has tackled the issues that have arisen in the monitoring and controlling the activities of the European driftnet fleet.

This chapter is divided up into four sub-sections. Section (i) reviews the use of driftnets in the North-East Atlantic and Mediterranean high seas tuna and swordfish fisheries as well as examining the Community's international obligation to restrict this type of fishing. Section (ii) examines **Article 9a** of Council Regulation (EEC) No. 3094/86[8] which during the period 1992–1998 prohibits the use by Community vessels of driftnet over 2.5 km in length. This is followed by a discussion of the Court of Justice decisions in several fishery cases regarding the application of Community law to vessels operating on the high seas area. In Section (iii) there is a summary of the initiatives the Community has taken to uphold its international enforcement responsibilities and a brief analysis of some practical legal issues which have arisen in the Member States as a consequence of the implementation of Council Regulation (EEC) No. 3094/86. There is also a brief assessment of the extra-territorial effect of US trade and environmental legislation on the enforcement of Community rules on driftnets in the Mediterranean high seas area. Section (iv) reviews legislation adopted in December 1998, Council Regulation No. 1239/98, to limit the use of driftnets and to facilitate diversification and conversion of vessels using this type of gear. The chapter concludes, in Section (v), with some comments on the Community's policy in relation to the high seas driftnet fishery and the prospect for improved enforcement prior to the prohibition of driftnets as from 2001.

Section (i) Driftnets: the international issue

1. *The high seas fisheries*[9]

In accordance with **Article 86** of the LOS Convention the high seas are the parts of the sea that are not included in the exclusive economic zone, in the territorial sea or in the internal waters of a State, or in the archipelagic waters of an archipelagic State. A fundamental principle of the LOS Convention is freedom of the high seas for all States, whether coastal or land-locked.[10] The concomitant of this principle is freedom to navigate on the high seas - and its corollary, the freedom to fish there.[11] The Community's competence to regulate fisheries in the area referred to as the high seas is now well established and has been subject to considerable judicial examination in the Court of Justice[12] and is discussed below solely in the context of the enforcement issues that have been highlighted as a result of the volatile nature of the high seas fishery in the North-East Atlantic. This fishery has produced an intensive gear conflict between the traditional pole and line Spanish vessels and the French, British and Irish driftnet fleet, as well as being the focus of international environmental concerns. It also offers an illustrative example of the issues that arise and the obstacles to be overcome if Member States are to comply with **Article 2(2)** of the Control Regulation and uphold their obligation to monitor, outside the Community fishery zone, the activities of their vessels in cases where such control is required to ensure compliance with Community rules applicable in those waters.[13]

As the North-East Atlantic and Mediterranean drift-net fisheries and the Community's international obligations which stem from United Nations Resolutions 44/225, 45/147 and 46/215 have posed a significant implementation and enforcement challenge for the European Community, it is thus appropriate to commence with a brief description of international developments, and the legal obligations which stem from the United Nations General Assembly Resolutions.

2. *International developments including United Nations Resolutions 44/225, 45/147 and 46/215*

Community fishery law does not evolve in a vacuum. Indeed, the Community regulation on the use of driftnets has been heavily influenced by legislative norms in international legal instruments and conventions for the conservation and management of the marine environment. The first efforts to deal with driftnets were initially taken on a regional basis. In May 1989, the parties to the international Convention for the high seas fisheries of the North Pacific Ocean, Canada, Japan, and the United States agreed to place restrictions on the activities of the Japanese driftnet fleet.[14] In July 1989, the South Pacific Forum adopted the Tarawa Declaration which 'recognised' that the use of driftnets was not consistent with international legal requirements in relation to high seas fisheries conservation, management and environmental principles.[15] In November 1989, a 'Castries Declaration', in similar terms, was adopted by the Organisation of Eastern Caribbean States.[16] Recommendations for preventing the expansion of driftnetting continued to be made by a number of international fishery organisations, notably the International Commission for the Conservation of the Atlantic Tunas (ICCAT) which drew up a resolution on preventing deep sea fishing with large scale driftnets in the Atlantic.[17] In the same vein the Wellington Convention of November 1990 bans the use of nets more than 2.5 km long in the South Pacific.[18] The definition of large-scale driftnets as nets more than 2.5 km in length appears to have set a trend which has since been followed in State practice.[19]

In response to growing public awareness and regional developments, the General Assembly of the United Nations in 1990 at its 44th session took a position on 'Large-scale driftnet fishing and its impact on the living marine resources of the world's oceans and seas' when it presented Resolution 44/225 of 15 March 1990.[20] It recommended that there should be agreement on several measures, including: the imposition of moratoria on all large-scale pelagic driftnet fishing by 30 June 1992; immediate action to reduce progressively such activities in the South Pacific with a view to their cessation by 1 July 1991; an immediate prohibition of further expansion of such activities on the high seas.[21] The moratoria and the prohibition could be lifted if effective conservation and management measures were taken to protect the resources and the regions concerned. Resolution 44/225 was followed by Resolution 45/147 of March 1991 which expressed 'deep concern' about reported attempts to increase driftnet fishing in the Atlantic and called upon States to ensure compliance with the ban on further expansion.[22] Concern was also expressed by the re-flagging of some vessels to avoid controls being imposed on driftnetting.[23] A third Resolution 46/215 of 20 December 1991 recommended a substantial reduction of driftnetting (50%) in the short term, and the setting of a revised timetable by which the moratorium presented in Resolution 44/215 of 15 March 1990 would be implemented.[24]

The question of gill nets was also discussed at the international conference on responsible fishing which led to the Cancun Declaration, May 1992, and as item 21 of the Agenda at the UN Conference of the Environment and Development (UNCED) held in Rio de Janeiro in 1992.[25] The latter called upon States to fully implement Resolution 46/215.

While it is not proposed to analyse in detail the aforementioned international legal developments,[26] nonetheless, they call for a number of observations. First of all, the UN Resolutions are recommendations and actually go beyond the high seas provisions of the LOS Convention. However, the 1989 and 1991 General Assembly Resolutions on high seas and their acceptance (thus far) by high seas fishing states do provide strong confirmation that conservation measures on the high seas are considered to be at least formally obligatory.[27] One commentator has noted that the UN resolutions on high seas driftnet fishing have played a key role in the establishment of the moratorium as binding customary international law.[28] While the legal significance to be attached to 'recommendatory resolutions' of the General Assembly themselves remain controversial and unsettled, the submission with regard to the driftnet resolutions is not that they are in themselves legally binding, but that they are the written embodiment of a particular measure which is independently legally binding.[29] As such, the United Nations Resolutions are important in that they embody the written detailed formulation of the measure, adopted by consensus and implemented by the international community.[30] Furthermore, although the UN Resolutions are based on questionable premises and without any supporting data, they are widely considered as a significant step in the development in the law of the sea.[31] However, in recommending a moratorium on large scale pelagic driftnet fishing, the UN Resolutions failed to specify how this might be implemented.

In any case, neither the exact legal significance of UN Resolutions nor the issue of implementation posed particular difficulty in the European Union where events were influenced by several extraneous factors. One of these was media attention (mainly as a result of the initiatives of Non Governmental Organisations), which focused on the threat to marine mammals and generated public awareness of the European driftnet fishery. Another was the awareness at Community level that the prohibition of driftnet fishing in other global areas would result in the driftnet fishing fleets moving to the Atlantic if this ocean remained unregulated.[32] This, combined with the external obligation to uphold an international commitment prompted the Community to take action.[33] One of the first initiatives was the decision taken by the Standing Technical Committee Fisheries (STCF) at its meeting in October 1990[34] to define large driftnets as nets more than 1 km in length.[35] At the Council meeting of Fisheries Ministers in December 1991, the Commission proposed limiting the length of driftnets used by Community vessel to 2.5 km in accordance with the Resolutions adopted by the General Assembly of the UN. The European Parliament endorsed the Commission proposal and issued a favourable opinion.[36] After lengthy discussions at the Council a compromise was reached whereby certain category of vessels were allowed to continue using driftnets up to 5 km.[37]

Section (ii) The law in the European Union during the period 1992–1998

1. *Article 9a of Council Regulation (EEC) No. 3094/86*[38]

Article 9a of Council Regulation (EEC) No. 3094/86[39] (the Technical Measures Regulation) which was introduced by **Article 1** paragraph 8 of Council Regulation No. 345/92 establishes the conditions under which Community fishing vessels may utilise driftnets during the period 1992–1998.[40] The reasons which justify the adoption of the prohibition are set out in the recital, which notes the following:

- on 22 December 1989 the General Assembly of the United Nations adopted Resolution 44/225 on large-scale pelagic driftnet fishing and its impact on the living marine resources of the world;
- the use of such nets has been the subject of discussions and resolutions in various international fora;
- the Council approved the Convention on the conservation of European wildlife and natural habitats (Berne Convention) by Council Decision 82/72/EEC;[41]
- the Community had signed the United Nations Convention on the Law of the Sea which requires all Members of the international community to cooperate in the conservation and management of the living resources of the high seas;
- the uncontrolled expansion and growth of driftnetting may entail serious disadvantages in terms of increased fishing effort and increased by-catches of species other than the target species;
- the Community should consider the interest aroused at international level and the concern expressed by ecological organisations and many fishermen including those of the Community;
- the provisions should be made for phases of adjustment for fishermen who are economically dependent on the use of driftnets, whilst limiting and analysing the ecological impact of corresponding fishing;
- the decision relating to driftnets should be capable of adjustment, where effective conservation and management measures are adopted on the basis of reliable statistical analyses;
- whereas management measures concerning fishing in the Baltic Sea should be adopted within the International Baltic Sea Fishery Commission.

As from 1 June 1992, **Article 9a** of Council Regulation No. 3094/86 prohibits Community vessels from keeping on board or using driftnets, the total length of which exceeds 2.5 km.[42]

A derogation was granted up to 31 December 1993 for certain categories of vessels which had fished for long-finned albacore tuna with driftnets in the North-East Atlantic during at least two years immediately preceding the coming into force of the regulation. In the event that the derogation was not extended the restriction of driftnets to 2.5 km would then apply to all Community vessels.[43]

There is also the requirement that driftnets longer than 1 km should remain attached to the vessels when fishing takes place outside the 12-mile coastal zone band.[44]

The scope of the regulation extends to all waters except the Baltic Sea, the Belts and the Sound under the sovereignty or jurisdiction of a Member State, and outside those waters, to all fishing vessels flying the flag of a Member State or registered in a Member State.[45]

2. *Commentary on Article 9a of Council Regulation (EEC) No. 3094/86*

Article 9a of Council Regulation (EEC) No. 3094/86 does not define a driftnet.[46] However, the length of net to which the regulation applies is the same as that referred to in the Wellington Convention.[47] This length appears to be generally accepted in State practice as the standard which defines 'large-scale' in terms of driftnets.[48] Indeed in *Etablissements Armand Mondiet SA* v. *Armement Islais SARL.* the Court of Justice rejected the argument that large-scale driftnets were nets longer than 40–50 km and pointed out that although the UN Resolution 44/225 did mention nets longer than 50 km, it did not give any definition of driftnets and contained no element warranting the conclusion that nets longer than 2.5 km were excluded from its scope.[49]

The purpose of the derogation which allowed certain categories of vessels to fish with nets up to 5 km in length up to 31 December 1993 was to allow a transitional period for vessels which had fished albacore with driftnets for a period of at least two years preceding the adoption of Council Regulation No. 345/92. This was availed of by 37 French vessels which fulfilled certain required criteria. Vessels flying the flag of the United Kingdom and Ireland did not benefit from the derogation because they only entered the fishery during the 1991 season. The derogation was not extended in 1994 because of the lack of scientific evidence showing the absence of any ecological risk as a consequence of using this type of gear.[50] Unusually, there appears to be a reversal of the burden of proof in this provision, as traditionally, the burden was on the scientific Community to demonstrate that damage is being done to a stock before restrictions are imposed on the industry. This provision in effect clearly indicates that unless established otherwise, driftnets of more than 2.5 km are considered harmful. This measure is thus an application of the precautionary principle.[51]

The requirement for vessels to remain attached to the nets would appear to be based on the original need to restrict vessels from fishing with more than one net and to reduce the threat from unattended nets to the safe navigation of other vessels. The condition of remaining attached to the net appears to have originated from the Mediterranean fishery where sea and weather conditions make compliance with this condition easier. In the North Atlantic, a number of reasons, including safety reasons, have been advanced as to why vessels may in some instances experience difficulty in complying with this requirement.[52] The requirement in the derogation for the submersion of the headlines of nets was also an effort to reduce the risk to navigation in the Mediterranean from this type of gear.[53]

3. *The Baltic Sea*

Driftnets in the Baltic do not come within the scope of **Article 9a** of Council Regulation No. 3094/86 (as consolidated in Council Regulation (EC) No. 88/98) and continue to be regulated by the Baltic Technical Regulation which allows vessels to have up to 21 km of driftnet on board.[54] There were several reasons why the Baltic Sea *et al.* were excluded from the general area of application of **Article 9a**. Ostensibly, it would appear that the 'Baltic Sea' is not considered 'high seas' as defined by UN Resolution No. 44/225 and that the resolution does not apply in areas where 'effective conservation and management measures' were taken on a regional basis. (The requirement of introducing conservation measures, however, is no longer contemplated by UN Resolution No. 46/215 which called for a total driftnet moratorium.) Conservation measures had been taken by the International Baltic Sea Fisheries Commission and implemented in Community law.[55] The majority of the members of the Baltic Sea Fishery Commission, however, are not members of the European Union and remain outside the remit of Community law and unaffected by the global move towards a driftnet moratorium. It could also be argued that the ecological risk presented by driftnets in the Baltic is less than that elsewhere.[56] Moreover, driftnetting in the Baltic is for the anadromous species, salmon, which is harvested in the coastal zone within national jurisdiction and consequently is less difficult to manage and control by national authorities.

The issue of the derogation in relation to the Baltic withstood the scrutiny of the Court of Justice in *Etablissements Armand Mondiet SA* v. *Armement Islais SARL*.[57] which is discussed in some detail below in the context of the *ratione territoriae and ratione materiae* of Community fishery law. One of the questions which the Court of Justice had to address when reviewing the validity of the Community measures on driftnets concerned the apparent difference in treatment between fishermen in the Baltic and those in other zones/areas.[58] The Court noted, with characteristic brevity, that the Community does not possess competence to regulate those seas and that the situation could therefore not be compared and discrimination could not occur.[59]

Article 9a of Council Regulation No. 3094/86 may be distinguished from the United Nations Resolutions by the expansive scope of the Community regulation which is not limited to the high seas of the world's oceans but also extends the prohibition to waters under Member State jurisdiction or sovereignty. This is important because the albacore and swordfish are vulnerable to harvesting in both the exclusive economic zone or the exclusive fisheries zone (EEZ/EFZ) and the high seas.[60] From a practical management and enforcement perspective, the limiting of the ban to the high sea areas *per se* would be ineffective if Community fishermen were allowed to fish with unlimited gear within the Community zone. If the Community has an interest in a particular conservation purpose in the high seas then it goes without question that this interest must also extend to the Community zone.[61] This is generally referred to as the 'principle of consistency', which requires that the measures that are applicable in the high seas area be consistent with those in the coastal state in the exclusive economic/exclusive fishery zone. It should be noted, however, that the UN Resolutions were essentially targeted at fishing practices in the Pacific which differ substantially from those in the North-East Atlantic and the Mediterranean.[62] Furthermore, virtually all the States concerned (mainly Pacific Rim) allow the use of driftnets and other fishing gear inside their national jurisdiction where the incidental catch rates of mammals and other fisheries species are equal to or, in many instances, higher than those in high seas driftnet fisheries.[63] This is one of the arguments frequently cited by proponents for the continuation of a limited driftnet fishery in the Atlantic.

4. *Does the scope of EC fisheries law extend to the high seas driftnet fishery in the North-East Atlantic and Mediterranean?*

Article 227(1) {299} of the EC Treaty states that the Treaty applies to a list of Member States, but does not offer any indication if the EC Treaty applies to maritime zones beyond the territorial sea. In determining the precise scope of Community powers the Court of Justice has tended to adopt a purposeful, rather than a literal, approach.[64] Nevertheless, several writers have expressed opposing views on the geographical scope of Community fishery law.[65] One view is that the EC Treaty has the same geographical scope as the jurisdiction of Member States in relation to the differing subject matters of the Treaty, unless the scope is expressly restricted. The Court confirmed this approach in the *Kramer* case[66] and stated that it should be made clear that although **Article 5** of the original regulation laying down a common structural policy for the fishing industry (Regulation 2141/70) was applicable only to a geographical limited area (the territorial sea and exclusive fishery zone):

> 'It none the less follows from **Article 102** of the Act of Accession, from **Article 1** of the said regulation and moreover from the very nature of things that the rule-making authority of the Community *ratione materiae* also extends - in so far as Member States have similar authority under public international law - to fishing on the high seas.'[67]

In *Commission* v. *Ireland*[68] the Court pointed out that the Community's powers are exercised primarily in, and in respect of, the territories to which the Treaties apply and that institutional acts adopted on the basis of the Treaties apply in principle to the same geographical areas as the Treaties themselves.[69] The Court did not have to address the issue whether this extended to the high seas area, but nonetheless, set down a clear guideline regarding the criteria which it used in assessing the field of application of Community regulation. The Court stated: '... Regulation ... must therefore be understood as referring to the limits of the field of application of Community law in its entirety, as that field may at any given time be constituted'.[70] Furthermore, as Advocate General Darmon pointed out in *Commission* v. *Spain*,[71] the judgement in *Commission* v. *Ireland*[72] 'clearly confirms the existence of parallelism in this area between internal and external powers, and does not in any event offer any support for the view that the Community has no independent authority to restrict fishing activities on the high seas'.[73] In drawing this conclusion he remarked that:

> 'although in practice States grant "*de facto* freedom to the individual operator" by not laying down rules in respect of conservation of stocks on the high seas. That observation, however ... does not in any way challenge the fundamental principle that the State is empowered, from the point of view of public international law, to impose restrictions on catches on the high seas.'[74]

The Court in *Commission* v. *Spain*[75] supported the view that the rule-making authority of the Community extended, to control of fishing on the high seas, in so far as the Member States had similar authority under international law. The Court cited the judgement in *Commission* v. *Ireland*,[76] to illustrate that there was no support whatsoever to uphold the argument that the only measures which the Community may take with regard to catches in waters outside the sovereignty or within the jurisdiction of the Member States of the EEC are those negotiated in international agreements and the adoption of measures designed to implement those agreements.[77]

From the enforcement perspective, the Court noted that because stocks move naturally inside or outside the demarcation line of the Community fisheries zone there is little doubt that the conservation policy would be undermined if it only applied to the Community zone.[78] Accordingly, the Community was entitled to require Member States to apply the control measures laid down by Council Regulations No. 2057/82 and 2241/87 (the 1982 and 1987 Control Regulations) to catches outside the Community fishing zone to catches subject to TAC or quota. As pointed out by one leading authority it may be deduced from the case law that, essentially the Community's competence in this matter is co-extensive with the legislative (or prescriptive) jurisdiction which its Member States have in fishery matters under international law.[79]

In *Anklagemydigheden* v. *Poulsen and Diva Navigation Corp.*[80] which concerned a ruling on a technical measure for the conservation of salmon (not **Article 9a** of Council Regulation (EEC) No. 3094/86), the Court held that Community legislation in respect of fish caught in certain areas not under the sovereignty or jurisdiction of the Member States (i.e. the high seas) may not be applied to vessels on the high seas registered in a non-member country, since in principle such vessels are only governed by its flag state.[81] The Court also held that the provisions could not be applied to a vessel sailing or on passage in the Exclusive Economic Zone of a Member State or crossing the territorial waters of a Member State in so far as the vessel is exercising the right of innocent passage and freedom of navigation in those areas.[82] Conversely, it may in principle be applied to such a vessel in the inland waters or in a port of a Member State.[83] It is implicit in the judgement that the Community enjoys, in respect to Member State vessels, the same legislative competence in respect of the high seas as afforded by international law to the flag state or the state of registration.[84]

The Community competence to regulate the high seas driftnet fishery was the subject of several preliminary questions referred to the Court of Justice by the Commercial Tribunal of La Roche-sur-Yon in the recent case *Etablissements Armand Mondiet SA* v. *Armement Islais SARL.*[85] This case resulted from a commercial dispute over the payment for two driftnets ordered by the respondent in August 1991. Allegedly, the reason cited for the cancellation of the said order was the adoption of Council Regulation No. 345/92 which limited the use of driftnets to 2.5 km. Thus, the repondent argued that he had no use for the nets. However, following the cancellation of the order, the first party *Mondiet*, brought a case before the Commercial Tribunal seeking payment for the driftnets. The Tribunal held that the adoption of Regulation No. 345/92 might constitute a case of *force majeure* exonerating the respondent from fulfilling his contractual obligation. As the resolution of the issue concerned the validity of the regulation the Tribunal referred several questions to the Court of Justice. There were 15 questions in total, the majority of which addressed the extent to which the Council was free to exercise its discretion in carrying out its legislative tasks. The first and second questions go to the heart of Community law competence to regulate for the high seas and indeed they offered an opportunity for the Court to clarify the geographical scope of the common fisheries policy which has been the subject of debate for some time. These questions were:

(1) 'Can Regulation No. 345/92 of 27 January 1992 amending for the eleventh time Regulation No. 3094/86 laying down certain technical measures for the conservation of fishery resources impose restrictions on the subject of the EC as regards freedom to fish on the high seas?'
(2) 'If the answer is no, can the Regulation prohibit, in areas under the jurisdiction of the Member States (exclusive economic zones and territorial seas), the keeping on board of driftnets of a certain length if that prohibition makes it impossible to fish with driftnets on the high seas?'

The Court, referring to its jurisprudence, decided that, in respect of the high seas, the Community has in matters relevant to its task the same legislative competence as afforded by international law to the flag State or to the States of registration.[86] In respect of fishing, that competence was established by the Geneva Convention of 29 April 1958 on Fishing and Conservation of the Living Resources of the High Sea which on this point codifies customary international law, and by LOS Convention of 10 December 1982.[87] The Court noted that although the latter had not entered into force (at the time), a large part of its provisions were considered as an expression of customary international law. **Article 6** of the Geneva Convention of 29 April 1958 recognised the interests of riparian states in the biological resources in the part of the high seas adjacent in the waters under their jurisdiction. Among others, **Articles 117** and **118** of LOS Convention obliged all members of the international community to co-operate in the conservation and management of the biological resources in the high seas.[88] In the Court's opinion this followed from the Community's competence to adopt measures for the conservation of fish stocks in the high seas in respect of a vessel flying the flag of a Member State or registered in a Member State.[89] As the answer to the first question was in the affirmative then *ipso facto* the Court was not obliged to assess the validity of the Community prohibition on the keeping on board of driftnets of over 2.5 km, in areas under the jurisdiction of the Member States (exclusive economic zones and territorial seas), this requirement was upheld.

The decision in *Etablissements Armand Mondiet SA* v. *Armement Islais SARL.* is important for several reasons. In the first instance, the Court's answers clarified the geographical scope (*de ratione loci*) of the CFP, particularly as **Article 227** {299} of the EC Treaty offers little guidance on this subject.[90] The decision also upheld the functional scope (*ratione materiae*) of the Community by reiterating the established principle that the utilisation and conservation of resources is within the exclusive competence of the Community.

The Court held that, *in matters relevant to its task*, the Community enjoys the same legislative competence in respect of the high seas as afforded by international law to the flag State or the State of registration. It may be thus deduced that the requirement in **Article 2(2)** of the Control Regulation for each Member State to monitor the activities of its vessels outside the Community fishery zone, and to ensure compliance with rules applicable in those waters is legally valid. It has been suggested that the Court's decision applies not only to the Community's competence in the sphere of fisheries, but equally governs its environmental international competences.[91] In other words, the Community is empowered to regulate such matters as dumping and oil pollution of the high seas.[92]

The second issue raised by this decision is the extent to which the Council is free to exercise its discretion in carrying out its legislative task. This issue, which is fundamental to the legal order of the Community, must be understood in the context of Treaty of European Union which has elevated the principle of subsidiarity to one of Community law.[93] If, for example, there is a challenge on some aspect of the legislative framework introduced by the Control Regulation on the grounds that the measures were not necessary by virtue of their scale and effects or that they offended the principle of proportionality as expressed in the second paragraph of **Article 3B** {5} of the EC Treaty, then, *Etablissements Armand Mondiet SA* v. *Armement Islais SARL.* may offer some useful guidance in anticipating how the Court will deal with such issues. The Court adhered to its previous case law when dealing with this particular issue and limited its enquiry to the question as to whether the impugned Regulation was the result of a manifest error or *détournement de pouvoir* or whether the authorities had manifestly exceeded their discretionary powers.[94] This approach which the Court had previously adopted in *Fedesa*[95] allowed it to conclude in the present case, that there was no reason to hold that the Council had committed a manifest error or exceeded its powers by limiting the applicability of the derogation to vessels which had fished for tuna with driftnets in the North-East Atlantic for at least two years prior to the coming into force of the Regulation.[96]

But the question arises, what if the Council of Ministers in exercising its discretion in carrying out its legislative task had, for example, extended the powers of Commission inspectors to carry out direct inspection on Member State vessels outside the Community zone? First of all, such a radical move would require a clear legislative base. There is little doubt that any proposal along these lines would be challenged on the grounds that the Council was exceeding its competence. There would also be the argument that such measures would be in breach of the principles of subsidiarity and proportionality.

Before examining the enforcement issues, the answer to the question which was posed at the start of this chapter – does Community fisheries law apply to the high seas? – is that the Court of Justice has unequivocally confirmed that in maritime zones outside Member State sovereignty or jurisdiction the Community has, in matters relating to the competences attributed to it, the same legislative powers at international level as those accorded to the flag State, or State of registration of a vessel.[97] In particular, in respect of fisheries, the Community has competence to adopt measures which apply to activities within the exclusive economic zone, the territorial sea, internal waters, and ports of the Member States,[98] and in so far as the Member States have similar authority under international law, to activities in the high seas.[99] The Community's competence must be exercised in accordance with the applicable rules of international law.[100]

2. *The specific problem of high seas fishing in the Mediterranean Sea*[101]

The geographical nature of the Mediterranean as a semi-enclosed sea connected by the narrow outlet of the Strait of Gibraltar and to the Black Sea by the Turkish Straits, surrounded by 22 countries with different territorial sea and coastal zone claims, and the absence of EEZs makes high seas fishing in the Mediterranean Sea complex and multi-faceted.[102] The provisions of Council Regulation No. 345/92 are applicable in the Mediterranean Sea, indeed UN Resolutions 44/225 and 45/197 specifically mention that the global moratorium on all large-scale pelagic driftnet fishing is fully implemented on the high seas of the world's oceans and seas, including *enclosed seas and semi-enclosed seas*. Driftnets are used for harvesting swordfish, albacore and other smaller types of tuna. There is a large Italian driftnet fleet located along the western and southern coasts of Sicily, Calabria and Campansia.[103] Fishing for swordfish with driftnets usually takes place from April to August while albacore fishing is conducted in the Spring and Autumn. Although catch data are incomplete, annual catches of swordfish using driftnets averages about 4000 tonnes while the albacore amount to approximately 700 tonnes.[104] The Italian fleet was estimated to consist of 682 vessels in 1990 but is reported to have decreased in size since then.[105]

Italian national legislation regulating the use of driftnets which precedes the adoption of Council Regulation No. 3094/86 and has been the subject of protracted legal proceedings initiated by ecological groups and fishermen's associations.[106] As noted by one commentator: 'The vicissitudes of the Italian legislation on driftnets are far from being a model of consistency. The Ministry and Parliament were probably in quite a difficult position, trying to strike a balance between the opposite pressures exerted by fishermen, environmentalists, administrative courts, and the public treasury. Finally, the EC acted as a *deus ex machina* in preventing further domestic trouble.'[107] The Italian driftnet problem is exacerbated by its scale (3,500 fishermen) and the fact that Italy, in common with the other Member States which border the Mediterranean, has no EEZ/EFZ. Thus there is the belief in Italy that an indiscriminate prohibition of driftnets would immediately cause Italian fishermen to be replaced by foreign fishermen, free from any obligation whatsoever.[108] In 1998, Spain instituted a 50-nautical-mile enforcement zone off its eastern Mediterranean coast. This initiative was designed to deter Italian fishing vessels from using driftnets in this zone. Spain's extension of jurisdiction only pertained to enforcement and does not constitute a claim to exclusive jurisdiction.

Section (iii) Enforcing the driftnet regulation

Introduction

This section examines the enforcement of **Article 9a** of Council Regulation No. 3094/86 during the fishing seasons in 1992–1996. It is followed by a brief analysis of the rôle of the Commission and the Member States during this period. Finally, the effect of extraterritorial United States environmental and trade legislation on the Community enforcement policy in the Mediterranean is assessed.

1. *Enforcement of Community legislation concerning the use of driftnets in the North-East Atlantic in 1992 and 1993*

Because of the heightened public awareness of the environmental issues at stake in the driftnet fishery and the need to ensure that the Community observed its obligations pursuant to the United Nations resolutions, the Commission focused on the activities of driftnet fleets and the requirement of Member States to enforce **Article 9a** of Council Regulation No. 3094/86. In particular the Commission asked the Member States to inform it of the specific control measures they intended introducing for this fishery. The Commission report on enforcement in the 1992 records that there were no replies.[109]

In 1993 Community inspectors carried out special inspection visits at sea and on land.[110] Spain was the only Member State to conduct a large-scale operation at sea including the provision of technical and medical assistance to the Spanish fleet. The Commission was able to conclude from the available reports in 1993 that there had been serious breaches of the rules and that the measures taken by Member States with fleets practising driftnetting on the high seas were largely inadequate. In effect, there were a large number of vessels violating the Community regulation by using nets far in excess of 2.5 km. Furthermore, the Commission concluded that the scale of irregularities revealed the inadequacy of the Member States' control measures.[111]

It was apparent by the end of the 1993 season that the remoteness of the fishing grounds made non-compliance easier and it was clearly evident that the Community system fell short of measures undertaken elsewhere to control oceanic driftnet fishing. In particular, the Commission noted that Japan, Korea and Taiwan had shown the way by utilising the appropriate technologies (monitoring by satellite), combined with management of authorised zones and controls at sea, supported by a licence regime and sanctions.[112] The Commission was able to conclude by the beginning of 1994 that there was an urgent requirement for a vigorous policy including the immediate and effective controls coupled with the complete prohibition of driftnets in the long run.[113]

2. *Enforcement of Community legislation concerning the use of driftnets in the North-East Atlantic in 1994*

Largely as a result of the experience gained in 1992 and 1993, the Commission, using the powers conferred by **Article 29(3)** of the Control Regulation required the Member States to provide detailed inspection plans for the 1994 season.[114] However, it was not until after events on the fishing grounds and an emergency meeting of National Control Experts on 29 July 1994 that all Member States subscribed to a clear interpretation of **Article 9a** of Council Regulation No. 3094/86 and gave a firm commitment to implement a vigorous control strategy.[115]

In retrospect, it is easy to conclude that the absence of adequate enforcement in the driftnet fishery was ultimately going to lead to problems on the fishing grounds. Early in the 1994 fishing season there were several minor incidents between the traditional pole and line fleet and driftnet vessels. It became apparent that there was an inherent risk, in the absence of authorised patrol vessels, that non driftnet vessels would attempt to apply Community law to driftnet vessels. This resulted in a crisis in the North-East Atlantic fishery which reached a hiatus in July 1994.[116] A meeting of control experts from the Member States was convened in Brussels on 29 July 1994.[117] At this meeting Member States gave firm commitments to maintain patrol vessels in the albacore fishery and to exchange information on a frequent basis regarding control activities undertaken.[118] All Member States gave a commitment to enforce strictly the maximum of 2.5 km of driftnet per vessel and agreed that reserve netting should not be carried on board fishing vessels.[119] It was also agreed that Commission inspectors would maintain an active presence on board patrol vessels of all Member States to ensure that all parties complied with their responsibilities.[120]

In 1994 it was fairly evident that the absence of an active enforcement presence at sea had contributed to the increase in tension between the respective fleets. This created a negative image for the CFP both in the eyes of the public and in third countries (many of which have fishery agreements with the EU).[121] During 1994 there was little inspection activity in the Mediterranean. In the Atlantic the 1994 albacore fishery revealed that effective control required a firm political commitment on the part of the Member States to apply the legal provisions and to invoke the full range of sanctions in the event of offence being detected. It also demonstrated that effective, albeit expensive, control and enforcement at sea could be attained even if the distance of the fishery from the Member States compounds the enforcement task.

3. *Enforcement of Community legislation concerning the use of driftnets in the North-East Atlantic in 1995–1996*

The 1995–1996 Commission Reports on the Enforcement of Community Legislation concerning the use of driftnets was able to note the main improvement attained in 1995–1996 compared to earlier years, these were stricter enforcement at sea and in port carried out by national control services, the absence of conflict at sea between the various fleets, a satisfactory level of co-operation between the Commission and the Member States, the common agreement on and understanding of the provisions of **Article 9a** of Council Regulation (EEC) No. 3094/86 and a reduction in the number of apparent infringements of the said article.[122]

A major innovation in terms of control in 1995 was the chartering by the Commission of a patrol vessel for use as an inspection platform. The vessel was deployed in the North-East Atlantic during the months of June to August and subsequently in the Mediterranean Sea. It was utilised by Community fisheries inspectors and those drawn from Member States (Spain, the United Kingdom, Italy, Ireland (observer role only) and Greece) as an inspection platform.[123] With respect to the role of the Commission vessel, Member States were invited to deploy inspectors on board in order that they could monitor and inspect the activities of their vessels at sea. The Community inspectors reported on general fleet activity and observed the inspections carried out in the presence of national inspectors. During the course of the season, regular meetings were carried out on board the Community vessels between the commanders of national patrol vessels. These meetings helped to diffuse tension on the fishing grounds and facilitated the sharing of information on the location and activities of the various fleets.

The Commission Report on the 1995 fishery records improved and increased levels of inspections conducted both in port and at sea, these revealed that the compliance level by the various fleets was almost universal.[124] Interestingly, only one fishing vessel was requested to return to port and cited for an apparent infringement of Community Regulations. The high level of compliance observed at sea was accredited to the more comprehensive and rigorous port inspections, the continuous presence of patrol vessels on the fishing grounds and the satisfactory level of cooperation between the Member States and the Commission. The only lacuna observed was the absence of enforcement of the rules with respect to keeping nets attached to fishing vessels. The Commission Services identified that the reluctance to enforce this particular requirement was due to safety concerns both for individual vessels and fleet generally. The Commission was also of the opinion that the presence of patrol vessels from different Member States helped to convince the various fleets that control was being exercised in a comprehensive and non-discriminatory manner and that apparent breaches of Community law were being investigated and dealt with.

The 1995/1996 Reports note that the costs incurred by the Member States throughout the season were considerable and disproportionate given the level of participation by fishing vessels using driftnets and the economic value of the quantities landed.[125] In order to reduce costs and increase efficiency, the Commission advocated that Member States give careful consideration to undertaking joint control efforts on the high seas and authorise their national inspectors to conduct inspections on their own vessels from inspection vessels of other Member States.

4. *Enforcement of Community legislation concerning the use of driftnets in the Mediterranean in 1995 and 1996*

The driftnet fishery in the Mediterranean differs in at least two respects from the one conducted in the North Atlantic: the control and enforcement challenge are different as are the biological considerations. The control challenge arises from the scale of fishing activity (± 600 vessels) and because Italian vessels operate in both the territorial sea and on the high seas.

The Commission Report on control in the driftnet fishery in 1995 records that the control regime put in place by national authorities in the Mediterranean remained insufficient and did not deter fishermen from using nets of an illegal length.[126] In reality, prior to 1996, there was little enforcement effort of significance mounted in the international zone. Nevertheless, the entry into force of the Community measures and the appearance of a Community patrol vessel in the Mediterranean in 1995 and 1996 improved matters.[127] The inspection vessel provided an opportunity for EC inspectors to monitor the activities of national inspectors from Italy and Greece which were on board. These patrols also provided the first opportunity for joint patrols with, and between, Member States and the Commission Inspectors, in the Mediterranean Sea.

Despite these Community initiatives, there were several allegations from non-governmental environmental and conservation groups regarding the impact of driftnet fishing in the Mediterranean and the absence of a credible control and enforcement system.[128] The impetus for an improved control regime in the Mediterranean Sea came not only from Community legislation and the activities of the Non-Governmental Organisations but also from legislative action and environmental policy developments in the United States.[129]

5. *The rôle of the Commission in enforcement*

Although limited from the implementation perspective the Commission's rôle in the scheme of enforcement of the driftnet fishery has been considerable and has involved a series of diplomatic, administrative and legislative actions. It is to the Commission that the Member States submit their detailed inspection plans. It is Commission officials who chair the meetings with national control experts who discuss the implementation and enforcement of Community legislation. The Commission's Legal Service advise the services of the Commission on the validity of Member States' interpretation and application of Community legislation. In the event of a crisis, it was the Commission which brokered a solution and mediated between the parties in dispute.[130] It is the Commission which may exclusively propose legislation which may amend Community law in respect to driftnets and it is the Commission who prepares and submits the monitoring and control reports to the Council of Ministers and the European Parliament.

Ultimately, the most serious action that the Commission can take, if the Member States have failed to fulfil their obligations, is to initiate an **Article 169** {226} action against a Member State.[131] Here the Commission plays a key rôle in the administrative stage of proceedings. The presence of EC inspectors on monitoring missions in the Member States lends itself to the informal phase of an **Article 169** {226} investigation when the Commission examines a possible breach of a legal duty, such as the failure to comply with Council Regulation No. 345/92. Inspectors' reports and observations from their missions allow the Commission to consider whether there is sufficient evidence to justify the commencement of formal proceedings.[132] Standard practice requires that this informal investigation be conducted with discretion. Indeed, the 29 July meeting in 1994 of national control experts is typical of the pivotal rôle the Commission takes after it has acquired evidence that Member States were not enforcing the restriction on the use of driftnets. In this instance, the Commission and Germany in their capacity as President of the European Union made a major effort to ascertain the facts and exert pressure on the Member States to comply with their obligations. The commencement of **Article 169** {226} EC Treaty proceedings against Member States for failing to mount an effective control campaign may have contributed to greater readiness to enforce Community rules. As Advocate General Roemer once commented in relation to **Article 169** {226} proceedings,[133] 'this procedure puts the Member States' prestige at issue: no one likes to be accused of having broken the law.'

One can conclude that the rôle of the Commission in the Community's driftnet policy is one in which certain elements of a federal approach are most apparent. It gives expression to the Community interest and a global environmental concern. The success of the Commission's rôle in achieving the uniform implementation of **Article 9a** Council Regulation No. 3094/86 is attributable to a number of factors which include the impartial pursuit of control and enforcement objectives but at the same time retaining the confidence of the Member States and remaining above national interests. Its most vital contribution has been the preparation of legislation and the formulation of the policy for the future of the driftnet fishery. In has also been successful in mediating between Member States and resolving differences without jeopardising the overall objective. In retrospect, it may be deduced that the proper implementation and effective enforcement of the limitation on the use of driftnets by the Community fleet would not have been achieved but for the vigilance and efforts of the Commission.

6. *The rôle of Member States*

The power to adopt fishery conservation and management measures rests with the Community institutions. Competence to apply such measures, however, is firmly in the hands of the Member States. Each Member State has its own national authorities and agencies who are responsible for implementing and enforcing Community regulations.[134] Moreover, the legislative framework within which these authorities operate, and the judicial process which they must follow in prosecuting suspected offenders of Council Regulation No. 3094/86 differ considerably. In the normal course of fishery patrols in Europe this does not pose a problem because each Member State is responsible for the enforcement of Community legislation in the waters under national jurisdiction or sovereignty. This accords with international law which specifies that the coastal state pursuant to the LOS Convention **Article 73(1)** is empowered to take such measures as may be necessary, including boarding, inspection, arrest and judicial process, of all vessels regardless of their flag once the vessel is located within their EFZ or EEZ. However, in high sea fisheries, Member States' competence is solely limited to inspection of vessels flying their flag. The strict adherence to flag state jurisdiction in the high seas area has resulted in the under utilisation of inspection resources. For example, although there were up to eight patrol vessels in the fishery at one period during the 1994 season,[135] each vessel was restricted to monitoring its own flag vessels. In some instances French patrol vessels had to monitor up to 60 driftnet vessels whereas the United Kingdom and Irish patrol vessels had less than 15 national vessels to inspect.[136] In 1995 the presence of a Community inspection vessel with Member State inspectors on board did not change the situation because national inspectors were still limited to inspections of their flag vessels and were not legally competent to inspect vessels of other Member States fishing in the albacore fishery. There were, however, frequent informal exchanges of information, including aerial surveillance information, between the respective inspection vessels. In 1995 when the albacore stock migrated into the Irish zone in the latter part of the season it was followed by the fishing fleet and the exclusive enforcement competence rested with the Irish authorities for the complete albacore fleet, which included 600 Spanish vessels and all the French, United Kingdom and Irish driftnet vessels.

As indicated above the limitation of enforcement competence to flag states when Member State vessels are fishing on the high seas has given rise to practical difficulties and is not cost effective. A possible solution to overcome these problems would be for the Community to promote the exchange of national inspectors or to implement joint inspection schemes,[137] or alternatively, to extend or share enforcement competence with other Member States. This would have the obvious advantage of improving cost effectiveness and efficiency as well as promoting the more uniform implementation of Community law. However, given the manner in which the devolved enforcement competence operates in the European Union much remains to be done before a common enforcement scheme can be put in place. A first step was taken in 1998, when the Council adopted an amendment to the Control Regulation which provides a legal base for inspectors from one Member State to accompany Commission inspectors while the latter are undertaking verification programmes in another Member State. Furthermore, Member States may also carry out, among themselves and on their own initiative, monitoring, inspecting and surveillance programmes concerning fisheries activities.

7. The effect of United States environmental and trade legislation on the implementation and enforcement of Community fishery law in the Mediterranean

US external policy on driftnets may be traced back to May 1989, when the parties to the International Convention for the High Seas Fisheries of the North Pacific Ocean, Canada, Japan and the United States, agreed to control Japanese driftnet fishing.[138] Bilateral Agreements to improve conservation were later entered into by the United States (pursuant to the Driftnet Impact Monitoring, Assessment and Control Act 1987) with Japan, Korea, and Taiwan. Similarly, concerning the salmon fishery in the Pacific, the United States and the Soviet Union agreed in principle to adopt sanctions against States defying driftnet limits.[139] No doubt bolstered by the success of their policy in the Pacific and driven on by increased awareness of global environmental matters at home, the United States has since focused its attention on other driftnet fisheries.

It has been suggested that American policy towards trade and the environment, like much of its trade policy, has reflected growing impatience with the multilateral approach, and that although the United States continues to participate in multilateral negotiations to protect the environment, there is increasing evidence that it has grown weary of the constraints of the process and the weakness of the resulting provisions, many of which provide for no enforcement at all.[140] The unilateral approach appeals to the United States because it is immediately effective and costs little to enforce, particularly when compared to the difficulties encountered in the multilateral process.

The significance of the implementation and enforcement of Community driftnet legislation must be appreciated against the background of the legislation adopted by the United States in 1990 to prohibit all use of high seas driftnet gear.[141] This legislation, the Driftnet Act Amendments 1990, declares that United States policy is to implement the moratorium called for in the United Nation resolutions and to 'secure a permanent ban on the use of destructive fishing practices, and in particular large-scale driftnets, by persons or vessels fishing beyond the exclusive economic zone of any nation'. The legislation requires the Secretary of State to negotiate international agreements to achieve this end and to provide for economic sanctions against States whose vessels use driftnets 'in a manner that diminishes the effectiveness of or is inconsistent with any international agreement governing large-scale driftnet fishing to which the United States is a party or otherwise subscribes'.[142] In July 1991, the United States House of Representatives adopted a resolution recording the wishes of Congress that the President, while seeking to achieve the UN moratorium, 'should work to achieve the United States policy of a permanent ban on large scale driftnet fishing, as set forth in the Fishery Conservation Amendments of 1990.'[143] Also in 1990, and more ominously for the European Community, the United States Senate adopted a bill requiring economic sanctions against any State using driftnets on the high seas. The Driftnet Fisheries Enforcement Act requires the Executive Branch (the Secretary of Commerce, the President) to ensure that the offending nation desists from driftnet fishing.[144]

These developments in the United States on the use of driftnets on the high seas must be viewed in the context of several other provisions in American municipal legislation which provide a legislative basis for the Unites States to impose unilaterally its environmental conservation and management policy upon other countries. For example, the Pelly Amendment Act,[145] the Marine Mammals Protection Act[146] and the Endangered Species Act Amendments[147] place mandatory embargoes on the importation of other fish and fish products if a country uses harvesting technology which takes an incidental by-catch of marine mammals or other protected species which are protected under American or international law. This ban may be avoided if the country wishing to export to the United States can show that the fish it wishes to export was not taken by using such gear as driftnets. The Pelly Amendment and the measures that rely on its provision for enforcement,[148] the Driftnet Impact, Monitoring and Assessment, and Control Act,[149] and the Driftnet Amendments of 1990,[150] use discretionary measures for enforcement. The United States President may impose bans on fish and fish products from a country[151] if the Secretary of Commerce has certified that the country's activities may affect international programmes, or may diminish the 'effectiveness' of international programmes to protect endangered or threatened species.[152] The Secretary of Commerce is directed to report to Congress annually on the use of large-scale driftnets on the high seas. Moreover the driftnet legislation provides for the imposition of a ban against countries that fail to reach agreement with the United States over the use of driftnets. Other than exerting pressure on other countries to achieve American environmental standards, the value of this municipal legislation is that it ensures that the United States fishing industry does not suffer a competitive disadvantage because of their adherence to tougher environmental standards to those applicable to their competitors. The legislation cited above has provided the basis for litigation in US Courts which has far reaching consequences for States involved in tuna fisheries.

In 1988, an environmental organisation successfully sued the Secretary of Commerce in a Federal District Court in the Northern District of California, and forced the United States to impose bans on yellowfin tuna from Mexico, Venezuela and Vanuatu, because the fishing fleets of these countries used fishing methods which involved the encircling of dolphins with purse-seine nets contrary to United States municipal law.[153] Subsequently, a GATT-related dispute arose between Mexico and the United States because Mexico contended that the embargo impugned several provisions of GATT agreements concerning restrictions of trade and that the American measures were disruptive to trade and protectionist in nature. In 1991 Mexico requested that the Council of GATT establish a Dispute Settlement Panel and in 1992 the Panel ruled in favour of Mexico stating that the United States was free to take domestic measures to protect the environment, but under GATT rules could not use discriminatory trade measures unilaterally to impose environmental laws outside its own jurisdiction.[154] In short, the GATT decision asserts that the only acceptable way to deal with extraterritorial environmental issues is through a multilateral framework. Nevertheless, in 1992 after intense debate within Congress, the Senate passed the International Dolphin Conservation Act of 1992, which called for a five-year moratorium on encircling dolphins with purse-seine nets.[155] It also made provision for the setting up of an International Dolphin Conservation Programme, membership of which was linked to the embargoes under the Marine Mammals Protection Act.[156] In the interim period, United States policy has achieved some success among ten Pacific Rim nations in reducing the mortality of dolphins, but on the other hand it has been criticised on the grounds that it applies restrictive laws to a resource that is beyond national jurisdiction while at the same time permitting the harvest of marine mammals taken incidentally in fisheries in its own EEZ.[157]

In view of the success of United States' environmental policy in the Pacific in curtailing the use of driftnets it was inevitable that domestic attention would focus on the Mediterranean Sea. In 1996, six environmental organisations successfully sued the Secretary of Commerce in the Court of International Trade in New York on the grounds that the Administration had failed to enforce the High Seas Driftnet Fisheries Enforcement Act and to certify Italy as a nation for which there is reason to believe that its nationals or vessels are conducting large-scale driftnet fishing beyond the exclusive economic zone of any nation.[158] Judge Thomas Aquilino noted that the 'Italians continue to engage in large-scale driftnet fishing ... in defiance of the law of their own country and the rest of the world'. As a result of this Court action and subsequent order, the Secretary of Commerce, in March 1996, threatened to impose sanctions on Italian fish products, worth around 1.8 trillion lira ($1.19 billion) if the Italian fleet continued to use driftnets in contravention of the UN resolutions and negotiations between the US President and the Government of Italy were not commenced within a period of 30 days and satisfactorily concluded within a further period of 90 days.[159]

It was evident that the Italian fleet continued to contravene the Community rules during 1996 and that it was necessary to take the threat of sanctions seriously.[160] On 26 July 1996, the United States and Italy concluded an agreement under which the Italian government would implement measures to stop driftnet fishing in the Mediterranean. Among the measures agreed were:[161] the Italian Government would step up its enforcement efforts both at sea and in port; expand its legislative authority to allow for the seizure of illegal driftnets; adopt a decree that prohibits virtually all Italian driftnet vessels from operating in bases in Sardinia in order to curtail their ability to conduct driftnet fishing in the Western Mediterranean; and the submission of two laws for Parliamentary approval, one to increase significantly penalties for illegal driftnet fishing, the other to allow Italy to ratify the International Commission for the Conservation of Atlantic Tunas (ICCAT) and that pending ratification the Government would apply ICCAT conservation rules provisionally.[162] As a long-term solution to the driftnet problem in the Mediterranean, the Italian Government presented, for approval by the European Union, a comprehensive three-year conversion plan, under which Italian driftnet vessels will either be removed from the national register or converted to other fishing methods, or find employment opportunities outside the fisheries sector. The cost of the plan, which covers the period 1996–1999, is estimated at $130 million (200 million ecu) and is to be shared between the EU and Italy. As part of the Italian–USA agreement the Italian Government proclaimed its full adherence to the UN moratorium. Indeed, much of the agreement depends on effective implementation, a point which the United States State Department emphasised as is evident from the statement issued at the conclusion of the agreement.[163]

Ostensibly, the extra-territoriality dimension of American trade and environmental policy has played a pivotal role in improving the prospects for effective enforcement and implementation of the UN resolutions in the Mediterranean Sea. The driftnet fishing in the Mediterranean takes place within 30 miles of the coast of Italy and in some early parts of the season within 30 miles of the Spanish Balearic Islands, in this regard it is important to emphasise that there is no 200-mile zone in the Mediterranean

Sea and the areas in question are within the potential exclusive jurisdiction of two Member States of the European Union. As noted by one American commentator,[164] the application of US law in the latent exclusive zones of the Mediterranean Sea, may effectively encourage States to extend their jurisdiction to the areas in question.

Even though neither Italy nor the Community has challenged the legality of the US decision at the World Trade Organisation, the threat of American sanctions and long-arm enforcement methods raises several important issues. In the past, the international system tended to treat such enforcement as matters of state administrative policy and not subject to international normative constraints. But increasingly the international system - or at least some states within the system - have been concerned with such extraterritorial enforcement. Particularly, as noted by one leading authority, unlike judicial enforcement, which has developed a body of jurisprudence and guidelines, international law relating to non-judicial enforcement is still primitive and inchoate.[165] There is now an urgent need for the development of norms to govern non-judicial enforcement measures, particularly, as states resent intrusion upon their territory, over-reaching upon their nationals and companies, and interference with their trade.[166] Although the issues raised be the extra-territoriality of US environmental and trade policy as it relates to the Mediterranean driftnet fishery continue to evolve it will be interesting to observe if the principles of 'reasonableness' and 'proportionality' will apply in the implementation phase of the new enforcement policy necessitated by the US–Italian agreement. Furthermore, it has been pointed out by several leading commentators that the reliance by the United States on unilateral trade sanctions in relation to the Pacific driftnet fishery has been counterproductive and has had undesirable side effects.[167] Finally, the United States measures do not accord with the FAO Code of Responsible Fisheries which provides, *inter alia*, that fish trade measures adopted by States to protect human or animal life or health, the interest of the consumer and of the environment, should not be discriminatory and should be in accordance with internationally agreed trade rules, in particular the principles, rights and obligations established in the Agreement on the Application of Sanitary and Phytosanitary Measures and the Agreement on Technical Barriers to Trade of the WTO.[168] It is perhaps appropriate to conclude this section by noting that it has been suggested elsewhere that in order to achieve conservation goals and to avoid potentially disruptive trade disputes, the United States Legislative and Executive Branches should work to integrate marine conservation and free trade policies rather than demand that the Judiciary fashion inflexible court orders.[169] In this context it may be significant that, outside the domain of the driftnet fishery, the United States has pursued a multilateral approach to combat the fishing activities of non-ICCAT members which have vessels operating in the Atlantic which fish for bluefin tuna and swordfish stocks. In particular it has been a strong advocate of the bluefin tuna and swordfish action plan which provides the means for multilateral trade measures to be imposed on non-member nations of ICCAT who are deemed to be conducting harmful fishing practices. In this regard ICCAT has adopted a measure requiring its members to impose trade measures against those nations determined to have vessels fishing in a manner that diminishes the effectiveness of ICCAT conservation measures for bluefin tuna. In the implementation of this measure, in December 1997, the United States banned all imports of bluefin tuna products from Panama, Honduras and Belize.

Section (iv) Measures to prohibit the use of driftnets by the Community fleet as from 2001

1. *The Commission proposal to prohibit driftnet fishing*

When the Council decided not to extend the derogation for certain categories of fishing vessels utilising driftnets in December 1993, it requested the Commission to bring forward a report on the use of driftnets by Community vessels.[170] This was followed by a proposal later in 1994 which envisaged the total phasing out of the use of large-scale driftnets by 31 December 1997 and the application of a strict enforcement regime in the interim period.[171] This proposal was the subject of protracted negotiations at successive Council meetings and several recommendations of the European Parliament. Attempts by the Commission to have the proposal adopted were frustrated by a blocking minority of Member States (France, the United Kingdom, Italy, and Ireland) which remained firmly attached to the view that the use of driftnets could be regulated and did not constitute an ecological risk to marine resources or mammals.

Discussion on the Commission's proposal commenced in earnest when the United Kingdom assumed the Presidency of the European Union on 1 January 1998.

The fisheries element of the Presidency's programme contained a commitment to secure agreements on the eventual phasing out of the use of large scale driftnets. Formally, the proposal sought to amend Council Regulation (EC) No. 894/97 (laying down certain technical measures for the conservation of resources) and in particular **Article 11**, which contained the existing rules on the use of large scale driftnets.[172] The Commission proposal of April 1994 was, in effect, taken by the United Kingdom Presidency, amended and presented as a compromise document to successive Fisheries Councils in March and June 1998. It was eventually adopted, as Council Regulation (EC) No. 1239/98, at the June Council 1998.[173]

2. *Community law prohibiting the use of driftnets – Council Regulation (EC) No. 1239/98*

With the exception of the Baltic Sea, the Belts and the Sound, driftnet fishing for 23 species is prohibited by Council Regulation (EC) No. 1239/98 as from 2001. The adopted text of Council Regulation (EC) No. 1239/98 differs in several respects from the Commission proposal. While the termination date for driftnet fishing is postponed until 2001, in the interim period a series of new conditions are imposed on fleets continuing to use this type of gear. Importantly, a fishing effort scheme is introduced to restrict the expansion of the driftnet fleet.

Before reviewing the elements in the regulation, mention should be made of two of the recitals. The fifth recital makes express reference to **Article 130r(2)** {174(2)} of the Treaty on European Union which falls under Title XVI {XIX} dealing with the environment. The said article states, *inter alia,* that Community policy on the environment shall be based on the precautionary principle and on the principles that preventive action should be taken to avoid damage to the environment. Recourse to this particular Treaty provision highlights the obligation on the Community to be mindful of the necessity of ensuring that resources including marine resources are not endangered through, in this case, the deployment of unselective fishing gear. The sixth recital refers to the Community's international obligations to contribute towards conservation and management of the biological resources of the oceans. Among the obligations included under this heading are those arising from the General Assembly Resolution of 21 January 1997 (which the Community supported), calling for full implementation of a global moratorium on all large-scale pelagic driftnets fishing on the high seas of the world's oceans and seas, including enclosed and semi enclosed seas. The 12th recital acknowledges the enforcement problem associated with the use of large scale driftnets and stresses that fishing using such nets should only take place when it can be and actually is controlled by inspection authorities.

Council Regulation No. 1239/98 inserts a new **Article 11** into the 1997 Technical Regulation. This stipulates that as of from 1 January 2002, no vessel may keep on board, or use for fishing, one or more driftnets intended for the capture of species listed in Annex VIII (ten species of tuna, marlins, sailfish, sauries and cephalopods, the list also includes dolphins and sharks). Annex VIII also prohibits the landing of the aforementioned species by Community vessels. Prior to the above date, vessels may continue to use nets whose individual or total length does not exceed more than 2.5 km. The combined effect of these two paragraphs is clear: vessels currently using large-scale driftnets must divest themselves of this gear and fish with alternative means such as longlines, trawls or pole and lines. Masters of vessels wishing to continue to utilise driftnets in the interim period must comply with a series of obligations:

- express prior authorisation must be obtained from the competent national authorities;
- gear may only be deployed according to certain rules;
- logbooks to record information on the total length of nets on board and used, the quantity of each species caught, including by-catches and discards at sea, and those pertaining to cetaceans, reptiles and sea-birds, the quantity of each species on board and the date and position of such catches must be retained;
- landing of species listed must be accompanied by a landing declaration indicating the dates and zones of catches;
- prior notice (at least two hours in advance) of intention to land must be communicated to the competent authorities of the Member State where the landing will be effected.

This combination of measures is designed in the first instance to improve both the quality and quantity of information on catches and by-catch and fishing effort of the various fleets using driftnets. Information on catches of species listed in the regulation should assist future management of the resources while data on effort should facilitate the monitoring of the scheduled reduction of the number of vessels participating in this particular fishery.

Article 11a(3) stipulates that the total number of vessels authorised (by national authorities to use driftnets) in 1998 shall not exceed 60% of the total number participating in the fishery during the period 1995–1997. Reductions attained in 1998 should be maintained up to the year 2001. Fleets using driftnets have already begun to diversify to other types of gear. French vessels have experimented with the use of trawls and to a lesser extent with long lines and trawl lines while Irish vessels also tested the effectiveness and efficiency of this type of gear. Fleet restructuring and diversification of gear use can be anticipated to accelerate in the near future. The full implementation of the conversion plan for the Italian fleet agreed by national authorities and the Commission seeks to encourage Italian fishermen to diversify away from certain fishing activities.[174] This plan seeks to encourage Italian vessel owners practising driftnet fishing to either cease the use of such nets definitively or diversify definitively towards the use of other types of gear. In return for diversification out of driftnets, participants in the plan are eligible for financial compensation payments over a three-year period 1997–1999. In the context of the Fisheries Minister meeting in June 1998, representatives of those Member States most affected by the phasing out of the use of large-scale driftnets pressed for a similar compensation scheme for their fleets. Recognising that conversion would entail financial burdens for the fleets concerned, the Council asked the Commission to bring forward proposals on a series of measures which could minimise anticipated economic hardship and facilitate reconversion generally.

In December 1998, the Council agreed to a range of financial measures designed to encourage diversification out of certain fishing activities. Under the terms of the Decision,[175] fishermen who are nationals of a Member State and work on board a fishing vessel flying the flag of Spain, France, Ireland, or the United Kingdom, using one or more driftnets intended for the capture of highly migratory species (those listed in the Annex to Council Regulation (EC) No. 1239/98) may receive compensatory payments up to a ceiling of €50,000 if they cease all economic activity before 1 January 2002. Alternatively, fishermen converting to another fishing activity or another sector may receive payments up to a limit of €20,000. Compensation payments are also available for individual fishing vessels provided they are more than ten years old. In order to avail of these two types of payments, fishermen must demonstrate that they have suffered a loss as a result of the ban on fishing with large scale driftnets.

In addition to the new provisions on the reporting of catches, discards and fishing effort, the new regulation sets out a number of conditions on how driftnets are to be deployed at sea during fishing activity. Heretofore, the masters of vessels were obliged to remain attached to the fishing vessel in order to reduce the possibility of nets being simply set and left unattended. The measure was also intended to establish clear responsibility for nets. In practice, as noted above, skippers experienced some difficulties in complying with this obligation particularly in the Atlantic Ocean where fishing conditions could be more severe than in the Mediterranean Sea. Under the new provisions, the masters of fishing vessel must keep their nets under constant visual observation and floating buoys, with radar reflectors, must be moored to each end of the netting so that its position can be determined at any time. The buoys must also be permanently marked with the registration letters and number of the vessel to which they belong. Close visual observation of the nets deployed should ensure that nets are not most at sea and thus reduce the danger of 'ghost fishing' sometimes associated with the use of driftnets.

Section (v) Conclusions

Since 1992 Community fisheries law enforcement has evolved considerably. At a first glance, the implementation of the amendment to Council Regulation No. 3094/86 would have been much easier from the enforcement perspective if it introduced a complete ban on driftnets. In retrospect, the toleration of nets up to 2.5 km has to be seriously questioned.[176] Firstly, because transitory measures are generally difficult to enforce and secondly, because it proved to be a contributory factor which in some instances led to non-compliance. Moreover, this approach was an avenue through which the use of driftnets was exploited, mainly because the fishery remained open and access for new participants was unlimited. The derogation granted to certain French vessels to use nets of 5 km in length up to the end of 1993, resulted in a different interpretation of the legislation and non compliance by the fleet well into the 1994 season. On the basis of experience of 1992–1996, the key to successful implementation of Community measures, in common with other areas of CFP, is immediate and effective enforcement.

But what constitutes immediate and effective controls? Two of the classic characteristics of fishery law enforcement require that the system should allow for credibility and efficiency of enforcement effort (economy in resources utilised or required).[177] The evolution of the Community enforcement policy on the use of driftnets can be usefully reviewed in the context of these criteria. The credibility of the Community enforcement effort was at issue during the 1992 and 1993 albacore seasons in the North-West Atlantic. Community driftnet fishermen flaunted the prescribed limit of 2.5 km and were not brought to justice. As there was no scheme of enforcement it was evident that questions would be raised about the Community's commitment to uphold their international environmental obligations. By 1997 circumstances had changed and the Community attained a satisfactory level of compliance in the North-East Atlantic driftnet fishery, albeit at a large cost in terms of resources and financial expenditure. Credibility is also the key to the successful implementation of the US–Italian Agreement on high seas driftnet fishing in the Mediterranean. The second criterion calls for efficiency of enforcement effort which in the high seas driftnet fishery is difficult to attain with the present control and enforcement structure in the European Union. The theatres of enforcement operations are international and, in most instances, well beyond the coastal zones of Member States. Even when deployed on duties in this fishery, national inspectors are limited to enforcing one technical measure. There is no possibility of offsetting the costs and increasing efficiency by monitoring, at the same time, the compliance with other regulatory measures because of the oceanic nature of the driftnet fishery.

The importance of effective controls at sea may be appreciated if **Article 9a** Council Regulation No. 3094/86 (and Council Regulation (EC) No. 1239/98) is schematically analysed by its constituent parts (Table 10.1).[178]

Given that the crucial rôle sea inspection plays and the problems and costs associated with it, how might flag States generally and the Community in particular enforce rules on the use of driftnets in the future? This task might be made easier if the Italian conversion plan is fully implemented and the trend, apparent in the French driftnet fleet, towards greater utilisation of other types of gear (mid-water trawls) continues. The introduction of a joint control scheme in ICCAT could also help to share the enforcement burden and provide the means for departing from the rigid flag State jurisdiction approach which has contributed to inefficiency and proved to be extremely costly over the years.

Table 10.1 Enforcement possibilities relating to the use of driftnets.

Item to be enforced	Enforced in port	Enforced at sea
Length of net onboard	Yes	Yes
Length of net in use	No	Yes
Technical conditions		
• submersion of the headline	No	Yes
• attachment to vessel	No	Yes

As is evident from the discussion in Section (iii) above, prior to the adoption of the current rules, control and inspection of vessels using large-scale driftnets was undermined by a number of factors which mitigated against efficient enforcement. Many of these factors were derived from the nature of the fishery: the migratory nature of species and hence the mobility of the fleets; the distance involved for control services; the disproportionate costs involved, and those factors which arose from the competitive/antagonistic relations between the various fleets. During several seasons this manifested itself in latent tension at sea, gear conflicts and a sense of injustice or harassment on the part of some fleets and vessel owners. The fact that this fishery attracted perhaps more than its share of media attention, and was closely observed by international ecological groups may also have had a major impact on the difficulties associated with enforcement.

While some of the factors mentioned above will not disappear in the period up to the year 2001, there are nevertheless grounds for optimism that enforcement will improve. The introduction of satellite-based monitoring systems by the Community and their application to those vessels using driftnets should reduce the difficulties and more particularly the cost of tracking and locating vessels operating a long distance from port.[179] The implementation of the catch and effort provisions will bring the driftnet fleets into line with other fleets harvesting migratory species on the high seas and in Community waters and thus remove or reduce the 'sense of grievance' sometimes expressed by those fleets not using driftnets. Acceleration of both the Italian conversion plan and measures to facilitate diversification should reduce the control burden further and make the enforcement task generally more manageable. Greater co-operation between national enforcement authorities can also play a rôle as can joint patrols at sea.[180] Over the period 1996–1998, there has been growing evidence that national authorities are more aware of the advantages of co-operation and are more prepared to move beyond the tentative approaches in the past. The range of specific conditions introduced for the transition period, prior to the termination of the fishery in 2001, should curtail scope for non-compliance and the highly international character of the fishery should result in a higher degree of vigilance on the part of the national control services. The prohibition of driftnets as from the year 2001 (other than in the Baltic Sea) will of course simplify an enforcement problem which in the past appeared to be almost intractable.

References

1 For a more specific definition, see, footnotes 19 and 46, *infra.*
2 For a discussion of the various actions and negotiations in the Pacific, see, Johnston, 'The Driftnetting Problem in the Pacific Ocean: Legal Considerations and Diplomatic Options', *Ocean Development and International Law* **21(1)**, (1990) 15. See also authorities cited in fn 26, *infra.*
3 Council Regulation (EEC) No. 1866/86, *OJ* L 162, 18.06.1986, p. 1. The Baltic technical regulation is described in Chapter 1, *ante.*
4 Communication from the Commission entitled, 'The Use of Large Scale Driftnets Under the Common Fisheries Policy', COM (94) 50 final, Brussels, 08.04.1994, Table 1, p. 22.
5 *Ibid.*, Table 3, p. 7.
6 The number of marine mammals has been considered by several authorities to be a matter of serious concern, see, Garcia, S. and Majkowski, J., 'State of High Seas Resources', *L.Sea Inst. Proc.*, (1992) 175–236.
7 Regarding the relationship between Community law and international law, see, *inter alios*: Groux, J., Manin, P., *European Communities in the International Order* (1985); Timmermans, C.W.A. and Völker, E.L.M., *Division of Powers between the European Communities and their Member States in the Field of External Relations* (1981); Meesen, 'The Application of Rules of Public International Law within Community Law' *CMLRev.* **13** (1976), 485. Specifically on Community fishery law see Churchill, R.R. and Foster N.G., 'European Community Law and Prior Treaty Obligations of Member States: The Spanish Fishermen's Cases' *International Comparative Law Quarterly* **36(3)** (1987), 504–524. Although somewhat out of date, see Koers, A.W., 'The European Economic Community and international fishery organizations', *LIEI* **1** (1984) 113–131. More recently two studies have made a significant contribution to the literature in this area, Frid, R., *The Relations Between the EC and International Organizations. Legal Theory and Practice*, (The Hague, 1995), see in particular Chapter 7 in relation to fisheries organizations; MacLeod, I., Hendry, I.D. and Hyett, S., *The External Relations of the European Communities*, (Oxford, 1996), Chapter 10.
8 *OJ* No. L42, 18.02.92, p. 15.
9 It is not proposed to present a detailed discussion on the High Seas fisheries regime, this is dealt with in all texts on the subject of international law of the sea, including *inter alios*: Churchill, R.R., Nordquist, M. and Simmons, K.R., (eds.) *New Directions on the Law of the Sea 1982: A Commentary'* (Vol 1–10, 1983–1990); Brown, E.D., '*The International Law of the Sea'*, Vol 1, Chapter 14, (Aldershot, 1994) For a more fishery-focused study, see, in particular, Burke, W.T., *The New International Law of Fisheries* (Oxford, 1994), 82–150; and, if not somewhat out of date, Koers, A.W., *The Enforcement of Fisheries Agreements on the High Seas: A Comparative Analysis of International State Practice*, Law of the Sea Inst. Occasional Paper No. 6 (1970). More recently, an invaluable contribution to the literature, is Treves, T. and Pineschi, L., (ed), *The Law of the Sea: the European Union and its Member States* (The Hague, 1996).
10 LOS Convention, **Article 87**.
11 LOS Convention, **Articles 87(a)** and **(e)**.
12 See Section (ii) *infra.*
13 Chapter 3, *ante.*
14 Large-scale pelagic driftnet fishing and its impact on the living marine resources of the world's oceans and seas, *Report of the Secretary-General*, UN GAOR, 46th Sess., at 3, UN Doc. A/46/615 (1991) and Large-scale pelagic driftnet fishing and its impact on the living marine resources of the world's oceans and seas, Report of the Secretary-General, UN GAOR, 47th Sess., at 3, UN Doc. A/47/487 (1992); Large-scale pelagic driftnet fishing and its impact on the living marine resources of the world's oceans and seas, *Report of the*

Secretary-General, UN GAOR, 50th Sess., UN Doc. A/50/553 (1995). See also, *The Regulation of Driftnet Fishing on the High Seas: Legal Issues, FAO Legislative Study No. 47*, (FAO, Rome, 1991).

15 Vol II, Doc.9.4; *LOS Bull.* **14** (December 1989), 29–30.

16 Vol II, Doc.9.15; *LOS Bull.* **14** (December 1989), 28.

17 ICCAT is an international organisation established by convention in 1966, which entered into force on 21 March 1969. Its area of competence encompasses all waters of the Atlantic Ocean, including the adjacent seas. Spain, Portugal, France and the United Kingdom (in respect to Bermuda) are members of ICCAT. In 1979 France, then the only Member State party to this Convention, requested the necessary amendments to be made to allow Community accession. It was not until quite recently, however, that the Community completed accession negotiations to ICCAT. In 1996 the Community was awaiting the last three ratification by Morocco, Gabon, Libya to become a full Member. By November 1997, the Community acceded to ICCAT.

18 Convention for the Prohibition of Fishing with Long Driftnets in the South Pacific, done at Wellington, 23 November 1989, reprinted in *Law of the Sea Bulletin* **14**, 38, (1989), *ILM* **29** 1449 (1990). Interestingly, the European Union had attended the negotiations in 1989 which concluded a convention which prohibited the use of driftnets in the southern Pacific. On the Convention and its developments, see, 'The Proposed "Driftnet Free Zone" in the South Pacific and the Law of the Sea Convention', **40**, (1991) *ICLQ* 184. For an outline of New Zealand Legislation to ratify the Convention, see Davidson, J.S., 'Driftnet Prohibition Act 1991', *IJMCL* 264–271. Certain sections of the New Zealand Act are much stricter than the Convention.

19 The Convention provides that a driftnet is 'a gillnet or other net or a combination of nets which is more than 2.5 km in length the purpose of which is to enmesh, entrap or entangle fish by drifting on the surface of or in the water', **Article 1(b)** of the Convention.

20 Large-scale pelagic driftnet fishing and its impact on the living marine resources of the world's oceans and seas, Report of the Secretary-General, UN GAOR, 46th Sess., at 3, UN, Doc. A/45/663 (1990).

21 *Ibid.*

22 Large -scale pelagic driftnet fishing and its impact on the living marine resources of the world's oceans and seas, Report of the Secretary-General, UN GAOR, 46th Sess., at 3, UN Doc. A/46/615 (1991) and Large-scale pelagic driftnet fishing and its impact on the living marine resources of the world's oceans and seas, Report of the Secretary-General, UN GAOR, 46th Sess., at 3, UN Doc. A/47/487 (1992).

23 *Ibid.*

24 *Ibid.*

25 For an interesting discussion of UNCED see Burke, W.T., 'UNCED and the Oceans', *Marine Policy* **17(6)**, (November 1993).

26 For a review of international practice see Burke, W.T. *The New International Law of Fisheries* (Oxford, 1994), 102–103, and by the same author 'The law of the sea concerning coastal state authority over driftnets in the high seas', *FAO Legislative Study* **47** (1991) 13–32; and Gurrish, 'Pressures to reduce by-catch on the High Seas: An Emerging International Norm', *Tulane Env. LJ* **5**, (1992) 473. See also Burke, W.T., Freeburg, M. and Miles, E.L., *The United Nations Resolutions on Driftnet Fishing: an Unsustainable Precedent for High Seas and Coastal Fisheries*, (School of Maritime Affairs, University of Washington, August 1993). For a study of US policy in the North Pacific see Song, Y.H., 'United States Ocean Policy: High Seas Driftnet Fisheries in the North Pacific Ocean', *Ocean Yearbook of International Law and Affairs* **11** (1991–92), 64–137. For international perspective, see, Shearer, I.A., 'High Seas: driftnets, highly migratory species and marine mammals', *Law of the Sea Institute Proceedings* (1992), 432–459. A Japanese Law Professor, Sumi, K., examines the driftnet issue as it pertains to

Japan in a paper entitled, 'The international legal issues concerning the use of driftnets with special emphasis on Japanese Practices and Responses', *FAO Legis. St* **47**, 62–63, (1991). On the South Pacific, see Hewison, G.J., 'The Convention for the Prohibition of Fishing with long driftnets in the South Pacific', *Case Western Journal of International Law* **25** (Summer 1993), 449–530. A number of incisive views on the subject of driftnets has been published in *The Living Resources*, in particular see Burke, W.T., 'Unregulated High Seas Fishing and Ocean Governance', 235–271; Carr, J., and Gianni, M., 'High Seas Fisheries, Large-Scale Drift Nets, and the Law of the Sea', 272–291; Sumi, K., 'The International Legal Issues Concerning the use of Driftnets With Special Emphasis on Japanese Practices and Responses', 292–309; Floit, C., 'Reconsidering Freedom of the High Seas: Protection of the Living Marine Resources on the High Seas', 310–326.

27 Burke, W., *The New International Law of Fisheries*, (Oxford, 1994), p.103.

28 Hewison, G., 'The Legally Binding nature of the Moratorium on Large-Scale High Seas Driftnet Fishing', *Journal of Maritime Law and Commerce*, **25**, (4), (1994), 557–579.

29 *Ibid.*

30 *Ibid.*

31 This is the second instance in which the General Assembly has been used to respond to problems beyond the national region of jurisdiction by recommending restrictions to be observed by states. The first one being the UN General Assembly recommendation of a moratorium on appropriation of the deep seabed for ocean mining purposes. Declaration of Principles Governing the Seabed and the Ocean Floor, and the subsoil thereof, Beyond the Limits of National Jurisdiction, GA Res. 2749 UN GAOR Supp. (No.28), at 24, UN Doc. No. A/8028(1970).

32 See Secretary General Report 1990, *op. cit.*, fn 20, p. 37. The same report records that the Government of Japan announced that it had taken measures effective from 15 August 1990 to prohibit Japanese vessels from using driftnets in the waters of the Atlantic Ocean, a similar ban for Taiwanese vessels was announced on 16 February 1990. Japan did not challenge Resolution 46/215 for apparently unrelated political reasons.

33 Indeed the Community actively supported the adoption of the United Nation's Resolutions, on this issue see, Commission reply to Parliamentary Questions No. 23/90, *OJ* C 246/7, 01.10.90

34 SEC (90)2498, STCF special meeting of November 1990 held in Charlottenlund.

35 For the definition and length of a driftnet, *op. cit.*, fns 19 and 46.

36 The Parliament again set out its views on the matter of driftnets at the part time session of 13 to 17 December 1993. In a resolution it proposed a ban on driftnets but considered that the Commission could, on the basis of a reasoned opinion from the Member States, authorise their use within the 12-mile limit, EP 177.124, Minutes 50 II, 17.12.1993.

37 This derogation was availed of by a number of French vessels and it expired on 31 December 1993, discussed *infra.*

38 *OJ* L42, 18.02.92, p. 15.

39 *Ibid.*

40 Council Regulation (EEC) No. 345/92 of 27.01.1992 amending for the eleventh time Regulation (EEC) No. 3094/86 laying down certain technical measures for the conservation of fishery resources, *OJ* L 42/15, 18.02.92. In 1997, the Council codified the technical measures in Council Regulation (EC) No. 894/97 of 29.04.1997 laying down certain technical measures for the conservation of fishery resources, OJ L 132/1, 23.05.1997. **Article 11** regulates the use of driftnets. In 1998, the Council adopted Council Regulation (EC) No. 1239/98 of 08.06.1998 amending Regulation (EC) No. 894/97 to prohibit the use of driftnets as from 1 January 2002, see Section (iv) *infra.*

41 *OJ* L 38, 10.02.1982, p. 1.

42 Council Regulation (EEC) No. 3094/86, *OJ* L 288, 11.10.1986, paragraph 1 of **Article 9a**.

43 *Ibid.*, paragraph 2.

44 *Ibid.*, paragraph 3 see discussion below on this requirement.

45 *Ibid.*, paragraph e.

46 New Zealand legislation and the Wellington Convention both provide a definition of a driftnet, see fns 18 and 19 *supra*. For other definitions in municipal law, see: Ireland - Sea Fisheries (International Waters) (Driftnet) Order 1994, para 4, SI 201 of 1994, which defines a driftnet as 'a wall of netting used in fishing which is free to move according to the wind and tide'; see also in the USA, Magnuson Fishery Conservation and Management Act (16 USC 1801–1882), the 1987 Driftnet Impact Monitoring, Assessment and Control Act (16 USC 1822).

47 *Ibid.*

48 The Standing Technical Committee Fisheries at its meeting in October 1990 defined large driftnets as meaning nets more than 1 kilometre in length, SEC (90)2498, STCF special meeting of November 1990 held in Charlottenlund. Interestingly, 1 km is the length of a prohibited net in the New Zealand Driftnet Prohibition Act 1991, Section 2, see *supra* fn 18.

49 Case C-405/92, [1993] ECR I-6133, paragraph 42, this case is discussed in detail *infra*.

50 See discussion *infra* Section (iv) regarding scientific evidence and the Commission proposal for a cessation of driftnet fishing. In April 1992, 33 French tuna fishermen brought an action in the Court of Justice under **Article 173** {230} of the EEC Treaty for the annulment of **Article 9a(2)** on the grounds that the derogation was too restrictive with regard to the duration of the transitional period and the length of net permitted. This issue was never decided because the Court ruled in Case C-131/92, *Thierry Arnaud and Others* v. *The Council*, [1993] ECR I-2573 that the fishermen's application was inadmissible on the grounds that the contested measure concerned the applicants only in their objective capacity as albacore fishermen and was thus not of direct and individual concern to the applicants, paras 12–18. To support the ruling the Court cited the previous decisions in Case 123/77, *UNICME* v. *Council*, [1978] ECR 845 and Case 26/86, *Deutz und Geldermann* v. *Council*, [1987] ECR 941.

51 The application of the precautionary principle approach to the regulation of the use of driftnets in Community legislation is discussed in the literature. There have been a number of definitions of the precautionary principle, which generally requires the taking of certain action or the cessation of certain action, even if there is no scientific evidence of a particular danger. It has found a new status in the 'FAO Conduct for Responsible Fishing', for a discussion see Gundling, 'The Status in International Law of the Principle of Precautionary Action', *IJMCL* **5** (1990), 23. See also, *inter alios*, MacDonald, J.M., 'Appreciating the precautionary principle as an ethical evolution in ocean management', *Ocean Development in International Law*, **26(3)** (1995), 255–286; Garcia, S.M., 'The Precautionary Principle: its implications in Capture Fisheries Management', *Ocean Development and International Law* **22** (1994), 99–125. More recently, McIntyre, O. and Mosedale, T., 'The Precautionary Principle as a Norm of Customary International Law', *Journal of Environmental Law*, **9(2)** (1997). For an informative discussion, see, Hey, H., 'The precautionary concept in environmental policy and law: Institutionalised caution', *Georgetown Inst. Env. Law Review* (1994), and by same author 'The precautionary approach. Implications of the revision of the Oslo and Paris Conventions', *Marine Policy* **15(4)**, (1991), 244–254.

52 The prudence of this requirement has been questioned, particularly if the crew of a fishing vessel wish to inspect their nets and free entangled mammals. On this point, see memorandum by the National Federation of Fishermen's Organisation submitted to House of Lords Select Committee on the European Communities, *Regulation of Driftnet Fishing, with evidence*, HL paper 77, 19 July 1994, session 1993–1994, Report 13, p. 17. Fishing vessels may also wish, in some instances, to steam alongside their nets in order to indicate the direction of set to passing shipping traffic. It may also be argued that driftnets pose the greatest threat to the safety of sailing vessels, although the driftnet fishing areas in the North-East

Atlantic are generally away from the busiest shipping routes used by commercial and leisure vessels.

53 This requirement makes the measurement task for enforcement officers more difficult, and has been criticised by fishermens organisations, see, *Regulation of Driftnet Fishing, with evidence, ibid.*

54 **Article 9a** of Council Regulation No. 1866/86, *OJ* L 162/1, 18.06.86, as amended by Council Regulation No. L 191/7, 22.07.88, allows the use of 600 nets at once per vessel, the length of each net not exceeding 35 metres measured in the gear's headrope. In addition to the permitted number of nets, not more than 100 reserve nets may be kept on board. This regulation has been codified, Council Regulation (EC) No. 88/98, *OJ* L 9/1, 15.01.1998. For the measures relating to driftnets, see **Article 9**.

55 See *Convention on Fishing and Conservation of the Living Resources in the Baltic Sea and the Belts*, [1983] *OJ* L 227/21, this convention was signed on 13 September 1973 by the States bordering the Baltic Sea, and established the Baltic Sea Fishery Commission. The Community became a party to the Convention in 1984, replacing Denmark and Germany. See Fitzmaurice, 'Common Market Participation in the Legal Regime of the Baltic Sea Fisheries' *Germ. YBIL* **33** (1990), 214–235. For a summary of control and fishery enforcement in the Baltic Sea see Nielsen, P., *Resource Management and Control in the Baltic Sea*, (Copenhagen, 1994), which reports on the Baltic Sea States meeting of Ministers of Fishery and the Baltic Fisheries Cooperation Committee seminar 24/25 May 1994, Gotland, Sweden.

56 According to the Swedish Ministry of Agriculture, no scientific evidence has been presented that large scale driftnets (21 km in the Baltic) had any significant effects on marine animals or sea birds in the Baltic Sea. A point supported by the Chairman of the International Council for the Exploration of the Sea (ICES), see House of Lords Select Committee on the European Communities, *Regulation of Driftnet Fishing, with evidence*, HL paper 77, 19 July 1994, session 1993–1994, report 13, p. 11. For a description of environmental issues as they pertain to the Baltic Sea see Fitzmaurice, M., *International Legal Problems of the Environmental Protection of the Baltic Sea*, (Dordrecht, London, Boston, 1992).

57 Case C-405/92, [1993] ECR I-6133.

58 *Ibid.*, preliminary question 12 referred by French commercial tribunal of Roche-sur Yon.

59 Case C-405/92, paragraphs 54–55.

60 This was particularly true in 1995 when the albacore was fished in the Irish EFZ for the latter part of the season.

61 The Community maritime zone is not a legal term and is used here to describe the combined sea area which falls within EEZs/EFZs of the Member States.

62 See discussion *infra* Section (iv) regarding the validity of scientific evidence.

63 See Burke, W., *The New International Law of Fisheries, op. cit.* fn 27, p. 106.

64 This accords with the basic rule of international law regarding the interpretation of treaties: 'A treaty shall be interpreted in good faith in accordance with the ordinary meaning to be given to the terms of the treaty in their context and in the light of its object and purpose', see Vienna Convention on the Law of Treaties, **Article 31**.

65 For a discussion of the Community's competence in maritime zones and activities in these zones, see all the articles in *Ocean Development and International Law* **23** (1992), 89–259, and especially Freestone, D., 'Some Institutional Implications of the Establishment of Exclusive Economic Zones by EC Member States,' 97 at 103–107; Churchill, R.R., 'EC fisheries and an EZ-Easy!', 145–163; Birnie, P., 'An EC Exclusive Economic Zone: Marine Environment Aspects, 193, at 200–206; Soons, A.H.A., 'Regulation of Marine Scientific Research by the European Community and its Member States, 259 at 262–267. See also, *inter alios*, Fleischer, C.A., 'L'accès aux lieux de pêche et la traité de Rome' *RMC* **141** (1971) 148–151; Koers, A.W., 'The External Authority of the EEC in regard to Marine Fisheries' *CMLRev* **14** (1977), 269 at 274–275.

66 Joined Cases 3, 4 and 6/76, *Kramer(Cornelius)*, [1976] ECR 1279.
67 *Ibid.*, paragraphs 30–33.
68 Case 61/77, *Commission* v. *Ireland*, [1978] ECR 417.
69 Case 61/77, paragraphs 44–47.
70 Case 61/77, paragraph 47. For a discussion of this case, see, Timmermans, C.W.A., *Social-economische wetgeving*. (1978), 582–589; Churchill, R., *European Law Review*, (1979), 391–396; Winkel, K. and Von Borries, R., *Common Market Law Review* (1978), 494–502.
71 Case C-258/89, [1991] ECR 3977.
72 Case 61/77, *Commission* v. *Ireland* [1978] ECR 417.
73 Case C-258/89, *Commission* v. *Spain*, [1991] I-3977, Opinion of Advocate-General Darmon, paragraph 46.
74 *Ibid.*, paragraph 56.
75 Case C-258/89, [1991] ECR 3977.
76 Case 61/77 *Commission* v. *Ireland* [1978] ECR 417.
77 Case C-258/89, [1991] ECR 3977, para 9.
79 Churchill R.R., *EC Fisheries Law*, pp. 67–81, *op. cit.*, fn 78. See Case 258/89, paragraph 12.
80 Case C-286/90, [1992] ECR 1-6019. Discussed Chapter 3 *supra*.
81 Case C-286/90, paragraph 22.
82 Case C-286/90, paragraphs 26–27.
83 Case C-286/90, paragraphs 28–29.
84 See Chapter 2, *ante*.
85 Case C-405/92, [1992]ECR I-6133, paragraph 12.
86 The Court referred to Joined Cases 3/76, 4/76, and 6/76, *Kramer*, [1976] ECR 1279; Case 61/77, *Commission* v. *Ireland*, [1977] ECR 937, Case C-286-89, *Poulsen et Diva Navigation*, [1992] ECR 1-6019.
87 Third Conference of LOS Convention, Official Documents, Vol XVII, 1984, Doc. A/Conf.62/122 and corrections, 157–231.
88 The Court cited its decision in Case C-286/90, *Poulsen et Diva Navigation*, [1992] ECR 1-6019, to support this point, Case C-405/92, [1992] ECR I-6133, paragraph 14.
89 Case C-405/92, [1992] ECR I-6133, paragraph 15.
90 There had been some dispute as regards the Community's competence to regulate over EEZs and the high seas, see the authorities cited, *op. cit.*, fn 65.
91 See, Somsen, H., case note, Case C-405/92, *Etablissements Armand Mondiet* v. *Armement Islais Sarl.*, [1992] ECR I-6133, in *European Environmental Law Review*, (1994), 117–121.
92 This is an important assertion which may be of benefit in assessing the inter-relationship between the European Union and the Member States at environmental meetings and negotiations, where foreign partners are occasionally uncertain about how responsibilities are allocated in the Community. Some of the legal and institutional aspects are fleetingly raised by Donoghue, J.E., 'EC Participation in Protection of the Marine Environment', *Marine Policy* (20) (1993), 515–518. The division of competence in mixed agreements has been ruled upon by the Court of Justice in Case C-25/94, *Commission of the European Communities* v. *Council of the European Union supported by United Kingdom of Great Britain and Northern Ireland*, [1996] ECR, I-1469, discussed in Chapter 2, *ante*.
93 On the application of principle of subsidiarity, see Chapters 2 and 3, *ante*.
94 Case C-405/92, *Etablissements Armand Mondiet SA* v. *Armement Islais SARL.*, [1992] ECR I-6133, paragraph 32.
95 Case C-331/88, *The Queen* v. *Minister of Fisheries and Food and Secretary of State for Health, ex parte: Fedesa and others*, [1990] ECR 1 4023.
96 Case C-405/92, *Etablissements Armand Mondiet SA* v. *Armement Islais SARL.*, [1992] ECR I-6133, paragraph 36.
97 Case C-405/92, *Etablissements Armand Mondiet SA* v. *Armement Islais SARL.*, [1992] ECR I-6133, at para 12.

98 Case C-286/90, *Anklagemydigheden* v. *Poulsen and Diva Navigation Corp.*, [1992] ECR I-6019, at paragraph 24.
99 Joined Cases 3/76, 4/76, and 6/76 *Kramer* [1976] ECR 127, at paragraphs 30–33; Case C-258/89 *Commission* v. *Spain*, [1991] ECR I-3977.
100 Case C-286/90, *Anklagemydigheden* v. *Poulsen and Diva Navigation Corp.*, [1992] ECR-6048 at paragraph 9. For the Court's application of the rules of the law of the sea, see paragraphs 25–27 of that judgement.
101 On the issue of high seas fishing in the Mediterranean Sea, see, in particular, Scovazzi, T., 'The Specific Problem of High Seas Fishing in the Mediterranean Sea', *Development Law Studies* **1**, (FAO, Rome, 1995).
102 *Ibid.*
103 For a detailed discussion of the Italian fisheries sector see *Regional Socio-economic Study in the Fisheries Sector – Italy*, Commission, DG XIV, XIV/418/92.
104 Communication from the Commission entitled, *The Use of Large Scale Driftnets Under the Common Fisheries Policy*, COM (94) 50 final, Brussels, 08.04.1994, p.5.
105 *Ibid.*
106 A concise summary of these measures is presented by Scovazzi T., *op. cit.*, fn 101, pp. 10–13.
107 *Ibid.*
108 *Ibid.* This would appear to be a legitimate concern because of the close proximity of other Mediterranean nations with fishing fleets which may readily access the high seas area which is adjacent to the Italian coastal zone. There are also some Japanese and Korean vessels which are active in international waters in the Mediterranean, although there appears to be little evidence that these vessels are using driftnets.
109 Communication from the Commission entitled, *The Use of Large Scale Driftnets Under the Common Fisheries Policy*, COM (94) 50 final, Brussels, 08.04.1994, p. 13.
110 This section is reproduced from the Commission Communication, *Ibid.*
111 A point noted in the evidence presented to the House of Lords Select Committee on the European Communities, *Regulation of Driftnet Fishing, with evidence*, HL paper 77, 19 July 1994, session 1993–1994, report 13, *passim.*
112 Following UN Resolutions 44/225 and 45/197, Japan, the Republic of Korea and Taiwan entered into arrangement with the United States and in the case of Japan, with Canada. The agreement provided for, *inter alia*, a scientific review of fishing in the North Pacific, the placing of satellite transponders on driftnet vessels to allow for real-time monitoring; time and area restrictions on the squid fishery to restrict the catching of salmonids; the deployment of Japanese, Taiwanese and Korean enforcement vessels on the fishing grounds; the exchange of observers; and the marking of gear. Report of Secretary-General United Nations 1991, *op. cit.* fn 14. With regard to the use of satellite surveillance to monitor the activities of vessels engaged in drift-net fishing, see, Chapter 11, *post.*
113 It was the Commission opinion that the fleet affected by the prohibition should be assisted by structural adjustment measures under the financial instruments for fisheries guidance (FIFG) to readjust to other types of fishing. These instruments are discussed in Chapter 1, *ante.*
114 Commission Paper, SEC (94) , Brussels, see also report in the *Financial Times*, 06.08.1994.
115 See statement and document issued by the Commission's *Service du Porte Parole*, 05.08.94, IP/94/785, entitled *Conclusions de la reunion du 29 Juillet sur la pêche au thon albacore (Commission; Espagne, Royaume-Uni, Ireland, et France).*
116 See, 'Tuna war threatened as Spanish attack EU rivals', *Financial Times*, 05.08.1994, p. 12; Reuters report 09/8/1994 and *Financial Times* 18.08.1994.
117 *Ibid.*
118 *Ibid.*
119 *Ibid.*

120 *Ibid.*
121 See, *inter alia*, leading article in the *Financial Times*, 10.08.1994, entitled 'Balancing the Scales', *The Economist*, article entitled 'Gunboat diplomacy in Europe's Seas', 13.08.1995.
122 Commission Paper, SEC (95) 2259, Brussels, 14.12.1995. See also Commission reply to Parliamentary Question E-103/96, 26.01.1996, *OJ* C 161/41, 05.06.96.
123 During the course of the campaign, the vessel was visited by the Fisheries Commissioner Bonino who participated in the monitoring of the inspection of vessels using driftnets.
124 *Op. cit.*, fn 122.
125 See also Commission reply to Parliamentary Question E-103/96, 26.01.1996, *OJ* C 161/41, 05.06.96.
126 Commission Paper, SEC (95) 2259, Brussels, 14.12.1995. See also Commission reply to Parliamentary Question E-103/96, 26.01.1996, *OJ* C 161/41, 05.06.96.
127 *Ibid.*
128 See written memorandum submitted by the non-governmental organisation 'Greenpeace' to the House of Lords Select Committee on the European Communities, *Regulation of Driftnet Fishing, with evidence*, HL paper 77, 19.07.1994, session 1993–1994, report 13, pp. 53–55.
129 Discussed *infra.*
130 The Commission has no legal competence in respect to private parties involved in, for example, a gear dispute or conflict on the fishing grounds. It may, nevertheless, bring senior political representatives in Member States together for bilateral discussions, as Commissioner Paliokrassas did in 1994 between Spanish and French Ministers after serious incidents in the driftnet fishery in 1994.
131 **Article 169** {226} of the EC Treaty of Rome actions are discussed in Chapter 8, *ante.*
132 The Commission usually initiates an administrative inquiry, as discussed in Chapter 6, to collect sufficient evidence. Member States are obliged to cooperate, pursuant to **Article 30** of the Control Regulation. There is also the broader Treaty obligation on Member States under **Article 5** {10} EC Treaty which obliges Member States to cooperate with the Commission in its investigation. Failure to do so may itself result in proceedings under **Article 169** {226}, on this point see Case 240/86, *Commission* v. *Greece*, [1988] ECR 1835.
133 Case 7/71, *Commission* v. *France*, [1971] ECR 1003, at 1026.
134 See discussion Chapters 2 and 8, *ante*, and the Commission Control Report 1996.
135 See report in the *Financial Times*, 06.08.1994, which records the presence of five EU inspectors monitoring the activities of the national inspectors in the fishery in 1994; and the Commission reply to Parliamentary Questions E-1696/94; E-1697/94, *OJ* C 30/9, 06.02.1995; E-1917/94, 06.09.1994, *OJ* C 36/13, 13.02.95.
136 See, discussion on the financial cost of fishery enforcement in the European Community, Chapter 9, *ante.*
137 There is a legal base for joint inspection programmes in **Article 34b(2)** of the Control Regulation, discussed in Chapter 3, *ante*. There is also budgetary provision for the Community to co-finance the cost of inspector exchanges, see Chapter 9, *ante.*
138 *Law of the Sea. Report of the Secretary-General, A/44/650*, 01.11.1989, at p. 33(115).
139 *Ibid.*, at pp. 33–34(115).
140 Hurlock, M.H., 'The GATT, United States law and the environment: a proposal to amend the GATT in light of the tuna/dolphin decision', *Columbia Law Review*, **29**, (1992) p. 2098–2161, and authorities cited therein.
141 See Burke W.T., *op. cit.*, fn 27, p. 272.
142 104 Stat. 4436, 4441(1990) (codified at 16 USC § 1826).
143 H.Con.Res. 113, 102nd Cong., 1st Sess. (9 July 1991).
144 MFCMA § 206, 16 USC § 1826 (1990).
145 22 USC § 1978 (1988).
146 16 USC § 1371 (a)(2)(E)(1988 & Supp. II 1990).

147 16 USC § 1537 (a)(2)(E)(1988 & Supp. II 1990).
148 16 USC § 1371 (a)(2)(E)(1988 & Supp. II 1990).
149 *Ibid.* § 1822 (1994).
150 16 USC § 1826(c) (1994).
151 See 22 USC § 1978(a)(2)(1988).
152 *Ibid.* § 1978(a)(2)(1988).
153 *Earth Island Institute* v. *Mosbacher*, 746 F. Supp.964 (N.D. Cal. 1990).
154 This decision has resulted in much literature in the United States, including, *inter alios*: Christensen, E., 'GATT Sets its Net on Environment Regulation: the GATT Panel Ruling on Mexican Yellowfin Tuna Imports and the Need for Reform of the International Trading System', *University of Miami Inter-American Law Review* **23**, (Winter, 1991–1992), 569–612; by same author, 'Making GATT Dolphin-Safe: Trade and Environment', *Duke Journal of Comparative and International Law*, **2**, (Spring 1992), 345–366; Manard, J.P., 'GATT and the Environment: the Friction Between International Trade and the World's Environment - the Dolphin and Tuna Dispute', *Tulane Environmental Law Journal* **5** (May, 1995), 373–428; Hurlock, M.H., 'The GATT, United States Law and the Environment: a Proposal to Amend the GATT in Light of the Tuna/Dolphin Decision', *Columbia Law Review* **29** (1992), 2098–2161; Blackwell, A., 'The Humane Society and Italian Driftnetters: Environmental Activists and Unilateral Action in International Environmental Law', *North Carolina Journal of International Law and Commercial Regulation* **23(2)** (1998), 2, 313–340.
155 Pub.L.No. 102–523, 106 Stat. 3425.
156 16 USC § 1371(a)(2)(E)(1988 & Supp. II 1990.
157 See Joseph, J., 'Tuna-Dolphin Controversy in the Pacific', *ODIL* **25**, p. 26. In 1996 the House of Representatives voted overwhelmingly to support a Bill implementing the Panama Declaration agreement on protecting dolphins in the eastern tropical Pacific, report in the *Journal of Commerce*, 5 August 1996. The Bill aimed to end the unilateral US embargo on tuna imports from Colombia, Costa Rica, Italy, Japan, Mexico, Panama, Vanuatu, and Venezuela, and would change the legal definition of 'dolphin safe' can labels from the present one (encirclement was not used to harvest tuna) to a new one (no dolphins were seen by accredited observers aboard to die in the harvest). Supporters of the bill argued that improvement in encirclement technology has reduced dolphin mortality to a negligible number. The Bill required US Senate approval in September 1996 before Congress adjourned prior to the Presidential election, the Senate failed to act and consequently the bill was never enacted.
158 *The Humane Society of the United States, Humane Society International, Defenders of Wildlife, Royal Society for the Prevention of Cruelty to Animals, Whale and Dolphin Conservation Society, and Earth Island Institute* v. *Ron Brown, Secretary of Commerce, and Warren Christopher, Secretary of State*, United States Court of International Trade, 920 F. Supp. 178, 183–84; Order, entered 18 March 1996, Court No. 95-05-00631. Interestingly, in this case the State Department argued that Italy should not be identified as a large-scale driftnet fishing nation, an argument that the Court rejected.
159 See *Inside US Trade*, 29.03.1996.
160 See report of press conference, Commissioner Bonino, Rome, 1 July 1996, TelexPress, 02/07/96; and report *Agence Europe*, 12/08/1996.
161 State Department Press Release 31.07.1996.
162 *Op. cit.*, fn 17.
163 The text of the statement states that: 'The Department of State is pleased with the commitment outlined in the agreement. We will, however, be monitoring implementation of the agreement closely to ensure that the commitments are carried out. Should illegal driftnet use persist, the United States Government will consider additional measures to solve the problem', *op. cit.*, fn 161.

164 Duff, A.J., 'Recent Applications of United States Laws to Conserve Marine Species Worldwide: Should Trade Sanctions be Mandatory?', *Ocean and Coastal Law Journal* **2(1)** (1997), 1–31.
165 Henkin, L., *International Law: Politics and Values*, (Dordrecht, 1995), p. 263.
166 *Ibid.*
167 See *inter alia*, Burke, W., 'Unregulated High Seas Fishing and Ocean Governance', and Chandler, M.P., 'Recent Developments in the Use of International Trade Restrictions as a Conservation Measure for Marine Resources', in Van Dyke, Zaelke, Hewison (eds) Freedom for the Seas in the 21st Century - Ocean Governance and Environmental Harmony (Washington DC, 1993), Chapters 17 and 21.
168 FAO Code of Responsible Fisheries, paragraph 11.2.4, (Rome, 1995).
169 *Op. cit.*, fn 164.
170 COM (94) 50 final of 08.03.1994.
171 *OJ* C 305, 31.10.1994.
172 Council Regulation (EC) No. 894/97 consolidated all technical measures adopted at Community level, and codified Council Regulation (EEC) No. 3094/86, *op. cit.*, fn 40, see Chapter 1, *ante.*
173 Council Regulation (EC) No. 1239/98 of 08.06.1998 amending Regulation (EC) No. 894/97 laying down certain technical measures for the conservation of fishery resources, *OJ* L 171 of 17.06.1998, p. 1.
174 Council Decision No. 97/292/EC of 28.04.1997 on a specific measure to encourage Italian fishermen to diversify out of certain fishing activities. *OJ* L 121/20, 13.05.97.
175 Council Decision No. 1999/27/EC of 17.12.1998 on a specific measure to encourage diversification out of certain fishing activities and amending Decision No. 97/292/EC, *OJ* L 8/22, 14.01.1999.
176 Some of the points in this paragraph are taken from the Communication from the Commission entitled, *The Use of Large Scale Driftnets Under the Common Fisheries Policy*, COM (94) 50 final, Brussels, 08.04.1994, *passim.*
177 See Burke, W., *The New International Law of Fisheries, op. cit.*, fn 27, p. 251.
178 *Ibid.*
179 See Chapter 11, *post.*
180 See Chapter 3, *ante.*

Chapter 11
New Developments in Fishery Law Enforcement

Introduction
This chapter examines two developments which will influence the future enforcement of fishery law. The first relates to the introduction of modern satellite technology to track fishing vessels. The second relates to three international legal instruments which, *inter alia,* address the issue of compliance with conservation and management measures, as well as providing a legal basis for more effective enforcement by flag States, port States and coastal States.

Section (i) Satellite technology and fishery law enforcement

Introduction
In the late 1970s the fisheries sector become one of first commercial user groups to use satellite technology to support and enhance the effectiveness of their industry. Today, many modern fishing vessels are equipped with satellite navigation systems, satellite communication and related equipment to receive and transmit satellite-generated images of sea surface temperature and other meteorological data. The fishing industry has reaped the benefits of modern technology. Contemporaneous with this development, satellite technology was used to survey the use of farm land for the purpose of monitoring compliance with the common agriculture policy, as well as a means to track protected and endangered species such as albatrosses as they follow their migratory routes across the Southern Ocean. The utility and versatility of satellite technology as an information gathering and surveillance tool is no longer disputed. It was thus somewhat inevitable that fishery managers would look at the use of satellite, computer and electronic technologies as an appropriate means to improve compliance with the rules which underpin fishery regulatory systems.

As is evident from the discussion in Chapter 2, the European Commission has long advocated the introduction of satellite tracking to monitor fishing vessels. Hence, it comes as no great surprise that the Control Regulation provides the legal base for the introduction of a satellite tracking system which will permit Member States to permanently monitor the activity of 4000 Community vessels by the year 2000. This is a particularly significant development because, for the fishery enforcement officer, one of the most difficult and practical problems is to verify the physical location and movement of vessels. A major part of the enforcement cost and effort is expended on simply finding out where vessels are fishing. Other than sighting reports from fishery patrol vessels and radio reports from fishing vessels, some Member States depend on aerial surveillance to monitor fishing activity within their exclusive fishery limits. The fact that several Member States have no airborne surveillance and the consequent reduction in the effectiveness of sea inspection was noted as a weakness by the Commission in national enforcement systems during the first phase of the CFP.[1] The technical and operational shortcomings of sea and aerial surveillance are numerous and add greatly to the expense of fishery enforcement. The introduction of modern technology will make surveillance simpler, safer, more effective and less expensive.

1. *Vessel monitoring systems*

A vessel monitoring system is any automated system capable of gathering information on the activities of fishing vessels. Information gathering may entail the use of coastal radar, electronic position recorders installed on board vessels, or alternatively the use of satellite communication systems to transmit data from the fishing vessel to a land-based earth station (LES). The LES distributes the data to a Fisheries Monitoring Centre (FMC) via the public telephone or data network. Vessel monitoring systems are generally store and forward systems, that is to say, it may take from a few minutes to several hours for the data sent from the fishing vessel to reach its final destination. A sophisticated vessel monitoring system may entail the use of earth observation satellites or satellites which rely upon synthetic aperture radar (SAR) to detect all vessels in a particular ocean region. The latter type of systems are commercially available and are passive systems in so far as they do not require the active or voluntary participation of the vessel in the surveillance scheme. However, it is not possible to obtain positive identification of the target vessels and consequently this is a major limitation from the fishery management point of view.

It is important to distinguish the different types of vessel monitoring system which may be used for fishery enforcement purposes. In the European Union, for example, the vessel monitoring systems involve the active participation of vessels in a scheme of surveillance through the installation of a tracking device on board the subject vessels and the transmission of data using communications satellites to Fisheries Monitoring Centres in the flag Member State. In such cases a vessel monitoring system has three main features.[2] Firstly, it requires to be capable of *data retrieval* from fishing vessels, the data as a minimum may include the vessel's position, speed, and course. The device installed on board the fishing vessel is euphemistically referred to as the 'blue box' (the colour of the box used in the initial test phase in the European Community). Essentially, the blue box contains a navigational positioning system such as a global positioning system (GPS) and a radio communications component supported by a satellite service provider.[3] The blue box transmits data from the vessel to a land-based earth station (LES) via a communications satellite. The precise configuration of the blue box may vary depending on which system is supplied. The second feature, is that the system must allow *data management* by the Fishery Monitoring Centre in the flag State. This requires the data received from the blue box to be matched to vessel details and analysed so as to determine whether the vessel is complying with the regulations. Information regarding suspected irregularities can be passed to the appropriate enforcement authorities. It is also envisaged that the system should be capable of providing statistical information on fishing effort such as the number of days a vessel spends in a given fishing effort area. Thirdly, the system must be capable of *data dis-*

tribution. Information retrieved from the system may need to be passed to the flag Member State of the vessel or the coastal Member State if the vessel is operating in the waters under the sovereignty or jurisdiction of a coastal State. Similarly, there may be a requirement to transmit the data to a regional organisation. In this regard it is essential to the integrity of the system that the data content and distribution would remain confidential as vessel data are of significant commercial and professional value.

A typical configuration for such a system is illustrated in Figure 11.1.

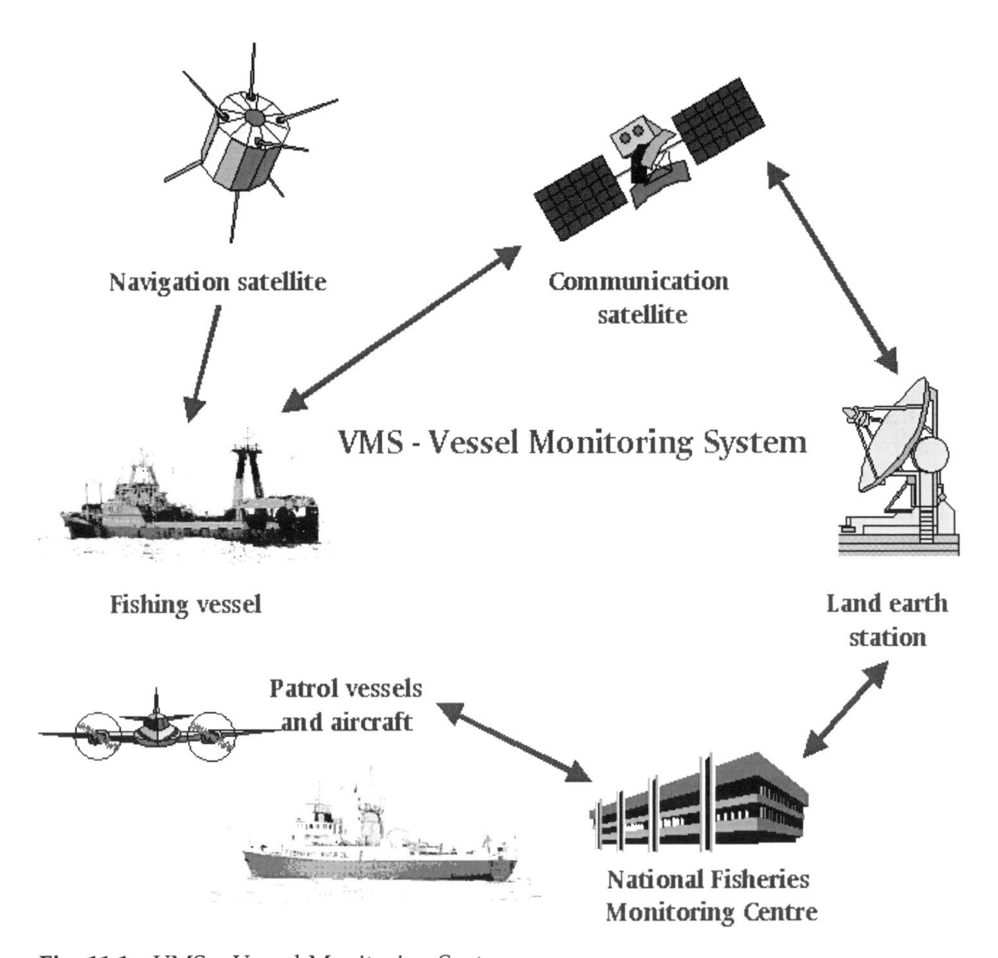

Fig. 11.1 VMS – Vessel Monitoring System.

2. *Continuous position monitoring of fishing vessels by a satellite-based surveillance system in the European Union*

The Commission proposal in 1992 (to improve the effectiveness of surveillance of fishing vessels) provided for the implementation of a satellite surveillance system. In the negotiations which led to the adoption of the new control structure in 1993 the proposal for satellite surveillance was rejected by the Member States. However, in the framework of an overall compromise it was agreed that the 1993 Control Regulation would place an obligation on the Council to decide, before 1 January 1996, if and to what extent a land-based or satellite continuous monitoring system would be installed on Community fishing vessels at a later date.[4] In the intervening period between the adoption of the 1993 Control Regulation and 1996, the feasibility of this type of technology for fishery enforcement was assessed by a pilot project carried out by Member States.[5]

3. *The pilot project to test the feasibility of satellite technology as a means to monitor fishing vessels*

The pilot project was in effect a feasibility study to examine the practical application and effectiveness of satellite systems for the surveillance of fishing vessels. Thirteen EU Member States (Belgium, Denmark, Germany, Greece, Spain, France, Ireland, Italy, the Netherlands, Portugal, Finland, Sweden and the United Kingdom) carried out tests to monitor 350 vessels.[6] Three different, commercially available, satellite-based vessel monitoring systems were tested and several Member States tested more than one of these systems.[7] Portugal carried out a unique programme entitled MONICAP. Finland and Sweden set up a joint pilot project with Denmark. Greece tested a marine radio data communication system, the operational range of which was curtailed to short distances between the subject vessels and the mother data retransmission patrol vessels. The United Kingdom conducted trials with an alternative system to satellite surveillance, referred to as an Automatic Position Recorder (APR), which was limited to storing data onboard the fishing vessel without transmitting the information in real-time to a land-based tracking station. The pilot project cost 10 million ecu which was financed from the Community fishery budget. The test projects in the Member States started in July 1994 and were completed by December 1995.

In the broader context of fishery enforcement in the European Community, the manner in which the pilot projects were established and completed is one of the first illustrations of close co-operation and co-ordination between various Member State enforcement agencies.[8] Each Member State was obliged to establish a Fisheries Monitoring Centre (FMC) to track the position of vessels included in the pilot project. The flag Member State received position data from vessels flying its flag at regular intervals. If the vessel was operating in the waters under the jurisdiction or sovereignty of another Member State, the flag State FMC re-transmitted the position data to the relevant coastal Member State. Denmark, Finland and Sweden operated a joint project, in which common hard- and software were installed in Copenhagen and data were distributed to Helsinki and Stockholm.

The report from the Commission to the Council and the European Parliament on the establishment of a satellite-based vessel monitoring system for Community fishing vessels, records that the pilot projects proved the reliability of real-time satellite position monitoring equipment on board fishing vessels and established that this type of technology will greatly enhance the efficiency and effectiveness of the traditional aerial, surface and land-based control resources.[9] The report recommended the phased introduction of a satellite monitoring system to monitor 10,000 Community vessels. In 1994 the cost for such a scheme was estimated to be in the region of 80 to 100 million ecu.[10]

4. Commission proposal for an operational vessel monitoring system

The Commission proposal for a Council Regulation amending the 1993 Control Regulation provided for the implementation by the Member States of a satellite-based monitoring system for tracking their respective vessels as well as the means to communicate vessels position reports to the appropriate Member State.[11] The proposal envisaged that the system would be introduced for all vessels exceeding 15 metres during the period 1997–1999, commencing with vessels that operate in the most sensitive fisheries such as the driftnet fisheries. In addition the proposal contained elaborate provisions to ensure co-operation between Member States.

At its meeting on 10 June 1996, the Council examined for the first time the Commission proposal and agreed to take a decision on the issue before the end of 1996. Subsequently, the Council formally approved an amendment to the 1993 Control Regulation to provide for the introduction of a satellite surveillance system.

5. Opinion of the European Parliament

In line with their previous call for a more robust and effective fishery enforcement regime, the European Parliament supported the Commission proposal to introduce a satellite surveillance system for Community fishing vessels.[12] Importantly, the Parliament advocated the financial participation by the European Union in the setting up of such a system. The Parliamentary report on the system stresses the importance of the system being applied fairly in all Member States and of not imposing an excessive administrative burden on fishermen. Interestingly, the Parliament recommended that each Member State impose the appropriate sanctions for failure to comply with the satellite surveillance system.[13] This recommendation was in line with the previous tough stance taken by the Parliament on the issue of sanctions. Furthermore, the Parliament was keen to include vessels using driftnets longer than one kilometre within the first phase of application of the Community Regulation. The latter point is of particular significance as the Community driftnet fleet generally operates outside the normal theatres of operation of Member State fishery inspection vessels which generally limit their patrols to the waters under the sovereignty and jurisdiction of their respective Member States.[14]

6. Council Regulation establishing a satellite-based surveillance system

On 14 April 1997 the Council formally approved and adopted an amendment to the 1993 Control Regulation in order to introduce a satellite-based vessel monitoring system for fishing vessels.[15] The regulation envisages the introduction of a satellite surveillance system in two phases. In the first phase, which commenced on 30 June 1998, vessels exceeding 20 metres between perpendiculars (24 metres overall) in the following categories are required to be equipped:

- vessels operating on the high seas, except in the Mediterranean Sea;
- vessels operating in the waters of third countries, provided provisions have been

made in Agreement with the relevant third country or countries for the application of a VMS to the vessels of such a country or countries operating in the waters of the Community;[16]
- vessels catching fish for reduction to meal and oil.[17]

In the second phase, which commences on 1 January 2000, all vessels exceeding 20 metres between perpendiculars (24 metres overall) are included in the system.[18] There is, however, an exception for vessels operating exclusively within 12 nautical miles of the baselines of the flag Member State, and for vessels which operate at sea for less than 24 hours.[19]

The devices fitted on board the fishing vessels shall enable the vessel to communicate by satellite to the flag State and to the coastal State concerned simultaneously its geographical position and in some instances the effort reports stipulated in the Control Regulation.[20] Furthermore, the onus of responsibility is placed on the master of the vessel to ensure that the devices are operating in accordance with the regulations.[21]

An obligation is placed on Member States to establish and operate Fisheries Monitoring Centres which will be equipped with the appropriate staff and resources to enable Member States to monitor the vessels flying their flag as well as the applicable vessels flying the flag of other Member States and third countries.[22] Further detailed rules for the implementation of the system are prescribed in a Commission Regulation.[23]

7. *Third-country vessels operating in the Community fishing zone*

From 1 January 2000, at the latest, all third-country fishing vessels exceeding 20 metres between perpendiculars (24 metres overall) which operate in the Community fishing zone are required to be equipped with a VMS position monitoring system approved by the Commission.[24]

8. *The Commission Application Regulation on a satellite-based vessel monitoring system*

In July 1997, the Commission adopted a detailed regulation prescribing the rules for the application of a satellite-based monitoring system.[25] The regulation specifies functional standards such as that devices installed on board the vessels are capable of transmitting:

- the vessel identification;
- the most recent geographical position of the vessel, with a position error which shall be less than 500 metres, with a confidence interval of 99%; and
- the date and time of the fixing of the said position of the vessel.

Moreover, the number of reports vessels have to make are specified in the regulation.[26] Significantly, vessels operating in the Community zone have to send messages every two hours, where as vessels operating in the Mediterranean and in the NAFO Regulatory area have only to send reports every 12 hours. Vessels in port have to transmit one position report per day and this ought to be sufficient to verify the accuracy of fishing effort or a days-at-sea regime.[27] Vessels using a system which does not permit polling (the facility to check the position and data pertaining to a vessel at any time) have an obligation to report their position every hour.[28] Interestingly, the difficult issue of transmission of data to the coastal Member State is overcome by specifying in the regulation that the data are to be transmitted to the coastal Member State and the flag Member State simultaneously. The regulation does not specify what constitutes a simultaneous transmission. This places an emphasis on a technical solution to the issue of data exchange and avoids resolving, or addressing, divisive issues pertaining to which Member State (i.e. the flag Member State or the coastal Member State) should be given priority to data access. In effect, both Member States should have information on the vessel position at the same time, thus the difficult issue of flag State competence *vis-à-vis* coastal State competence does not arise. In this regard, it is important to note that the issue of priority on data access is academic provided that enforcement personnel in the appropriate Member State have access to vessel position reports in sufficient time to take the necessary enforcement action or inspections if this should be required. The simultaneous data transmission requirement is also consistent with the fishing effort regime which requires effort reports to be sent to both the flag Member States and the coastal Member States if a vessel is operating in waters under the sovereignty and jurisdiction of a coastal Member State.[29] The regulation also places responsibility on the master of fishing vessels to ensure that the devices function correctly and stipulates that the obligation is on the master to have defective equipment repaired expeditiously. Indeed, vessels in most instances are only allowed one month to repair defective devices. During such periods the master is obliged to report the position of the vessel by other communication means such as by radio. Finally, there is a provision which provides that the Commission may have access to satellite data on specific request to the Member States.

9. *External programmes on satellite tracking*

Other than playing a leading rôle in the establishment of satellite surveillance as a mandatory enforcement tool, the Community has continued to participate in the work of various international fisheries organisations which are researching the use of this type of technology in the monitoring of vessels. Some of these organisations, such as the Commission for the Conservation of Antarctic Marine Living Resources (CCAMLR) and International Commission for the Conservation of Atlantic Tunas (ICCAT), are considering the introduction of satellite monitoring for international fisheries. The Community has taken the initiative by funding and implementing several external VMS programmes. These programmes have been established as a result of obligations arising from participation in regional fishery organisations such as North Atlantic Fisheries Organisation (NAFO), the North-East Atlantic Fisheries Commission (NEAFC), as well as within the framework of bilateral fisheries agreements with third countries. These programmes include the following.

(i) NAFO Pilot Project for satellite tracking (1996–1997)

NAFO Contracting Parties agreed to implement, during the period from 1 January 1996 to 31 December 1998, a Pilot Project to provide for satellite tracking devices on

35% of their respective vessels fishing in the NAFO Regulatory Area in the North-West Atlantic. As a NAFO Contracting Party, the European Community participates in this pilot project. In practice, Spain and Portugal track vessels operating in the NAFO Regulatory Area and send position data to the European Commission in Brussels which forwards the data to the NAFO Secretariat in Halifax. The NAFO Secretariat is obliged to make this information available to all NAFO Contracting Parties with inspection vessels operating in the NAFO Regulatory Area.

(ii) Fisheries agreement between the Kingdom of Morocco and the EU
In 1995, the European Union and Morocco concluded a four-year fisheries agreement that allows mainly Spanish fishing vessels to fish in Moroccan waters.[30] This agreement has significant fishery enforcement provisions which include the establishment of a pilot project for satellite monitoring. Vessel tracking in the Moroccan fisheries zone will allow a direct control of the provisions concerning fishing effort and geographical restrictions.

(iii) Fisheries agreement between the Islamic Republic of Mauritania and the EU
In 1996 the EU and Mauritania concluded an Agreement in the sea fisheries sector.[31] The Agreement stipulates that pending the implementation of a national satellite monitoring system for fishing vessels of similar type operating in Mauritania's fishing zone, both Parties agree to implement a bilateral satellite tracking project for Community vessels. Similar to the Moroccan Agreement, vessel tracking in the Mauritania fisheries zone will allow a direct control of the provisions concerning fishing effort and geographical restrictions.

10. Satellite monitoring – outside the European Union

Outside the European Union several third countries have taken their own initiatives to introduce satellite technology as a means to improve regulatory compliance. In April 1990, an international agreement between the United States of America, Canada and Japan required that satellite monitoring be placed on 100% of the Japanese squid and large-mesh driftnet fishing vessels operating in the North Pacific in 1990.[32] Similar agreements were reached between the United States and Korea and the United States and Taiwan. The United States has monitored the operations of nearly 800 fishing vessels on the high seas. Several American longline vessels have been equipped with satellite surveillance devices in the Western Pacific pelagic longline fishery and a scheme was introduced for two specific fisheries in the New England region (groundfish and scallops). In 1994, Australia officially implemented its first Vessel Monitoring System for 30 vessels in a deep sea Orange Roughy trawl fishery. Further developments include extending the coverage of the system to tuna longliners and the northern prawn fishery. Since 1994, New Zealand has required certain categories of vessels to carry and operate vessel monitoring equipment. Several other countries worldwide, namely Argentina, Canada, Chile, French overseas territories (French Polynesia and New Caledonia), Indonesia, Iran, Japan, Morocco, Norway, Peru, Russia and Taiwan have been conducting trials.

11. Satellite monitoring – international conventions and agreements

Several international agreements and conventions regulating fishing activity contain provisions regarding the use of satellite surveillance systems for fishery enforcement. These include the United Nations Implementation Agreement for Straddling Stocks and Highly Migratory Species which provides that the flag State is obliged to take measures to monitor and control the fishing activities and related activities of vessels flying its flag by, *inter alia,* the development and implementation of vessel monitoring systems, including, as appropriate, satellite transmitter systems, in accordance with any national programmes and those which have been subregionally, regionally or globally agreed among the States concerned.[33] There are provisions in both the Agreement to Promote Compliance with International Conservation and Management Measures by Fishing Vessels on the High Seas (Compliance Agreement) and the Code of Conduct for Responsible Fisheries[34] which require the flag State to monitor the activities of its vessels, although in both the latter two Agreements there are no express provisions stating that there is a requirement for the development of a satellite system. The Bering Sea Convention also requires each contracting party to monitor its fishing vessels that fish for pollock in the convention area by using satellite technology while vessels operate in the Bering Sea. Moreover, as noted above, several regional organisations such as CCAMLR (Antarctic), ICCAT (Atlantic tuna), Forum Fisheries Agency (South Pacific) and NAFO (Northwest Atlantic), are in the pilot project phase to test the technology. The ICCAT project is based on the European Community model and involves the active participation of 25 Contracting Parties. It is the most elaborate pilot project to date and will involve the monitoring of 250 vessels.

12. Commentary on the concept of satellite surveillance of fishing vessels

The attractiveness of a satellite-based continuous monitoring system is very apparent.[35] The system adopted for the Community fleet has no doubt been influenced by the success of satellite monitoring programme schemes introduced elsewhere in the world, particularly, as in recent years, many nations outside the European Community have resorted to technological innovations such as monitoring vessel locations by satellite as a means of improving the enforcement and effectiveness of their fishery management regimes. Furthermore, because the renewed emphasis on 'closed boxes' and 'sensitive areas' (such as the Shetland Box and the Mackerel Box)[36] as conservation tools, as well as the complex access regime that implemented as a consequence of the Iberian Accession readjustment in 1995, there is an urgent need for a more sophisticated approach to enforcement. In this regard, it is beyond dispute that the satellite surveillance system will complement and improve existing enforcement tools. However, the use of this particular type of satellite technology will not provide conclusive evidence of anything other than the real-time information (i.e. the position, course and speed of a vessel). In short, it does not affect the requirement to undertake inspections at sea but perhaps will replace or reduce, in due course, the requirement for aerial surveillance.

The full value of the system will largely depend on how it is integrated with existing control mechanisms in the Member States.[37] The system, nonetheless, may have some tangible benefits for quota management, particularly as the present system for quota allocation is based on an ICES division quota allocation system which is frequently circumvented by the misreporting of fishing areas. It may also be advantageous in monitoring distant water fleets, particularly those operating on the high seas, which the EU has international obligations to control. The surveillance of Community vessels operating in the Southern Ocean as well as vessels fishing for highly migratory species such as tuna in the Indian ocean, come within the scope of the system. This is an important element because the European Union is occasionally criticised by the international community for not taking sufficient measures to control the activities of Community vessels fishing outside the Community maritime area. It thus appears that the Community has taken a significant initiative by introducing satellite technology as an enforcement tool. The scale of this initiative is evident when it is considered that international fora such as CCAMLAR and ICCAT have been slow to introduce mandatory satellite surveillance systems for the respective fisheries.

A positive aspect of the vessels monitoring system introduced by the Community is that the issue of who should bear the cost of the system has been resolved, with the Community underwriting a significant amount of the cost of the devices to be installed on board the fishing vessels and in the flag Member States. Interestingly, vessels which are obliged to install the devices in order to comply with national legislation (which exceeds the Community legislative requirement) are also eligible for the same financial support as vessels which come within the mandatory scope of the regulation.[38] The latter is a persuasive inducement to the industry to participate on a voluntary basis in the scheme.

One outstanding issue on the satellite monitoring system, however, is its cost-effectiveness. The cost of the system will depend on the number of participating vessels and on the system(s) selected by the Member States. (i.e. the annual cost of monitoring a fleet of 4000 vessels is likely to be of the order of €8 million). This may be compared to the cost of a boarding and inspection of a fishing vessel at sea which is in the region of €2500 or the cost of a new offshore patrol vessels which is about €30 million.[39] Furthermore, it ought to be pointed out that costs may be substantially reduced if Member States and fishermen work together to choose the least expensive system that achieves the control and surveillance objectives. The benefits from the system will be derived from its utility and effectiveness as an enforcement tool to address the shortcomings in the enforcement of the CFP. In this respect the system is the only enforcement means that provides continuous information on the location of fishing vessels. This allows Member States to monitor directly the compliance with all provisions related to geographical restrictions. Benefits from satellite technology will further be achieved through the *synergy* with the conventional control means, in particular the improvement of the aerial and marine surveillance. Information provided by the system will improve the deployment of aircraft and patrol vessels. Less time will be spent searching for fishing vessels, more time will be devoted to inspection. The shore-based inspectors will benefit from the information provided by satellite surveillance as efficiency will be increased, since the inspection officers will have advance notification of possible illegal or unauthorised landings and transhipment, which have been very difficult to combat using the traditional enforcement tools. Satellite surveillance also offers valuable information with which the data in logbooks may be verified including the cross-checking of the catch area against positions recorded in the logbook. Further scope for improving management measures is provided by the facility introduced by satellite surveillance to collect more comprehensive statistics on fishing activity. Satellite monitoring also has a deterrent effect. Fishermen will be less inclined to mis-report their position and their activity, as they will be aware that the authorities are continuously monitoring their position. This form of preventive enforcement is very beneficial, however, it is difficult to quantify in monetary terms.

The principal weakness of VMS is that it requires the active participation of vessels in the system. If a vessel is not fitted with a tracking device then it cannot be monitored. However, this weakness may be mitigated by combining VMS with other surveillance technology such as aerial surveillance data or data from radar satellites. Indeed an integrated control and monitoring system would combine data from several sources including VMS, aerial surveillance, coastal radar and SAR satellite images.

13. Legal issues pertaining to the introduction of a satellite-based vessel monitoring system

In their reports on the pilot project Member States did not undertake an elaborate assessment of the legal and evidential issues pertaining to the data acquired from a satellite-based vessel monitoring system.[40] It should nevertheless be pointed out that the vessels participated in the pilot projects on a voluntary basis and that many issues such as the legal consequences of tampering or misuse of the devices fitted on board the vessels for the purpose of cloning (the unauthorised use of a legitimate unit to mislead the authorities as to the correct location of vessels) or of ghosting (the unauthorised disabling of a unit) did not arise during the period of the pilot project. It is nonetheless clear from the Community Regulation that the full onus of responsibility for the operation of the equipment, and the obligation to ensure that data are transmitted, rests with the master of the vessel.[41] Appropriate sanctions for the misuse of equipment will be dealt with in national legislation in the Member States. In this regard the issue of jurisdiction is the same as it is in respect to other Community fishery regulations, that is to say, it follows the traditional international law jurisdictional framework which distinguishes flag State powers from those of the coastal State. Member States may thus make breaches of the Satellite Regulation an offence under their criminal or administrative law, and may do so in respect to vessels flying their flag wherever they may be and in respect to vessels flying the flag of other Member States or third countries when such vessels are fishing in the territorial sea or EEZ/EFZ. Furthermore, Member States may only enforce such measures against vessels flying their flag when such vessels operate in the territorial sea, EEZ/EFZ, and on the high seas. Similarly, the coastal Member State may enforce the provisions with respect to all vessels which are subject to the regulation when such vessels operate in the waters under the sovereignty or jurisdiction of the coastal Member State.

In 1999, the Council adopted a list of types of behaviour which seriously infringe the rules of the Common Fisheries Policy. The list includes tampering with the satellite-based vessel monitoring system.

The use of VMS type of technology for fishery enforcement raises many issues regarding the evidential value of the data acquired from the systems, particularly, whether the data obtained from the system will be admissible in judicial proceedings against vessels. In this regard the admissibility and evidential value of the data provided by a satellite-based vessel monitoring system could to some extent differ in each Member State, principally because Member States have widely diverging rules of evidence. This is also the case with other types of surveillance evidence, e.g. photographs taken during aerial surveillance flights. In this regard, it is interesting to note that probative value of evidence obtained from aerial surveillance is inadmissible in a number of jurisdictions as the sole proof of a fishery violation. This may imply that certain Member States will have to introduce rules under their national law in order to enable the evaluation of evidence obtained by a satellite-based system. On this point, it is important to recall that experience in the Member States and in other countries such as New Zealand and Norway has shown that the use of aerial surveillance by aircraft has played only a limited rôle in fishery enforcement, principally because aircraft (even helicopters) can provide only limited information on the vessels observed fishing. This information, which usually consists of the identity of the vessel, its position, course and speed, is normally passed on to other inspection services ashore or at sea. Surface patrol vessels may then pursue further investigations and, where possible, conduct boarding operations. The restricted rôle and lack of probative value of aerial surveillance evidence is because the information provided does not allow the enforcement authorities to certify what type of activity the vessel sighted is engaged in, even if the net may be visible coming out of the water with the catch visible. In the latter case, the skipper might rebut accusations of illegal activity by claiming that the nets were being cleaned or checked at sea. The problem is further complicated by international legal rights such as the freedom of navigation in the EEZ/EFZ or the right to innocent passage in the territorial sea. Furthermore, under national law in some Member States, the power to redirect a vessel, or to route a vessel to port, is vested only in sea fishery inspection officers or naval inspection officers after they have formed a reasonable suspicion that an offence has been committed.[42] This may necessitate a thorough inspection of the vessel by a boarding party. In this regard, aerial surveillance of fishing vessels does not yield sufficiently precise elements of proof of an infringement, because the presence of vessels in a particular area alone can be based on many considerations such as weather conditions, innocent passage, mechanical breakdown etc. These are some of the factors which will have to be taken into consideration when Member States assess the probative value of evidence acquired from satellite surveillance systems. In all probability, Member States which rely upon criminal courts may question the *prima facie* value of satellite data for the purposes of obtaining a conviction for a serious offence such as illegal fishing in a prohibited area. On the other hand, Member States which rely upon civil courts or administrative fora to deal with fishery offences may require a less rigorous burden of proof and in such instances satellite data may be adequate evidence to establish an offence and as a means to identify the perpetrator.

Another issue relates to which parties have proprietary interests in the data acquired from the surveillance system. The Council Regulation does not deal with data access rights for vessels owners or operators. Data access rights thus remain covered by other provision in Community law and the appropriate national legislation in the Member States. Legal safeguards regarding data confidentiality are also dealt with in **Article 37** of the Control Regulation, discussed in Chapter 3. In this regard, rules on confidentiality, professional secrecy and data protection as laid down in Directive 95/46/

EC of the European Parliament and of the Council of 24 October 1995 on the protection of individuals with regard to the processing of personal data and on the free movement of such data will apply.[43]

14. Conclusion

With the introduction of mandatory satellite monitoring the European Community has taken a major initiative to improve the monitoring and control of fishing vessels. The functional nature of Community regulations which specify the fundamental requirements of the system and ought to facilitate rapid implementation of the scheme in the Member States. Indeed, by setting down the minimum standard for data formats and communications protocols, as well as by loosely defining the architecture of the system, the Community has established a blueprint for satellite monitoring which may be followed by other regional organisations and third countries. Ultimately, satellite tracking will be seen as a tangible step towards strengthening the control and monitoring of Community fishing vessels and in the long-term, this could make a significant contribution to the conservation of fishery resources. The introduction of a mandatory position monitoring system approved by the Commission for third-country vessels operating in the Community fishing zone will also improve the monitoring and control of such vessels.

Section (ii) International legal instruments having an impact on high sea fishing

Introduction

The issue of enforcement is particularly significant when viewed against the background of three international instruments which address the issue of compliance by fishing vessels with conservation and management measures. These are:

(1) The Agreement to Promote Compliance with International Conservation and Management Measures by Fishing Vessels on the High Seas (Compliance Agreement);[44]
(2) The Agreement for the implementation of the provisions of the United Nations Convention on the Law of the Sea of 10 December 1982, relating to the conservation and management of Straddling Fish Stocks and Highly Migratory Fish Stocks (the 1995 Straddling Stocks Agreement);[45]
(3) The Code of Conduct for Responsible Fishing.[46]

As these instruments constitute a new framework which will, *inter alia*, regulate high seas fisheries, it is proposed briefly to outline some of their salient features in relation to enforcement. It must be pointed out that only the 'Compliance Agreement' is confined to high seas fisheries, it, nonetheless, forms a part of the 'Code of Conduct' which addresses all fisheries. Similarly, the conservation requirement contained in the 'Straddling Stocks Agreement' also apply, *mutatis mutandis*, within waters under national jurisdiction of coastal States.

1. *The Compliance Agreement*

Since the early 1990s the FAO has undertaken the negotiation of a treaty to promote compliance and in particular to deter re-flagging of vessels as a means of avoiding compliance with applicable conservation and management rules for fishing activities on the high seas. During the FAO negotiations it became apparent that re-flagging was only one part of a more difficult enforcement problem: the limitations of exclusive flag State jurisdiction over flag vessels when operating on the high seas.[47]

The Compliance Agreement is an integral part of the Code of Conduct for Responsible Fisheries and is the only part of the Code which is legally binding on parties which have lodged instruments of acceptance. The objective of the Compliance Agreement is broad in so far as it imposes on all States whose vessels operate on the high seas several obligations aimed at ensuring that the activities of their vessels comply with international management and conservation goals. Significantly, the Compliance Agreement does not expand on enforcement powers for non flag States, either through at-sea boarding and inspection or through dispute settlement, against flag States that remain irresponsible.[48] It relies on the circulation of information on high seas fishing operations to regional fisheries bodies and to individual parties, many of which are coastal states.[49] It has been suggested that the latter may be in a stronger position to take effective compliance control action, given that most vessels operating on the high seas also operate for part of the time in coastal State waters.[50] Although on the substantive enforcement issues it is apparent that the practical benefit of the Compliance Agreement is attenuated, it should, nonetheless, be seen as providing the basis for an effective enforcement and compliance regime for vessels which have benefited from change of flag in order to circumvent international measures for the conservation and management measures by appropriate and effective means co-ordinated at international level. Thus the Compliance Agreement is a significant initiative to combat the practice of vessel owners reflagged existing, and in some cases new, vessels in countries which are not members of regional organisations as a means to avoid high seas conservation and management obligations. The Compliance Agreement received the unequivocal support of the European Community and is viewed as a significant step towards improving international co-operation and the multilateral process.[51] The Compliance Agreement requires 25 instruments of acceptance to come into force and as of 1999 only thirteen States and the European Community have submitted instruments of acceptance.

2. *The 1995 United Nations Agreement for the Conservation and Management of Straddling Fish Stocks and Highly Migratory Fish Stocks*

Arguably, one of the most significant developments in the international law of fisheries is the Agreement for the implementation of the provisions of the United Nations Convention on the Law of the Sea of 10 December 1982, relating to the Conservation and Management of Straddling Fish Stocks and Highly Migratory Fish Stocks.[52]

Although each of the substantive articles of the Straddling Stocks Agreement merit separate treatment in their own right, it is not within the remit of this chapter to examine in detail this elaborate document which comprises of 49 articles and 2 annexes.[53] Nonetheless, it calls for a number of general observations. Firstly, the Agreement contains extensive provisions on the management and conservation as well as providing a mechanism to resolve the compatibility problem between EEZs and the high seas.[54] There are also several elements in the Straddling Stocks Agreement which mirror measures in the Compliance Agreement such as duties of monitoring, control, and surveillance of vessels and the regulation of high seas transhipment of catch.[55] Secondly, the Agreement broadens coastal State authority over foreign vessels on the high seas that have previously engaged in such unauthorised fishing within the EEZ.[56] Thirdly, there are extensive provisions on the issue of compliance and enforcement in Part VI of the Agreement. The flag State is to co-operate with the coastal State in taking enforcement action, and may authorise authorities of the coastal State to board and inspect the vessel on the high seas.[57] These articles are quite complex and vest extensive boarding and inspection powers in all parties to the Agreement that are members of regional fishery organisations.[58] In some instances, if fishery violations are serious, the vessel may be directed to port for further investigation.[59] However, inspecting states may not take other enforcement action without the consent of the flag State.[60] Interestingly, with respect to a vessel which is suspected to be without nationality, a State may board and inspect the vessel and take other action as may be appropriate in accordance with international law.[61] As noted by one authority, 'the enforcement provisions reaffirm and enhance coastal State responsibility for EEZ enforcement, elaborate further on the concepts of flag State responsibility and the duty to cooperate, and give non-flag States, in the context of regional agreements, a range of authority to board and inspect vessels of other parties to the agreement.'[62] Port State authority is also considerably enhanced.[63] Finally and most importantly, the Straddling Stocks Agreement provides for a compulsory and binding dispute settlement procedure in relation to disputes occurring on the high seas.[64]

Significantly, while UNCLOS vested almost exclusive enforcement competence in the flag State, the Straddling Stocks Agreement is the first major international instrument which provides a comprehensive inventory of responsibilities for the flag State with regard to the fishing activities of vessels on the high seas, ranging from issuance of fishing licences and monitoring of fishing activities to enforcement of sanctions. Similar to the provisions in the Control Regulation,[65] the Agreement imposes an obligation on the flag State to carry out investigations and judicial proceedings 'expeditiously' and apply sanctions of such severity that they will secure compliance, discourage future violations, and deprive the benefactors from any benefits accruing from illegal activities. It exceeds the Control Regulation in so far as the Agreement requires that punitive measures shall include the refusal, withdrawal, or suspension of authorisation to serve as a master or as an officer on a fishing vessel. Indeed, this sanction is severe and appears to exceed the type and range of sanctions presently prescribed in the municipal laws of the Member States. Overall, it may be argued, nonetheless, that the Agreement is an incontrovertible acknowledgement of the fact that effective fisheries conservation and management cannot be achieved unless there is appropriate enforcement and monitoring mechanism which ensures compliance with the rules.

One commentator has suggested that there are several enforcement provisions in the 1995 Straddling Stocks Agreement which exceed the established norms of customary and conventional international law.[66] Because of the significance of this assertion in relation to the scope of the 1995 Straddling Stocks Agreement it is a useful exercise to mention these provisions. Firstly, extrapolating from the general duty of conservation in customary international law,[67] the question is posed as to whether the implementing measures prescribed in **Articles 18** and **19** of the Agreement (namely, strengthening the duties of the flag State) should also be considered as rules of customary international law. If such a supposition is accepted then failure by a non-party to the Agreement to comply with the requirements in **Articles 18** and **19** may be assimilated to an abuse of rights. That is to say, such a failure would contravene **Article 300** of UNCLOS as well as **Article 34** of the 1995 Straddling Stocks Agreement, both of which require state parties to fulfil their obligations and exercise their rights in a manner which do not constitute an abuse of rights.[68] Secondly, pointing to the qualified exclusive right of flag States in respect of vessels flying their flag,[69] one author notes that the provisions in **Articles 20** and **21**, which allow the coastal State to take enforcement action on the high seas against vessels which have engaged in authorised fishing within areas under national jurisdiction, are conventional rules and thus only apply to parties to the 1995 Straddling Stocks Agreement.[70] Such measures may not, therefore, be enforced against non-parties to the Agreement unless they have consented to their application. Thirdly, bearing in mind the right of innocent passage and freedom of navigation, the authority notes that **Article 21(14)** of the 1995 Agreement is not consistent with established international law in so far as that this provision allows a State party, which is a member of the relevant regional organisation, to board and inspect a vessel flying the flag of another State party when such a vessel is transiting (on passage) in an area under national jurisdiction during the same fishing trip. This right is triggered when there are clear grounds to believe that the vessel in question has violated the high seas conservation and management measures of the regional organisation concerned.[71] Fourthly, the improved port State enforcement measures with respect to vessels which have violated international high seas conservation measures, exceeds the provisions in UNCLOS, and is thus binding on only those States who are party to the Agreement.

In the overall context of fisheries law, a major point which requires to be emphasised is that **Articles 21** and **22** establish an unprecedented and far-reaching exception to flag-State jurisdiction. This exception allows a State party to the Agreement (and at some time a member of regional fisheries organisation) to board and inspect fishing vessels flying the flag of another State (which is neither a party to the Agreement nor a member of a regional fisheries organisation) in order to ensure compliance with the management measures adopted by the regional fisheries organisation in question. That is to say, a member of a regional organisation such as NAFO may take enforcement action within the NAFO Regulatory Area against any vessel that flies the flag of a State that is a party to the Straddling Stocks Agreement even if such a State is not a Contracting Party pursuant to the NAFO Convention. Furthermore, as stated above **Article 21** prescribes detailed rules for the boarding and inspection of vessels, where there are clear grounds for believing that a vessel has violated conservation and management measures. In addition, in some circumstances it allows the inspecting State to secure evidence and take the vessel suspected of a violation to the nearest port. As a safeguard against abuses of this provision, **Article 22** provides a code of conduct for boarding parties, and **Articles 21(12)** and **21(18)** provide that the decision of a State to fulfil its flag State responsibilities supersedes the action taken by the inspection State. Aside from these specific safeguards and the right of flag State preemption, **Article 21(18)** provides that States are liable for damage or loss resulting from actions taken that are unlawful or exceed those reasonably required to implement enforcement.

While the Straddling Stocks Agreement acknowledges the primacy of the rôle for the flag State, the provisions in **Articles 21** and **22**, nevertheless, constitute a major erosion of the exclusive nature of flag State responsibility and inspection jurisdiction to control the activities of vessels which fly their flag. The provisions place a clear requirement on the flag State to take action, but in the absence of such action, there is a clear mandate for other State parties of a regional organisation to take effective enforcement action. Interestingly, there are no similar provisions in Community law which allow for a Member State to inspect the vessels of another Member State on the high seas, a point which was raised in Chapter 10. Furthermore, there are no provisions in Community fishery law regarding the use of force to effect fishery enforcement operations. Indeed, the use of force which in some instances is regulated in the municipal law of the Member States is generally considered counter-productive and entirely inappropriate as a means to achieve harmonious and effective enforcement in marine fisheries. Under the Straddling Stocks Agreement the use of force is to be avoided unless the inspectors are endangered or obstructed from the conduct of their duties, and if force is employed at all, it is not to exceed a level that is 'reasonably required in the circumstances'.[72]

It has been argued that the Straddling Stocks Agreement fails to address two serious problems which contributed to the underlying cause of the fishery dispute between Canada and the European Union in 1995.[73] Firstly, while the Straddling Stocks Agreement places an emphasis on the rôle of the regional organisation such as NAFO in the formulation and enforcement of regulation, it does not adequately address voluntary compliance or the consentual framework which are features of such organisations. Secondly, by placing primary responsibility on the flag State for the prosecution and sanctioning of violations of regional fishery rules, the Straddling Stocks Agreement may fail to address the problem of *under-enforcement* which has been a feature in the operation of regional organisations in the past.

Reaction of the European Union to the Straddling Stocks Agreement

The European Commission, which had sole responsibility for representing the interests of the Member States at the United Nations Conference, had sought to uphold the balance between the rights and obligations of the coastal State and the interest of State(s) with vessels fishing on the high seas, as well as the promotion of a multilateral solution process. Although the agreement refers to a dispute settlement procedure for disputes of a technical nature,[74] and places an emphasis on the rôle of regional or subregional organisations, there is evidence that the balance between the conflicting interest of the coastal State and the high seas fishing State may have been tilted in favour of the former.[75] Of particular concern, is the extension of enforcement powers and action that is authorised when the flag State is unwilling or unable to monitor the activities of its vessels operating in high seas fisheries.[76] The Agreement was the subject of detailed analysis and evaluation in a communication from the Commission to the Council in November 1995.[77] Ministers agreed to sign the Agreement at the

Fisheries Council on 10 June 1996. Subsequently, on 27 June 1996, the European Community together with several of its members signed the Agreement. At the time of writing (1999), only 24 States have deposited instruments of ratification or accession and 30 States will be required to ratify before the bring the Agreement into force. Among NAFO Contracting Parties, only Iceland, Norway, the Russian Federation Canada, and the United States have deposited instruments of ratification, although Japan, and Korea have signed the Agreement.

There is no doubt that those parties that have signed are aware that a strong international law framework as well as compulsory dispute settlement mechanisms may be the only remedies capable of preventing coastal State(s) from unilateral or bilateral extension of jurisdiction that could be very harmful to the larger interests of the UNLOS Convention as a whole.

3. *The Code of Conduct for Responsible Fisheries*

The Compliance Agreement is one part of a twin-track approach to enforcement.[78] The second part is the development of an international code of conduct for responsible fisheries under the aegis of the FAO. In 1995, after two years of intensive negotiations, the Code was adopted by the Conference of the FAO at its 28th session.[79] The Code is voluntary and more broad-based than the Straddling Stocks Agreement. The Code covers all fisheries, regardless of whether they fall within coastal State jurisdiction or not. It also applies to the activities of the Community fleet and will provide clear guidelines and direction which will be assimilated into all future developments of the common fisheries policy.[80] In this regard, the Code sets out principles and international standards of behaviour for responsible practices with a view to ensuring the effective conservation, management and development of living aquatic resources, with due respect for the ecosystem and biodiversity.[81]

The Code consists of 12 Articles which cover the capture, processing and trade of fish and fishery products, fishing operations, aquaculture, fisheries research and the integration of fisheries into coastal area management.[82] There are several important provisions in the Code which will influence the development of a coherent and effective enforcement structure to help States to establish or to improve the legal and institutional framework required for the exercise of responsible fisheries, and in the formulation and implementation of appropriate measures to attain a sustainable resource. In this regard, the Code, in accordance with international law, places an obligation on States to ensure compliance with conservation and management measures. In particular, there is an obligation on States to establish effective mechanisms to monitor and control the activities of fishing vessels and fishing support vessels.[83] Furthermore, States are obliged to ensure that the activities of their vessels do not undermine the effectiveness of conservation and management measures taken in accordance with international law and adopted at the national, subregional, regional or global levels.[84] Significantly, the Code states that States should establish, within their respective competence and capacities, effective mechanisms for fisheries monitoring, surveillance, control and enforcement to ensure compliance with their conservation and management measures, and those adopted by subregional or regional organisations or arrangements.[85] The provisions in the Code in relation to sanctions are very similar to those in the Control Regulation, in so far as the Code stipulates that sanctions should be sufficient in severity to be effective, including sanctions which allow for the refusal, withdrawal or suspension of authorisation to fish in the event of non-compliance with conservation and management measures in force.[86] In common with both Community law and the 1995 Straddling Stocks Agreement the Code requires States, in conformity with their national law, to implement enforcement measures including, where appropriate, observer programmes, inspection schemes and vessel monitoring systems.[87]

3. *Conclusions*

From the aforementioned brief description of the new international instruments it is clear that the high seas fisheries regime has undergone fundamental change, this change is most apparent in the new regime that pertains to straddling stocks and highly migratory species. It may be suggested that this change was necessitated by the fact that the relevant UNCLOS provisions (in particular, **Article 63(2)**) lacked sufficient detail.[88] Whatever the initiator of change, the new high seas fishing order is important for EC fisheries law for several reasons. Firstly, because the activities of a significant section of the Community fleet will be subject to the new international framework. Secondly, because EC fisheries law does not evolve in a vacuum. It is tempered by world events and is bound by international legal instruments which, if not adopted into Community law, provide the benchmarks for Community standards.

References

1 Report from the Commission to the Council and the European Parliament on monitoring and implementation of the CFP, SEC (92) 394 of 06.03.1992, p. 7. See Chapter 2, *ante*.

2 See, *inter alia*: Verborgh, J., 'Blueprint for a Satellite-based System for the Monitoring of Fishing Activities'; and Springer, S.C., 'Monitoring High-seas Fishing Operation by Satellite', *Fisheries Enforcement Issues*, (OECD, Paris, 1991). For a general discussion of the use of satellites in fisheries enforcement, see Bailey, J., 'Marine Boundary Enforcement from Space: Satellite Technology and Fisheries Jurisdiction' in Blake, G.H. (eds.), *The Peaceful Management of Transboundary Resources*, (London, 1994). See also Molenaar, E.J., *Monitoring, Control and Surveillance of Fisheries Within the Framework of the Indian Ocean Commission*, paper presented at the Workshop on Fisheries, Mauritius, 2–3 February 1999.

3 There are several satellite communications service providers which have systems capable of monitoring fishing vessels. INMARSAT has several geostationary satellites which cover four ocean regions with near global coverage (±75° latitude). It is primarily used for maritime communications and safety. EUTELTRACS operated by the European Organisation of Telecommunications by Satellite (EUTELSAT) also uses geostationary satellites and covers western Europe from about 1,000 miles west of Ireland to Moscow. It is primarily a satellite tracking system. Argos uses satellites belonging to the National Oceanic and Atmospheric Administration (NOAA) which are on polar orbits and have near global coverage during a twenty-four hour period. It is primarily a satellite tracking system and has been extensively used in both the maritime and environmental fields.

4 Council Regulation 2847/93, *OJ* L 261/1, **Article 3(1)**. (Repealed.)

5 *Ibid*. **Article 3(2)**. (Repealed.)

6 Commission Regulation (EC) No. 897/94 laid down detailed rules for the pilot projects, *OJ* L 104, 23.04.1994, p.18.
7 For precise details on the systems tested see COM(96) 232 final, 96/0140(cns), 28.05.1996. This document also presents the Commission proposal for a satellite monitoring system.
8 Chapter 3, *ante.*
9 *Op. cit.*, fn 7.
10 Expenditure was eligible under Council Decision No. 95/527/EC on a Community financial contribution towards certain expenditure incurred by the Member States implementing the monitoring and control systems applicable to the Common Fisheries Policy, *OJ* L 301, 14.12.1995, p. 30; Amended in *OJ* L 302, 15.12.1995, p. 45.
11 *Op. cit.*, fn 7.
12 Opinion delivered on 13 December 1996, *OJ* C 20/382, 20.01.97.
13 On sanctions see discussion in Chapter 6, *ante.*
14 The enforcement difficulties associated with the use of driftnets is examined in Chapter 10, *ante.*
15 Council Regulation (EC) No. 686/97 of 14.04.1997 amending Regulation No. 2847/93 establishing a control system applicable to the common fisheries policy, *OJ* L 102/1, 19.04.97.
16 This is an example of the principle of reciprocity which is a common feature in international fishery agreements.
17 Council Regulation (EC) No. 686/97, **Article 3(1)**.
18 *Ibid.*, **Article 3(2)**. Amended by Council Regulation No. 2846/98, *OJ* L 358, 31.12.1998, p. 5. **Article 1(3)**.
19 *Ibid.* **Article 3(3)**. Reciprocity only applies during the first phase of VMS. After the year 2000 all third-country vessels over 24 metres overall length are required to have a VMS approved by the Commission. See fn 24 *infra.*
20 *Ibid.*, **Articles 3(5)**, and **19**, Council Regulation No. 2847/93, *OJ* L 261, 20.10.1993, p. 1.
21 *Ibid.*, **Article 3(6)**.
22 *Ibid.*, **Articles 3(7)–3(9)**.
23 *Ibid.*, **Article 3(10)**.
24 Council Regulation (EC) No. 2846/98, *OJ* L 358, 31.12.1998, p. 5. **Article 1(16)** inserting a new **Article 28c** in Council Regulation No. 2847/93.
25 Commission Regulation No. 1498/97 of 29.07.1997 laying down detailed rules for the application of Council Regulation No. 2847/93 as regards satellite-based vessel monitoring systems, *OJ* L 202/18, 30.07.97.
26 *Ibid.*, Annex 1.
27 See Chapter 4, Section (ii), *ante.*
28 *Op. cit.* fn 25, **Article 3(3)**.
29 See Chapter 4, Section (ii), *ante.*
30 Council Regulation No. 150/97 of 12.12.1996 on the conclusion of an Agreement on co-operation in the sea fisheries sector between the European Community and the Kingdom of Morocco and laying down provisions for its implementation, *OJ* L 30, 31.01.1997, Chapter VI.
31 Council Decision of 26 November 1996 on the conclusion of an Agreement in the form of an Exchange of Letters concerning the provisional application of the Agreement on co-operation in the sea fisheries sector between the European Community and the Islamic Republic of Mauritania, (96/731/EC), *OJ* L 334/16, 23.12.1996, Chapter VII.
32 Springer S., *op. cit.* fn 2, examines the successful implementation of a satellite programme under a tripartite International Driftnet Fishery Agreement negotiated between United States, Canada, and Japan. The cost of the system was funded by the participating flag States. Details of another trial programme are presented in a study by the same author, *Fishing Vessel Tracking: Application for Fisheries Management and Enforcement*, (Maryland,

1991). The United States has proposed specifications for consideration by the international community, require the monitoring system to be, *inter alia*, tamper-proof, automatic and continuous, accurate to 400 metres, supplying vessel data (position, course, speed), allow for vessel interrogation facility, two-way message network, and elaborate data management facility on the shore side.

33 United Nations Implementation Agreement for Straddling Stocks and Highly Migratory Species, **Article 18(3)(g)(iii)**. There are several other provisions which refer to satellite systems, including several Articles in Annex I which specify requirements for the collection and sharing of data.

34 See Section (ii), *infra*.

35 See Verborgh, J., *op. cit.*, fn 2.

36 For a discussion of the success of these fishery management tools see, Holden, *op. cit.*, fn 1, Chapter 1, *ante*.

37 Additional benefit is gained if the system complies with the Global Maritime Distress and Safety System (GMDSS) and thus allows for improved search and rescue capabilities.

38 Council Regulation No. 686/97, **Article 5**, *op. cit.* fn 15.

39 See Annual Report and Accounts 1995–1996, Scottish Fishery Protection Agency 1995–1996, (HMSO, 1996).

40 *Op. cit.* fn 1.

41 Council Regulation No. 686/97, **Article 6**, *op. cit.*, fn 15.

42 A typical example of national legislation in relation to the powers of an inspector are set out in the Irish Fisheries Consolidated Act 1959 (as amended). In Ireland, fishery protection officers have extensive powers, and may, pursuant to Chapter II and Chapter III of the Fishery Act, stop vessels and board them, examine and take copies of the documentation pertaining to that vessel, question the master, search the vessel, take the names and address of the crew. Most importantly if the fishery officer suspects that a fishery offence is committed, he may detain the vessel and commence the process for investigating and prosecuting the suspected violator. The issue of how reasonable this suspicion must be has not been ruled on specifically in a fishery case, it must therefore follow the general rule of law in this regard.

43 *OJ* L 281, 23.11.1995, p. 31.

44 The FAO conference adopted at its 27th session in November 1993 the Agreement to Promote Compliance with International Conservation and Management Measures by Fishing Vessels on the High Seas.

45 There was a request at the United Nations Conference on Environment and Development (UNCED) to convene a conference on straddling fish stocks and highly migratory fish stocks. On 4 August 1995, following three years of negotiations, the United Nations Conference adopted without a vote a new agreement to regulate these stocks. The Agreement will enter force 30 days after ratification or accession by 30 signatories, **Article 40**.

46 There was a call for the establishment of an International Code of Conduct for Responsible Fisheries in the Declaration of Cancún of May 1992.

47 Moore, G., 'The Food and Agriculture Organisation of the United Nations Compliance Agreement', *IJCML*, 412–415.

48 **Article (iii)** of the Compliance Agreement sets out flag state responsibilities.

49 **Articles (v)** and **(vi)** of the Compliance Agreement.

50 *Op. cit.*, fn 47.

51 The Compliance Agreement was accepted by the Community on the 25 June 1996, Council Decision 96/428/EC, *OJ* L 177/24, 16.07.1996. On the issue of legislative competence to vote at the FAO meeting see Case C-25/94 *Commission* v. *Council* [1996] I-1469, discussed Chapter 2, *ante*.

52 UN Document A/Conf.164/33, 03.08.1995.

53 The Agreement has been subject to a number of incisive articles elsewhere, see, *inter alia*: Juda, L., 'The 1995 United Nations Agreement on Straddling Fish Stocks and Highly Migratory Fish Stocks: A Critique', *ODIL*, **28** (1997), 147–166; Tahindro, A., 'Conservation and Management of Transboundary Fish Stocks: Comments in Light of the Adoption of the 1995 Agreement for the Conservation and Management of Straddling Fish Stocks and Highly Migratory Fish Stocks', *ODIL*, **28** (1997), 1–58, Barston, R., 'United Nations Conference on Straddling and Highly Migratory Fish Stocks', *Marine Policy* **19(2)** (1995), 159–166; Balton, D.A., 'Strengthening the Law of the Sea: the New Agreement on Straddling Stocks and Highly Migratory Fish Stocks', *ODIL*, **27** (1996), 125–151; Anderson D.H., 'The Straddling Stocks Agreement of 1995 - An Initial Assessment', *International and Comparative Law Quarterly*, **45** (1996), 463–475.

54 1995 Straddling Stocks Agreement, **Article 7**.

55 The 1995 Straddling Stock Agreement obliges states to become members of regional organisations, **Articles 8(3)** and **8(4)**.

56 *Ibid.* **Article 20(6)**.

57 *Ibid.*, without prejudice to the right of *'hot pursuit'* pursuant to UNCLOS **Article 111**.

58 The procedures for boarding and inspections are set out in **Article 22**.

59 1995 Straddling Stocks Agreement, **Article 21(8)**. A serious violation is defined in **Article 21(11)**.

60 *Ibid.*, **Article 21(12)**.

61 1995 Straddling Stocks Agreement, **Article 17**. Unfortunately, the Agreement does not specify the action.

62 Balton, D.A., *supra*, footnote 53, p. 141.

63 1995 Straddling Stocks Agreement, **Article 23**.

64 *Ibid.*, Part VIII, **Articles 27–32**.

65 Chapter 6, *ante*.

66 Tahindro, A., *op. cit.*, fn 53.

67 Based on **Article 117** of UNCLOS and the decision of the International Court in the 1974 *Fisheries Jurisdiction* cases.

68 Tahindro, A., *op. cit.*, fn 53, pp. 35–36. In this regard it may be suggested that this argument is cyclical and that it is only state practice which will establish the value of **Articles 18** and **19** as customary norms.

69 Subject to **Articles 110** and **111** of UNCLOS.

70 Tahindro, A., *op. cit.*, fn 53, pp. 38–39.

71 *Ibid.*, p. 39.

72 1995 Straddling Stocks Agreement **Article 22(1)(f)**.

73 See *inter alia*, Kedziora, D.M., 'Gunboat Diplomacy in the Northwest Atlantic: the 1995 Canada-EU Fishing Dispute and the United Nations Agreement on Straddling and High Migratory Fish Stocks', *Northwestern Journal of International Law and Business*, **17** (1996–1997), 1132–1162; Curran, P. and Long, R., 'Fishery Law, Unilateral Enforcement in International Waters: the Case of the *Estai*', *Irish Journal of European Law* **5(2)** (1996), 123–163.

74 1995 Straddling Stocks Agreement, Part VIII.

75 See de Yturrigaga, J.A., *The International regime of Fisheries From UNCLOS 1982 to the Presential Sea*, (The Hague, 1997), 209–226.

76 The head of the European delegation, Commissioner Bonino, noted that the European Community was *'très préoccupée par la possibilité de l'usage de la force, en raison de la formulation très vague' (du texte)*. Reported by Reuters, 3 and 4 August 1995.

77 See, Communication from the Commission to the Council on the signature of the Agreement for the implementation of the provisions of the United Nations Convention on the law of the Sea of 10 December 1982 relating to the conservation and management of straddling stocks and highly migratory fish stocks, COM (95) 591 final, 22.11.1995.

78 FAO Doc. CCRF (Rev. 1), 28.09.1995.
79 See, Edeson, W.R., 'The Code of Conduct for Responsible Fisheries, An Introduction', *IJMCL* **11(2)** (1996), 233–238.
80 FAO Code of Conduct for Responsible Fisheries, **Article 1.4**.
81 FAO Code of Conduct for Responsible Fisheries, Introduction.
82 FAO Code of Conduct for Responsible Fisheries, **Article 1.3**.
83 *Ibid.*, **Article 6**.
84 *Ibid.*, **Article 7**.
85 *Ibid.*, **Article 7.1.7**.
86 *Ibid.*, **Article 7.7.2**.
87 *Ibid.*, **Article 7.7.3**.
88 See Brown, E.D., *The Law of the Sea*, Vol. 1, p. 228. A useful discussion of developments in the 1990s is provided by Hey, E., 'Global Fisheries Regulations in the First Half of the 1990s', *IJMCL* **11(4)** (1996), 459–490.

Chapter 12
Conclusions

Introduction

There is general consensus that the unilateral extension of exclusive coastal state jurisdiction over the living resources within a 200-mile zone in 1976, and the conclusion of the 1982 Convention on the Law of the Sea, are milestones in the evolution of the international law of fisheries. Similarly, the adoption of the common policy for fisheries in 1983 was an event of major significance in the development of fisheries law in the European Community. In 1983 these developments provided a basis for optimism and progress. In the intervening years, many of these expectations remain unrealised and a number of the substantive issues associated with these developments are still unresolved. Fisheries in the European Community have suffered from the same defects and shortcomings as other global fisheries. Thus, while 95% of total marine catches come from waters which are encompassed by national jurisdiction,[1] fisheries have not produced maximum sustainable benefits and in general continue to be exploited to the threshold of extinction.[2] Moreover, from the global perspective, extended coastal State jurisdiction has not resolved fundamental issues pertaining to conservation and the management of fisheries which have continued to suffer from inadequate and inefficient regulatory and enforcement structures.[3] Indeed, in recent years a common feature in fishery management has been the increased reliance on regulatory intervention as a means to stay the decline in resources.

Similarly since 1983, one of the predominant characteristics of fisheries law in the European Community is the extraordinarily large volume of legislation, adopted for different purposes at different times. This legislation operates in different areas of the fishery policy with varying degrees of effectiveness. It may be contended, however, that if one excludes the adoption of application regulations, the apogee of Community legislative action in the life of the present policy appears to have been attained in 1993. Furthermore, despite the initial difficulties encountered in adopting the legislative framework for such a disparate group of interests as exist in sea fisheries in the first phase of the CFP, the revised policy introduced in 1993 has become less diffuse and consequently more focused. In this regard, although the CFP will no doubt continue to evolve in the light of experience and circumstance, its future is not likely to depend on any new policy redirection and subsequent legislative action, but rather on the effective enforcement of the legislation in place.[4] Indeed, it is easy to conclude that the plethora of interrelated regulations which govern the conservation, structural, and market aspects of the policy will be rendered meaningless unless they are effectively enforced. Henceforth, it will no longer be sufficient to look at any substantive area of fishery law, without examining the seminal issues of enforcement and compliance. In this context, the aim of these conclusions is to summarise some of the themes which have appeared either expressly or implicitly in preceding chapters.

Section (i) The rôle of fishery enforcement in the CFP

Since its inception, Community fisheries law has embraced and upheld the various legal principles which have circumscribed the policy. Indeed, what is remarkable, from the discussion of the historical evolution of the CFP,[5] is that, although the policy developed in the 1980s and 1990s into an intricate web of secondary legislation guided by established norms such as the principle of equal access, the principle of relative stability, the *acquis communautaire* in relation to the accession of new Member States, no equivalent empirical norm evolved on which to develop a *communautaire* enforcement policy. The normative bifurcated approach to the prescription of policy on the one hand, and the execution and enforcement of Community rules on the other, was most apparent in relation to the principle of equal access - a remnant of the controversial principle in international law of freedom of fishing - which the Community adopted as a guiding axiom in the establishment of the CFP. It may also be pointed out, that this approach was in marked contrast to the development in public international law, and the acceptance by the majority of States elsewhere, of exclusive coastal state jurisdiction and property rights over the resources in the EFZ/EEZ. In the European Community, however, the *communautaire* approach to resource allocation and resource management manifested by the common rules for the conservation, structural and market aspects of the policy was not matched by, or linked to, a federal system of, or *communautaire* approach to, fishery enforcement. As explained in Chapter 2, the power to enforce Community fishery law delineated along the established lines of separation in customary and convention law which distinguish the rights and obligations of the flag State from those of the coastal State. It may also be recalled that there were several reasons for the cleavage between the common resource management policy agreed and implemented at a Community level, and the power to enforce the common rules which was vested exclusively in the Member State. Firstly, there were external constraints imposed on Member States by the hierarchy of customary and convention norms which governed the EEZ/EFZ regime. These mitigated against any impetus to redefine an enforcement policy exclusively for the benefit of implementing and enforcing a Community fisheries regime in a more effective manner.[6] Secondly, there were internal constraints as a result of the uniqueness of the Community legal order, such as the almost exclusive autonomy of Member States to deal through their criminal and administrative procedures with infringements of fishery law, albeit such infringements were invariably infringements of Community law. Furthermore, notwithstanding the principle of direct effect, the Community did not have a secure legal base in the Treaty, or perhaps the requisite political conviction, to pursue actively the development of formal collaboration between Member States, nor indeed to oblige Member States to act in a uniform manner in the enforcement of Community fishery law. Significantly, this model of enforcement for the fishery policy was in marked contrast to the express legal bases in **Articles 85–86** {81–82} of the EC Treaty and Council Regulation No. 17/62 which provided for a direct rôle in enforcement for the Community institutions in ensuring that the competition policy was uniformly and effectively applied by all parties.[7] Thus, in the first phase of the CFP one may conclude that, despite the almost exclusive prescriptive jurisdiction of the Community in the domain of fisheries and the development of a number of broader principles by the Court of Justice in respect of Member State obligations to deal with non-compliance of

Community law as well as the supervisory rôle of Community institutions,[8] responsibility for the practical enforcement of Community fisheries law rested almost exclusively with the Member States. In substance and effect, fishery law enforcement in the Community accorded with the custom and convention model in international law and thus remained securely within the exclusive remit of the flag and the coastal State authorities.[9] In short, the great strength of the CFP, i.e. a common management plan for the resource, had been effectively undermined by reliance on the devolved enforcement structure in the Member States.

Within a short period of agreement on a common policy for fisheries, however, it was evident that there were certain detracting forces inhibiting the attainment of what had become by necessity the principal and over-riding policy objective which was the sustainable management of the resource. Today, even with the benefit of hindsight, it is not clear whether it was the patent failure of the resource management plan to work satisfactorily, or the absence of adequate regulatory compliance by the industry, which were the principal contributory factors to the spiralling trend towards stock over-exploitation and resource depletion.[10] Whatever the reason, throughout the early phase of the policy, the management and regulation of fisheries continued to be increasingly *communuataire* in approach, a progression which was frequently abetted by an expanding body of Court of Justice jurisprudence in cases relating to the CFP.[11] Increased regulation inevitably resulted in the placing of further constraints on the industry. These constraints encompassed rules to regulate, *inter alia*: quotas; the use of particular types of fishing gear; access to particular fishing zones; the modification of existing vessels and the building of new fishing vessels; and the marketing of fishery products. Increased regulation, however, hightened the need for effective enforcement of Community rules. Furthermore, problems which stemmed from the devolved enforcement structure became increasingly manifest, a principal weakness being the absence of accountability by the Member States in the utilisation of quotas,[12] and major disparities in the uniform application of Community rules because of widespread differences between national law enforcement processes and procedures.

The Community experience in the management of fisheries during the first ten years of the CFP revealed that there was an urgent requirement to improve the enforcement of fishery law at the end of that period. In retrospect, it is possible to identify and enumerate the factors which led to a revised enforcement structure for the second period of the policy. Firstly, the CFP did not attain many of its initial objectives, and by 1993 it was evident that there was a major gap between the law in theory, which was to ensure the effective and uniform enforcement of Community rules, and law in practice, which revealed that non-compliance with the regulatory framework was widespread. Lack of uniformity in the application of regulations in the Member States was perceived as not only inflicting damage on the credibility of the CFP as a common policy and distorting the operation of the single market, but was also undermining some well-established principles in the Community legal order. Secondly, the pressing requirement for an improved enforcement structure stemmed mainly from the failure of Member States to effectively enforce and observe their regulatory obligations during the period 1983–1992. Thus, the need for more effective enforcement arose out of a situation which the Member States and the industry had themselves created and, consequently, neither party vigorously opposed the adoption of the new regulatory framework in 1993.[13] Thirdly, the failure of the CFP to deliver long-term sustainability of fisheries, placed the twin issues of compliance and enforcement high on the agenda which had to be dealt with by the mid-term policy review in 1992.[14] Revision of the enforcement structure was thus not an exercise to achieve greater European integration but had become an essential prerequisite to ensure the survival of fish stocks. Finally, it was also clear by 1992 that the international law of fisheries, which through convention and custom imposed responsibility on the basis of flag State and coastal State jurisdiction, lacked, from a global perspective, the substantive and procedural provisions to ensure an adequate level of enforcement and compliance with resource management schemes.[15] This view was expressed at various international fora where concern had long been expressed regarding the clear signs of over-exploitation of important fish stocks and damage to eco-systems. Many parties thus sought to place the issue of enforcement at the top of the international fisheries agenda.[16] In Europe, by 1992, there was a shared recognition that a more robust enforcement regime was needed for the second ten-year phase of the CFP.

Section (ii) Confronting the enforcement dilemma

This book attempts to evaluate the new enforcement regime introduced in 1993 and its underlying legal instrument - the Control Regulation. The central question posed is whether this instrument will contribute to a more integrated and harmonious approach to the enforcement of fishery law in the second half of the 1990s and beyond. The conclusion reached on the basis of the provisions examined in Chapters 3–7 is that the instrument, despite a modest start in the first phase of application during the period 1994–1998,[17] certainly has the potential to make a significant contribution to such a development. It reflects the global consensus in the European Community that enforcement is now the substantive issue to be addressed in the overall fishery conservation and management plan. This consensus did not exist in 1983 and has only come about after critical reappraisal which led to a more comprehensive Community enforcement policy. Three of the more commendable features of this instrument are: the development of a scheme of enforcement covering all aspects of the policy from producer to consumer;[18] the extension of the scope of enforcement to cover the activities of the Community fleet fishing outside the Community maritime area;[19] and the emphasis on the introduction of satellite technology and integrated computerised systems to reduce the enforcement deficit.[20]

The framework that the Control Regulation introduces may, on the one hand, be subject to a number of criticisms. Firstly, it is perceived as being overtly interventionist yet stops short of prescribing minimum inspection standards or a compulsory level of control in the Member States.[21] Secondly, it must continue to rely on the same substantive Treaty provisions **Articles 169–171** {226–228} to ensure compliance by Member States with their Community obligations.[22] This defect, however, is mitigated by the soft law approach to law enforcement and structural reform.[23]

It may also be argued that the Control Regulation does not adequately address the important rôle the industry could play in self-regulation, nor does it provide a legal base for the development of an alternative to the traditional scheme of enforcement which places an emphasis on non-compliance and the importance of effective investigation, prosecution and the imposition of sanctions, as opposed to promoting compliance by the industry with the regulatory framework. In this respect, it could be contended that the Control Regulation reflects a preference for authoritive prescription which regulates by means of legal and administrative instrument. In contrast to the fishery enforcement model in use in Norway or indeed the Community competition policy, the Community fishery enforcement model does not entail any reallocation of responsibility to the industry or encourage self-regulation or user participation in the enforcement process. The latter are frequently suggested by political and social scientists as a panacea for the crises that may occur in governing society.[24] On the other hand, this omission does not imply that Member States should not support transparent and accountable self-regulatory schemes for central elements of the CFP nor that the Community would look unfavourably at such developments. Moreover, the ostensible aim of the Control Regulation is to establish the obligations on Member States, and the latter have sufficient scope within the principle of subsidiarity to adopt enforcement models which entail co-responsibility and self-regulation in addition to authoritative prescription. It is thus evident that there are several possibilities outside the scope of the Control Regulation for Member States to implement schemes of self-regulation which may entail mandatory participation by the industry. In this regard, one Member State, the Netherlands, has taken several initiatives through the medium of private law to involve the industry in the enforcement process.[25]

On a more global perspective, with the introduction of the 1993 Control Regulation we may also trace a gradual progression away from the traditional distinction between flag State and coastal State enforcement jurisdiction which was a predominant feature in the implementation of the CFP during the period 1983–1993. Indicators of this trend include: the functional, if non-executive, rôle played by the Commission in fishery enforcement matters (other than in licensing where the Commission in some instances may institute proceedings and invoke sanctions);[26] the requirement of simultaneous transmission of fishing effort reports to the flag State and the coastal State under the fishing effort regime and the corresponding requirement regarding the transmission of satellite data;[27] the prescription of Community standards relating to proceedings and sanctions in cases of fishery violations;[28] and the emphasis placed in the Control Regulation on co-operation and co-ordination of the enforcement efforts in the Member States.[29] Further proof of this move is the inclusion of a provision in the EU–Morocco Fishery Agreement which allows Moroccan observers to accompany Spanish national inspectors in Spanish ports while the latter carry out inspections of vessels flying the Spanish flag and fishing under the aforesaid Agreement.[30]

This trend of moving away from the exclusive remit of the flag State to take enforcement action also reflects the new orientation of global enforcement strategies outside the Community legal order as is evident from the agreement on the Straddling Stocks Agreement and other international instruments.[31] In particular, the provisions in the 1995 Straddling Stocks Agreement which allows for high seas enforcement of fisheries law by non-flag State breaks with the tradition of the exclusive jurisdiction of the flag State as reflected in the 1958 High Seas Convention and the 1982 UNCLOS.[32] Similarly, outside the European Community, there have been several bilateral agreements which authorise States to conduct non-flag State inspections of fishing vessels in specific fisheries on the high seas. Because the task in fishery management is thus by its very nature inter-jurisdictional, it may be argued, from this perspective, that the established normative relationship between the coastal state and exclusive resource responsibility makes little sense.

Section (iii) Contemplating the future of fishery law enforcement in the European Union

The common fisheries policy has been in operation for 14 years and the enforcement structure introduced by the 1993 Control Regulation is, at the time of writing, nearly six years old. Significantly, the debate regarding the future direction of the policy in the *post*-2003 period is already underway and the subject of enforcement should not be overlooked. At this stage it is not possible to contemplate the outcome of policy discussions in the run up to 2002. As far as enforcement is concerned its future development is likely to be shaped by a number of factors. Firstly, the subject of enforcement impinges on several areas outside the domain of the CFP and thus the new policy will not only reflect recent progress in the international law of fisheries but will also be influenced by other developments in the landscape of Community law. Particularly, it will be affected by developments in inter-Member State co-operation in justice and home affairs provisions pursuant to Title VI {VI} of the TEU and amending Treaties.[33] Moreover, it will also be affected by regulations pertaining to frontier control, data protection, and the prevention of fraud, all of which are areas which bridge the divide between the first and third pillars of the European Union. Secondly, the debate on the new policy is unlikely to focus on the issues pertaining to resource allocation and the principle of relative stability. Consequently, any call for the introduction of individual vessel quotas or user-based property rights in selected stocks and the concomitant freedom of exchange will be strongly contested in some quarters. Thirdly, the architecture of the management policy may be modified substantially from the present CFP to include: the determination of TACs on a multi-annual basis for the main groups of species; greater reliance on the precautionary principle; increased use of sensitive areas and biological stops as a management tool; greater emphasis on protecting the integrity of the marine eco-system and ensuring that due regard is paid to the wider environmental implications when considering and implementing fishery management plans. Furthermore, it is almost certain that the management of fishing effort will be a predominant feature in the revised policy and will be closely linked to the satellite surveillance system for fishery enforcement.[34] Fourthly, the external policy in fisheries will continue to increase in significance and will be adapted to suit circumstances in particular third-world countries and linked into the wider framework of development aid. In the case of international waters and the recent important evolution regarding the rôle of regional organisations the Community will play a major rôle in ensuring that the agreed international norms and conventions are upheld. In particular, the Community will endeavour to see that the present balance between coastal State and flag State will constitute a stable and permanent basis for the regulation of high seas fisheries.[35]

Given, as we have seen, that the hard law option of **Article 169** {226} proceedings is unwieldy and considering that formal co-operative enforcement structures are locked away from Community institutional interference by Treaty,[36] it will be difficult in the short term to develop through Community law precise enforcement processes and procedures to ensure compliance with Community law in the Member States. Furthermore, Member States have displayed a persistent reluctance to rely on **Article 170** {227} to ensure that other Member States enforce Community law.[37] Consequently, increased use may have to be made of secondary or intermediate instruments such as codes of practice, recommendations, guidelines, resolutions, declaration of principles, standards and best practices.[38] It may be argued that this approach is particularly suitable for achieving greater harmonisation and assimilation of Member States' enforcement practices and standards. Indeed, the principal attraction of this approach is that Member States are able to undertake obligations and simultaneously retain a degree of autonomy over their actions and resources. As noted by one commentator in the context of environmental law, the soft law approach allows states to tackle problems collectively at a time when they do not want to shackle their freedom of action.[39]

It should not be forgotten when assessing the future direction of enforcement that fisheries management in the European Community has gained considerably from the political stability of the Community. In particular, the non-partisan rôle played by Community institutions in resolving inter-Member State disputes should not be undervalued.[40] Indeed, the most positive feature in the present structure is that all Member States play an active part in the decision-making process which helps to avoid automatic adherence to the dichotomy between coastal and flag State prerogatives which characterises international fishery relations and global enforcement issues. The strength of the Community enforcement system, cumbersome as it may appear, is that the present system ensures that resource and enforcement disputes can be settled amicably, as is evident in the case study examined above in relation to the driftnet fishery.[41] In the emerging perspective of community fishery law the interest of the Member State is no longer the sole factor which shapes the global objectives of fishery management. Neither is the status of inter-state dispute settlement forums a significant or particular obstacle for the European Union. Indeed, it may be argued that **Article 170** {227} of the EC Treaty actions provide a unique mechanism for dispute resolution in so far as it offers Member States a ready made mechanism to challenge any other Member State that is suspected of failing to comply with Community obligations. However, as only one fishery case has been brought before the Court of Justice under **Article 170** {227}, it is suggested that national governments, if left to their own devices, will rarely initiate enforcement proceedings.[42] Nevertheless, **Articles 169–171** {226–228} enforcement proceedings provide additional security that Member States do not inadvertently omit to fulfil their international obligations.[43] This is particularly significant because of the new-found emphasis on regional organisations in the Straddling Stocks Agreement and the new-found importance of consensus in the resolution of difficulties in the management of fisheries.[44] The Community must be able to accommodate participation in regional and global organisations without the complications arising from differences in enforcing Community law in the Member States. In this respect, the European Union has an institutional structure suitable for resolving difficulties and dealing with trans-national problems. A significant point is that the Court of Justice is vested with mandatory jurisdiction and thus does not suffer from the same defects as a variety of international dispute resolution fora. Consequently, and in conclusion, it may be argued that the rôle of Community institutions in the enforcement process will be central to the evolution of the new policy in the *post*-2003 period.

In the international arena, the Community will continue to work within the LOS Convention framework and co-operate with third countries at the regional and sub-regional level to ensure the adoption and enforcement of regulations which will ensure the sustainable development of resources. Indeed, from the global perspective, fisheries enforcement and compliance are now recognised as universal tasks which require a global partnership with a solid legal foundation in order to build international co-operation. Evidence to support this view may be deduced from several international instruments, such as: the emphases in the UN Driftnet Resolutions on effective compliance;[45] the identification in Chapter 17 of Agenda 21 of the lack of effective enforcement to deal with the wider concerns of the marine ecosystem; the obligation pursuant to the Compliance Agreement for States to exchange information, including evidential material which enables the flag State to identify vessels which have undermined international conservation and management measures;[46] and the emphasis in the Code of Conduct for Responsible Fisheries on the establishment of effective enforcement structures at regional or sub-regional level.[47]

In relation to trade, the Community is unlikely to pursue the United States option of trade sanctions as a tool to achieve improved compliance with international conservation and management options. Significantly, during the EU–Canada dispute in 1995, the Community resisted the temptation to let a fishery altercation escalate into a trade dispute.[48] Furthermore, in the context of the driftnet fishery, the Community has taken issue with the United States approach of linking compliance with the United Nations Driftnet Resolution with the threat of economic sanctions.[49] Indeed, the Community approach accords with the view that the compatibility of trade measures and their relationship with fishery conservation measures is a question to be determined through the dispute settlement procedure within the World Trade Organisation (WTO).[50] More specifically, the Community approach accords with the FAO Code of Conduct for Responsible Fisheries which prescribes that fish trade measures adopted by States to protect human or animal life or health, and the interests of consumers or the environment, should not be discriminatory and should be in accordance with internationally agreed trade rules, in particular the principles, rights and obligations established in the Agreement on the Application of Sanitary and Phytosanitary Measures and the Agreement on Technical Barriers to Trade of the WTO.[51] In conclusion, the Community will play a major rôle in the development of the multilateral co-operative approach to fishery enforcement in the international setting.

Section (iv) A greater rôle for the industry in the enforcement process

It is increasingly evident that non-compliance with fishery law is most prevalent when basic enforcement structures do not exist and there are strong financial incentives to violate Community regulations. Other than removing the cause of fraud, or the opportunity to commit fraud, by increasing the level of law enforcement there appear to be few viable alternative methods to ensure the effectiveness of Community law. On the other hand, there is some evidence in environmental law that a different approach has been adopted to encourage environmentally responsible behaviour. This approach entails the use of fiscal and economic incentives and disincentives such as taxes, subsidies and grants towards the pursuit of desirable practices. The objective is to induce greater compliance by making it cost-effective to comply with the law and by making it cost-prohibitive to engage in illegal practices. As noted by one authority, those who abate cheaply will abate most.[52] At present, there is no direct commercial or economic incentive for the industry or the Member States to comply with Community fisheries regulations. In contrast, it may be recalled that in the United States the campaign to introduce *Dolphin safe* labels on products which had been captured with environmentally-friendly fishing gear was a major factor in influencing international fishing practices.[53] Similarly, an eco-label system awarded to products which accord with environmental standards pursuant to Council Regulation No. 880/92 has been effective in promoting responsible behaviour.[54] This regulation does not apply to fishery products, however, there appears to be considerable scope for the industry to introduce similar measures to promote responsible fishing practices by European fishermen.

It is, of course, possible to argue that there are other approaches to fisheries enforcement which might be more desirable, and more efficacious than present ones. In particular there are several incontrovertible reasons why the industry ought to be made a fuller partner in the enforcement process. When considering the logic of this move it should be borne in mind that the fishing industry is the net benefactor of a successful fisheries policy. Conversely, the industry is also the ultimate loser if the policy fails. Furthermore, to provide an enforcement rôle for the industry is merely to acknowledge that protecting marine resources is not exclusively a problem for lawyers, scientists, and policy administrators. In this regard one needs to add a word of caution, increased participation by the industry in the enforcement process will not in itself remove the requirement for formal enforcement structures and procedures. It will, however, improve the collective awareness of the need for effective enforcement and remove some of the traditional barriers which have existed between the industry and those responsible for undertaking enforcement tasks.

Several practical steps may be taken to provide the industry with a greater rôle in the enforcement process. Firstly, industry participants require formal instruction on the regulations which underpin the CFP. An appropriate and effective way to achieve this is to introduce a mandatory professional certificate for vessel operators to be obtained by examination on the relevant aspects of the policy. This could become part of the professional training of current and prospective captains. Indeed, the endorsements of such certificates in the event of conviction for fishery offences would offer a tangible deterrent to would-be offenders of Community regulations.[55] Secondly, the industry also needs to be made continually aware of the long-term damage to fisheries as a result of non-compliance. Greater awareness through the media and industry-focused seminars could be a means of disseminating the appropriate information. Thirdly, the industry should also be encouraged to report violations of Community law. In particular, the industry ought to be encouraged to report the activities of non-licensed operators and illegal practices.[56] Fourthly, the industry ought to be encouraged to establish observer programmes similar to those that exist under the EU–Morocco and EU–Mauritania Agreements and under the NAFO Scheme of Enforcement. The provision of financial assistance for the industry to promote voluntary participation in the scheme of satellite surveillance is a positive steps in this direction.[57] Fifthly, the rôle of sanctions in fishery law needs to be reassessed.

Participation by the industry in the enforcement process could place greater emphasis on compliance as one of the objectives of regulation, and give less prominence to the concept of the punitive sanction as the law's main response to illegal activities in marine fisheries. As a consequence, the characterisation of Community fishery law enforcement will no longer be solely one of authoritative prescription backed by sanction, but of resource trusteeship and concern for common interests at a collective level. In the long-term, only this approach will lead to a diminution of regulation and supervision.

References

1 See, *Report of the FAO World Conference on Fisheries Management and Development* (Rome, FAO, 1984), Appendix D, p. 1.
2 The FAO Committee reported, at its 20th session in 1993, that 69% of the world's marine stocks for which data are available were either fully exploited, overexploited, depleted, or very slowly recovering from overfishing and were therefore in need of urgent corrective conservation and management measures, see, FAO Fisheries Dept., *State of World Fisheries*, **8**.
3 A view first documented in 1934 by Daggett, A., 'The Regulation of Maritime Fisheries by Treaty', *American Journal of International Law* **28** (1934), 693–717. The recent experience in the Canadian Atlantic fisheries and in the European North Sea would support this view, see, *inter alia*: Charles, A.T., 'The Atlantic Canadian Ground-Fishery: Roots of a Collapse', *Dalhousie Law Journal*, **18** (1995) 65–83.
4 See discussion on the future development of the CFP *infra*.
5 Chapter 1, *ante*.
6 On this point, see discussion on division of enforcement competence in Chapter 2, *ante*.
7 Council Regulation No. 17: First Regulation implementing **Articles 85** {81} and **86** {82} of the Treaty, *OJ* 013, 21.02.1962, p. 204. See, Temple Lang, J., 'Community Antitrust Law and National Regulatory Procedures', *Annual Proceedings of the Fordham Corporate Law Institute* **24** (1997–1998), 297–334; 'European Community Constitutional Law and the Enforcement of Community Antitrust Law', *Annual Proceedings of the Fordham Corporate Law Institute*, **20** (1993–1994), 524–604.
8 In relation to general principles and the subject of sanctions, the deliberations of the Court in, *inter alia*, Case 68/88, *Commission* v. *Greece* [1989] ECR 1095; and Case 240/90, *Germany* v. *Commission*, [1992] ECR I-5381, 5431-2, are pertinent, as is **Article 209A** {280} of the Treaty of European Union. On this point, see Chapters 2 and 3, *ante*.
9 As noted in Chapter 2, the Community played an active part, however, in the International Scheme of Enforcement under the NAFO Convention which provided for a multilateral inspection scheme. Similarly, the Community is a Contracting Party to the Convention for the Conservation of Antarctic Marine Living Resources (CCAMLR) and is thus obliged to comply with the CCAMLR inspection scheme which also provides for multilateral inspection. Up to 1997, some Community Members, namely, Spain, France, Portugal and the United Kingdom (in respect of Bermuda) were contracting parties of ICCAT and were thus obliged to implement the International Convention for the Conservation of Atlantic Tunas (ICCAT) scheme of enforcement. In respect to the latter, it needs to be stressed that the scheme of enforcement is flag State based and there is no provision for non-flag State inspection of vessels operating on the high seas, see Chapter 2, *ante*. The European Union at the time of writing has acceded to ICCAT and there is a new draft new scheme of enforcement in preparation.
10 As noted in the introduction to this chapter, extended coastal-State jurisdiction proved to be

no guarantee of effective management and conservation of resources. For a further discussion of this point see, *inter alia*, Johnston, D.M., 'Is Coastal State Fishery Management a Success or Not?', *Ocean Development and International Law* **22** (1991), 199–208; Schrank, W., 'Extended Fisheries Jurisdiction', *Marine Policy* **19** (1995), 285–299; Churchill, R.R., 'Shared Fisheries Management in the European Community', *RECIEL* **2** (1993), 260–269; and by the same author 'Fisheries in the European Community: Sustained Development or Sustained Mismanagement' in Couper, A. and Gold, E. (eds.) *The Marine Environment and Sustainable Development: Law, Policy, and Science*, (Law of the Sea Institute, 1993), 140–177.

11 See discussion of Cases relating to unilateral national measures discussed in Chapter 3, *ante*.

12 A typical example of this was the difficulties experienced in Member States when dealing with the phenomenon of quota hopping Chapter 7, *ante*.

13 See, discussion of the legislative history of the 1993 Control Regulation, Chapter 3, *ante*.

14 Chapter 2, *ante*.

15 As mentioned in Chapter 2, UNCLOS contains relatively few provisions on enforcement within national jurisdiction (**Articles 73, 49, 21**, and **25**), and even less with regard to enforcement outside the coastal State jurisdiction (**Articles 66, 110–111**, and perhaps **116**).

16 It is important to note, however, despite the concerns voiced by several States, the freedom to fish did not undergo an express modification in the 1992 global negotiation on the development of the environment (UNCED). The fisheries component of Agenda 21 expressly endorses the fisheries provisions of UNCLOS, albeit without reference to freedom of fishing. However, as is evident from the discussion in Chapters 10 and 11, the proliferation of international instruments in the 1990s, namely, the UN Resolutions on large-scale pelagic driftnet fishing, Chapter 17 of Agenda 21, the FAO Compliance Agreement, the 1995 Straddling Stocks Agreement, the FAO Code of Responsible Fisheries, and the Convention on Biological Diversity, all support the view that the UNCLOS framework does not provide all the answers to the conservation and management of marine resources. See, further, Hey, E., 'Global Fisheries in the First Half of the 1990s', *IJMCL* **11(4)** (1996), 359–490.

17 Chapter 8, *ante*.

18 Chapter 3, *ante*.

19 *Ibid*.

20 Chapter 11, *ante*.

21 For a different approach see for example the requirements for a mandatory level of control of products eligible for export refunds under the CAP, discussed by Huyzer, M., 'The Responsibility of the Member States for Implementation and Enforcement of Common Market and Price Policy', Vervaele (eds.) *Administrative Law Application and Enforcement of Community Law*, (Boston, 1994), p. 22.

22 Chapter 8, *ante*.

23 *Ibid*.

24 For further discussion of this point, in relation to fisheries see, *inter alios*, Dubbink, W. and van Vliet, M., 'Market Regulation Versus Co-management', *Marine Policy* **20(6)** (1996), 499–516; Garza-Gil, D. Iglesia-Malvido, C., Suris-Regueiro, J.C. and Varele-Lafuente, M., 'The Spanish Case Regarding Fishing Regulation', *Marine Policy* **20(3)** (1996), 249–259. For a wider perspective, see Stone, C.D., *Where the Law Ends: The Social Control of Corporate Behavior*, (New York, 1975); Pressman, J. and Wildavski, A, *Implementation: How Great Expectations in Washington are Dashed in Oakland*, (Berkeley, 1973).

25 A recent research study, the Pioneer programme, has been undertaken in the Centre for Law Enforcement and European Integration, University of Utrecht, the Netherlands, which examines the interaction between Community law and national administrative economic and criminal law. Several case studies have been examined including Foodstuffs and the Enforcement of the Hygiene Directive 93/43/EEC in the Netherlands. On the topic of private law in domain of sea fisheries, see Berg, A., 'Shifting Boundaries Between Public and

Private Law Enforcement: the Case of Fisheries Self-enforcement', unpublished paper presented at the Conference on Compliance and Enforcement of European Law, Utrecht, 2–3 October 1997.

26 Chapter 5, *ante.*

27 Chapters 3 and 4, *ante.*

28 Chapter 6, *ante.*

29 Chapter 3, *ante.*

30 Mutual observation for shore based controls, Chapter VI, EU-Morocco Fisheries Agreement, *OJ* L 30, Vol. 40, 31.01.1997.

31 Chapter 11, *ante.*

32 See **Article 22** of the 1958 Convention on the High Seas, and **Article 110** of UNCLOS. See however the argument presented in Chapter 11 as to whether the provisions in Straddling Stocks Agreement exceed customary and convention international law.

33 Chapter 8, *ante.*

34 Chapter 1, *ante.*

35 See, *inter alia*, 'Communication from the Commission to the Council and the European Parliament,' *Fishing on the High Seas, a Community Approach*, Sec (92) 565 final; Proposal for A Council Decision on the ratification by the European Community of the Agreement for the implementation of the provisions of the United Nations Convention on the Law of the Sea of 10 December 1982 relating to the conservation and management of straddling stocks and highly migratory species, *OJ* C 367, Vol. 39, 05.12.1996, pp. 24–25.

36 See, Chapter 8, *ante.*

37 *Ibid.*

38 *Ibid.*

39 See, Birnie, P.W., *International Law and the Environment*, (Oxford, 1992), p.28.

40 See discussion on the rôle of the Commission in respect to the driftnet fishery, Chapter 10, *ante.*

41 See, Chapter 10, *ante.*

42 See Chapter 8, *ante.*

43 However, the weaknesses of these provisions are discussed in Chapter 8, *ante.*

44 See, Chapter 11, *ante.*

45 See, Chapter 10, *ante.*

46 See Chapter 11, *ante.*

47 *Ibid.*

48 *Ibid.*

49 See Chapter 10, *ante.*

50 GATT panels have dealt with unilateral trade measures taken by States to enforce conservation and management measures. See further, Kinsbury, B., 'The Tuna-Dolphin Controversy, the World Trade Organisation, and the Liberal Project to Reconceptualize International Law', *Yearbook of International Environmental Law* **5** (1994), 1–40; Weiss, F., 'The Second Tuna GATT Panel Report', *Leiden Journal of International Law* **8** (1995), 135–150.

51 FAO Code of Conduct of Responsible Fisheries, **Article 11.2.4**.

52 Scannell, Y., *Environmental and Planning Law*, (Dublin, 1995), 36–38.

53 For a stimulating discussion, from the environmental perspective, of a market based solution to marine habitat protection as an alternative to litigation or the legislative process, see Alker, S.C., 'The Marine Mammal Protection Act: Refocusing the Approach to Conservation', *UCLA Law Review*, **44**, (1996), pp. 527–577.

54 *OJ* L 99/92, 11.04.1992.

55 Under the present scheme of law in several Member States, vessel operators are not penalised because fishery offences are processed as actions *in rem*, and consequently no conviction is recorded against the captain of the vessel in the event of the successful prosecution

of the vessel under his/her command. In contrast there are provisions in the Straddling Stocks Agreement which provide that punitive measures shall include the refusal, withdrawal, or suspension of authorisation to serve as a master or officer of a fishing vessel.

56 The success of this approach in other policy areas may be gauged, for example, from the number of complaints lodged by industry participants regarding alleged violations of Community environmental law, see Chapter 8, *ante*.

57 See, Chapter 11, *ante*.

Selected Bibliography

Adams, M. (1984) *The Role of the US Navy in Domestic Law Enforcement* (Naval War College Paper).

Alker, S.C. (1996) 'The Marine Mammal Protection Act: Refocusing the Approach to Conservation', *UCLA Law Review*, Vol. 44, No. 2, pp. 527–77.

Allain, R.J. (1982) 'A Study of Aerial Fisheries Surveillance in Certain CECAF Coastal States', FAO CECAF/TECH/82/46, Rome.

Anand, R.P. (1982) 'The Politics of a New Legal Order for Fisheries', *ODIL*, Vol. II, No. 3/4, pp. 265–95.

Andersen, P. (1994) 'Economic Theory and Fisheries Law Enforcement: A Comment', *Regulation of Fisheries: Legal Economic and Social Aspects*, University of Tromø, Norway, pp. 82–5.

Anderson, D. (1992) 'The Straits of Dover and the Southern North Sea: Some Recent Legal Developments', *IJMCL*, Vol. 7, No. 2, pp. 85–97.

Anderson, D.H. (1995) 'Legal Implications of the Entry into Force of the UN Convention on Law of the Sea', *ICLQ*, Vol. 44, pp. 313–26.

Anderson, D.H. (1996) 'The Straddling Stocks Agreement of 1995 - an Initial Assessment', *ICLQ*, Vol. 45, pp. 463–75.

Anderson, L.G. (1989) 'Enforcement Issues in Selecting Fisheries Management Policy', *Marine Resources Economics*, Vol. 6, pp. 261–77.

Anderson, L.G. and Lee, D.R. (1986) 'Optimal Governing Instrument, Operation Level and Enforcement in Natural Resource Regulation: the Case of the Fishery', *American Journal of Agriculture Economics*, Vol. 68, No. 3, pp. 678–90.

Armas Pfirter, F.M. (1995) 'Straddling Stocks and Highly Migratory Stocks in Latin American Practice and Legislation: New Perspectives in Light of Current International Negotiations', *ODIL*, Vol. 26, pp. 127–50.

Armstrong, A.J. (1991) 'Development of International/Regional Co-operation in Fisheries, Monitoring, Control and Surveillance', Commonwealth Secretariat, London.

Attard, D. (1987) *The Exclusive Economic Zone in International Law* (Clarendon Press, Oxford).

Balton, D.A. (1996) 'Strengthening the Law of the Sea: the New Agreement on Straddling Stocks and Highly Migratory Fish Stocks', *ODIL*, Vol. 27, No. 1–2, pp. 125–51.

Barav, A. (1975) 'Failure of Member States to Fulfil their Obligations under Community Law', *CMLRev* 385, Vol. 12.

Barston, R.P. (1995) 'United Nations Conference on Straddling and Highly Migratory Fish Stocks', *Marine Policy*, Vol. 19, No. 2, pp. 159–66.

Berg, A. (1996) 'Enforcement of the Common Fisheries Policy, with Special Reference to the Netherlands', in Harding (ed.), *Enforcing European Community Rules* (Dartmouth Publishing Company, Aldershot).

Berg, A. and Vervaele, J. (1994) *Fisheries Legislation in the Member States: Issues of Enforcement and Co-operation in the Member States,* Study commissioned by the European Commission, Utrecht.

Bergin, A. (1988) 'Fisheries Surveillance in the South Pacific', *Ocean and Coastal Management,* Vol. 19, No. 3, pp. 467–91.

Biais, G. (1995) 'An Evaluation of the Policy of Fishery Resources Management by TACs in European Community Waters from 1983 to 1992', *Aquatic Living Resources,* Vol. 8, No. 3, pp. 241–51.

Bieber, R. & Monar, J. (eds) (1995) *Justice and Home Affairs in the European Union* (Presses interuniversitaires européennes, Brussels).

Birnie, P. (1992) 'An EC Exclusive Economic Zone: Marine Environment Aspects', *ODIL,* Vol. 23, pp. 200–06.

Birnie, P. (1997) 'Are Twentieth-Century Marine Conservation Conventions Adaptable to Twenty-First Century Goals and Principles?: Part 1', *IJMCL,* Vol. 12, No. 3, pp. 307–39.

Birnie, P. (1997) 'Are Twentieth-Century Marine Conservation Conventions Adaptable to Twenty-First Century Goals and Principles?: Part 11', *IJMCL,* Vol. 12, No. 4, pp. 488–532.

Birnie, P. and Boyle, A.E. (1995) *Basic Documents on International Law and the Environment* (Clarendon Press, Oxford).

Boss, D. (1984) 'European Fisheries. Situation and outlook after the Adoption of the Common Fisheries Policy', *MAK Toplaterne. Diesel Engine Journal,* No. 053, pp. 23–32.

Boyle, A.E. (1991) 'Saving the World? Implementation and Enforcement of International Environmental Law through International Institutions', *Journal of Environmental Law,* Vol. 3, No. 2, pp. 229–45.

Boyle, A.E. (1992) *International Law and the Environment* (Clarendon Press, Oxford).

Blackwell, A. (1998) 'The Humane Society and Italian Driftnetters: Environmental Activists and Unilateral Action in International Environmental Law', *North Carolina Journal of International Law and Commercial Regulation,* Vol. 23, No. 2, pp. 313–40.

Blake, G.H., Hildesley, W., Pratt, M., Ridley, R. and Schofield, C. (eds) (1995) *The Peaceful Management of Transboundary Resources* (Graham & Trotman, Nijhoff).

Bridge, J.W. (1984) 'Procedural Aspects of the Enforcement of EC Law through the Legal Systems of the Member States', *ELRev.,* Vol. 9, No. 1, pp. 28–42.

Brown, E.D. (1972) 'British Fisheries and the Common Market', *Current Legal Problems,* Vol. 25, pp. 37–73.

Brown, E.D. (1994) *The International Law of the Sea,* Vols I and II (Dartmouth Publishing Company, Aldershot).

Brownlie, I. (1990) *Principles of Public International Law* (Oxford University Press).

Burke, W.T. (1982) 'US Fisheries Management and the New Law of the Sea', *AJIL,* Vol. 76, pp. 24–55.

Burke, W.T. (1989) 'Fishing in the Berring Sea Donut: Straddling Stocks and the New International Law of Fisheries', *Ecology Law Quarterly,* Vol. 16, No. 1, pp. 285–310.

Burke, W.T. (1993) 'UNCED and the Oceans', *Marine Policy,* Vol. 17, pp. 519–33.

Burke, W.T. (1994) *The New International Law of Fisheries: UNCLOS 1982 and Beyond* (Clarendon Press, Oxford).

Burke, W.T., Freeberg M. and Miles, E.L. (1994) 'The United Nations Resolutions on Driftnet Fishing – an Unsustainable Precedent for High Seas and Coastal Fisheries', *ODIL*, Vol. 25, No. 2, pp. 127–86.

Burke, W.T., Legatski R. and Woodhead, W. (1975) *National and International Law Enforcement in the Ocean* (University of Washington Press, Seattle & London).

Burrows, F. (1987) *Free Movement in European Community Law* (Clarendon Press, Oxford).

Caddy, J.F. (1996) *Resources and Environmental Issues Relevant to Mediterranean Fisheries Management* (Rome).

Caddy, J.F. and Griffiths R. (1990) *Recent Trends in the Fisheries and Environment in the General Fisheries Council for the Mediterranean (GFCM) Area,* (FAO, General Fisheries Council for the Mediterranean).

Campiglio, L. (1994) *The Environment after Rio: International Law and Economics* (Graham & Trotman, London).

Canadian Dept. of External Affairs (1993) *Global Overview of Straddling and Highly Migratory Fish Stock: The Non-Sustainable Nature of High Seas Fisheries* (Halifax).

Canfield, J.L. (1993) 'Recent Developments in Berring Sea Fisheries Conservation and Management', *ODIL*, Vol. 24, pp. 269–74.

Carlin, F. (1996) 'The Data Protection Directive: the Introduction of Common Privacy Standards', *European Law Review,* Vol. 21, No. 1, pp. 65–70.

Carroz, J.E. and Roche, A.G. (1968) 'International Policing of High Seas Fisheries', *Canadian Yearbook of International Law,* Vol. 6, pp. 61–8.

Cataldi, G. (1992) 'The EEC and Fisheries: Some Recent Developments', *Proceedings from the 26th Annual Conference, Law of the Sea Institute,* pp. 539–63 (LOSI).

Charles, A.T. (1992) 'Fishery Conflicts – A Unified Framework', *Marine Policy,* Vol. 16, No. 5, pp. 379–93.

Charles, A.T. (1995) 'The Atlantic Canadian Ground-Fishery: Roots of a Collapse', *Dalhousie Law Journal,* Vol. 18, No. 1, pp. 65–83.

Charney, J.I. (1994) 'The Marine Environment and the 1982 United Nations Convention on the Law of the Sea', *International Lawyer,* Vol. 28, No. 4, pp. 879–902.

Christensen, E. (1991/92) 'GATT Sets Its Net On Environment Regulation: The GATT Panel Ruling On Mexican Yellowfin tuna imports and the Need for Reform of the International Trading System', *Univ. of Miami Inter-American Law Rev.,* Vol. 23, pp. 596–612.

Christensen, E. (1992) 'Making GATT Dolphin-Safe: Trade and Environment', *Duke Journal of Comparative and International Law,* Vol. 2, pp. 345–66.

Churchill, R.R. (1977) 'The EEC Fisheries Policy – Towards a Revision', *Marine Policy,* Vol. 1, pp. 26–36.

Churchill, R.R. (1980) 'Revision of the EEC's Common Fisheries Policy Part II', *European Law Review,* Vol. 5, pp. 3–37.

Churchill, R.R. (1986) *EC Fisheries Law* (Kluwer Academic Publishers, Dordrecht).

Churchill, R.R. (1987) 'The EEC's Contribution to "State" Practice in the Field of Fisheries', *Law of the Sea Conference Proceedings 1985,* pp. 557–68.

Churchill, R.R. (1988) 'The EEC's Fisheries Management System: A Review of the First Five Years of Operation', *CMLRev.,* Vol. 25, pp. 369–89.

Churchill, R.R. (1990) 'Quota Hopping: The Common Fisheries Policy Wrongfooted?', *CML Rev.,* Vol. 27, No. 2, pp. 209–47.

Churchill, R.R. (1991) 'Fisheries in the European Community - Sustainable Development or Sustained Mismanagement?', *Proceeedings of the 25th Annual Conference of the Law of the Sea Institute,* pp. 557–68 (Malmo, Sweden).
Churchill, R.R. (1992) 'EC Fisheries and an EZ - Easy', *ODIL,* Vol. 23, pp. 145–63.
Churchill, R.R. (1993) 'Fisheries Issues in Maritime Boundary Delimitation', *Marine Policy,* Vol. 17, No. 1, pp. 44–57.
Churchill, R.R. (1993) 'Shared Fisheries Management in the European Community', *RECIEL,* Vol. 2, pp. 260–69.
Churchill, R.R. (1996) 'Enforcement of the Common Fisheries Policy, with Special Reference to the United Kingdom', in Harding (ed.), *Enforcing European Community Rules,* pp. 83–102 (Dartmouth Publishing Company, Aldershot).
Churchill, R.R. (1997) 'Decision to Defer Accession to the LOS Convention: A Convincing Move?', *IJMCL,* Vol. 12, No. 1, pp. 110–21.
Churchill, R.R and Foster, N.G. (1987) 'Double Standards in Human Rights? The Treatment of Spanish Fishermen by the European Community', *ELRev,* Vol. 12, No. 06, pp. 430–43.
Churchill, R.R and Foster, N.G. (1987) 'European Community Law and Prior Treaty Obligations of Member States: the Spanish Fishermen's Cases', *ICLQ,* Vol. 36, No. 3, pp. 504–24.
Churchill, R.R. and Lowe, A.V. (1997) *The Law of the Sea,* 2nd ed. (Manchester University Press, Manchester).
Churchill, R.R. and Orebech, P. (1993) 'The European Economic Area and Fisheries', *IJMCL,* Vol. 8, No. 4, pp. 453–69.
Clingan, T.A. (1994) *The Law of the Sea: Ocean Law and Policy* (Austin & Winfield, San Francisco).
Coffey, C. (1995) *Introduction to the Commmon Fisheries Policy: An Environmental Perspective,* Institute for European Environmental Policy Background Briefing No. 2. (London).
Collins, A. and O'Reilly, J. (1990) 'The Application of Community Law in Ireland 1973-1989', *CMLRev.,* Vol. 27, No. 2, pp. 315–39.
Coppel, J. (1993) *Individual Enforcement of Community Law: the Future of the Francovich Remedy,* EUI working papers. Law 06 (Florence).
Crean, K. and Symes, D. (1994) 'The Discard Problem: Towards a European Solution', *Marine Policy,* Vol. 18, No. 5, pp. 422–34.
Crean, K. and Symes, D. (eds) (1996) *Fisheries Management in Crises* (Blackwell Science Ltd, Oxford).
Curran, P. and Long, R. (1996) 'Fishery Law, Unilateral Enforcement in International Waters: the Case of the *Estai*', *Irish Journal of European Law,* Vol. 5, No. 2, pp. 123–63.
Curtin, D. (1993) 'The Constitutional Structure of the Union: a Europe of Bits and Pieces', *CMLRev.,* Vol. 30, No. 1, pp. 17–69.
Curtin, D. and O'Keefe, D. (1992) *Constitutional Adjudication in European Community and National Law. Essays for the Hon. Mr. Justice T.F. O'Higgins* (Butterworth Law (Ireland), Dublin).
Cuthbert, J.A.M. (1991) *Scotland's Fishing Interests in the Light of the Common Fisheries Policy of the European Community,* Unpublished thesis (Glasgow).
Daggett, A. (1934) 'The Regulation of Maritime Fisheries by Treaty', *AJIL,* Vol. 28, pp. 693–717.

Dahmani, M. (1987) *The Fisheries Regime of the Exclusive Economic Zone* (Kluwer Academic Publishers, Dordrecht).

Daintith, T. (eds) (1995) *Implementing EC Law in the UK: Structures for Indirect Rule* (Chancery Wiley Law Publications, London).

Dashwood, A. and White, R. (1989) 'Enforcement Actions under Article 169 and 170 EEC Treaty', *ELRev*, Vol. 14, pp. 388–413.

Davies, P.G.G. (1995) 'The EC/Canadian Fisheries Dispute in the Northwest Atlantic', *ICLQ*, Vol. 44, pp. 927–39.

Day, D. (1995) 'Tending the Achilles heel of NAFO: Canada Acts to Protect the Nose and Tail of the Grand Banks', *Marine Policy*, Vol. 19, pp. 257–70.

Delmas-Marty, M. (1997) 'Union européenne et droit pénal', *Cahiers de droit européen*, No. 5-6, pp. 607–53.

Delmas-Marty, M., Summers, M. and Mongin, G. (ed.) (1996) *What kind of Criminal Policy for Europe?* (Kluwer Law International, The Hague).

Den Boer, M. (1995) 'Police Co-operation in the TEU: Tiger in a Trojan Horse', *CMLRev*, Vol. 32, pp. 555–78.

Derham, P.J. (1985) 'The Problems of Quota Management in the European Community Concept', in Food and Agriculture Organization of the United Nations, Expert Consultation on the Regulation of Fishing Effort, *FAO Fisheries Report No.289*, Supplement 3, pp. 241–50.

Derham, P.J. (1987) 'The Implementation and Enforcement of Fisheries Legislation', in Ulfstein (ed.), *Proceedings of the European Workshop on the Regulation of Fisheries: Legal, Economic and Social Aspects*, pp. 71–81 (Strasbourg).

Dias, A.K. and Begg, M. (1994) 'Environmental Policy for Sustainable Development of Natural Resources. Mechanisms for Implementation and Enforcement', *Natural Resources Forum*, United Nations, Oxford, Vol. 18, No. 4, pp. 275–86.

Donoghue, J.E. (1993) 'EC Participation in the Protection of the Marine Environment', *Marine Policy*, Vol. 20, pp. 515–18.

Duff, J.A. (1997) 'Recent Applications of United States Laws to Conserve Marine Species Worldwide: Should Trade Sanctions be Mandatory', *Ocean and Coastal Law Journal*, Vol. 2, pp. 1–31.

Dunlap, W.V. (1994) 'Canada Asserts Jurisdiction Over High Seas Fisheries', *Boundary and Security Bulletin*, Vol. 2, pp. 63–9.

Dunlap, W.V. (1995) 'Bering Sea, the Donut Hole Agreement', *IJMCL*, Vol. 10, pp. 114–26.

Dunlap W.V. (1995) 'Environmental Jurisdiction on the High Seas: The Special Case of the Arctic Ocean', in Blake (ed.), *The Peaceful Management of Transboundary Resources*, pp. 319–49 (Kluwer Law International).

Edeson, W. (1996) 'The Code for Responsible Fisheries: An Introduction', *IJMCL*, Vol. II, No. 2, pp. 233–8.

Eser, A. (1992) *Principles and Procedures for a New Transnational Criminal Law* (Freiburg im Breisgau, Max-Planck-Institut).

European Commission (1986) *Guidelines and Initiatives for the Development of the Common Fisheries Policy*, COM Document 1986/0302 final (Brussels).

European Commission (1986) *Report on the Enforcement of the Common Fisheries Policy*, COM Document 1986/0301 final (Brussels).

European Commission (1991) *Report 1991 on the Common Fisheries Policy*, SEC Document 1991/2288 final (Brussels).

European Commission (1992) *Report on Monitoring Implementation of the Common Fisheries Policy,* SEC/92/0394 (Brussels).
European Commission (1993) *The New Components of the Common Fisheries Policy and their Practical Implementation,* COM Document 1993/0664 final (Brussels).
European Commission (1993) *Maintenance, Extension and Creation of Databases Appropriate to the Implementation of the Common Fisheries Policy,* COM Document 1993/0501 final (Brussels).
European Commission (1994) *The Use of Large Scale Driftnets under the Common Fisheries Policy,* COM Document 1994/50 final (Brussels).
European Commission (1994) *Enforcement of Community Legislation Concerning the Use of Driftnets in 1994,* SEC Document (94) 2003.
European Commission (1995) *Report from the Commission to the Council and the European Parliament on the Community's Financial Contribution Towards Expenditure Incurred by Member States for the Purpose of Ensuring Compliance with the Common Fisheries Policy,* COM Document 1995/243/II/final (Brussels).
European Commission (1995) *Implementation of Technical Measures in the Common Fisheries Policy,* COM Document 1995/0669 final (Luxembourg).
European Commission (1996) *Report to the Council and the European Parliament on the Establishment of a Satellite-Based Vessel Monitoring System for Community Fishing Vessels,* COM Document 1996/0323 final (Luxembourg).
European Commission (1996) *Communication from the European Commission to the Council and the European Parliament, Fisheries Agreements, Current Situation and Perspectives,* COM(96) 488 final (Brussels).
European Commission (1997) *Report from the Commission on the Monitoring the Common Fisheries Policy,* COM Document 1997/0226 final (Luxembourg).
European Commission (1998) *Improving the Implementation of the Common Fisheries Policy: an Action Plan,* SEC Document 1998/949 final (Brussels).
European Commission (1998) *The Common Fisheries Policy,* Directorate-General XIV-Fisheries, CU-12-98-441-EN-C (Luxembourg).
European Commission (1998) *Fisheries Monitoring Under the Common Fisheries Policy,* COM/98/0092 final (Luxembourg).
European Commission (1998) *Commission Working Document on Enforcement of European Consumer Legislation,* SEC Document 1998/0527 final (Brussels).
European Court of Auditors (1994) *Court of Auditors' Special Report No. 3/93 Concerning the Implementation of the Measures for the Restructuring, Modernising and Adaptation of the Capacities of Fishing Fleets in the Community Together with the Commission's Replies,* OJ C2, 4.1.1994 (Luxembourg).
European Parliament (1985) *Report on the Enforcement of the Common Fisheries Policy,* Author: O'Hagan C.T.S, EP Document 1985/0162 A2 (Luxembourg).
European Parliament (1985) *Report on Budgetary Control with Regard to the Measures Taken under the Common Fisheries Policy,* Author: Battersby R.C., EP Document 1985/0034 A2 (Luxembourg).
European Parliament (1988) *Report on Measures Intended to Develop the Social Aspects of the Common Fisheries Policy,* Author: Morris D., EP Document 1987/0310 A2 (Luxembourg).
European Parliament (1991) *Interim Report on the Common Fisheries Policy and the Adjustment to be Made,* Author: Pery N., EP Document 1991/0335 A3 (Luxembourg).

European Parliament (1992) *Interim Report on the Common Fisheries Policy and the Adjustment to be Made,* 02, Author: Pery N., EP Document 1992/0175 A3 (Luxembourg).

European Parliament (1993) *European Community Fisheries Agreements with third Countries and Participation in International Fisheries Agreements,* Research Author Mathers S. (Luxembourg).

European Parliament (1994) *Report on the Commission Proposal for a Council Decision on Accession of the EC to the Agreement to Promote Compliance with International Conservation and Management Measures by Fishing Vessels on the High Seas,* EPDocument A4-86/94 (Luxembourg).

European Parliament (1997) *Report on the Common Fisheries Policy after the Year 2002,* Author: Fraga EstÈvez C., EP Document 1997/0298 A4 (Luxembourg).

European Parliament (1997) *The Common Fisheries Policy Beyond 2002: Alternative Options to the TACS and Quotas System for the Conservation and Management of Fisheries Resources,* Authors Angelidis A., Rodgers P., Valatin G. (Luxembourg).

European Parliament (1997) *Report on International Fisheries Agreements,* Author: Duncan Crampton P., EP Document 1997/0149 A4 (Luxembourg).

European Parliament (1998) *Report on the Commission Working Document 'Improving the Implementation of the Common Fisheries Policy: An Action Plan',* EP Document 1998/0462 A4 (Luxembourg).

Evans, A.C. (1979) 'The Enforcement of Article 169 EEC: Commission Discretion', *ELRev,* Vol. 4, pp. 449–55.

Farnell, J. and Elles, J. (1984) *In Search of a Common Fisheries Policy* (Gower Technical, Aldershot).

Fauteux, P. (1993) 'L'initiative juridique canadienne sur la pêche en haute mer', *Canadian Yearbook of International Law,* Vol. 31, pp. 38–87.

Fauteux, P. (1994) 'L'Organisation des Pêches de L'Atlantique Nord Ouest et le conflit Canada-CEE', *Revue de l'INDEMER,* Vol. 2, pp. 65–89.

Fidell, E.R. (1976) 'Fisheries Legislation: Naval Enforcement', *J.Mar.L.*and *Comm.,* Vol. 7, pp. 351–66.

Findlater, J. (1992) *Legal Aspects of Commercial Sea-Fishing in the EC* (Irish Centre for European Law, Dublin).

Fisher, R. (1981) *Improving Compliance with International Law* (University Press of Virginia, Charlottesville).

Fitzmaurice, M. (1990) 'Common Market Participation in the Legal Regime of the Baltic Sea Fisheries', *GYBIL,* Vol. 33, pp. 214–35.

Fitzmaurice, M. (1992) *International Legal Problems of the Environmental Protection of the Baltic Sea* (Martinus Nijhoff, Dordrecht).

Fleischer, C.A. (1971) 'L'accès aux lieux de pêche et la traité de Rome', *RMC,* Vol. 141, pp. 148–51.

Fleischer, C.A. (1977) 'The Right of a 200-mile Exclusive Economic Zone or a Special Fishery Zone', *SDLR,* Vol. 14, pp. 548–83.

Fleischer, C.A. (1988) 'The New Regime of Maritime Fisheries', *Collected Courses of the Hague Academy of International Law,* Vol. 2, T. 209, pp. 95–222 (Dordrecht).

Fleischmann, A. (1995) 'Personal Data Security: Divergent Standards in the European Union and the United States', *Fordham International Law Journal,* Vol. 19, pp. 143–80.

Flum (eds.) (1987) *The Commission's Agricultural Proposals for Implementing the Single Act. Own-Initiative Opinion of the Section for Agriculture and Fisheries,* Economic and Social Committee (Brussels).

Food and Agriculture Organization of the United Nations (FAO) (1982) 'The United Nations Convention on the Law of the Sea: Impacts on Tuna Regulation', *FAO Legislative Study,* No. 26 (Rome).

Food and Agriculture Organization of the United Nations (FAO) (1985) 'Expert Consultation in the Regulation of Fishing Effort', *FAO Fisheries Report,* No. 289, Suppl. 3 (Rome).

Food and Agriculture Organization of the United Nations (FAO) (1991) 'The Regulation of Driftnet Fishing on the High Seas: Legal Issues', *FAO Legislative Study,* Vol. 47 (Rome).

Food and Agriculture Organization of the United Nations (FAO) (1994) 'A Global Assessment of Fisheries by-Catch and Discards', *FAO Fisheries Technical Paper* 339 (Rome).

Food and Agriculture Organization of the United Nations (FAO) (1996) 'Introducing Monitoring, Control and Surveillance. A Tool for Fisheries Management', *FAO Fisheries Technical Document 338* (Rome).

Food and Agriculture Organization of the United Nations (FAO) (1996) 'Coastal State Requirements For Foreign Fishing', *FAO Legislative Study,* No. 57 (Rome).

Food and Agriculture Organization of the United Nations (FAO) (1997) 'Review of the State of World Fishery Resources: Marine Fisheries', *Fisheries Circular, No. 920* (Rome).

Francis, P., Davies, P. and Jupp, V. (1997) *Policing Futures: the Policy, Law Enforcement and the Twenty-First Century* (St Martins Press, Basingstoke).

Franckx, E. (1992) 'EC Maritime Zones, the Delemitation Aspect', Vol. 23, pp. 239–58.

Freestone, D. (1994) 'The Road from Rio: International Environmental Law after the Earth Summit', *Journal of Environmental Law,* Vol. 6, No. 2, pp. 193–218.

Freestone, D. (1995) 'The Effective Conservation and Management of High Seas Living Resources: Towards a new Regime?', *Canterbury Law Review,* Vol. 5, pp. 341–62.

Freestone, D. (1997) 'Some Institutional Implications of the Establishment of Exclusive Economic Zones by EC Member States', *ODIL,* Vol. 23, No. 2–3, pp. 97–114.

Freestone, D. and Hey, E. (1996) *The Precautionary Principle and International Law: the Challenge of Implementation* (Kluwer Law International, The Hague).

Freestone, D. and Makuch, Z. (1996) 'The New International Environmental Law of Fisheries: the 1995 United Nations Straddling Stocks Agreement', *Yearbook of International Environmental Law,* Vol. 7, pp. 3–51.

Frid, R. (1995) *The Relations Between the EC & International Organisations – Legal Theory and Practice* (Kluwer Law International, The Hague).

Furlong, W.J. (1991) 'The Deterrent Effect of Regulatory Enforcement in the Fishery', *Land Economics,* Vol. 67, No. 1, pp. 116–29.

Ganz, G. (1987) *Quasi-Legislation: Recent Development in Secondary Legislation* (London).

Garcia, S.M. (1994) 'The Precautionary Principle: its Implications in Capture Fisheries Management', *Ocean and Coastal Management,* Vol. 22, pp. 99–125.

Garron, R. (1971) 'Le Marché Commun De La Pêche Maritime', *Collection de droit maritime et des transports,* Part III (Paris).

Garza-Gil, D. (1996) 'The Spanish Case Regarding Fishing Regulation', *Marine Policy,* Vol. 20, pp. 249–59.

Gerardu, J. and Wasserman, C. (1996) 'Fourth International Conference on Environmental Compliance and Enforcement: Proceedings', *International Conference on Environmental Compliance and Enforcement,* 2 Vols (4th, 1996, Chiang Mai, Thailand).

Gold, E. (1993) *The Marine Environment and Sustainable Development Law, Policy, and Science* (Law of the Sea Institute).

Goldberg, R.M. (1975) 'Ends and Means: the Role of Enforcement Analysis in International Fisheries Regulations', in Knight H.G. (ed.), *The Future of International Fisheries Management* (Cust. Publishing Company, St. Paul Minn).

Gormley, L. (1986) 'The Application of Community Law in the United Kingdom, 1976–1985', *CMLRev,* Vol. 23, No. 2, pp. 287–323.

Groux, J. and Manin, P. (1985) *European Communities in the International Order* (European Communities, Luxembourg).

Guldenmund, R. and Westeroun Van Meeteren, L. (1996) 'Towards an Administrative Sanctioning System in the Common Agriculture Policy', in Harding (ed.), *Enforcing European Community Rules* (Dartmouth Publishing Company, Aldershot).

Gurrish, J.A. (1992) 'Pressures to Reduce By-Catch on the High Seas: An Emerging International Norm', *Tulane Environmental Law Journal,* Vol. 5, p. 473.

Handoll, J. (1995) *Free Movement of Persons in the EU* (Chancery Wiley Law Publications, Chichester).

Hannesson, R. (1995) 'Fishing on the High Seas', *Marine Policy,* Vol. 19, No. 5, pp. 371–7.

Harding, C. (1982) 'The Impact of Article 177 of the EEC Treaty on the Review of Community Action', *Yearbook of European Law 1981,* pp. 93–113 (Clarendon Press, Oxford).

Harding, C. (1992) *European Community Investigations & Sanctions* (Leicester University Press, Leicester).

Harding, C. (1992) 'Who Goes to Court in Europe? An analysis of Litigation against the European Community', *ELR,* Vol. 17, No. 2, pp. 105–25.

Harding, C. (1993) *European Community Investigations and Sanctions: The Supranational Control of Business delinquency* (Leicester University Press, Leicester).

Harding, C. (1997) 'Member State Enforcement of European Community Measures: the Chimera of "Effective" Enforcement', *Maastricht Journal of European and Comparative Law,* Vol. 4, No. 1, pp. 5–24.

Harding, C. and Swart, B. (ed.) (1996) *Enforcement of Community Rules* (Dartmouth Publishing Company, Aldershot).

Hart, J. (1976) *The Anglo Icelandic War of 1972–1973,* Research Series, No. 29 (Institute of International Studies, Berkeley).

Hartley, T.C. (1986) 'Direct Applicability and Direct Effect: Two Distinct and Different Concepts in Community Law', *CMLRev,* Vol. 9, p. 425.

Hartley, T.C. (1998) *The Foundations of European Community Law,* 4th ed. (Oxford University Press, Oxford).

Hatcher, A., Thébaud, O., Jaffry, S. and Bennett, E. (1998) *Compliance with Fishery Regulation,* CEMARE Report 51, Research study funded by the European Commission.

Hayashi, M. (1993) 'The Management of Transboundary Fish Stocks under the LOS Convention', *IJMCL,* Vol. 8, No. 2, pp. 245–61.

Hayashi, M. (1995) 'The Role of the United Nations in Managing the World's Fisheries', in Blake, G.H. (ed.), *The Peaceful Management of the Transboundary Resources* (Kluwer Law International).

Henkin, L. (1995) *International Law: Politics and Values* (Martinus Nijhoff, Dordrecht).

Hewison, G. (1993) 'The Convention for the Prohibition of Fishing with Long Driftnets in the South Pacific', *Case Western Journal of International Law*, Vol. 25, pp. 449–530.

Hewison, G. (1993) 'The Precautionary Approach to Fisheries Management: An Environmental Perspective', *IJMCL*, Vol. 11, No. 3, pp. 301–32.

Hewison, G. (1994) 'The Legally Binding Nature of the Moratorium on Large Scale High Seas Driftnet Fishing', *Journal of Maritime law and Commerce*, Vol. 25, No. 4, pp. 557–79.

Hey, E. (1989) *The Regime for the Exploitation of Transboundary Marine Fisheries Resources: the United Nations Law of the Sea Convention Cooperation Between States* (Martinus Nijhoff, Dordrecht).

Hey, E. (1991) 'The Precautionary Approach. Implications of the Revision of the Oslo and Paris Conventions', *Marine Policy*, Vol. 15, No. 4, pp. 244–54.

Hey, E. (1996) 'Global Fisheries in the First Half of the 1990s', *IJMCL*, Vol. 11, No. 4, pp. 459–90.

Hiester, E. (1976) 'The Legal Position of the European Community with regard to the Conservation of the Living Resources of the Sea', *Legal Issues of European Integration*, Vol. 55, pp. 55–79.

Holden, M.J. and Garrod, D. (1996) *The Common Fisheries Policy: Origin, Evaluation and Future*, 2nd ed. (Fishing News Books, Oxford).

Houghton, R.G. (ed.) (1981) *Analysis of Methods Used to Determine Fishing Capacity and Establishment of a Method Suitable for Community Needs* (EC Commission, Brussels).

House of Lords (1984) *The Common Fisheries Policy, with Minutes of Evidence*, Select Committee of the European Communities, Session 1984/1985. Report 01 (London).

House of Lords (1992) *Implementation and Enforcement of Environmental Legislation*, Select Committee of the European Communities, Session 1991/1992. Report 09 (London).

House of Lords (1992) *The Common Fisheries Policy, with Minutes of Evidence*, Select Committee of the European Communities, Session 1992/1993. Report 02 (London).

House of Lords (1993) *Regulation of Drift Net Fishing, with Evidence*, Select Committee on the European Communities, Session 1993/1994. Report 13 (London).

Howarth, W. (1990) 'The Single European Market and the Problems of Fish Movement', *ELRev*, Vol. 15, pp. 34–50.

Hurlockm, M.H. (1992) 'The GATT, United States Law and the Environment: a Proposal to Amend the GATT in Light of the Tuna/Dolphin Decision', *Columbia Law Review*, Vol. 29, pp. 2098–161.

Ijstra, T. (1992) 'Development of Resource Jurisdiction in the EC's Regional Seas: National EEZ Policies of EC Member States in the Northeast Atlantic, the Mediterranean Sea and the Baltic Sea', *ODIL*, Vol. 23, Nos 2–3, pp. 165–92.

Jacobson, J.L. (1995) 'Managing Marine Living Resources in the Twenty-First Century: The Next Level of Ocean Governance?', in Nordquist, M. and Moore, J.N. (ed.), *Entry into Force Law of the Sea Convention*, pp. 311–22 (Martinus Nijhoff, Dordrecht).

Joerges, C. and Jürgen, N. (1997) 'From Intergovernmental Bargaining to Deliberative Political Processes: The Constitutionalisation of Comitology', *ELJ*, Vol. 3, No. 3, pp. 273–99.

Johnston, D.M. (1987) *The International Law of Fisheries. A Framework for Policy-Oriented Inquiries* (Kluwer Academic Publishers).
Johnston, D.M. (1991) 'Is Coastal State Fishery Management Successful or Not?', *ODIL*, Vol. 22, pp. 199–208.
Johnston, D.M (1991) 'The Driftnetting Problem in the Pacific Ocean: Legal Considerations and Diplomatic Options', *ODIL*, Vol. 22, No. 2, p. 15.
Johnston, B. (ed.) (1977) 'Canada's 200 Mile Zone: The Problem of Compliance', *ODIL*, Vol. 4, pp. 67–107.
Joseph, J. (1994) 'Tuna-Dolphin Controversy in the Pacific', *ODIL*, Vol. 25, p. 26.
Joyner, C.C. (1993) 'Chile's Presential Sea Proposals: Implications for Straddling Stocks and the International Law of Fisheries', *ODIL*, Vol. 24, No. 1, pp. 99–21.
Juda, L. (1997) 'The 1995 United Nations Agreement on Straddling Fish Stocks and Highly Migratory Fish Stocks: A Critique', *ODIL*, Vol. 28, pp. 147–66.
Karagiannakos, A. (1995) *Fisheries Management in the European Union* (Ashgate Publishing Company, Aldershot).
Kerameus, K. (1997) 'Enforcement in the International Context', *Collected Courses of the Hague Academy of International Law*, Vol. 264, pp. 179–410.
Kelly, J.M. (1992) *A Short History of Western Legal Theory* (Clarendon Press, Oxford).
Kinsbury, B. (1994) 'The Tuna Dolphin Controversy, the World Trade Organisation, and the Liberal Project to Reconceptualize International Law', *Yearbook of International Environmental Law*, Vol. 5, pp. 1–40 (Clarendon Press, Oxford).
Kitzinger, U. (1973) *Diplomacy and Persuasion: How Britain Joined the Common Market* (Thames and Hudson, London).
Koers, A.W. (1970) *A Comparative Analysis of International State Practice*, Law of the Sea Inst. Univ of Rhode Island (Occasional Paper No. 6, June).
Koers, A.W. (1973) *International Regulation of Marine Fisheries: a Study of Regional Fisheries Organisations* (Fishing News Books, London).
Koers, A.W. (1977) 'The External Authority of the EEC in Regard to Marine Fisheries', *CMLRev*, Vol. 14, pp. 269–301.
Koers, A.W. (1979) 'Participation of the European Economic Community in a New Law of the Sea Convention', *American Journal of International Law*, Vol. 73, pp. 426–43.
Koers, A.W. (1983) 'The Fisheries Policy', *Thirty Years of Community Law*, Commission of the European Communities (Luxembourg).
Koers, A.W. (1983) 'Fisheries Management: the Case of the EEC and the Capabilities Required', *Proceedings of a Seminar in Jakarta.*
Koers, A.W. (1984) 'The European Economic Community and International Fishery Organisations', *Legal Issues of European Integration*, Vol. 1, pp. 113–31.
Kolte, L. (1988) 'The Community Budget: New principles for Finance, Expenditure, Planning and Budgetary Disciple', *CMLRev*, Vol. 25, pp. 487–501.
Kunzlik, P. (1997) 'The Enforcement of EU Environmental Law: Article 169, the Ombudsman and the Parliament', *European Environmental Law Review*, Vol. 6, No. 2, pp. 46–52.
Kwiatkowska, B. (1989) *The 200 mile Exclusive Economic Zone in the New Law of the Sea* (Martinus Nijhoff, Dordrecht).
Kwiatkowska, B. (1991) 'Creeping Jurisdiction Beyond 200 miles in the Light of the 1982 Law of the Sea Convention and State Practice', *ODIL*, Vol. 22, No. 2, pp. 157–73.

Kwiatkowska, B. (1993) 'The High Seas Fisheries Regime: at a Point of No Return?', *IJMCL*, Vol. 8, No. 3, pp. 327–55.
Lasok, D. and Lasok, K.P.E. (1998) *Law and Institutions of the European Union*, 7th ed. (Butterworths, London).
Leigh, M. (1983) *European Integration and the Common Fisheries Policy* (Croom Helm, London).
Lewis, C. (1996) *Remedies and Enforcement of European Community Law* (Sweet & Maxwell, London).
Leonard, L. (1944) *International Regulation of Fisheries* (Carnegie Endowment, Washington).
Lodge, J. (1983) *Institutions and Policies of the European Community* (Pinter, London).
Lugt, M. (1994) 'Enforcement of European Food Law in the Netherlands and its Future', *European Food Law Review*, Vol. 4/9, pp. 391–412.
Lugten, G.L. (1995) 'Fisheries War for the Halibut', *Environmental Policy and Law*, Vol. 25, No. 4–5, pp. 223–9.
MacLeod, I., Hendry, I.D. and Hyett, S. (1996) *The External Relations of the European Communities: A Manual of Law and Practice* (Oxford University Press, Oxford).
MacRory, R. (1992) 'The Enforcement of Community Environmental Laws, Some Critical Issues', *CMLRev.*, Vol. 29, No. 1, pp. 347–69.
Manard, J.P. (1995) 'GATT and the Environment: the Friction Between International Trade and the World's Environment – the Dolphin and Tuna Dispute', *Tulane Environmental Law Journal*, Vol. 5, pp. 373–478.
MacDonald, J.M. (1995) 'Appreciating the Precautionary Principle as an Ethical Evolution in Ocean Management', *ODIL*, Vol. 26, pp. 255–86.
Mazany, R.L. (1993) 'The Economics of Crime and Law Enforcement', in Charles A.T. (ed.), *Fishery Enforcement: Economic Analysis and Operational Models* (Oceans Institute of Canada, Halifax).
McDorman, T.L. (1994) 'Stateless Fishing Vessels, International Law and the U.N. High Seas Fisheries Conference', *Journal of Maritime Law and Commerce*, Vol. 25, No. 4, pp. 531–55.
McGinley, J. (1991) *Ireland's Fishery Policy* (Teelin, Co. Donegal, Croaghlin Press).
McKenna, P. (1996) *Report on the Commission Report on Monitoring the Common Fisheries Policy – the Commission Report*-3 European Parliament, PE 217.749 final (European Parliament).
Meltzer, E. (1994) 'Global Overview of Straddling and Highly Migratory Fish Stocks: The Non Sustainable Nature of High Seas Fish Stocks', *ODIL*, Vol. 25, No. 3, pp. 255–344.
Mendrinou, M. (1996) 'Non-Compliance and the European Commission's Role in Integration', *Journal of European Public Policy*, Vol. 3, No. 1, pp. 1–22.
Miles, E. (ed.) (1989) 'Pressure on the United Nations Convention on the Law of the Sea of 1982 arising from New Fisheries Conflicts: The Problem of Straddling Stocks', *ODIL*, Vol. 20, pp. 344–49.
Miles, E. (ed.) (1989) *Management of World Fisheries: Implications of Extended Coastal State Jurisdiction* (University of Washington Press, London).
Milliman, S.R. (1986) 'Optimal Fishery Management in the Presence of Illegal Activity', *Journal of Environmental Economics and Management*, Vol. 13, pp. 363–87.
Mlimuka, A.K.L.J. (1995) 'The Influence of the 1982 United Nations Convention on

Law of the Sea on State Practice: the Case of Tanzanian Legislation Establishing the Exclusive Economic Zone', *ODIL*, Vol. 26, pp. 57–73.
Molenaar, E.J. (1999) 'Monitoring, Control and Surveillance of Fisheries within the Framework of the Indian Ocean Commission', *Paper presented at the workshop on fisheries, organised by the Indian Ocean Commission*, Mauritius, 2–3 February 1999.
Monace, F.R. (1995) 'Europol: The Culmination of the European Union's International Police Cooperation Efforts', *Fordham International Journal*, Vol. 19, pp. 247–308.
Moore, G. (1980) 'National Legislation for the Management of Fisheries under Extended Coastal State Jurisdiction', *JMLC*, Vol. 11, pp. 153–82.
Moore, G. (1987) 'Enforcement Without Force: New Concepts in Compliance Control for Foreign Fishing Operations', *The Law of the Sea: Essays in the Memory of Jean Carroz*, pp. 159–69.
Moore, G. (1988) *Coastal State Requirements for Foreign Fishing*, FAO Legis. St. No. 21 Rev. 3.
Moore, G. (1993) 'Enforcement Without Force: new Techniques in Compliance Control for Foreign Fishing Operations based on Regional Cooperation', *Law of the Sea Institute Proceedings*, Vol. 26, pp. 335–44.
Mortelmans, K.J.M. (1997) 'Compliance and Enforcement of European Community Law', *Paper delivered at PIONIER Conference, Utrecht*, 2–3 October 1997.
Munir, A.E. (1991) *Fisheries after Factortame* (Butterworths, London).
Murray, L.E. (1994) 'Maritime Enforcement. The Canadians Federal Government's Marine Fleets and Navy's Mission', *Marine Policy*, Vol. 18, No. 6, pp. 521–29.
Nielsen, J.R. (1994) 'Participation in Fishery Management Policy Making, National and EC Regulation of Danish Fishermen', *Marine Policy*, Vol. 18, No. 1, pp. 29–40.
Nielsen, J.R. and Vedsmand, T. (1997) 'Fishermen's Organisations in Fisheries Management: Perspectives for Fisheries co-Management Based on Danish Fisheries', *Marine Policy*, Vol. 21, No. 3, pp. 277–88.
Noirfalisse, C. (1992) 'The Community System of Fisheries Management and the Factortame Case', *Yearbook of European Law*, Vol. 12, pp. 325–51.
Nordic Council of Ministers (1993) 'Information Technology in Fisheries Management and Control', *Workshop in Reykjavik, 14–15 September 1993.*
Nordquist, M. (ed. in chief) (1983–1990) *New Direction in the Law of the Sea 1982: A Commentary*, Vols 1–4, 1983–1990 (Martinus Nijhoff, The Hague).
Nordquist, M.H. and Moore, J.N. (ed.) (1994) *Entry into Force of the Law of the Sea Convention* (Martinus Nijhoff, The Hague).
O'Connell, D.P. (Shearer, I.A.) (1982–1984) *The International Law of the Sea*, Vols I and II (Clarendon Press, Oxford).
O'Connell, M.E. (1992) 'Enforcing the New International Law of the Environment', *German Yearbook of International Law*, Vol. 35, pp. 293–332.
O'Reilly, J. (1992) 'Judicial Review and the Common Fisheries Policy in Community Law', in Curtin D. and O'Keefe D. (eds), *Constitutional Adjudication in European Community and National Law*, pp. 51–65 (Butterworths Law (Ireland), Dublin).
O'Reilly Hinds, L. (1995) 'Crises in Canada's Atlantic Sea Fisheries', *Marine Policy*, Vol. 19, pp. 271–83.
Oda, S. (1973) 'New Directions in the International Law of Fisheries', *AJIL*, Vol. 17, pp. 84–90.
Oda, S. (1977) *The Law of the Sea in our Time*, Vols I and II (Sijthoff, Leyden).

Oda, S. (1983) 'Fisheries Under the United Nations Convention on the Law of the Sea', *AJIL*, Vol. 77, No. 4, pp. 739–55.
Oda, S. (1989) *International Control of Sea Resources* (Martinus Nijhoff, Dordrecht).
OECD (1993) *The Use of Individual Quotas in Fisheries Management* (Organisation for Economic Cooperation and Development, Paris).
OECD (1994) *Fisheries Enforcement Issues* (Organisation for Economic Cooperation and Development, Paris).
OECD (1994) *Measures de réglementation dans les pêches* (Organisation for Economic Cooperation and Development, Paris).
Oliver, P. (1987) 'Enforcing Community Rights in the English Courts', *Modern Law Review*, Vol. 50, No. 7, pp. 881–907.
Oliver, P. (1996) *Free Movement of Goods in the European Community, under articles 30 to 36 of the Rome Treaty* (Sweet & Maxwell).
Olmi, G. (1972) 'Agriculture and Fisheries in the Treaty of Brussels of January 22, 1972', *CMLRev*, Vol. 9, pp. 293–321.
Orrego Vicuña, F. (1992) 'The "Presential Sea": Defining Coastal States' Special Interests in High Seas Fisheries and Other Activities', *German Yearbook of International Law*, Vol. 35, pp. 264–92.
Orrego Vicuña, F. (1993) 'Towards an Effective Management of High Seas Fisheries and the Settlement of Pending Issues of the Law of the Sea', *ODIL*, Vol. 24, pp. 81–92.
Orrego Vicuña, F. (1998) *The Changing International Law of High Seas Fisheries* (Cambridge University Press, New York).
Oude Elferink, A.G. (1995) Fisheries in the Sea of Okhotsk High Seas Enclave - the Russian Federation's Attempts at Coastal State Control, *IJMCL*, Vol. 10, No. 1, pp. 1–18.
Payoyo, P.B. (1994) 'Implementation of International Conventions Through Port State Control: an Assessment', *Marine Policy*, Vol. 18, No. 5, pp. 379–92.
Peterson J. (1994) 'Subsidiarity: a Definition to Suit any Vision', *Political Affairs*, Vol. 47, No. 1.
Posner, R.A. (1997) 'Social norms and the Law: an Economic Approach', *American Economic Review* (Paper and Proceedings), Vol. 87, No. 2, pp. 365–9.
Power, V. (1992, 1st Suppl. 1994) *EC Shipping Law* (Lloyd's of London Press, London).
Pyle, D.J. (1983) *The Economics of Crime and Law Enforcement* (London).
Rayfuse, R. (1998) 'Enforcement of High Seas Fisheries Agreements: Observation and Inspection under the Convention on the Conservation of Antarctic Marine Living Resources', *ODIL*, Vol. 13, No. 4.
Robb, D. (1983) 'Access Conditions and Compliance Control', *FAO Fisheries Report*, No. 293, pp. 157–73.
Savini, M.J. (1979) 'The New International Law of Fisheries Emerging from Bilateral Agreements', *Marine Policy*, Vol. 3.
Sherlock, A. and Harding, C. (1991) 'Controlling Fraud within the European Community', *ELR*, Vol. 16, No. 3, pp. 20–36.
Schrank, W. (1995) 'Extended Fisheries Jurisdiction', *Marine Policy*, Vol. 19, pp. 285–99.
Scovazzi, T. (1985) 'Explaining Exclusive Fishery Jurisdiction', *Marine Policy*, Vol. 9, pp. 120–5.
Scovazzi, T. (1995) 'The Specific Problem of High Seas Fishing in the Mediterranean

Sea', *Development Law Studies* (Food and Agriculture Organization of the United Nations).
Sevenster, H.G. (1992) 'Criminal Law and EC Law', *CMLRev*, Vol. 29, No. 1, pp. 29–70.
Shackleton, M. (1986) *The Politics of Fishing in Britain and France* (Ashgate Publishing Company).
Shackleton, M. (1990) *Financing the European Community* (Council on Foreign Relations).
Sharp, P. (1987) 'Small State Foreign Policy and International Regimes: the case of Ireland and the European Monetary System and the Common Fisheries Policy', *Millenium Journal of International Studies*, Vol. 16, No. 1, pp. 55–72.
Shearer, A. (1986) 'Problems of Jurisdiction and Law Enforcement Against Delinquent Vessels', *ICLQ*, Vol. 35, No. 2, pp. 320–43.
Shearer, A. (1992) 'High Seas: Driftnets, Highly Migratory Species and Marine Mammals', *Law of the Sea Institute Proceedings*, pp. 432–59.
Simmonds, K.R. (1989) 'The European Economic Community and the New Law of the Sea', *Collected courses of the Hague Academy of International Law*, Vol. 6, No. 218, pp. 9–164.
Snyder, F. (1985) *Law of the Common Agricultural Policy* (Sweet and Maxwell, London).
Snyder, F. (1993) 'The Effectiveness of European Community Law: Institutions, Processes, Tools and Techniques', *Modern Law Review*, Vol. 56, No. 1, pp. 19–54.
Snyder, F. (1993) 'Soft Law and Institutional Practice in the European Community', European University Institute (EUI) Working Paper.
Sohn, L.B. (1995) 'Settlement of Law of the Sea Disputes', *IJMCL*, Vol. 10, No. 2, pp. 205–17.
Song, Y.H. (1991–1992) 'United States Ocean Policy: High Seas Driftnet Fisheries in the North Pacific Ocean', *Ocean Yearbook of International Law and Affairs*, Vol. 11, pp. 64-137.
Song, Y.H. (1995) 'The EC's Common Fisheries Policy in the 1990s', *ODIL*, Vol. 26, No. 1, pp. 31–7.
Song, Y.H. (1997) 'The Canada–European Union Turbot Dispute in the Northwest Atlantic: An Application of the Incident Approach', *ODIL*, Vol. 20, No. 3, pp. 269–313.
Song, Y.H. (1998) 'The Common Fisheries Policy of the European Union: Restructuring of the Fishing Fleet and the Financial Instrument for Fisheries Guidance', *IJMCL*, Vol. 13, No. 4, pp. 537–77.
Soons, A.H.A. (1994) 'Regulation of Marine Scientific Research by the European Community and its Member States', *ODIL*, Vol. 25, pp. 259–67.
Steel, D.IA. (1984) 'Fisheries Policy and the EEC: the Democratic Influence', *Marine Policy*, Vol. 8, pp. 350–53.
Steiner J. (1992) 'EC Law Problems of Enforcement', *Collected Courses of the Academy of European Law*, Vol. 3, book 1, pp. 241–88 (Martinus Nijhoff, Florence).
Steiner J. (1993) 'From Direct effect to *Francovich*: Shifting Means of Enforcement of Community Law', *ELR*ev., Vol. 18, No. 1, pp. 3–22.
Steiner, J. (1995) *Enforcing EC Law* (Blackstone Press, London).
Stigler, J.G. (1978) 'The Optimum Enforcement of Laws', *Journal of Political Economy*, Vol. 78, pp. 526–36.
Sullivan, M.S. (1977) 'The Case in International Law for Canada's Extension of Fisheries Jurisdiction Beyond 200 miles', *ODIL*, Vol. 20, No. 3, pp. 203–68.

Sutinen, J.G. (1988) 'Enforcement Economics in Exclusive Economic Zones', *GEOJOURNAL,* Vol. 16, No. 3, Special Issue, 'Marine Resource Economics', pp. 273–81.

Sutinen, J.G. and Andersen, P. (1985) 'The Economics of Fisheries Law Enforcement', *Land Economics,* Vol. 61, No. 4, pp. 387–97.

Sutinen, J.G. and Hennessey, T.M. (1986) 'Enforcement: the Neglected Element in Fishery Management', in Miles, E., Pealy, R. and Stokes, R. (eds), *Natural Resource Economics and Policy Applications* (Seattle).

Sutinen, J.G., Rieser, A. and Gauvin, J.R. (1990) 'Measuring and Explaining Noncompliance in Federally Managed Fisheries', *ODIL,* Vol. 21, pp. 335–72.

Svein Jentoft (1989) 'Fisheries Co-Management: Delegating Government Responsibility to Fishermen's Organisations', *Marine Policy,* Vol. 13, No. 2, pp. 137–54.

Symes, D. (ed.) (1997) *Property Rights and Regulatory Systems in Fisheries* (Blackwell Science Ltd, Oxford).

Symes, D. (1998) *Northern Waters: Management Issues and Practice* (Fishing News Books, Oxford).

Symmons, C.R. (1993) *Ireland and the Law of the Sea* (Round Hall Ltd, Dublin).

Symmons, C.R. (1997) 'Ireland and the Law of the Sea', in Treves (ed.), *The Law of the Sea, The European Union and its Member States,* pp. 261–316 (Kluwer Law International, The Hague).

Tahindro, A. (1997) 'Conservation and Management of Transboundary Fish Stocks: Comments in Light of the Adoption of the 1995 Agreement for the Conservation and Management of Straddling Fish Stocks and Highly Migratory Fish Stocks', *ODIL,* Vol. 28, pp. 1–58.

Teasdale, A.L (1993) 'Subsidiarity in Post Maastricht Europe', *Political Quarterly,* Vol. 64, No. 2, April–June.

Temple Lang, J. (1981) 'Community Antitrust Law - Compliance and Enforcement', *CMLRev,* Vol. 18, No. 3, pp. 335–62.

Temple Lang, J. (1990) 'Community Constitutional Law: Article 5 EEC Treaty', *CMLRev,* Vol. 27, No. 4, pp. 645–81.

Temple Lang, J. (1993) 'European Community Constitutional Law and the Enforcement of Community Antitrust Law', *Annual Proceedings of the Fordham Corporate Law Institute,* Vol. 20, pp. 525-604.

Temple Lang , J. (1998) 'Community Antitrust Law and National Regulatory Procedures', *Annual proceedings of the Fordham Corporate Law Institute,* Vol. 24, pp. 297–334.

Temple Lang, J. and Gallagher, E. (1995) 'The Role of the Commission and Qualified Majority Voting: a Unique Relationship with the Council', *Occasional paper, Institute of European Affairs* (Dublin).

Timmermans, C.W.A. and Völkers, E.L.M (1981) *Division of Powers Between the European Communities. and their Member States in the Field of External Relations* (Deventer, Kluwer).

Toth, A. (1994) 'Is Subsidiarity Justifiable?', *European Law Review,* Vol. 19, No. 3, pp. 268–85.

Tyler, T.R. (1990) *Why People Obey the Law* (New Haven and London, Yale University Press).

Treves, T. (1996) 'The Proceedings Concerning Prompt Release of Vessels and Crews Before the International Tribunal for the Law of the Sea', *IJMCL,* Vol. 11, pp. 179–200.

Treves, T. and Pineschi, L. (ed.) (1996) *The Law of the Sea, The European Union and its Member States* (Kluwer Law International, The Hague).

Ulfstein, G. (1983) '200 mile Zones and Fisheries Management', *Nordisk tidsskrift for international ret,* Vol. 52, No. 3–4, pp. 3–33.

Ulfstein, G., Andersen, P. and Churchill, R.R. (1987) *Proceedings of the European Workshop on the Regulation of Fisheries: Legal Economic and Social Aspects,* University of Tromso, Norway, 2–4 June 1986 (Strasbourg).

Usher, J.A. (1985) 'The Scope of Community Competence. Its Recognition and Enforcement', *Journal of Common Market Studies,* Vol. 24, No. 2, pp. 121–36.

Van Dyke, J.M. (1995) 'Modifying the 1982 Law of the Sea Convention: New Initiatives on Governance of High Seas Fisheries Resources: the Straddling Stocks Negotiations', *IJMCL,* Vol.10, No. 2, pp. 219–27.

Van Dyke, J.M. (1996) 'The Straddling and Migratory Stocks Agreement and the Pacific', *IJMCL,* Vol. 11, No. 3, pp. 406–15.

Van Mensbrugghe, Y. (1975) 'The Common Market Fisheries Policy of the Law of the Sea', *Netherlands Year Book of International Law,* Vol. 6, pp. 199–228.

Vervaele, J. (1991) 'La lutte contre la fraude communautaire: une mise ? l'Èpreuve de la loyauté communautaire des Etats membres?', *Revue de Droit Pénal et de Criminologie,* No. 6, pp. 569–85.

Vervaele, J. (1992) *Fraud against the Community: the Need for Community Fraud Legislation* (Kluwer Law International, Deventer).

Vervaele, J. (ed) (1994) *Administrative Law Application and Enforcement of Community Law in the Netherlands* (Kluwer Law International, Deventer).

Vervaele, J. (1995) 'Principles of Justice, de-Regulation and re-Regulation and the Enforcement of Community Law in the EU', in Paasivirta (ed.), *Principles of Justice and the Law of the European Union* (European Union European Commission, Luxembourg).

Verveale J., Betlem, G., de Lange, R. and Veldman, A.G. (ed.) (1999) *Compliance and Enforcement of European Community Law* (Kluwer Law International, The Hague).

Vignes, D. (1994) 'La Convention sur le droit de la mer répond-elle à l'attente?', *Studia diplomatica,* Vol. 47, No. 6, pp. 29–57.

Vos, E. (1997) 'The Rise of Committees', *ELJ,* Vol. 3, No. 3, pp. 210–29.

Wägenbaur, R. (1996) 'How to Improve Compliance with European Community Legislation and the Judgments of the European Court of Justice', *Fordham International Law Journal,* Vol. 19, No. 3, pp. 936–50.

Wallace, H. and Young, A.R. (1997) *Participation and Policy-Making in the European Union* (Oxford University Press, Oxford).

Wallace, R. (1984) 'Special Economic Dependency and Preferential Rights in Respect of Fisheries: Characterisation and Articulation within the European Communities', *CMLRev.,* Vol. 21, No. 3, pp. 523–37.

Weatherill, S. and Beaumont, P. (1991, 2nd ed. 1995) *EC Law: The Essential Guide to the Legal Workings of the European Community* (Penguin Books, London).

Weiler, J. (1991) 'The Transformation of Europe', *YLJ,* Vol. 100, p. 2465.

Weiler, J. (1991) 'EPC and the Single Act: From Soft Law to Hard Law?', in Holland (ed.), *The Future of European Political Co-operation* (Macmillan Press).

Weiss, F. (1995) 'The Second Tuna GATT Panel Report', *Leiden Journal of International Law,* Vol. 8, pp. 135–50.

Wellens, K.C. and Borchardt, G.M. (1989) 'Soft Law in European Community Law', *ELRev,* Vol. 14, No. 5, pp. 267–321.

Wilke, M. and Wallace, H. (1990) *Subsidiarity: Approaches to Power-Sharing in the European Community,* Discussion Papers No. 27, London, Royal Institute of International Affairs.

Winkel, K. (1977) 'Equal Access Community Fishermen to the Member States Fishing Grounds', *CMLRev,* Vol. 14, pp. 329–37.

Williams, C. (1996) 'Fisheries in the Future: Sustainability or Extinction?', *Marine Policy,* Vol. 20, No. 1, pp. 91–5.

William, H.G. and Lugten, G. (1994) 'Fishing for Complements', *Environmental Policy and Law,* Vol. 24/5, pp. 254–60.

Williams, R. (1994) 'The European Commission and the Enforcement of Environmental Law: an Invidious Position', *Yearbook of European Law,* Vol. 14, pp. 351–99.

Wise, M. (1984) *The Common Fisheries Policy of the European Community* (Methuen, London).

Young, O.R. (1979) *Compliance and Public Authority* (Johns Hopkins University Press, Baltimore).

Yturriaga, J.A. de (1995) 'El Conflicto Pesquero Canadiense-Comunitario:! Y Luego Dicen Que el Pescadoes Cairo', *Revista De Instituciones Europeas,* Vol. 22, No. 2, pp. 511–31.

Yturriaga, J.A. de (1997) *The International Regime of Fisheries From UNCLOS 1982 to the Presential Sea* (Martinus Nijhoff, The Hague).

Zangl, P. (1989) 'The Inter-Institutional Agreement on the Improvement of the Budgetary Procedure', *CMLRev,* Vol. 26, p. 675.

Zerdick, T. (1995) 'European Aspects of Data Protection: What Rights for the Citizen?', *Legal Issues of European Integration,* No. 2, pp. 59–86.

COUNCIL REGULATION (EC) No 1447/1999

of 24 June 1999

establishing a list of types of behaviour which seriously infringe the rules of the common fisheries policy

THE COUNCIL OF THE EUROPEAN UNION,

Having regard to the Treaty establishing the European Community, and in particular Article 37 thereof,

Having regard to the proposal from the Commission ([1]),

Having regard to the opinion of the European Parliament ([2]),

Whereas:

(1) according to Article 31(2a) of Council Regulation (EEC) No 2847/93 of 12 October 1993 establishing a control system applicable to the common fisheries policy([3]), the Council, acting on the basis of Article 37 of the Treaty, may draw up a list of types of behaviour which seriously infringe the rules of the common fisheries policy;

(2) in order to improve transparency in the common fisheries policy, Member States should provide the Commission with information on instances of such behaviour and action taken by Member States;

(3) Article 37 of Regulation (EEC) No 2847/93 provides for the protection of certain personal data in the framework of the control system applicable to the common fisheries policy;

(4) the abovementioned list is to be coherent with similar provisions adopted by international fisheries organisations;

(5) for certain measures laid down by this Regulation it is appropriate to provide for detailed rules of implementation,

HAS ADOPTED THIS REGULATION:

Article 1

The types of behaviour which seriously infringe the rules of the common fisheries policy as referred to in Article 1 of Regulation (EEC) No 2847/93 shall be those listed in the Annex hereto.

Article 2

1. The Member States shall notify the Commission on a regular basis of the instances of behaviour referred to in Article 1 that have been discovered and shall provide it with all information regarding action taken by the administrative and/or judicial authorities.

2. The Commission shall make the information it receives pursuant to paragraph 1 available to the European Parliament, the Council and the Advisory Committee of Fisheries.

3. The information notified under paragraph 1 and made available under paragraph 2 shall be treated in accordance with the provisions of Article 37 of Regulation (EEC) No 2847/93.

4. Detailed rules for the implementation of this Article, shall be laid down in accordance with the procedure in Article 36 of Regulation (EEC) No 2847/93.

Article 3

This Regulation shall enter into force on the seventh day following that of its publication in the *Official Journal of the European Communities.*

This Regulation shall be binding in its entirety and directly applicable in all Member States.

Done at Luxembourg, 24 June 1999.

For the Council
The President
J. TRITTIN

([1]) OJ C 105, 15.4.1999, p. 3.
([2]) Opinion delivered on 4 May 1999 (not yet published in the Official Journal).
([3]) OJ L 261, 20.10.1993, p. 1. Regulation as last amended by Regulation (EC) No 2846/98 (OJ L 358, 31.12.1998, p. 5).

ANNEX

LIST OF TYPES OF BEHAVIOUR WHICH SERIOUSLY INFRINGE THE RULES OF THE COMMON FISHERIES POLICY

A. Failure to cooperate with the authorities responsible for monitoring

- Obstructing the work of fisheries inspectors in the exercise of their duties in inspecting for compliance with the applicable Community rules.
- Falsifying, concealing, destroying or tampering with evidence which could be used in the course of inquiries or judicial proceedings.

B. Failure to cooperate with observers

- Obstructing the work of observers in carrying out their duties, laid down by Community Law, of observing compliance with the applicable Community rules.

C. Failure to observe the conditions to be met when fishing

- Fishing without holding a fishing licence, a fishing permit or any other authorisation required for fishing and issued by the flag Member State or by the Commission.
- Fishing under cover of one of the abovementioned documents the content of which has been falsified.
- Falsifying, deleting or concealing the identification marks of the fishing vessel.

D. Failure to comply during fishing operations

- Using or keeping on board prohibited fishing gear or devices affecting the selectiveness of gear.
- Using prohibited fishing methods.
- Not lashing or stowing fishing gear the use of which is prohibited in a certain fishing zone.
- Directed fishing for, or keeping on board of species from, stocks subject to a moratorium or a prohibition of fishing.
- Unauthorised fishing in a given zone and/or during a specific period.
- Failure to comply with the rules on minimum sizes.
- Failure to comply with the rules and procedures relating to transhipment and fisheries operations involving joint action by two or more vessels.

E. Failure to comply in connection with resources for monitoring

- Falsifying or failing to record data in logbooks, landing declarations, sales notes, takeover declarations and transport documents or failure to keep or submit these documents.
- Tampering with the satellite-based vessel monitoring system.
- Deliberate failure to comply with the Community rules on remote transmission of movements of fishing vessels and of data of fishery products held on board.
- Failure of the master of the fishing vessel of a third country or his representative to comply with the applicable control rules when operating fishing in Community waters.

F. Failure to comply in connection with landing and marketing of fishery products

- Landing of fishery products not respecting the Community rules on control and enforcement.
- Stocking, processing, placing for sale and transporting fishery products not meeting the marketing standards in force and, in particular, those concerning minimum sizes.

Index